DEGREES AND RADIANS (1.1, 2.1)

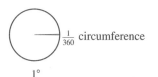

$\frac{1}{360}$ circumference

1°

1 radian

$$\frac{\theta_{deg}}{180°} = \frac{\theta_{rad}}{\pi \text{ rad}}$$

Degrees $\times \dfrac{\pi}{180}$ = Radians   Radians $\times \dfrac{180}{\pi}$ = Degrees

TRIGONOMETRIC FUNCTIONS (2.3, 2.6)

$\sin x = \dfrac{b}{R}$ $\csc x = \dfrac{R}{b}$

$\cos x = \dfrac{a}{R}$ $\sec x = \dfrac{R}{a}$

$\tan x = \dfrac{b}{a}$ $\cot x = \dfrac{a}{b}$

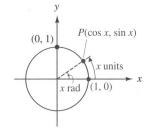

$R = \sqrt{a^2 + b^2} > 0$

(x in degrees or radians)
For x any real number and T any trigonometric function,
$\quad T(x) = T(x \text{ rad})$
For a unit circle:

$P(\cos x, \sin x)$

$(0, 1)$

x units

x rad   (1, 0)

SPECIAL VALUES (2.5)

θ	$\sin \theta$	$\csc \theta$	$\cos \theta$	$\sec \theta$	$\tan \theta$	$\cot \theta$
0° or 0	0	N.D.	1	1	0	N.D.
30° or $\pi/6$	1/2	2	$\sqrt{3}/2$	$2/\sqrt{3}$	$1/\sqrt{3}$	$\sqrt{3}$
45° or $\pi/4$	$1/\sqrt{2}$	$\sqrt{2}$	$1/\sqrt{2}$	$\sqrt{2}$	1	1
60° or $\pi/3$	$\sqrt{3}/2$	$2/\sqrt{3}$	1/2	2	$\sqrt{3}$	$1/\sqrt{3}$
90° or $\pi/2$	1	1	0	N.D.	N.D.	0

N.D. = Not defined

PYTHAGOREAN THEOREM (1.3, C.2)

$a^2 + b^2 = c^2$

SPECIAL TRIANGLES (2.5)

30°–60° triangle   45° triangle

NUMBER OF SIGNIFICANT DIGITS (A.3)

Form of Number	Number of Significant Digits
x, with no decimal point	Count the digits of x from left to right, starting with the first digit and ending with the last nonzero digit.
x, with a decimal point	Count the digits of x from left to right, starting with the first non-zero digit and ending with the last digit (which may be zero).

ACCURACY OF CALCULATED VALUES (A.3)

The number of significant digits in a calculation involving multiplication, division, powers, and/or roots is the same as the number of significant digits in the number in the calculation with the smallest r significant digits.

ACCURACY FOR TRIANGLES (1.3)

Angle to Nearest	Significant Digits for Sid
1°	2
10′ or 0.1°	3
1′ or 0.01°	4
10″ or 0.001°	5

(Continued inside back cover)

D1410291

M. Lourdes Emanuelli

ANALYTIC TRIGONOMETRY

WITH APPLICATIONS

RECEIVED
FEB 0 7 1997
C B R
Vol. IV

ANALYTIC TRIGONOMETRY

WITH APPLICATIONS

Sixth Edition

Raymond A. Barnett
Merrit College

Michael R. Ziegler
Marquette University

PWS PUBLISHING COMPANY

I(T)P **An International Thomson Publishing Company**
Boston • Albany • Bonn • Cincinnati • Detroit • London • Madrid • Melbourne • Mexico City
New York • Paris • San Francisco • Singapore • Tokyo • Toronto • Washington

PWS PUBLISHING COMPANY
20 Park Plaza, Boston, MA 02116-4324

Copyright © 1995 by PWS Publishing Company, a division of International Thomson Publishing Inc.
Copyright © 1992, 1988, 1984, 1980, 1976 by Wadsworth, Inc.

All rights reserved. No part of this book may be reproduced, stored in a retrieval
system, or transmitted in any form or by any means—electronic, mechanical,
photocopying, recording, or otherwise—without the prior written permission of
PWS Publishing Company.

International Thomson Publishing
The trademark ITP is used under license.

For more information, contact:

PWS Publishing Co.
20 Park Plaza
Boston, MA 02116

International Thomson Publishing Europe
Berkshire House 168-173
High Holborn
London WC1V 7AA
England

Thomas Nelson Australia
102 Dodds Street
South Melbourne, 3205
Victoria, Australia

Nelson Canada
1120 Birchmont Road
Scarborough, Ontario
Canada M1K 5G4

International Thomson Editores
Campos Eliseos 385, Piso 7
Col. Polanco
11560 Mexico D.F., Mexico

International Thomson Publishing GmbH
Konigswinterer Strasse 418
53227 Bonn, Germany

International Thomson Publishing Asia
221 Henderson Road
#05-10 Henderson Building
Singapore 0315

International Thomson Publishing Japan
Hirakawacho Kyowa Building, 31
2-2-1 Hirakawacho
Chiyoda Ku, Tokyo 102
Japan

Library of Congress Cataloging-in-Publication Data

Barnett, Raymond A.
    Analytic trigonometry with applications / Raymond A. Barnett,
Michael R. Ziegler. — 6th ed.
        p.   cm.
    Includes indexes.
    ISBN 0-534-94344-6
    1. Trigonometry, Plane.   I. Ziegler, Michael R.   II. Title.
QA533.B32   1994
516.3'4—dc20

94-25418
CIP

Sponsoring Editor: David Dietz
Editorial Assistant: Julia Chen
Production Coordinator: Robine Andrau
Marketing Manager: Marianne C. P. Rutter
Manufacturing Coordinator: Marcia A. Locke
Interior/Cover Designer: Julia Gecha

Interior Illustrator: Scientific Illustrators
Typesetter: Ruttle, Shaw & Wetherill, Inc.
Cover Printer: Henry N. Sawyer Company
Text Printer/Binder: Quebecor Printing/Hawkins

Printed and bound in the United States of America
96 97 98 — 10 9 8 7 6 5 4 3 2

C O N T E N T S

☆ Sections marked with a star may be omitted without loss of continuity.

IDENTITIES   187

INVERSE TRIGONOMETRIC FUNCTIONS; TRIGONOMETRIC EQUATIONS   241

ADDITIONAL TRIANGLE TOPICS; VECTORS   293

7 POLAR COORDINATES; COMPLEX NUMBERS   359

A COMMENTS ON NUMBERS   398

B FUNCTIONS AND INVERSE FUNCTIONS   413

C PLANE GEOMETRY: SOME USEFUL FACTS   428

A *nalytic Trigonometry with Applications,* Sixth Edition, has benefited from the generous response from users of the earlier editions. Prerequisites for the book are one and a half to two years of high school algebra and one year of high school geometry or their equivalents.

We have taken great care to produce a book that students can actually read, understand, and enjoy. To gain student interest quickly, the text moves directly into trigonometric concepts and applications and reviews essential material from prerequisite courses only as needed. Almost every concept is illustrated by an example followed by a matching problem (with the answers near the end of a section) to encourage an active rather than passive involvement in the learning process. Concept development proceeds from the concrete to the abstract (e.g., right triangle definitions of the trigonometric functions precede the unit circle definitions of these functions). There are enough applications from different fields to convince even the most skeptical student that trigonometry is really useful.

◆ New Features in the Sixth Edition

- A new **full-color design** and **expanded art program** make the book more visually appealing and pedagogically effective. Being able to ''see'' mathematics helps many students.
- Many new **applications** have been added and older applications have been made current. Some of the new applications **model real-world data** (e.g., Exercise 3.3, Problems 41 and 42). More **multistep applications** have been included that involve the use of concepts from preceding sections as well as concepts from the current section (e.g., Exercise 1.4, Problems 21, 22, 27, 28, 37, and 38; Exercise 1.5, Problems 31 and 32). A number of the multistep problems require the establishment of general relationships among variables and constants before specific numerical values are assigned (e.g., Exercise 1.4, Problems 21 and 22; Exercise 1.5, Problem 32).
- **Cumulative reviews** after Chapters 3, 5, and 7 are new. These review sections include problems corresponding to every chapter in the text up to that point. **Chapter summaries** now include section-by-section summaries of all important concepts and formulas. Chapter review exercise sets include many more applications. Review problems are purposely not identified by section numbers of relevant sections; however, the answers to these problems are. The inclusion of **optional sections** in the chapter reviews and in the cumulative reviews is new; this material is clearly marked with a star (☆) so that it can be omitted without loss of continuity, if desired.
- **Graphing calculator problems** are interspersed in context where appropriate throughout the book (e.g., Section 3.1, Exercise 3.1, Exercise 3.2). These problems are often exploratory in nature and are designed to add additional

insight into the concepts being developed. All graphing calculator material is identified by either a graphing calculator icon or by a ''*c*'' in a box and can be omitted without loss of continuity, if desired.

- **Caution** warnings, identified by an icon, alert students to potential problem areas (e.g., Sections 1.1 and 1.3).
- **Historical remarks** have been added or expanded where appropriate (e.g., Sections 1.2 and 2.1).
- **Titles** identifying the subject area or problem type have been added to examples; and the **numbering system** for examples, figures, and tables has been simplified by the start of the numbers anew in each section.

◆ Specific Changes in the Sixth Edition

Chapter 1   Right Triangle Ratios

Section 1.1 on angles, degrees, and arcs was completely rewritten. Now *angle* is defined only as a rotation. The old Section 1.2 on *significant digits* was simplified and moved to Appendix A. Section A.3, along with a table in the front endpapers of the book, guides *computational accuracy* throughout the text.

Chapter 2   Trigonometric Functions

To provide more focus, a separate subsection on *reference triangles and angles* was added to Section 2.5. Also, problems specifically on reference triangles and angles were added to Exercise 2.5. Section 2.6 on *circular functions* has been partly rewritten and expanded and more problems were added to the exercise set.

Chapter 3   Graphing Trigonometric Functions

This chapter was fine-tuned. Additional exercises and applications were added, including *modeling problems using real data* (e.g., Exercise 3.3, Problems 41 and 42). *Optional graphing calculator problems* are now distributed throughout the chapter.

Chapter 4   Identities

This chapter was fine-tuned. The process of *verifying an identity* was made more explicit. Optional graphing calculator problems are now distributed throughout the chapter.

Chapter 5   Inverse Trigonometric Functions; Trigonometric Equations

A subsection on *angle ranges in degree and radian measure* was added to Section 5.1. Problems involving degree measure were added to the exercise set. Optional graphing calculator problems are now distributed throughout the chapter.

Chapter 6   Additional Triangle Topics; Vectors

Two sections from the previous edition on the *law of sines* were completely rewritten and combined into one section. The section on the *law of cosines* was also rewritten and is now Section 6.2. The rewriting has provided *clearer strategies for solving oblique triangles,* given ASA, AAS, SSA, SAS, and SSS. There are now additional simpler applications and multistep applications for both the law of sines and the law of cosines.

Chapter 7 Polar Coordinates; Complex Numbers

By popular demand, the *"cis" notation* was deleted from the chapter. Optional graphing calculator problems are now distributed throughout the chapter.

◆ Important Features Retained in the Sixth Edition

- The focus of this book is student comprehension. An **informal style** is used for exposition, definitions, and theorems. Precision, however, is not compromised.
- The subject matter is related to the real world through many carefully selected **realistic applications** from many different fields. These applications are uniformly distributed throughout the book.
- To gain reader interest quickly, the text moves directly into trigonometric concepts and applications. **Review material** from prerequisite courses is either integrated in certain developments (particularly in Chapters 4 and 5) or can be found in the appendixes. The material can be reviewed as needed by the student or taught in class by the instructor.
- The book includes **more than 2,000 carefully selected and graded problems**. Problems in most exercise sets are divided into A, B, and C groupings. The A problems are easy and routine, the B problems are more challenging but still emphasize mechanics, and the C problems are a mixture of more difficult mechanics and theory problems. In short, the text is designed so that an average or struggling student will be able to experience success and a very capable student will be challenged. **Answers** to most of the odd-numbered problems and almost all of the chapter and cumulative review exercises are included at the end of the book.
- The content **satisfies the requirements for many succeeding courses**, including calculus, analytic geometry, physics, and technical mathematics courses.
- The **trigonometric functions** are defined first in terms of angle domains, using degree and radian measure side by side, and then in terms of real number domains. All of this is done early in the book and is reinforced throughout. By the end of the course, students should be relatively comfortable with all three modes and should be able to shift from one to the other without difficulty.
- **Historical remarks** are included where appropriate to provide perspective (see Sections 1.1 and 1.3).

◆ Student Aids

- Most concepts are illustrated by an **example** followed by a **matched problem** (with the answer near the end of the section) to encourage an active rather than passive involvement in the learning process.
- **Annotation** of examples and developments is found throughout the book to help students through critical stages (see Sections 1.2, 1.4, 4.1, 4.2, 5.3).
- **Chapter reviews** are included at the end of each chapter and **cumulative reviews** are included after chapters 3, 5, and 7.

- **Answers** to almost all chapter and cumulative review exercises and most odd-numbered problems from the other exercises are in the back of the book so that a student can easily check his or her progress.
- **Dashed "think boxes"** are used to indicate steps that are usually performed mentally after a concept or procedure is understood (e.g., Sections 1.1 and 2.4).
- **Functional use of color** guides students through critical steps and clarifies figures.
- **Generic calculator steps** are included in the early chapters of the book (e.g., Sections 1.1 and 1.3).
- **Formulas and symbols** (keyed to sections in which they are first introduced) and the metric system are summarized on the front and back endpapers of the book for convenient reference.
- A perforated **Quick Reference Card** minimizes the need for page turning by putting key equations and graphs at students' fingertips.

◆ Pedagogical Use of Color

Color is used not only to make the text more attractive, but more importantly it is used functionally to improve communications. For example, color is used in

1. Commentary that accompanies a solution process (e.g., Sections 1.3 and 4.2).
2. Graphing to visually separate the various parts of a graph (e.g., Chapters 3 and 7).
3. Boxed highlighted material to distinguish Assumptions/Definitions, Theorems, and Strategies/Processes (e.g., Sections 1.3, 4.2, and 6.5).

ASSUMPTIONS/DEFINITIONS

THEOREMS

STRATEGIES/PROCESSES

◆ Ancillaries

1. An **Instructor's Answer Manual** contains most of the answers that have been excluded from the text.

2. A **Student's Solutions Manual**, containing worked-out solutions to all chapter review and cumulative review exercises and all odd-numbered problems in the book, is available for students to purchase.

3. A **Trigonometry Videotape Series** reviews key topics in the text and features professional math instructors. The videotapes are free to qualified adopters; students may visit your math laboratory to view the tapes of their choice. These tutorial tapes, produced by Educational Video Resources, can significantly improve students' comprehension and performance in trigonometry.

4. **Test Items**, an improved printed and bound test bank, is available free to adopters.

5. **EXPTest** and **ExamBuilder** are improved computer-generated test systems for Windows, DOS, and Macintosh, and are available to instructors without cost.

6. A set of full-color acetate **Transparencies** depicting key concepts from trigonometry is available free to adopters.

7. **DERIVE Notebook Software** for the IBM provides students using the DERIVE computer algebra system with the opportunity to study, graph, and solve selected exercises from the text. It is available for purchase.

8. **INVESTIGATE Tutorial Software** in IBM and Macintosh formats allows students to practice the skills taught in the textbook. For topics in the text, the tutorial provides multiple-choice problems and instruction. The student is presented with solutions to incorrectly answered problems. This diagnostic program keeps track of right and wrong responses and can report to the instructor on students' progress.

9. A graphing calculator supplement, **Trigonometry Activities for the TI-82 and TI-85 Graphing Calculators** by Cynthia Dennis and Linda Neal, coordinates the use of graphing calculators with text topics in a set of exploratory learning activities that follows the organization of the text. It is available for students to purchase.

ACKNOWLEDGMENTS

In addition to the authors, the publication of a book requires the effort and skills of many people. We would like to extend particular thanks to some very capable people who had major responsibilities in the preparation of this book for publication: Robine Andrau, for her skill and expertise in guiding the project through the production process; Stephen J. Merrill and Robert E. Mullins of Marquette University, for their careful accuracy check of the manuscript; Fred Safier, City

College of San Francisco, for his careful checking of exercise sets and preparation of the solutions manual that accompanies this text; technical typist Rita Zubli, who also helped to prepare the solutions manual; George Alexander, University of Wisconsin—Madison, for his work on the test bank that accompanies this text; Laurel Technical Services for its work on the Instructor's Answer Manual and Student's Solutions Manual; Cynthia Dennis, San Jacinto College North, and Linda Neal, San Jacinto College South, for their preparation of an effective graphing calculator supplement; and George Morris and his staff at Scientific Illustrators for their effective illustrations.

We also wish to thank the many users and reviewers of the fifth and previous editions for their helpful suggestions and comments. In particular, we wish to thank

George Alexander
University of Wisconsin—Madison

Charlene Beckman
Grand Valley State University

Ronald Bzoch
University of North Dakota

Robert A. Chaffer
Central Michigan University

Rodney Chase
Oakland Community College

Barbara Chudilowsky
De Anza College

Richard L. Francis
Southeast Missouri State University

Al Giambrone
Sinclair Community College

Frank Glaser
California State Polytechnic University—Pomona

Jack Goebel
Montana College of Mineral Science and Technology

Deborah Gunter
Langston University

Pamala Heard
Langston University

Steven Heath
Southern Utah State College

Robert Henkel
Pierce College

John G. Henson
Mesa College

Norma F. James
New Mexico State University—Las Cruces

Ted Laetsch
University of Arizona

Vernon H. Leach
Pierce College

Lauri Lindberg
Pierce College

LaVerne McFadden
Parkland College

John Martin
Santa Rosa Junior College

Eldon Miller
University of Mississippi

Georgia Miller
Millsaps College

Margaret Ullmann Miller
Skyline College

Kathy Monaghan
American River College

Karen Murany
Oakland Community College

Harald M. Ness
University of Wisconsin

Nagendra N. Pandey
Montana College of Mineral Science and Technology

Howard Penn
El Centro College

Gladys Rockin
Oakland Community College

Michael Seery
Scottsdale Community College

M. Vali Siadat
Richard J. Daley College

Thomas Spradley
American River College

Lowell Stultz
Kalamazoo Valley Community College

Richard Tebbs
Southern Utah State College

Roberta Tugenberg
Southwestern College

Steve Waters
Pacific Union College

Bobby Winters
Pittsburgh State University

Thomas Worthington
Grand Rapids Junior College

Because of the careful independent checking by three competent college mathematics instructors, the authors and publisher believe this book to be substantially error-free. If any errors remain, the authors would be grateful if corrections were sent to them care of Precalculus Editor, PWS Publishing Company, 20 Park Plaza, Boston, MA 02116-4324.

Raymond A. Barnett
Michael R. Ziegler

U se of calculators is emphasized throughout this book. Many brands and types of scientific calculators are available and can be found starting at about $10. Graphing calculators are more expensive; but, in addition to having most of the capabilities of a scientific calculator, they have very powerful graphing capabilities. Your instructor should help you decide on the type and model best suited to this course and the emphasis on calculator use he or she desires.

Scientific Calculator (left) and Graphing Calculator (right)
Courtesy Texas Instruments Incorporated

Whichever calculator you use, it is essential that you read the user's manual for your calculator. A large variety of calculators are on the market, and each is slightly different from the others. Therefore, take the time to read the manual. The first time through do not try to read and understand everything the calculator can do—that will tend to overwhelm and confuse you. Read only those sections pertaining to the operations you are or will be using; then return to the manual as necessary when you encounter new operations.

In the beginning chapters of the text, calculator steps for calculations are shown. These are only aids. Try the calculation first, and then use the aid if you get stuck.

It is important to remember that a calculator is not a substitute for thinking. It can save you a great deal of time in certain types of problems, but you still must understand basic concepts so that you can interpret results obtained through the use of a calculator.

RIGHT TRIANGLE RATIOS

f you were asked to find your height, you would no doubt take a ruler or tape measure and measure it directly. But if you were asked to find the area of your bedroom floor in square feet, you would not be likely to measure the area directly by laying 1 ft squares over the entire floor and counting them. Instead, you would probably find the area **indirectly** by using the formula $A = ab$ from plane geometry, where A represents the area of the room and a and b are the lengths of its sides, as indicated in the figure.

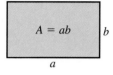

In general, **indirect measurement** is a process of determining unknown measurements from known measurements by a reasoning process. How do we measure quantities such as the volumes of containers, the distance to the center of the earth, the area of the surface of the earth, and the distances to the sun and the stars? All these measurements are accomplished indirectly by the use of special formulas and deductive reasoning.

The Greeks in Alexandria, during the period 300 B.C.–A.D. 200, contributed substantially to the art of indirect measurement by developing formulas for finding areas, volumes, and lengths. Using these formulas, they were able to determine the circumference of the earth (with an error of only about 2%) and to estimate the distance to the moon. We will examine these measurements as well as others in the sections that follow.

It was during the early part of this Greek period that trigonometry, the study of triangles, was born. Hipparchus (160–127 BC), one of the greatest astronomers of the ancient world, is credited with making the first systematic study of the indirect measurement of triangles.

1.1 ANGLES, DEGREES, AND ARCS

- ♦ **Angles**
- ♦ **Degree Measure of Angles**
- ♦ **Angles and Arcs**
- ♦ **Approximation of Earth's Circumference**
- ♦ **Approximation of Sun's Diameter**

Angles and the degree measure of an angle are the first concepts introduced in this section. Then, a very useful relationship between angles and arcs of circles is developed. This simple relationship between an angle and an arc will enable us to measure indirectly many useful quantities such as the circumference of the earth (a classic Greek accomplishment) and the sun's diameter. (If you are rusty on certain geometric relationships and facts, refer to Appendix C as needed.)

♦ Angles

Central to the study of trigonometry is the concept of angle. An **angle** is formed by rotating a half line, called a **ray,** around its endpoint. One ray k, called the

initial side of the angle, remains fixed; a second ray *l*, called the **terminal side** of the angle, starts in the initial side position and rotates around the common endpoint *P* in a plane until it reaches its terminal position. The common endpoint *P* is called the **vertex.** (See Figure 1.)

FIGURE 1
Angle *θ*: A ray rotated around its endpoint

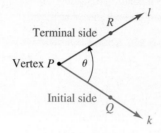

We may refer to the angle in Figure 1 in any of the following ways:

Angle *θ*	∠ *θ*
Angle *P*	∠ *P*
Angle *QPR*	∠ *QRP*

The symbol ∠ denotes *angle*. The Greek letters theta (*θ*), alpha (*α*), beta (*β*), and gamma (*γ*) are often used to name angles.

There is no restriction on the amount or direction of rotation in a given plane. When the terminal side is rotated counterclockwise, the angle formed is **positive** (see Figures 1 and 2(a)); when it is rotated clockwise, the angle formed is **negative** (see Figure 2(b)). Two different angles may have the same initial and terminal sides, as shown in Figure 2(c). Such angles are said to be **coterminal.** In this chapter we will concentrate on positive angles; we will consider more general angles in detail in subsequent chapters.

FIGURE 2

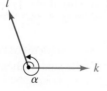

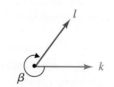

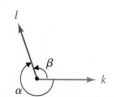

(a) Positive angle *α*:   (b) Negative angle *β*:   (c) *α* and *β* are
    Counterclockwise   Clockwise rotation   coterminal
    rotation

◆ Degree Measure of Angles

To compare angles of different sizes a standard unit of measure is necessary. Just as a line segment can be measured in inches, meters, or miles, an angle is measured in *degrees* or *radians*. (We will postpone our discussion of radian measure of angles until Section 2.1.)

DEGREE MEASURE OF ANGLES

An angle formed by one complete revolution of the terminal side in a counterclockwise direction has measure 360 degrees, written 360°. An angle of one degree measure, written 1°, is formed by $\frac{1}{360}$ of one complete revolution in a counterclockwise direction.

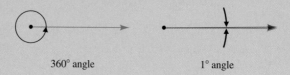

360° angle   1° angle

Angles of measure 90° and 180° represent $\frac{90}{360} = \frac{1}{4}$ and $\frac{180}{360} = \frac{1}{2}$ of complete revolutions, respectively. A 90° angle is called a **right angle** and a 180° angle is called a **straight angle.** An **acute angle** has angle measure between 0° and 90°. An **obtuse angle** has angle measure between 90° and 180°. (See Figure 3.)

FIGURE 3
Special angles

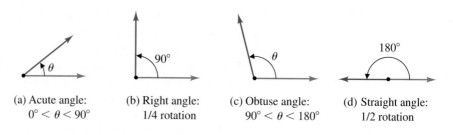

(a) Acute angle: (b) Right angle: (c) Obtuse angle: (d) Straight angle:
$0° < \theta < 90°$ 1/4 rotation $90° < \theta < 180°$ 1/2 rotation

REMARK   In Figure 3 we used θ in two different ways: to name an angle and to represent the measure of an angle. This usage is common; the context will dictate the interpretation.   ◇

Two positive angles are **complementary** if the sum of their measures is 90°; they are **supplementary** if the sum of their measures is 180° (see Figure 4).

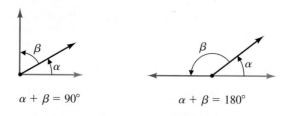

$\alpha + \beta = 90°$ $\alpha + \beta = 180°$

(a) Complementary angles   (b) Supplementary angles

FIGURE 4

A degree can be divided using decimal notation. For example, 36.25° represents an angle of degree measure 36 plus one-fourth of 1 degree. A degree can also be divided into minutes and seconds just as an hour is divided into minutes and seconds. Each degree is divided into 60 equal parts called minutes ('), and each minute is divided into 60 equal parts called seconds ("). Thus,

   5°12′32″

is a concise way of writing 5 degrees, 12 minutes, and 32 seconds.

EXAMPLE 1   Converting Degrees and Minutes to Minutes

How many minutes are in 5°45′?

SOLUTION   5°45′ = (5 · 60 + 45)′ = 345′   5 × 60 + 45 =   ◆

MATCHED PROBLEM 1*   How many seconds are in 2°20′13″?

Most scientific calculators compute with degrees in decimal form, and many of these calculators can directly convert a degree-minute-second representation into a decimal equivalent and vice versa. Look for keys of the form DMS and  DD , or their equivalents. As you probably guessed, DMS represents *Degrees-Minutes-Seconds* and DD represents *Decimal Degrees*.

 Caution   Consult the owner's manual for your particular calculator. Most graphing calculators use a different sequence of steps from those used by scientific calculators.
   ◇

CONVERSION ACCURACY

If an angle is measured to the nearest second, the converted decimal form should not go beyond three decimal places.

EXAMPLE 2   From Degree-Minute-Second Form to Decimal Form

Convert 12°6′23″ to decimal degrees.

---

* Answers to matched problems are located at the end of each section just before the exercise set.

SOLUTION   **Method I Multi-Step Conversion**

Since

$$6' = \left(\frac{6}{60}\right)^\circ \qquad \text{and} \qquad 23'' = \left(\frac{23}{3600}\right)^\circ$$

then

$$12°6'23'' = \left(12 + \frac{6}{60} + \frac{23}{3600}\right)^{\circ*}$$

$$\approx 12.106° \qquad\qquad \text{\textit{To three decimal places}}$$

Method II Single-Step Calculator Conversion

Here we use a scientific calculator with a $\boxed{\text{DD}}$ key or its equivalent. $12°6'23''$ is to be entered in the form

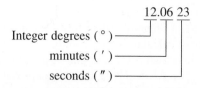

Integer degrees (°)
minutes (′)
seconds (″)

◆

Enter	Press	Display	Round to 3 Decimals
12.0623	DD	12.10638889	12.106

Thus,

$$12°6'23'' = 12.106°$$

⚠ Caution   When entering single-digit minutes or seconds, be sure to place a zero to the left of each; for example, $15°4'7''$ should be entered in the form 15.0407. Look at Example 2 again.   ◇

MATCHED PROBLEM 2   Convert $128°42'8''$ to decimal degrees.

128° 42′ 8″ (math) ▶DEC

---

* Dashed "think boxes" indicate steps that can be performed mentally once a concept or procedure is understood.

It is also useful in some types of problems to convert decimal degree forms to degree-minute-second forms. The next example illustrates the process.

 **EXAMPLE 3**

From Decimal Form to Degree-Minute-Second Form

Convert 35.413° to degree-minute-second form.

SOLUTION **Method I Multi-Step Conversion**

$$
\begin{aligned}
35.413° &= 35°(0.413 \cdot 60)' \\
&= 35°24.78' \\
&= 35°24' (0.78 \cdot 60)'' \\
&\approx 35°24'47''
\end{aligned}
$$

Method II Single-Step Calculator Conversion

Here we use a scientific calculator with a $\boxed{\text{DMS}}$ key or its equivalent. (Most graphing calculators use a different sequence of steps.) ◇

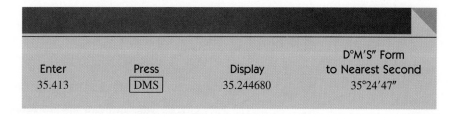

Enter	Press	Display	D°M′S″ Form to Nearest Second
35.413	$\boxed{\text{DMS}}$	35.244680	35°24′47″

Thus, 35.413° = 35°24′47″.

MATCHED PROBLEM 3 Convert 72.103° to degree-minute-second form.

◆ **Angles and Arcs**

Given an arc *RQ* of a circle with center *P*, the angle *RPQ* is said to be the **central angle** that is **subtended** by the arc *RQ* (and vice versa); see Figure 5.

FIGURE 5
Central angle *RPQ* subtended by arc *RQ*

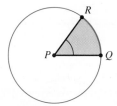

It follows from the definition of degree that a central angle subtended by an arc $\frac{1}{4}$ the circumference of a circle has degree measure 90; $\frac{1}{2}$ the circumference of a circle, degree measure 180, and the whole circumference of a circle, degree measure 360. In general, to determine the degree measure of an angle θ subtended by an arc of *s* units for a circle with circumference *C* units, use the following proportion:

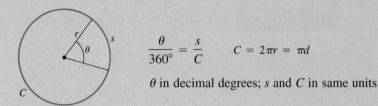

PROPORTION RELATING CENTRAL ANGLES AND ARCS

$$\frac{\theta}{360°} = \frac{s}{C} \qquad C = 2\pi r = \pi d$$

θ in decimal degrees; s and C in same units

If we know any two of the three quantities s, C, or θ, we can find the third by using simple algebra. For example, in a circle with circumference 72 in., the degree measure of a central angle θ subtended by an arc of length 12 in. is given by

$$\theta = \frac{12}{72} \cdot 360° = 60° \qquad \boxed{12}\ \boxed{\div}\ \boxed{72}\ \boxed{\times}\ \boxed{360}\ \boxed{=}$$

REMARK   Note that "an angle of 6°" means "an angle of degree measure 6," and "$\theta = 72°$" means "the degree measure of angle θ is 72".   ◇

◆ Approximation of Earth's Circumference

The early Greeks were aware of the proportion relating central angles and arcs, which Eratosthenes (240 BC) used in his famous calculation of the circumference of the earth. He reasoned as follows: It was well-known that at Syene (now Aswan), during the summer solstice, the noon sun was reflected on the water in a deep well (this meant the sun shone straight down the well and must be directly overhead). Eratosthenes reasoned that if the sun rays entering the well were continued down into the earth, they would pass through its center (see Figure 6). On the same day at the same time, 5,000 stadia (approx. 500 mi) due north, in Alexandria, sun rays crossed a vertical pole at an angle of 7.5° as indicated in

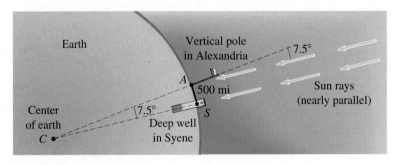

FIGURE 6
Estimating the earth's circumference

Figure 6. Since sun rays are very nearly parallel when they reach the earth, Eratosthenes concluded that $\angle ACS$ was also 7.5°. (Why?)*

Even though Eratosthenes' reasoning was profound, his final calculation of the circumference of the earth requires only elementary algebra:

$$\frac{s}{C} = \frac{\theta°}{360°}$$

$$\frac{500 \text{ mi}}{C} \approx \frac{7.5°}{360°}$$

$$C \approx \frac{360}{7.5}(500 \text{ mi}) = 24,000 \text{ mi} \quad \boxed{360} \; \boxed{\div} \; \boxed{7.5} \; \boxed{\times} \; \boxed{500} \; \boxed{=}$$

The value calculated today is 24,875 mi.

The diameter of the earth (D) and the radius (R) can be found from this value using the formulas $C = 2\pi R$ and $D = 2R$ from plane geometry.[†] [*Note:* $\pi = C/D$ for all circles.]

◆ EXAMPLE 4   Arc Length

How large an arc is intercepted by a central angle of 6.23° on a circle with radius 10 cm? (Use $\pi \approx 3.14$ and compute the answer to two decimal places.)

SOLUTION   Since

$$\frac{s}{C} = \frac{\theta°}{360°} \quad \text{and} \quad C = 2\pi R$$

then

---

* If line p crosses parallel lines m and n, then angles α and β have the same measure.

† The constant π has a long and interesting history; a few important dates are listed below:

1650 B.C.	Rhind Papyrus	$\pi \approx \frac{256}{81} = 3.16049\ldots$
370 B.C.	Euclid	$3 < \pi < 4$
240 B.C.	Archimedes	$3\frac{10}{71} < \pi < 3\frac{1}{7}$ ($3.1408\ldots < \pi < 3.1428\ldots$)
A.D. 264	Liu Hui	$\pi \approx 3.14159$
A.D. 470	Tsu Ch'ung-chih	$\pi \approx \frac{355}{113} = 3.1415929\ldots$
A.D. 1674	Leibniz	$\pi = 4(1 - \frac{1}{3} + \frac{1}{5} - \frac{1}{7} + \frac{1}{9} - \frac{1}{11} + \cdots)$
		$\approx 3.1415926535897932384626$
		(This and other series can be used to compute π to any decimal accuracy desired.)
A.D. 1761	Johann Lambert	Showed π to be irrational (π as a decimal is nonrepeating and nonterminating)

$$\frac{s}{2\pi R} = \frac{\theta°}{360°}$$ *Replace C with* $2\pi R$

$$\frac{s}{2(3.14)(10 \text{ cm})} \approx \frac{6.23°}{360°}$$

$$s \approx \frac{2(3.14)(10 \text{ cm})(6.23)}{360} \approx 1.09 \text{ cm}$$

| 2 | × | 3.14 | × | 10 | × | 6.23 | ÷ | 360 | = |

◆

MATCHED PROBLEM 4   How large an arc is intercepted by a central angle of $50.73°$ on a circle with radius 5 m? (Use $\pi \approx 3.14$ and compute the answer to two decimal places.)

◆   **Approximation of Sun's Diameter**

◆   EXAMPLE 5   **Sun's Diameter**

If the distance from the earth to the sun is about 93,000,000 mi, find the approximate diameter of the sun if it subtends an angle of $\frac{1}{2}°$ on the surface of the earth. (Use $\pi \approx 3.14$.)

SOLUTION   For small central angles in circles with very large radii, the **intercepted arc** (arc opposite the central angle) and its **chord** (the straight line joining the end points of the arc) are approximately the same length (see Figure 7). We thus use the intercepted arc to approximate its chord in many practical problems, particularly when the length of the intercepted arc is easier to compute. We apply these ideas to finding the diameter of the sun to the nearest 10,000 mi as follows:

$$\frac{s}{2\pi R} = \frac{\theta°}{360°}$$

$$s = \frac{2\pi R\theta}{360} \approx \frac{2(3.14)(93,000,000 \text{ mi})(0.5)}{360} \approx 810,000 \text{ mi}$$

◆

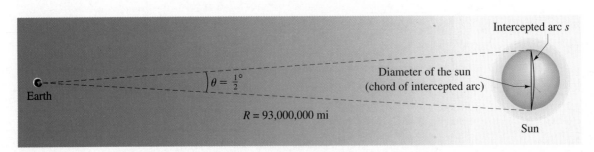

Intercepted arc s

Diameter of the sun
(chord of intercepted arc)

$\theta = \frac{1}{2}°$

Earth

$R = 93,000,000$ mi

Sun

FIGURE 7

MATCHED PROBLEM 5   If the moon subtends an angle of about $\frac{1}{2}°$ on the surface of the earth when it is 239,000 mi from the earth, estimate its diameter. (Use $\pi \approx 3.14$ and round to the nearest 10 mi.)

Answers to   **1.** 8,413″   **2.** 128.702°   **3.** 72°6′11″   **4.** 4.42 m
Matched Problems   **5.** 2,080 mi

EXERCISE 1.1

Throughout the text the mechanical problems in most exercises sets are divided into A, B, and C groupings: The A problems are easy and routine, the B problems are more challenging but still emphasize mechanics, and the C problems are a mixture of difficult mechanics and theory.

A *Indicate the number of degrees in the angle formed by the terminal side rotating counterclockwise through*

1. $\frac{1}{2}$ revolution   **2.** $\frac{1}{4}$ revolution

3. $\frac{1}{8}$ revolution   **4.** $\frac{1}{3}$ revolution

5. $\frac{2}{3}$ revolution   **6.** $\frac{1}{12}$ revolution

Identify the following angles as acute, right, obtuse, or straight. If the angle is none of these, say so.

7. 123°	**8.** 18°	**9.** 180°
10. 90°	**11.** 45°	**12.** 91°
13. 270°	**14.** 225°	

Convert to the indicated quantities:

15. 10°33′ = ?′   **16.** 40°27′ = ?′   **17.** 1°10′12″ = ?″

18. 3°2′25″ = ?″   **19.** 72′ = ?°?′   **20.** 84″ = ?′?″

B *Change to decimal degrees accurate to three decimal places:*

21. 43°21′4″   **22.** 61°52′11″   **23.** 2°12′47″

24. 23°5′21″   **25.** 103°17′41″   **26.** 228°40′51″

Change to degree-minute-second form.

27. 13.633°   **28.** 22.767°   **29.** 83.017°

30. 74.023°   **31.** 187.204°   **32.** 204.333°

Two angles, α and β, have degree measures as indicated. Write α < β or α > β as appropriate.

33. α = 27°9′17″
    β = 27.163°

34. α = 123°16′5″
    β = 123.254°

35. α = 12.807°
    β = 12°47′13″

36. α = 68.916°
    β = 68°55′42″

37. α = 248°7′3″
    β = 248.132°

38. α = 341°4′9″
    β = 341.118°

In Problems 39–46, find C, θ, s, or r as indicated, using π ≈ 3.14. Refer to the figure.

Figure for 39–46

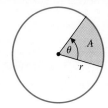

39. C = 1,000 cm,   θ = 36°,   s = ?   (Exact)

40. s = 12 m,   C = 108 m,   θ = ?   (Exact)

41. s = 25 km,   θ = 20°,   C = ?   (Exact)

42. C = 740 mi,   θ = 72°,   s = ?   (Exact)

C **43.** r = 5,400,000 mi,   θ = 2.6°,   s = ?
    (To nearest 10,000 mi)

44. s = 38,000 cm,   θ = 45.3°,   r = ?
    (To nearest 1,000 cm)

45. θ = 12°31′4″,   s = 50.2 cm,   C = ?
    (To nearest 10 cm)

46. θ = 24°16′34″,   s = 14.23 m,   C = ?
    (To one decimal place)

In Problems 47–50, find A or θ as indicated, using π ≈ 3.14. Refer to the figure: The ratio of the area of a sector of a circle (A) to the total area of the circle (πr^2) is the same as the ratio of the central angle of the sector (θ) to 360°. That is,

Figure for 47–50

$$\frac{A}{\pi r^2} = \frac{\theta}{360°}$$

47. r = 25.2 cm,   θ = 47.3°,   A = ?
    (To nearest unit)

48. r = 7.38 ft,   θ = 24.6°,   A = ?
    (To one decimal place)

49. r = 12.6 m,   A = 98.4 m^2,   θ = ?
    (To one decimal place)

50. r = 32.4 in.,   A = 347 in.^{2},   θ = ?
    (To one decimal place)

Applications

Biology: Eye   *In Problems 51–54, find r, θ, or s as indicated, using* π ≈ 3.14. *Refer to the following: The eye is roughly spherical with a spherical bulge in front called the cornea. (Based on the article, "The Surgical Correction of Astigmatism" by Sheldon Rothman and Helen Strassberg in* The UMAP Journal, *Vol. V No. 2, 1984.)*

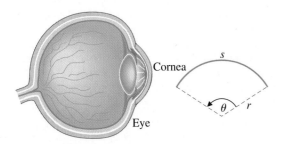

Figure for 51–54

51. $s = 11.5$ mm, $\theta = 118.2°$, $r = ?$
(To two decimal places)

52. $s = 12.1$ mm, $r = 5.26$ mm, $\theta = ?$
(To one decimal place)

53. $\theta = 119.7°$, $r = 5.49$ mm, $s = ?$
(To one decimal place)

54. $\theta = 117.9°$, $s = 11.8$ mm, $r = ?$
(To two decimal places)

In Problems 55–58, find the distance on the surface of the earth between each pair of cities to the nearest mile, given their respective latitudes. Latitudes are given to the nearest 10′. Note that each chosen pair of cities has approximately the same longitude (i.e., lies on the same north-south line). Use π ≈ 3.14 *and r* ≈ 3,960 *miles for the earth's radius. The figure shows the situation for San Francisco and Seattle. [Hint: C = 2πr]*

∗**55. Geography/Navigation**   San Francisco, CA, 37°50′N; Seattle, WA, 47°40′N∗

∗**56. Geography/Navigation**   Phoenix, AZ, 33°30′N; Salt Lake City, UT, 40°40′N

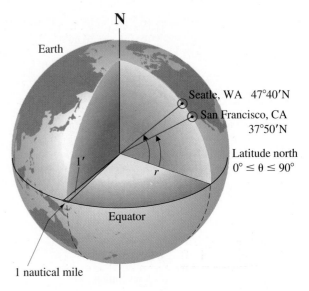

Figure for 55–62

∗**57. Geography/Navigation**   Dallas, TX, 32°50′N; Lincoln, NE, 40°50′N

∗**58. Geography/Navigation**   Buffalo, NY, 42°50′N; Durham, NC, 36°0′N

For Problems 59–62, find the distance on the earth's surface, to the nearest nautical mile, between each pair of cities. A nautical mile is the length of 1′ of arc on the equator or any other circle on the surface of the earth having the same center as the equator. See the figure above. Since there are 60 × 360 = 21,600 minutes in 360°, the length of the equator is 21,600 nautical miles.

59. Geography/Navigation   San Francisco, CA, 37°50′N; Seattle, WA, 47°40′N

60. Geography/Navigation   Phoenix, AZ, 33°30′N; Salt Lake City, UT, 40°40′N

61. Geography/Navigation   Dallas, TX, 32°50′N; Lincoln, NE, 40°50′N

62. Geography/Navigation   Buffalo, NY, 42°50′N; Durham, NC, 36°0′N

63. Diameter of a Distant Object   A round object 1,000 m away subtends an angle of 2°. Approximately what is its diameter (in meters)? (Use π ≈ 3.14 and compute answer to one decimal place.)

∗**64. Satellite Telescopes**   In the Orbiting Astronomical Observatory launched in 1968, ground personnel could direct telescopes in the vehicle with an accuracy of 1′ of arc. They claimed this corresponds to hitting a 25¢ coin at a distance of 100 yd (see the figure on p. 13). Show that

---

* The most difficult problems are double-starred (∗∗); moderately difficult problems are single-starred (∗); easier problems are not marked.

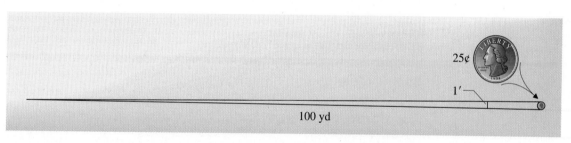

25¢

1'

100 yd

Figure for 64

this claim is approximately correct; that is, show that the length of an arc subtended by 1' at 100 yd is approximately the diameter of a quarter, 0.94 in.

65. Sun Diameter Approximate the diameter of the sun by using the average sun–earth distance of 92,956,000 mi (measured indirectly in 1958 by radar reflection off Venus) and the fact that the diameter of the sun subtends an angle of 32' on the surface of the earth (Use $\pi \approx 3.142$ and compute answer to the nearest 1,000 mi.)

∗∗66. Moon Distance and Diameter Hipparchus calculated the distance from the earth to the moon to be about 33.5 earth diameters. Using Eratosthenes' calculation for the circumference of the earth (24,000 mi), calculate the distance to the moon in miles. (Use $\pi \approx 3.14$.) Estimate the diameter of the moon if it subtends an angle of $\frac{1}{2}°$ on the surface of the earth. (Calculate the distance to the nearest 10,000 mi and the diameter to the nearest 100 mi.)

1.2 SIMILAR TRIANGLES

♦ **Euclid's Theorem and Similar Triangles**

♦ **Applications**

Properties of similar triangles, stated in Euclid's theorem below, are central to this section and form a cornerstone for the development of trigonometry. Euclid (300 B.C.), a Greek mathematician who taught in Alexandria, was one of the most influential mathematicians of all time. He is most famous for writing *The Elements,* a collection of thirteen books on geometry. In the Western World, next to the Bible, *The Elements* is probably the most studied book of all time.

The ideas on indirect measurement presented here are included to help you understand basic concepts. More efficient methods will be developed in the next section.

Since calculators routinely compute to 8 or 10 digits, one could easily believe that a computed result is far more accurate than warranted. Generally, a final result cannot be any more accurate than the least accurate number used in the calculation. Regarding calculation accuracy, we will be guided by Appendix A.3 on significant digits throughout the text.

♦ Euclid's Theorem and Similar Triangles

In Section 1.1, problems were included that required knowledge of the distances from the earth to the moon and the sun. How can inaccessible distances of this type be determined? Surprisingly, the ancient Greeks made fairly accurate calculations of these distances as well as many others. The basis for their methods is the following elementary theorem of Euclid:

EUCLID'S THEOREM

If two triangles are similar, their corresponding sides are proportional.

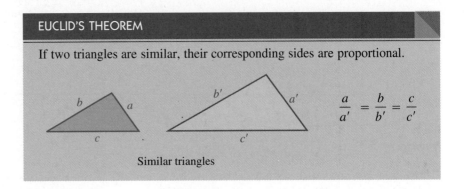

$$\frac{a}{a'} = \frac{b}{b'} = \frac{c}{c'}$$

Similar triangles

REMARK   Recall from plane geometry (see Appendix C.2) that the sum of the measures of the three angles of any triangle is always 180° and that two triangles are similar if two angles of one triangle have the same measure as two angles of the other. If the two triangles happen to be right triangles then they are similar if an acute angle in one has the same measure as an acute angle in the other. Take a moment to draw a few figures and to think about these statements.   ◇

◆   **Applications**

We now solve some elementary problems using Euclid's theorem.

EXAMPLE 1   Height of a Tree

A tree casts a shadow of 32 ft at the same time a vertical yardstick (3.0 ft) casts a shadow of 2.2 ft (see Figure 1). How tall is the tree?

FIGURE 1

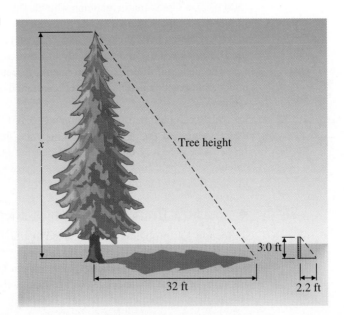

Tree height

x

3.0 ft

32 ft

2.2 ft

SOLUTION   The parallel sun rays make the same angle with the tree and the yardstick. Since both triangles are right triangles and have an acute angle of the same measure, the triangles are similar. The corresponding sides are proportional, and we can write

$$\frac{x}{3.0 \text{ ft}} = \frac{32 \text{ ft}}{2.2 \text{ ft}}$$

$$x = \frac{3.0}{2.2} (32 \text{ ft})$$

$$= 44 \text{ ft} \qquad \text{To two significant digits} \qquad \blacklozenge$$

MATCHED PROBLEM 1   A tree casts a shadow of 31 ft at the same time a 5.0 ft vertical pole casts a shadow of 0.56 ft. How high is the tree?

 **EXAMPLE 2**   Length of an Air Vent

Find the length of the proposed air vent indicated in Figure 2.

FIGURE 2

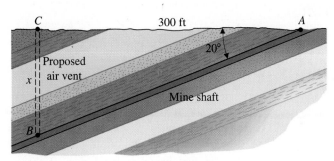

Air vent

SOLUTION   We make a careful scale drawing of the main shaft relative to the proposed air vent as follows: Pick any convenient length, say 2.0 in., for $A'C'$; copy the 20° angle CAB and the 90° angle ACB using a protractor (see Figure 3). Now measure $B'C'$ (approx. 0.7 in.), and set up an appropriate proportion:

$$\frac{x}{0.7 \text{ in.}} = \frac{300 \text{ ft}}{2.0 \text{ in.}}$$

$$x = \frac{0.7}{2.0} (300 \text{ ft})$$

$$\approx 100 \text{ ft} \qquad \text{To one significant digit}$$

FIGURE 3

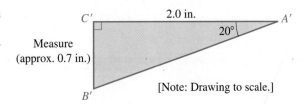

[*Note:* The use of scale drawings for finding indirect measurements is included here only to demonstrate basic ideas. A more efficient method will be developed in Section 1.3.] ◆

MATCHED PROBLEM 2   Suppose in Example 2 that $AC = 500$ ft and $\angle A = 30°$. If in a scale drawing $A'C'$ is chosen to be 3 in. and $B'C'$ is measured as 1.76 in., find BC, the length of the proposed mine shaft.

      Now let us calculate the distance to the moon using Euclid's elementary theorem and a refinement of Eratosthenes' estimate of the circumference of the earth. Suppose stations A and B are established 4,150 mi apart at convenient locations on earth (see Figure 4). At a certain time on a particular night, the moon is directly overhead at A, and the observed angle between a vertical pole at B and the moon is 60.8°. We now have enough information to find the distance from the earth to the moon!

FIGURE 4
Distance from earth to moon

      We first calculate the radius of the earth EB using $C = 2\pi R$, with $C = 24,900$ mi and $\pi \approx 3.14$:

$$R = \frac{C}{2\pi} \approx \frac{24,900}{2(3.14)} \approx 3,960 \text{ mi}$$

Next, we calculate the angle α:*

$$\alpha = \frac{s}{C}(360°) = \frac{4,150 \text{ mi}}{24,900 \text{ mi}}(360°) = 60.0°$$

Since $\angle EBM$ is the supplement of β ($\angle EBM + \beta = 180°$), we have

$$\angle EBM = 180° - \beta = 180° - 60.8° = 119.2°$$

Triangle EBM is determined!
      We now carefully draw a similar triangle $E'B'M'$ to scale, choosing a convenient value for $E'B'$ (see Figure 5).

_____

* More will be said about the correspondence between angle accuracy and side accuracy for triangles in the next section.

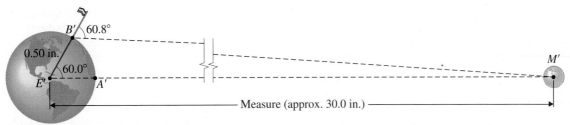

[Note: Drawing to scale.]

FIGURE 5

We then measure $E'M'$ (approx. 30.0 in.) and solve for EM using an appropriate proportion:

$$\frac{EM}{E'M'} = \frac{EB}{E'B'}$$

$$EM = \frac{E'M'}{E'B'}(EB) \approx \frac{30.0 \text{ in.}}{0.50 \text{ in.}}(3{,}960 \text{ mi})$$

$$\approx 240{,}000 \text{ mi} \qquad \text{To two significant digits}$$

Answers to Matched Problems
  1. 280 ft (to two significant digits)
  2. 300 ft (to one significant digit)

EXERCISE 1.2

Round each answer to an appropriate accuracy; use the guidelines from Appendix A.3.

A *Given two similar triangles, as in the figure, find the unknown length indicated.*

FIGURE for 1–6

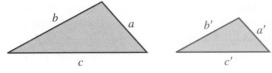

  1. $a = 5$, $b = 15$, $a' = 2$, $b' = ?$
  2. $b = 3$, $c = 21$, $b' = 1$, $c' = ?$
  3. $c = ?$, $a = 12$, $c' = 18$, $a' = 2.0$
  4. $a = 51$, $b = ?$, $a' = 24$, $b' = 8.0$
  5. $b = 52{,}000$, $c = 18{,}000$, $b' = 8.0$, $c' = ?$
  6. $a = 630{,}000$, $b = ?$, $a' = 15$, $b' = 0.50$

B *Given the similar triangles in the figure, find the indicated quantities to two significant digits.*

Figure for 7–18

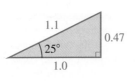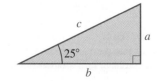

  7. $a = 10.1$ m, $b = ?$, $c = ?$
  8. $a = 5.0$ mi, $b = ?$, $c = ?$
  9. $b = 51$ in., $a = ?$, $c = ?$
  10. $b = 32$ cm, $a = ?$, $c = ?$
  11. $c = 2.0 \times 10^5$ km, $a = ?$, $b = ?$
  12. $c = 5.0 \times 10^{-7}$ m, $a = ?$, $b = ?$
  13. $a = 23.4$ m, $b = ?$, $c = ?$
  14. $a = 63.19$ cm, $b = ?$, $c = ?$
  15. $b = 2.489 \times 10^9$ yd, $a = ?$, $c = ?$
  16. $b = 1.037 \times 10^{13}$ m, $a = ?$, $c = ?$
  17. $c = 8.39 \times 10^{-5}$ mm, $a = ?$, $b = ?$
  18. $c = 2.86 \times 10^{-8}$ cm, $a = ?$, $b = ?$

C *Find the unknown quantities. (If you have a protractor, make a scale drawing and complete the problem using your own measurements and calculations. If you do not have a protractor, use the quantities given in the problem.)*

19. Suppose in the figure that $\angle A = 70°$, $\angle C = 90°$, and $a = 101$ ft. If a scale drawing is made of the triangle by choosing a' to be 2.00 in. and c' is then measured to be 2.13 in., estimate c in the original triangle.

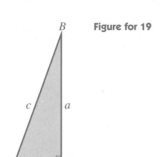

Figure for 19

20. Repeat Problem 19, choosing $a' = 5.00$ in. and c', a measured quantity, to be 5.28 in.

Applications

21. **Tennis**   A ball is served from the center of the baseline into the deuce court. If the ball is hit 9 ft above the ground, travels in a straight line down the middle of the court, and the net is 3 ft high, how far from the base of the net will the ball land if it just clears the top of the net? (See the figure.) Assume all figures are exact and compute the answer to one decimal place.

22. **Tennis**   If the ball in Problem 21 is hit 8.5 ft above the ground, how far away from the base of the net will the ball land? Assume all figures are exact and compute the answer to two decimal places. (Now you can see why tennis players try to spin the ball on a serve so that it curves downward.)

23. **Indirect Measurement**   Find the height of the tree in the figure given that $AC = 24$ ft, $CD = 2.1$ ft, and $DE = 5.5$ ft.

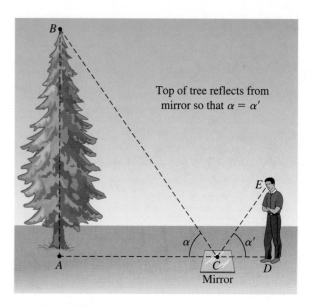

Top of tree reflects from mirror so that $\alpha = \alpha'$

Figure for 23 and 24

24. **Indirect Measurement**   Find the height of the tree in the figure given that $AC = 25$ ft, $CD = 2$ ft 3 in., and $DE = 5$ ft. 9 in.

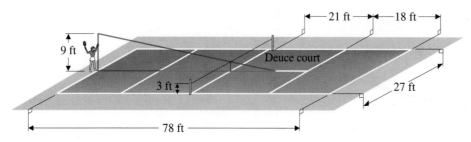

Figure for 21 and 22

Problems 25 and 26 are optional for those who have pro-tractors and can make scale drawings.

25. Astronomy The following figure illustrates a method that is used to determine depths of moon craters from observatories on earth. If sun rays strike the surface of the moon so that ∠ *BAC* = 15° and *AC* is measured to be 4.0 km, how high is the rim of the crater above its floor?

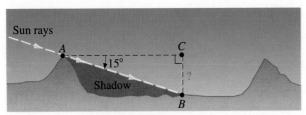

Depth of moon crater

Figure for 25

26. Astronomy The figure illustrates how the height of a mountain on the moon can be determined from earth. If sun rays strike the surface of the moon so that ∠ *BAC* = 28° and *AC* is measured to be 4.0×10^3 m, how high is the mountain?

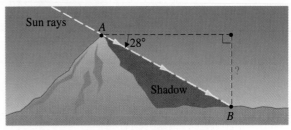

Height of moon mountain

Figure for 26

1.3 TRIGONOMETRIC RATIOS AND RIGHT TRIANGLES

- ◆ **Trigonometric Ratios**
- ◆ **Complementary Angles and Cofunctions**
- ◆ **Calculator Evaluation**
- ◆ **Solving Right Triangles**

If in a right triangle we are given the measure of two sides, or the measure of an acute angle and one side, then the triangle is determined. Our problem is to find the measures of the remaining sides and acute angles. This is called **solving a triangle.** In this section we show how this can be done without using scale drawings. The concepts introduced here will be generalized extensively as we progress through the book.

◆ Trigonometric Ratios

We see in Figure 1 that there are six possible ratios of the sides of a right triangle that can be computed for each angle θ.

FIGURE 1
Six possible ratios of sides

$$\frac{b}{c} \quad \frac{c}{b} \quad \frac{a}{c} \quad \frac{c}{a} \quad \frac{b}{a} \quad \frac{a}{b}$$

These ratios are referred to as **trigonometric ratios,** and because of their importance, each is given a name: sine (sin), cosine (cos), tangent (tan), cosecant (csc), secant (sec), and cotangent (cot). And each is written in abbreviated form as follows:

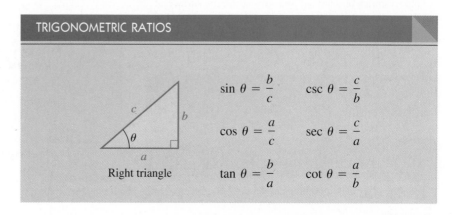

TRIGONOMETRIC RATIOS

Right triangle

$$\sin \theta = \frac{b}{c} \qquad \csc \theta = \frac{c}{b}$$

$$\cos \theta = \frac{a}{c} \qquad \sec \theta = \frac{c}{a}$$

$$\tan \theta = \frac{b}{a} \qquad \cot \theta = \frac{a}{b}$$

Side b is often referred to as the **side opposite** angle θ, a as the **side adjacent** to angle θ, and c as the **hypotenuse.** Using these designations, the ratios become:

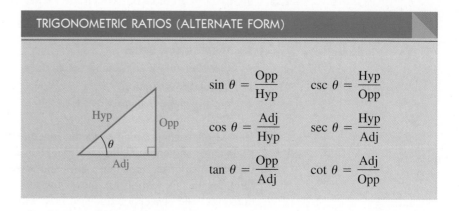

TRIGONOMETRIC RATIOS (ALTERNATE FORM)

$$\sin \theta = \frac{\text{Opp}}{\text{Hyp}} \qquad \csc \theta = \frac{\text{Hyp}}{\text{Opp}}$$

$$\cos \theta = \frac{\text{Adj}}{\text{Hyp}} \qquad \sec \theta = \frac{\text{Hyp}}{\text{Adj}}$$

$$\tan \theta = \frac{\text{Opp}}{\text{Adj}} \qquad \cot \theta = \frac{\text{Adj}}{\text{Opp}}$$

These trigonometric ratios should be learned, because they will be used extensively in the work that follows. It is also important to note that the right angle in a right triangle can be oriented in any position and that the names of the angles are arbitrary, but the hypotenuse is always opposite the right angle.

For all right triangles, no matter how large or small, it follows from Euclid's theorem concerning similar triangles (Section 1.2) that for a given acute angle θ, $\sin \theta$ will always be the same. This is easy to see, since if the measure of the acute angle θ is the same in two right triangles, the triangles are similar. And if they are similar, their corresponding sides are proportional (see Figure 2). If we can compute $\sin \theta$ accurately for one triangle, then we have $\sin \theta$ for all triangles similar to the original! The same reasoning applies to the other five trigonometric ratios.

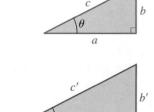

FIGURE 2

Similiar triangles: $\sin \theta = \dfrac{b}{c} = \dfrac{b'}{c'}$

◆ Complementary Angles and Cofunctions

What is the meaning of the prefix *co-* in cosine, cosecant, and cotangent? The *co-* refers to a complementary angle relationship. Recall that two positive angles are complementary if their sum is 90°. The two acute angles in a right triangle are complementary. (Why?)* Referring to the definition of the six trigonometric ratios, we see that:

COMPLEMENTARY RELATIONSHIPS

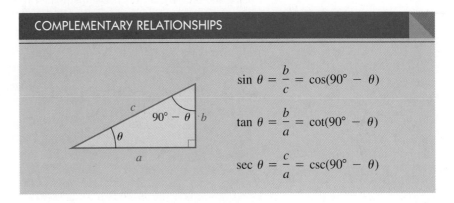

$$\sin \theta = \frac{b}{c} = \cos(90° - \theta)$$

$$\tan \theta = \frac{b}{a} = \cot(90° - \theta)$$

$$\sec \theta = \frac{c}{a} = \csc(90° - \theta)$$

Thus, the sine of θ is the same as the cosine of the complement of θ (which is $90° - \theta$ in the triangle shown), the tangent of θ is the cotangent of the complement of θ, and the secant of θ is the cosecant of the complement of θ. The trigonometric ratios cosine, cotangent, and cosecant are sometimes referred to as the **cofunctions** of sine, tangent, and secant, respectively.

◆ Calculator Evaluation

For the trigonometric ratios to be useful in solving right triangle problems, we must be able to find each for any acute angle. Scientific and graphing calculators can approximate (almost instantly) these ratios to eight or ten significant digits. Scientific and graphing calculators generally use different sequences of steps. Consult the user's manual for your particular calculator. *We will use calculators throughout this book.*

Most scientific calculators have a choice of three trigonometric modes: degree (decimal), radian, or grad. Our interest now is in **degree mode.** Later we will discuss radian mode in detail. The grad mode (of special interest in certain areas of engineering) will not be considered in this book.

⚠ Caution   Refer to the user's manual accompanying your calculator to determine how it is to be set in degree mode, and set it that way. *This is an important step and should not be overlooked.* Many errors can be traced to calculators being set in the wrong mode. ◇

---

* Since the sum of the measures of all three angles in a triangle is 180°, and a right triangle has one 90° angle, the two remaining acute angles must have measures that sum to $180° - 90° = 90°$. Thus, the two acute angles in a right triangle are always complementary.

If you look at the function keys on your scientific calculator, you will find three keys labeled

| sin | | cos | | tan |

These keys are used to find sine, cosine, and tangent ratios, respectively. The calculator also can be used to compute cosecant, secant, and cotangent ratios using the reciprocal* relationships, which follow directly from the definition of the six trigonometric ratios.

RECIPROCAL RELATIONSHIPS FOR $0° < \theta < 90°$

$$\csc \theta \sin \theta = \frac{c}{b} \cdot \frac{b}{c} = 1 \qquad \text{thus} \qquad \csc \theta = \frac{1}{\sin \theta}$$

$$\sec \theta \cos \theta = \frac{c}{a} \cdot \frac{a}{c} = 1 \qquad \text{thus} \qquad \sec \theta = \frac{1}{\cos \theta}$$

$$\cot \theta \tan \theta = \frac{a}{b} \cdot \frac{b}{a} = 1 \qquad \text{thus} \qquad \cot \theta = \frac{1}{\tan \theta}$$

⚠ Caution   When using reciprocal relationships, many students tend to associate cosecant with cosine and secant with sine: just the opposite is correct.   ◇

◆ **EXAMPLE 1**   Evaluating Trigonometric Ratios

Evaluate to four significant digits using a scientific calculator:

(A)   sin 23.72°   (B)   tan 54°37′
(C)   sec 49.31°   (D)   cot 12°52′

SOLUTIONS   (A)   Set calculator in degree mode and use the | sin | key.

   sin 23.72° = 0.4023   *use* | sin | *key*

(B)   Convert to decimal degrees and proceed as in part (A).

   tan 54°37′ = tan(54.61666. . .)°
            = 1.408   *Use* | tan | *key*

---

* Recall that two numbers a and b are **reciprocals** of each other if $ab = 1$; then we may write $a = 1/b$ and $b = 1/a$.

(C) Use the reciprocal relationship sec $\theta = 1/\cos \theta$. $\cos(49.31)\ \boxed{x^{-1}}$

$\quad$ sec $49.31° = 1.534$ *Use* $\boxed{\cos}$ *and* $\boxed{1/x}$ *or* $\boxed{x^{-1}}$ *keys*

(D) Convert to decimal degrees and proceed as in part (C).

$\quad$ cot $12°52' = \cot(12.8666. . .)°$ *Convert to decimal degrees*

$\qquad\qquad\quad = 4.378$ *Use* $\boxed{\tan}$ *and* $\boxed{1/x}$ *or* $\boxed{x^{-1}}$ *keys*

$\diamond$

MATCHED PROBLEM 1   Evaluate to four significant digits using a scientific calculator:

(A)   cos $38.27°$ (B)   sin $37°44'$
(C)   cot $49.82°$ (D)   csc $77°53'$

Now we reverse the process illustrated in Example 1. Suppose we are given

$$\sin \theta = 0.3174$$

How do we find θ? That is, how do we find the acute angle θ whose sine is 0.3174? The solution to this problem is written symbolically as either

$$\theta = \arcsin 0.3174$$ *"arcsin" and "sin^{-1}" both represent the same thing*

or

$$\theta = \sin^{-1} 0.3174$$

Both of these expressions are read "θ is the angle whose sine is 0.3174."

⚠ Caution   It is important to note that $\sin^{-1} 0.3174$ does not mean $1/(\sin 0.3174)$; the -1 "exponent" is a superscript that is part of a function symbol. More will be said about this in Chapter 5, where a detailed discussion of these concepts is given.

$\diamond$

We can find θ directly with a scientific calculator as follows: The $\boxed{\sin^{-1}}$ key or the two key combination $\boxed{\text{inv}}$ $\boxed{\sin}$, both short for **inverse sine,** takes us from a trigonometric sine ratio back to the corresponding acute angle in decimal degrees (if the calculator is in degree mode). Thus, if $\sin \theta = 0.3174$, then we can write either $\theta = \arcsin 0.3174$ or $\theta = \sin^{-1} 0.3174$. We choose the latter, and proceed as follows:

$\qquad\qquad \theta = \sin^{-1} 0.3174$ *Use* $\boxed{\sin^{-1}}$ *or* $\boxed{\text{inv}}$ $\boxed{\sin}$ *keys*
$\qquad\qquad\quad = 18.506°$ *To three decimal places*
$\qquad\qquad\quad$ or $18°30'22''$ *To nearest second*

✔Check   sin $18.506° = 0.3174$ *Use* $\boxed{\sin}$ *key*

◆   **EXAMPLE 2**   Finding Inverses

Find each acute angle θ to the accuracy indicated:

(A) $\cos \theta = 0.7335$ (To three decimal places)
(B) $\theta = \tan^{-1} 8.207$ (To nearest minute)
(C) $\theta = \arcsin 0.0367$ (To nearest 10′)

SOLUTIONS   First, set calculator in degree mode.

(A)   If $\cos \theta = 0.7335$, then

   $\theta = \cos^{-1} 0.7335$ Use $\boxed{\cos^{-1}}$ or $\boxed{\text{inv}}$ $\boxed{\cos}$ keys
   $\quad = 42.819°$ To three decimal places

(B) $\theta = \tan^{-1} 8.207$ Use $\boxed{\tan^{-1}}$ or $\boxed{\text{inv}}$ $\boxed{\tan}$ keys
   $\quad = 83.053$
   $\quad = 83°3′$ To nearest minute

(C) $\theta = \arcsin 0.0367$ Use $\boxed{\sin^{-1}}$ or $\boxed{\text{inv}}$ $\boxed{\sin}$ keys
   $\quad = 2.103°$
   $\quad = 2°10′$ To nearest 10′   ◆

MATCHED PROBLEM 2   Find each acute angle θ to the accuracy indicated:

(A) $\tan \theta = 1.739$ (To two decimal places)
(B) $\theta = \sin^{-1} 0.2571$ (To nearest 10″)
(C) $\theta = \arccos 0.0367$ (To nearest minute)

   We postpone any further discussion of $\cot^{-1}$, $\sec^{-1}$, and $\csc^{-1}$ until Chapter 5. The preceding discussion will handle all our needs at this time.

◆   **Solving Right Triangles**

Recall that earlier in this section we said: If in a right triangle we are given the measure of two sides, or the measure of an acute angle and one side, then the triangle is determined. Our problem is to find the measures of the remaining sides and acute angles. This is called **solving a triangle.** We are now ready to solve right triangles. Solving right triangles is best illustrated through examples. We note at the outset that accuracy of the computations involved is governed by the following table (which is also reproduced inside the front cover for easy reference):

Angle to nearest	Significant digits for side measure
1°	2
10′ or 0.1°	3
1′ or 0.01°	4
10″ or 0.001°	5

if you measure the angle to the nearest 1°, you need to have two signif digits

When we use the equal sign ($=$) in the following computations, it should be understood that equality holds only to the number of significant digits justified by this table. The approximation symbol ($\approx$) is only used when we want to emphasize the approximation.

◆ **EXAMPLE 3**   Solving a Right Triangle

Solve the right triangle in Figure 3.

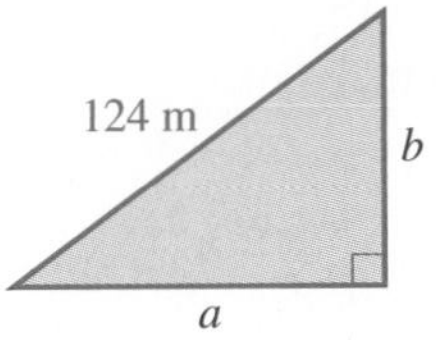

SOLUTION   *Solve for the complementary angle.*

$$90° - \theta = 90° - 35.7° = 54.3° \qquad \text{Remember, 90° is exact.}$$

Solve for b.

Since $\theta = 35.7°$ and $c = 124$ m, we look for a trignonometric ratio that involves θ and c (the known quantities) and b (an unknown quantity). Referring to Definition 1, we see that both sine and cosecant involve all three quantities. We choose sine, and proceed as follows:

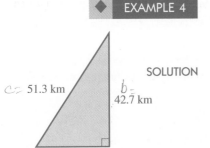

124 m b

a

FIGURE 3

$$\sin \theta = \frac{b}{c}$$

$$\begin{aligned} b &= c \sin \theta \\ &= (124 \text{ m})(\sin 35.7°) \\ &= 72.4 \text{ m} \end{aligned}$$

Solve for a.

Now that we have b, we can use the tangent, cotangent, cosine, or secant to find a. We choose the cosine. Thus,

$$\cos \theta = \frac{a}{c}$$

$$\begin{aligned} a &= c \cos \theta \\ &= (124 \text{ m})(\cos 35.7°) \\ &= 101 \text{ m} \end{aligned}$$

◆

MATCHED PROBLEM 3   Solve the triangle in Example 3 with $\theta = 28.3°$ and $c = 62.4$ cm.

◆ **EXAMPLE 4**   Solving a Right Triangle

Solve the right triangle in Figure 4 for θ and $90° - \theta$ to the nearest $10'$ and for a to three significant digits:

SOLUTION   *Solve for θ.*

$c = 51.3$ km

$b = 42.7$ km

a

FIGURE 4

$$\sin \theta = \frac{b}{c} = \frac{42.7 \text{ km}}{51.3 \text{ km}}$$

$$\sin \theta = 0.832$$

Given the sine of θ, how do we find θ? We can find θ directly using a scientific calculator as discussed in Example 2.

The $\boxed{\sin^{-1}}$ key or the combination $\boxed{\text{inv}}$ $\boxed{\sin}$ takes us from a trigonometric ratio back to the corresponding angle in decimal degrees (if the calculator is in degree mode).

$$\begin{aligned} \theta &= \sin^{-1} 0.832 \qquad \text{\textit{Use} }\boxed{\text{inv}}\ \boxed{\sin}\ \text{\textit{or}}\ \boxed{\sin-1}\ \text{\textit{keys}}\\ &= 56.3° \qquad\quad\ \ (0.3)(60) = 18' \approx 20'\\ &= 56°20' \qquad\quad \text{\textit{To nearest} }10' \end{aligned}$$

Solve for the complementary angle.

$$\begin{aligned} 90° - \theta &= 90° - 56°20'\\ &= 33°40' \end{aligned}$$

Solve for a.

Use cosine, secant, cotangent, or tangent. We will use tangent:

$$\tan \theta = \frac{b}{a}$$

$$\begin{aligned} a &= \frac{b}{\tan \theta}\\ &= \frac{42.7 \text{ km}}{\tan 56°20'}\\ &= 28.4 \text{ km} \end{aligned}$$

◆

✔Check   We check by using the Pythagorean theorem.* (See Figure 5.)

$$28.4^2 + 42.7^2 \overset{?}{=} 51.3^2 \qquad \text{\textit{Compute both sides to three significant digits.}}$$
$$2,630 \overset{\checkmark}{=} 2,630$$

FIGURE 5

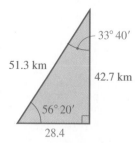

MATCHED PROBLEM 4   Repeat Example 4 with $b = 23.2$ km and $c = 30.4$ km.

In this section we have concentrated on technique. In the next section we will consider a large variety of applications involving the techniques discussed in this section.

_____

* Pythagorean theorem: A triangle is a right triangle if and only if the sum of the squares of the two shorter sides is equal to the square of the longest side:

$$a^2 + b^2 = c^2$$

Answers to
Matched Problems

1. (A) 0.7851 (B) 0.6120 (C) 0.8445 (D) 1.023
2. (A) 60.10° (B) 14°53'50" (C) 87°54'
3. 90° − θ = 61.7°, b = 29.6 cm, a = 54.9 cm
4. θ = 49°40', 90° − θ = 40°20', a = 19.7 km

EXERCISE 1.3

A *Consider the figure and identify each of the given ratios without looking back in the text. For example, csc θ = c/b.*

Figure for 1–12

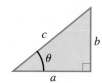

1. cos θ   **2.** sin θ   **3.** tan θ
4. cot θ   **5.** sec θ   **6.** csc θ

Consider the figure and identify each of the ratios by name without looking back in the text. For example, c/a is sec θ.

7. b/c   **8.** a/c   **9.** b/a   **10.** a/b   **11.** c/b   **12.** c/a

Find each to three significant digits.

13. sin 25.6°   **14.** cos 36.4°   **15.** tan 35°20'
16. cot 12°40'   **17.** sec 44.8°   **18.** csc 18.3°
19. cos 72.9°   **20.** sin 63.1°   **21.** cot 54.9°
22. tan 48.3°   **23.** csc 67°30'   **24.** sec 51°40'

B *Find each acute angle θ to the accuracy indicated.*

25. sin θ = 0.8032   (To two decimal places)
26. tan θ = 3.144   (To two decimal places)
27. θ = arccos 0.7153   (To nearest 10')
28. θ = arcsin 0.1152   (To nearest 10')
29. θ = tan⁻¹ 1.948   (To nearest minute)
30. θ = cos⁻¹ 0.5509   (To nearest minute)
31. θ = sin⁻¹ 0.3772   (To nearest second)
32. θ = tan⁻¹ 0.7765   (To nearest second)

Solve the right triangle (labeled as in the figure at the beginning of the exercise) given the information in each problem.

33. θ = 58°40', c = 15.0 mm
34. θ = 62°10', c = 33.0 cm

35. θ = 83.7°, b = 3.21 km
36. θ = 32.4°, a = 42.3 m
37. θ = 71.5°, b = 12.8 in.
38. θ = 44.5°, a = 2.30 × 10⁶ m
39. a = 22.4 cm, 90° − θ = 33°40'
40. c = 3.45 in., 90° − θ = 17°50'
41. b = 63.8 ft, c = 134 ft (Angles to nearest 10')
42. b = 22.0 km, a = 46.2 km (Angles to nearest 10')
43. b = 132 mi, a = 108 mi (Angles to nearest 0.1°)
44. a = 134 m, c = 182 m (Angles to nearest 0.1°)

In Problems 45–48, verify each statement for the indicated values.

45. (sin θ)² + (cos θ)² = 1
   (A) θ = 11° (B) θ = 6.09° (C) θ = 43°24'47"
46. (sin θ)² + (cos θ)² = 1
   (A) θ = 34° (B) θ = 37.281° (C) θ = 87°23'41"
47. sin θ − cos(90° − θ) = 0
   (A) θ = 19° (B) θ = 49.06° (C) θ = 72°51'12"
48. tan θ − cot(90° − θ) = 0
   (A) θ = 17° (B) θ = 27.143° (C) θ = 14°12'33"

C *In Problems 49–56 solve the right triangles (labeled as in the figure at the beginning of the exercise).*

49. θ = 37.46°, b = 5.317 cm
50. θ = 29.83°, c = 4.032 m
51. a = 23.82 mi, θ = 83°12'
52. a = 6.482 m, θ = 35°44'
53. b = 42.39 cm, a = 56.04 cm
   (Angles to nearest 1')
54. a = 123.4 ft, c = 163.8 ft
   (Angles to nearest 1')
55. b = 35.06 cm, c = 50.37 cm
   (Angles to nearest 0.01°)
56. b = 5.207 mm, a = 8.030 mm
   (Angles to nearest 0.01°)

57. Show that $(\sin \theta)^2 + (\cos \theta)^2 = 1$, using Definition 1 and the Pythagorean theorem.

58. Without looking back in the text, show that for each acute angle θ:

(A) $\csc \theta = \dfrac{1}{\sin \theta}$ (B) $\cos(90° - \theta) = \sin \theta$

59. Without looking back in the text, show that for each acute angle θ:

(A) $\cot \theta = \dfrac{1}{\tan \theta}$ (B) $\csc(90° - \theta) = \sec \theta$

60. Without looking back in the text, show that for each acute angle θ:

(A) $\sec \theta = \dfrac{1}{\cos \theta}$ (B) $\cot(90° - \theta) = \tan \theta$

Problems 61–66 refer to the figure, where O is the center of a circle of radius 1, θ is the acute angle AOD, D is the intersection point of the terminal side of angle θ with the circle, and EC is tangent to the circle at D.

61. Geometric Interpretation of Trigonometric Ratios
Show that

(A) $\sin \theta = AD$ (B) $\tan \theta = DC$ (C) $\csc \theta = OE$

62. Geometric Interpretation of Trigonometric Ratios
Show that

(A) $\cos \theta = OA$ (B) $\cot \theta = DE$ (C) $\sec \theta = OC$

63. Geometric Interpretation of Trigonometric Ratios
Explain what happens to each of the following as the acute angle θ approaches 90°:

(A) $\sin \theta$ (B) $\tan \theta$ (C) $\csc \theta$

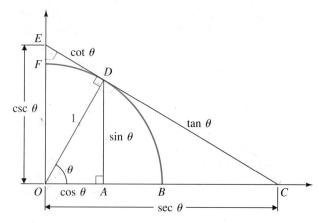

Figure for 61–66

64. Geometric Interpretation of Trigonometric Ratios
Explain what happens to each of the following as the acute angle θ approaches 90°:

(A) $\cos \theta$ (B) $\cot \theta$ (C) $\sec \theta$

65. Geometric Interpretation of Trigonometric Ratios
Explain what happens to each of the following as the acute angle θ approaches 0°:

(A) $\cos \theta$ (B) $\cot \theta$ (C) $\sec \theta$

66. Geometric Interpretation of Trigonometric Ratios
Explain what happens to each of the following as the acute angle θ approaches 0°:

(A) $\sin \theta$ (B) $\tan \theta$ (C) $\csc \theta$

1.4   RIGHT TRIANGLE APPLICATIONS

Now that you know how to solve right triangles, we can consider a variety of interesting and significant applications.

◆   EXAMPLE 1   Mine Shaft Application

Solve the mine shaft problem in Example 2, Section 1.2. See Figure 1.

FIGURE 1

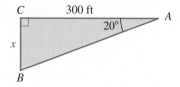

SOLUTION

$$\tan \theta = \frac{\text{Opp}}{\text{Adj}}$$

$$\tan 20° = \frac{x}{300 \text{ ft}}$$

$$x = (300 \text{ ft})(\tan 20°) = 100 \text{ ft} \quad \text{To one significant digit} \quad \blacklozenge$$

MATCHED PROBLEM 1   Solve the mine shaft problem in Example 1 if $AC = 500$ ft and $\angle A = 30°$.

Before proceeding further, we introduce two new terms: **angle of elevation** and **angle of depression.** An angle measured from the horizontal upward is called an angle of elevation; one measured from the horizontal downward is called an angle of depression (see Figure 2).

FIGURE 2

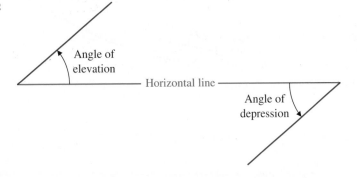

◆ EXAMPLE 2   Depth of the Grand Canyon

From an aerial photograph, a portion of the Grand Canyon is found to be 8.72 km wide. From one rim of the canyon, the angle of depression to the bottom of the other side of the canyon is measured to be 6.6°. How deep is the canyon? (Assume both rims are the same altitude above sea level.)

SOLUTION   We first sketch a picture (Figure 3) and label the known parts (change kilometers to meters).

FIGURE 3

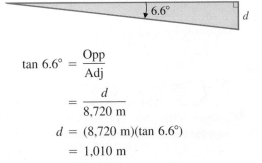

$$\tan 6.6° = \frac{\text{Opp}}{\text{Adj}}$$

$$= \frac{d}{8,720 \text{ m}}$$

$$d = (8,720 \text{ m})(\tan 6.6°)$$

$$= 1,010 \text{ m} \qquad \blacklozenge$$

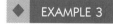   The horizontal shadow of a vertical tree is 23.4 m long when the angle of elevation of the sun is 56.3°. How tall is the tree?

Astronomy

If we know that the distance from the earth to the sun is approximately 93,000,000 mi, and we find that the largest angle between the earth-sun line and the earth-Venus line is 47°, how far is Venus from the sun? (Assume that the earth and Venus have circular orbits around the sun—see Figure 4.)

FIGURE 4

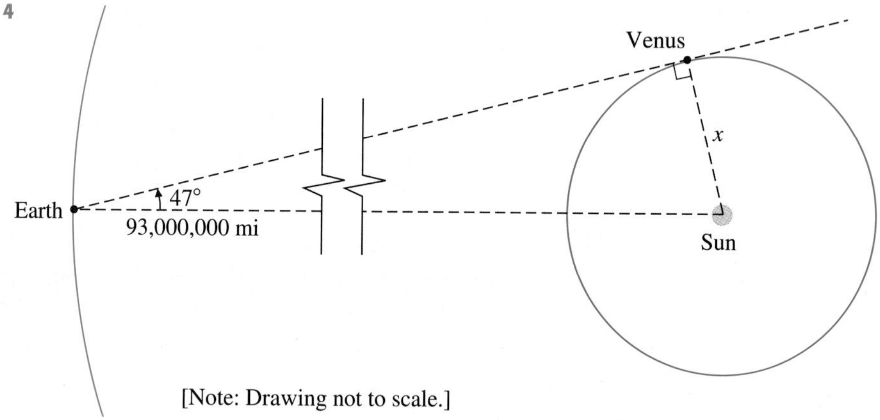

Earth ·—···47°
93,000,000 mi

Venus

x

Sun

[Note: Drawing not to scale.]

SOLUTION   The earth-Venus line at its largest angle to the earth-sun line must be tangent to Venus's orbit. Thus, from plane geometry, the Venus-sun line must be at right angles to the earth-Venus line at this time. The sine ratio involves two known quantities and the unknown distance from Venus to the sun. To find *x* we proceed as follows:

$$\sin 47° = \frac{x}{93,000,000}$$

$$x = 93,000,000 \sin 47°$$

$$= 68,000,000 \text{ mi}$$

◆

MATCHED PROBLEM 3   If the largest angle that the earth-Mercury line makes with the earth-sun line is 28°, how far is Mercury from the sun? (Assume circular orbits.)

Coastal Piloting

A boat is cruising along the coast on a straight course. A rocky point is sighted at an angle of 31° from the course. After continuing 4.8 mi, another sighting is taken and the point is found to be 55° from the course (see Figure 5). How close will the boat come to the point?

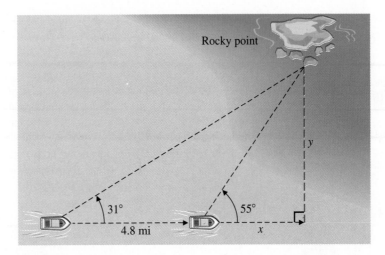

FIGURE 5

SOLUTION   Referring to Figure 5, y is the closest distance that the boat will be to the point. To find y we proceed as follows. From the small right triangle we note that

$$\cot 55° = \frac{x}{y}$$

$$x = y \cot 55° \tag{1}$$

Now, from the large right triangle, we see that

$$\cot 31° = \frac{4.8 + x}{y}$$

$$y \cot 31° = 4.8 + x \tag{2}$$

Substituting equation (1) into (2), we obtain

$$y \cot 31° = 4.8 + y \cot 55°$$

$$y \cot 31° - y \cot 55° = 4.8$$

$$y(\cot 31° - \cot 55°) = 4.8$$

$$y = \frac{4.8}{\cot 31° - \cot 55°}$$

$$y = 5.0 \text{ mi} \qquad \blacklozenge$$

MATCHED PROBLEM 4   Repeat Example 4 after replacing 31° with 28°, 55° with 49°, and 4.8 mi with 5.5 mi.

Answers to   **1.** 300 ft (to one significant digit)   **2.** 35.1 m
Matched Problems **3.** 44,000,000 mi   **4.** 5.4 mi

EXERCISE 1.4

Applications

A **1. Construction**   A ladder 8.0 m long is placed against a building as indicated in the figure. How high will the top of the ladder reach up the building?

Figure for 1

8.0 m

61°

2. Construction   In Problem 1, how far is the foot of the ladder from the wall of the building?

3. Boat Safety   Use the information in the figure to find the distance *x* from the boat to the base of the cliff.

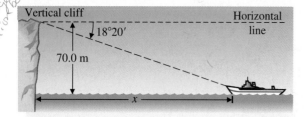

Vertical cliff   Horizontal line

18°20′

70.0 m

x

Figure for 3

4. Boat Safety   In Problem 3, how far is the boat from the top of the cliff?

5. Geography on the Moon   Find the depth of the moon crater in Problem 25, Exercise 1.2.

6. Geography on the Moon   Find the height of the mountain on the moon in Problem 26, Exercise 1.2.

7. Flight Safety   A glider is flying at an altitude of 8,240 m. The angle of depression from the glider to the control tower at an airport is 15°40′. What is the horizontal distance (in kilometers) from the glider to a point directly over the tower?

8. Flight Safety   The height of a cloud or fog cover over an airport can be measured as indicated in the figure. Find *h* in meters if *b* = 1.00 km and α = 23.4°.

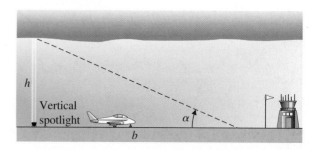

h

Vertical spotlight

α

b

Figure for 8

9. Space Flight   The figure shows the reentry flight pattern of a space shuttle. If at the beginning of the final approach the shuttle is at an altitude of 3,300 ft and its ground distance is 8,200 ft from the beginning of the landing strip, what glide angle must be used for the shuttle to touch down at the beginning of the landing strip?

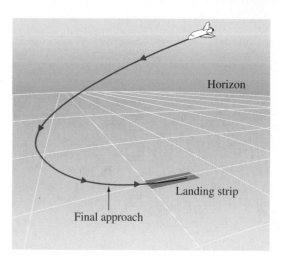

Horizon

Landing strip

Final approach

Figure for 9

10. Space Flight If at the beginning of the final approach the shuttle in Problem 9 is at an altitude of 3,600 ft and its ground distance is 9,300 ft from the beginning of the landing strip, what glide angle must be used for the shuttle to touch down at the beginning of the landing strip?

B **11. Architecture** An architect who is designing a two story house in a city with a 40°N latitude wishes to control sun exposure on a south-facing wall. Consulting an architectural standards reference book, she finds that at this latitude the noon summer solstice sun has a sun angle of 75° and the noon winter solstice sun has a sun angle of 27° (see the figure).

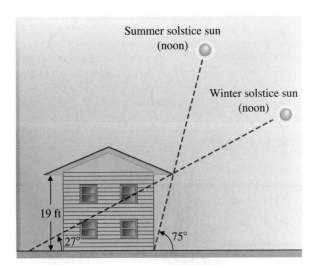

Figure for 11

(A) How much roof overhang should she provide so that at noon on the day of the summer solstice the shadow of the overhang will reach the bottom of the south-facing wall?

(B) How far down the wall will the shadow of the over-hang reach at noon on the day of the winter solstice?

12. Architecture Repeat Problem 11 for a house located at 32°N latitude, where the summer solstice sun angle is 82° and the winter solstice sun angle is 35°.

13. Geometry What is the altitude of an equilateral triangle with side 4.0 m? [An equilateral triangle has all sides (and all angles) equal.]

14. Geometry The altitude of an equilateral triangle is 5.0 cm. What is the length of a side?

15. Geometry Find the length of one side of a nine-sided regular polygon inscribed in a circle with radius 8.32 cm (see the figure).

all sides are of equal length

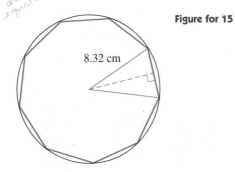

Figure for 15

8.32 cm

16. Geometry What is the radius of a circle inscribed in the polygon in Problem 15? (The circle will be tangent to each side of the polygon and the radius will be perpendicular to the tangent line at the point of tangency.)

17. Lightning Protection A grounded lightning rod on the mast of a sailboat produces a cone of safety as indicated in the figure. If the top of the rod is 67.0 ft above the water, what is the diameter of the circle of safety on the water?

90.0°

Figure for 17

18. Lightning Protection In Problem 17, how high should the top of the lightning rod be above water if the diameter of the circle on the water is to be 100 ft?

19. Diagonal Parking To accommodate cars of most sizes, a parking space needs to contain an 18 ft by 8.0 ft rectangle as shown in the figure. If a diagonal parking space makes an angle of 72° with the horizontal, how long are the sides of the parallelogram that contain the rectangle?

Figure for 19

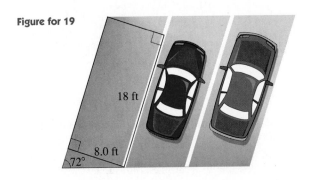

18 ft

8.0 ft

72°

20. Diagonal Parking Repeat Problem 19 using 68° instead of 72°.

21. Earth Radius A person in an orbiting spacecraft (see the figure) *h* mi above the earth sights the horizon on the earth at an angle of depression of *α*. (Recall from geometry that a line tangent to a circle is perpendicular to the radius at the point of tangency.) We wish to find an expression for the radius of the earth in terms of *h* and *α*.

(A) Express cos *α* in terms of *r* and *h*.
(B) Solve the answer to part (A) for *r* in terms of *h* and *α*.
(C) Find the radius of the earth if *α* = 22°47′ and *h* = 335 mi.

22. Orbiting Spacecraft Height A person in an orbiting spacecraft sights the horizon line on earth at an angle of depression *α*. (Refer to the figure in Problem 21.)
(A) Express cos *α* in terms of *r* and *h*.
(B) Solve the answer to part (A) for *h* in terms of *r* and *α*.
(C) Find the height of the spacecraft *h* if the sighted angle of depression *α* = 24°14′ and the known radius of the earth, *r* = 3,960 mi, is used.

23. Navigation Find the radius of the circle that passes through points *P*, *A*, and *B* in part (a) of the figure. [*Hint:* The central angle in a circle subtended by an arc is twice any inscribed angle subtended by the same arc—see figure (b)]. If *A* and *B* are known objects on a maritime navigation chart, then a person on a boat at point *P* can locate the position of *P* on a circle on the chart by sighting the angle *APB* and completing the calculations as sug-

Figure for 23

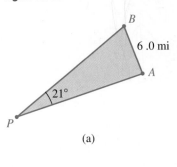

B

6.0 mi

A

21°

P

(a)

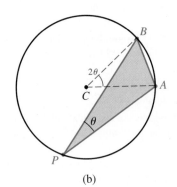

B

2*θ*

C

A

θ

P

(b)

Figure for 21

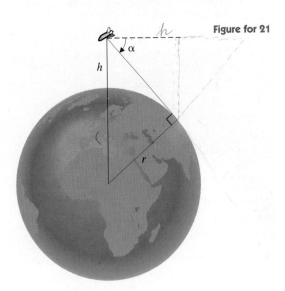

h

α

h

r

gested. By repeating the procedure with another pair of known points, the position of the boat on the chart will be at an intersection point of the two circles.

24. **Navigation**   Repeat Problem 23 using 33° instead of 21° and 7.5 km instead of 6.0 mi.

25. **Geography**   Assume the earth is a sphere (it is nearly so) and that the circumference of the earth at the equator is 24,900 mi. A **parallel of latitude** is a circle around the earth at a given latitude that is parallel to the equator (see the figure). Approximate the length of a parallel of latitude passing through San Francisco, which is at a latitude of 38°N. See the figure, where θ is the latitude, R is the radius of the earth, and r is the radius of the parallel of latitude. In general, show that if E is the length of the equator and L is the length of a parallel of latitude at a latitude θ, then $L = E \cos \theta$.

Figure for 25

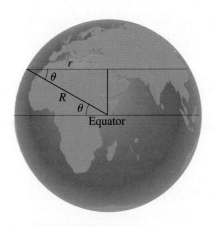
Equator

26. **Geography**   Using the information in Problem 25 and the fact that the circumference of the earth at the equator is 40,100 km, determine the length of the Arctic Circle (66°33′N) in kilometers.

27. **Precalculus: Lifeguard Problem**   A lifeguard sitting in a tower spots a distressed swimmer, as indicated in the figure. To get to the swimmer, the lifeguard must run some distance along the beach at rate p, enter the water, and swim at rate q to the distressed swimmer.
   (A) Express the total time T it takes the lifeguard to reach the swimmer in terms of θ, d, c, p, and q. (In a course in calculus, students are asked to find θ to make the time minimum.)
   (B) Find T (in seconds) if $\theta = 77°$, $d = 380$ m, $c = 76$ m, $p = 6.5$ m/sec, and $q = 1.4$ m/sec.

(C) How far did the lifeguard run on the sand under the conditions in part (B)?

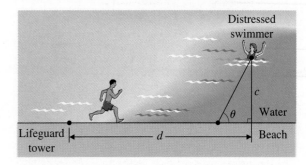

Figure for 27

28. **Precalculus: Lifeguard Problem**   Refer to Problem 27:
   (A) Express the total distance D covered by the lifeguard from the tower to the distressed swimmer in terms of d, c, and θ.
   (B) Find D for $d = 320$ m, $c = 64$ m, and $\theta = 68°$.

29. **Precalculus: Pipeline**   An island is 4 mi offshore in a large shallow bay. A water pipeline is to be run from a water tank on the shore to the island, as indicated in the figure. If the pipeline costs $20,000 per mile in the ocean and on the island and $10,000 per mile along the shore, what will be the cost (to the nearest thousand dollars) of the pipeline for $\theta = 30°$? (In a course in calculus, students are asked to find θ so that the cost is minimum.)

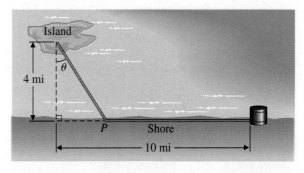

Figure for 29

30. **Precalculus: Pipeline**   Repeat Problem 29 with:
   (A) $\theta = 15°$ (B) $\theta = 45°$

31. Surveying Use the information in the figure to find the height y of the mountain.

$$\tan 42° = \frac{y}{x} \qquad \tan 25° = \frac{y}{1.0 + x}$$

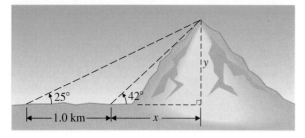

Figure for 31

32. Surveying

(A) Using the figure, show that

$$h = \frac{d}{\cot \alpha - \cot \beta}$$

(B) Use the results in part (A) to find the height of the mountain in Problem 31.

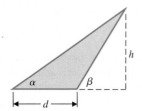

Figure for 32

33. Surveying From the sunroof of Janet's apartment building, the angle of depression to the base of an office building is 51.4° and the angle of elevation to the top of the

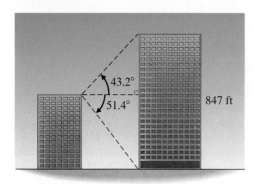

Figure for 33

office building is 43.2° (see the figure). If the office building is 847 ft high, how far apart are the two buildings and how high is the apartment building?

34. Surveying

(A) Using the figure, show that

$$h = \frac{d}{\cot \alpha + \cot \beta}$$

(B) Use the results in part (A) to find the distance between the two buildings in Problem 33.

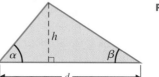

Figure for 34

35. Precalculus: Physics In physics one can show that the velocity (v) of a ball rolling down an inclined plane (neglecting air resistance and friction) is given by

$$v = g(\sin \theta)t$$

where g is the gravitational constant and t is time (see the figure).

Galileo's experiment

Figure for 35

Galileo (1564–1642) used this equation in the form

$$g = \frac{v}{(\sin \theta)t}$$

so he could determine g after measuring v experimentally. (There were no timing devices available then that were accurate enough to measure the velocity of a free-falling body. He had to use an inclined plane to slow the motion down, and then he was able to calculate an approximation for g.) Find g if at the end of 2.00 sec a ball is traveling at 11.1 ft/sec down a plane inclined at 10.0°.

36. Precalculus: Physics In the preceding problem find g if at the end of 1.50 sec a ball is traveling at 12.4 ft/sec down a plane inclined at 15.0°.

C **37.** **Geometry** In part (a) of the figure, *M* and *N* are midpoints to the sides of a square. Find the exact value of sin *θ*. [*Hint:* The solution utilizes the Pythagorean theorem, similar triangles, and the definition of sine. Some useful auxiliary lines are drawn in part (b) of the figure.]

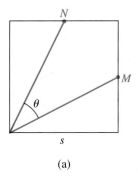

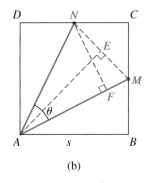

(a)   (b)

Figure for 37

38. **Geometry** Find *R* in the figure. The circle is tangent to all three sides of the isosceles triangle. (An isosceles triangle has two sides equal.) [*Hint:* The radius of a circle and a tangent line are perpendicular at the point of tangency. Also, the altitude of the isosceles triangle will pass through the center of the circle and will divide the original triangle into two congruent triangles.]

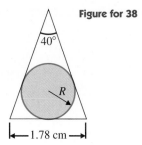

Figure for 38

CHAPTER 1 SUMMARY

1.1
ANGLES, DEGREES, AND ARCS

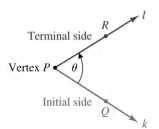

FIGURE 1

An **angle** is formed by rotating a half line, called a **ray,** around its endpoint. See Figure 1: One ray *k*, called the **initial side** of the angle, remains fixed; a second ray *l*, called the **terminal side** of the angle, starts in the initial side position and is rotated around the common endpoint *P* in a plane until it reaches its terminal position. The common endpoint is called the **vertex.** An angle is **positive** if the terminal side is rotated counterclockwise and **negative** if the terminal side is rotated clockwise. Different angles with the same initial and terminal sides are called **coterminal.** An angle of **one degree** is $\frac{1}{360}$ of a complete revolution in a counterclockwise direction. Names for special angles are noted in Figure 2.

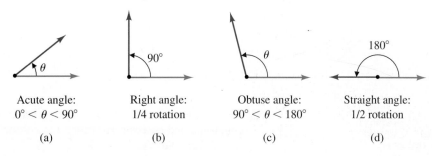

Acute angle: $0° < θ < 90°$

(a)

Right angle: 1/4 rotation

(b)

Obtuse angle: $90° < θ < 180°$

(c)

Straight angle: 1/2 rotation

(d)

FIGURE 2

Two positive angles are **complementary** if the sum of their measures is 90°; they are **supplementary** if the sum of their measures is 180°.

Angles can be represented in terms of **decimal degrees,** or in terms of **minutes** ($\frac{1}{60}$ of a degree) and **seconds** ($\frac{1}{60}$ of a minute). Calculators can be used to convert from decimal degrees to degrees-minutes-seconds and vice versa. The **arc length** s of an arc **subtended** by a **central angle** θ in a circle of radius r (Figure 3) satisfies

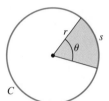

$$\frac{\theta}{360°} = \frac{s}{C} \qquad C = 2\pi r = \pi d$$

θ in decimal degrees; s and C in same units

FIGURE 3

1.2
SIMILAR TRIANGLES

The properties of similar triangles stated in **Euclid's theorem** are central to the development of trigonometry: If two triangles are similar, their corresponding sides are proportional. See Figure 4.

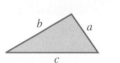
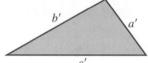

$$\frac{a}{a'} = \frac{b}{b'} = \frac{c}{c'}$$

Similar triangles

FIGURE 4

1.3
TRIGONOMETRIC RATIOS AND RIGHT TRIANGLES

The six **trigonometric ratios** for the angle θ in a right triangle (see Figure 5) with **opposite side** b, **adjacent side** a, and **hypotenuse** c are:

$$\sin \theta = \frac{b}{c} = \frac{\text{Opp}}{\text{Hyp}} \qquad \csc \theta = \frac{c}{b} = \frac{\text{Hyp}}{\text{Opp}}$$

$$\cos \theta = \frac{a}{c} = \frac{\text{Adj}}{\text{Hyp}} \qquad \sec \theta = \frac{c}{a} = \frac{\text{Hyp}}{\text{Adj}}$$

$$\tan \theta = \frac{b}{a} = \frac{\text{Opp}}{\text{Adj}} \qquad \cot \theta = \frac{a}{b} = \frac{\text{Adj}}{\text{Opp}}$$

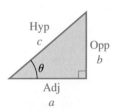

Right triangle

FIGURE 5

The complementary relationships shown at the top of p. 39 illustrate why cosine, cotangent, and cosecant are called the **cofunctions** of sine, tangent, and secant, respectively. See Figure 6.

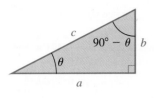

FIGURE 6

Complementary Relationships

$$\sin \theta = \frac{b}{c} = \cos(90° - \theta)$$

$$\tan \theta = \frac{b}{a} = \cot(90° - \theta)$$

$$\sec \theta = \frac{c}{a} = \csc(90° - \theta)$$

Reciprocal Relationships

$$\csc \theta = \frac{1}{\sin \theta}$$

$$\sec \theta = \frac{1}{\cos \theta}$$

$$\cot \theta = \frac{1}{\tan \theta}$$

Solving a right triangle involves finding the measures of the remaining sides and acute angles when given the measure of two sides or the measure of one side and one acute angle. Accuracy of these computations is governed by the following table:

Angle to nearest	Significant digits for side measure
1°	2
10' or 0.1°	3
1' or 0.01°	4
10" or 0.001°	5

1.4
RIGHT TRIANGLE APPLICATIONS

An angle measured upward from the horizontal is called an **angle of elevation** and one measured downward from the horizontal is called an **angle of depression.**

CHAPTER 1 REVIEW EXERCISE

Work through all the problems in this chapter review and check answers in the back of the book. Answers to all review problems are there, and following each answer is a number in italics indicating the section in which that type of problem is discussed. Where weaknesses show up, review appropriate sections in the text.

A **1.** $2°1'20" = \,?"$

2. An arc of $\frac{1}{6}$ the circumference of a circle subtends a central angle of how many degrees?

3. Given two similar triangles, as shown in the figure, find a if $c = 20{,}000$, $a' = 2$, and $c' = 5$.

Figure for 3

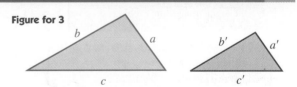

4. Change $36°23'$ to decimal degrees (to two decimal places).

5. If an office building casts a shadow of 31 ft at the same time a vertical yardstick (36 in.) casts a shadow of 2.0 in., how tall is the building?

6. For the triangle shown here, identify each ratio:
(A) sin θ (B) sec θ (C) tan θ
(D) csc θ (E) cos θ (F) cot θ

Figure for 6

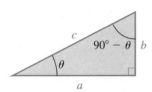

7. Solve the right triangle in Problem 6, given $c = 20.2$ cm and $\theta = 35.2°$.

B **8.** Find the degree measure of a central angle subtended by an arc of 8.00 cm in a circle with circumference 20.0 cm.

9. If the minute hand of a clock is 2.00 in. long, how far does the tip of the hand travel in exactly 20 min? (Use $\pi \approx 3.14$.)

10. Convert 74.273° to degree-minute-second form.

11. Write $\alpha < \beta$ or $\alpha > \beta$ as appropriate for $\alpha = 32°47'18''$ and $\beta = 32.783°$.

12. For the triangles in Problem 3, find b to two significant digits if $a = 4.1 \times 10^{-6}$ mm, $a' = 1.5 \times 10^{-4}$ mm, and $b' = 2.6 \times 10^{-4}$ mm.

13. For the triangle in Problem 6, identify by name each of the following ratios relative to angle θ:
(A) a/c (B) b/a (C) b/c
(D) c/a (E) c/b (F) a/b

14. Solve the right triangle in Problem 6, given $\theta = 62°20'$ and $a = 4.00 \times 10^{-8}$ m.

15. Find each θ to the accuracy indicated.
(A) tan $\theta = 1.662$ (To two decimal places)
(B) $\theta = $ arccos 0.5607   (To nearest 10')
(C) $\theta = \sin^{-1} 0.0138$ (To nearest second)

16. Solve the right triangle in Problem 6, given $b = 13.3$ mm and $a = 15.7$ mm. (Find angles to the nearest 0.1°.)

17. Find the angles in Problem 16 to the nearest 10'.

18. If an equilateral triangle has a side of 10 ft, what is its altitude to two significant digits?

C 19. A curve of a railroad track follows an arc of a circle of radius 1,500 ft. If the arc subtends a central angle of 36°, how far will a train travel on this arc? (Use $\pi \approx 3.14$.)

20. Find the area of a sector with central angle 36.5° in a circle with radius 18.3 ft. Compute your answer to the nearest unit.

21. Solve the triangle in Problem 6, given $90° - \theta = 23°43'$ and $c = 232.6$ km.

22. Solve the triangle in Problem 6, given $a = 2,421$ m and $c = 4,883$ m. (Find angles to the nearest 0.01°.)

23. Use a calculator to find csc 72.3142° to four decimal places.

Applications

24. Precalculus: Shadow Problem   A person is standing 20 ft away from a lamppost. If the lamp is 18 ft above the ground and the person is 5 ft 6 in. tall, how long is the person's shadow?

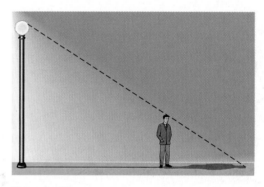

Figure for 24

25. Construction   The front porch of a house is 4.25 ft high. The angle of elevation of a ramp from the ground to the porch is 10.0°. (See the figure.) How long is the ramp? How far is the end of the ramp from the porch?

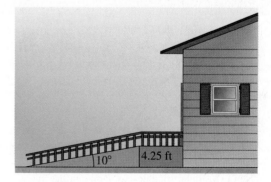

Figure for 25

check did it diff

26. Medicine: Stress Test   Cardiologists give stress tests by having patients walk on a treadmill at various speeds and inclinations. The amount of inclination may be given as an angle or as a percentage (see the figure). Find the angle of inclination if the treadmill is set at a 4% incline. Find the percentage of inclination if the angle of inclination is 4°.

Figure for 26

a Angle of inclination: θ

Percentage of inclination: $\dfrac{a}{b}$

θ

b

27. Geography/Navigation   Find the distance (to the nearest mile) between Green Bay, WI, with latitude 44°31′N and Mobile, AL, with latitude 30°42′N. (Both cities have approximately the same longitude.) Use $\pi \approx 3.14$ and $r \approx 3{,}960$ mi for the earth's radius.

∗28. Precalculus: Balloon Flight   The angle of elevation from the ground to a hot air balloon at an altitude of 2,800 ft is 64°. What will be the new angle of elevation if the balloon descends straight down 1,400 ft?

∗29. Surveying   Use the information in the figure to find the length x of the island.

∗30. Precalculus: Balloon Flight   Two tracking stations 525 m apart measure angles of elevation of a weather balloon to be 73.5° and 54.2°, as indicated in the figure. How high is the balloon at the time of the measurements?

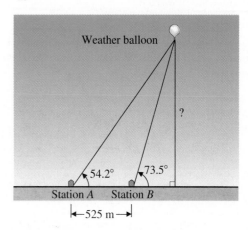

Weather balloon

?

54.2°   73.5°

Station A Station B

|←525 m→|

Figure for 30

NO

∗31. Navigation: Chasing the Sun   Your flight is westward from Buffalo, NY, and you notice the sun just above the horizon. How fast would the plane have to fly to keep the sun in the same position? (The latitude of Buffalo is 42°50′N, the radius of the earth is 3,960 mi, and the earth makes a complete rotation about its axis in 24 hrs.)

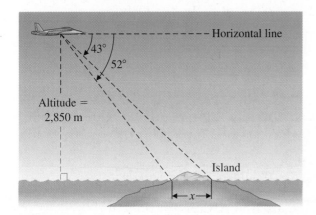

Horizontal line

43°

52°

Altitude = 2,850 m

Island

←x→

Figure for 29

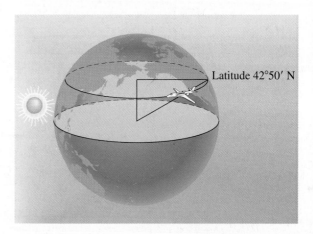

Latitude 42°50′ N

Figure for 31

32. Solar Energy   A truncated conical solar collector is aimed directly at the sun as shown in part (a) of the figure. An analysis of the amount of solar energy absorbed by the collecting disk requires certain equations relating the quantities shown in part (b) of the figure. Find an equation that expresses

(A) β in terms of α

(B) r in terms of α and h

(C) $H - h$ in terms of r, R, and α

(Based on the article, "The Solar Concentrating Properties of a Conical Reflector" by Don Leake in *The UMAP Journal*, Vol. 8 No. 4, 1987.)

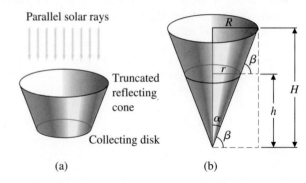

(a)   (b)

Figure for 32

TRIGONOMETRIC FUNCTIONS

☆ Sections marked with a star may be omitted without loss of continuity.

T he trigonometric ratios we studied in Chapter 1 provide a powerful tool for indirect measurement. It was for this purpose only that trigonometry was used for nearly 2,000 years. Surveying, map-making, navigation, construction, military uses, and astronomy catalyzed the extensive development of trigonometry as a tool for indirect measurement.

A turning point in trigonometry occurred after the development of the rectangular coordinate system (credited mainly to the French philosopher-mathematician René Descartes, 1596–1650). The trigonometric ratios, through the use of this system, were generalized into trigonometric functions. This generalization increased their usefulness far beyond the dreams of those originally responsible for this development. The Swiss mathematician Leonhard Euler (1707–1783), probably the greatest mathematician of his century, made substantial contributions in this area. (In fact, there were very few areas in mathematics in which Euler did not make significant contributions.)

Through the demands of modern science, the periodic nature of these new functions soon became apparent, and they were quickly put to use in the study of various types of periodic phenomena. The trigonometric functions began to be used on problems that had nothing whatsoever to do with angles and triangles.

In this chapter we generalize the concept of trigonometric ratios along the lines just suggested. Before we undertake this task, however, we will introduce a new measure of angle called the *radian*.

2.1   DEGREES AND RADIANS

- **Degree and Radian Measure of Angles**
- **Angle in Standard Position**
- **Arc Length and Sector Area**

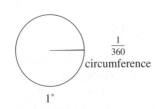

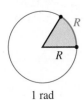

FIGURE 1
Degree and radian measure

◆ Degree and Radian Measure of Angles

In Chapter 1 we defined an angle and its degree measure. Recall that a central angle in a circle has angle measure $1°$ if it subtends an arc $\frac{1}{360}$ of the circumference of the circle. Another angle measure that will be of considerable use to us is *radian measure*. A central angle subtended by an arc of length equal to the radius of the circle is defined to be an angle of **radian measure 1** (see Figure 1). These definitions of angle measure are independent of the size of the defining circle.

Thus, when we write $\theta = 2°$, we are referring to an angle of degree measure 2; when we write $\theta = 2$ rad, we are referring to an angle of radian measure 2.

It follows from the definition that the radian measure of a central angle θ subtended by an arc of length s is found by determining how many times the length of the radius R, used as a unit length, is contained in the arc length s. In terms of a formula, we have the following:

RADIAN MEASURE OF CENTRAL ANGLES

$$\theta = \frac{s}{R} \text{ radians (rad)}$$

Also,

$$s = R\theta$$

[*Note:* s and R must be in the same units.]

Because of their importance, the formulas in the box should be understood before proceeding further.

What is the radian measure of a central angle subtended by an arc of 32 cm in a circle of radius 8 cm?

$$\theta = \frac{32 \text{ cm}}{8 \text{ cm}} = 4 \text{ rad}$$

REMARK   *Radian measure is a unitless number.* The units in which the arc length and radius are measured cancel; hence, we are left with a "unitless," or pure, number. For this reason, the word *radian* is often omitted when we are dealing with the radian measure of angles unless a special emphasis is desired.   ◇

What is the radian measure of an angle of 180°? A central angle of 180° is subtended by an arc $\frac{1}{2}$ of the circumference of the circle. Thus, if C is the circumference of a circle, then $\frac{1}{2}$ of the circumference is given by

$$s = \frac{C}{2} = \frac{2\pi R}{2} = \pi R \qquad \text{and} \qquad \theta = \frac{s}{R} = \frac{\pi R}{R} = \pi \text{ rad}$$

Hence, 180° corresponds to π rad. This is important to remember, since the radian measures of many special angles can be obtained from this correspondence. For example, 90° is 180°/2; therefore, 90° corresponds to $\pi/2$ rad. Since 360° is twice 180°, 360° corresponds to 2π rad. Similarly, 60° corresponds to $\pi/3$ rad, 45° to $\pi/4$ rad, and 30° to $\pi/6$ rad. These special angles and their degree and radian measures will be referred to frequently throughout this book. We summarize these special correspondences here for ease of reference.

Radians	$\pi/6$	$\pi/4$	$\pi/3$	$\pi/2$	π	2π
Degrees	30	45	60	90	180	360

In general, we can use the following proportion to convert degree measure to radian measure and vice versa.

RADIAN–DEGREE CONVERSION FORMULAS

$$\frac{\theta_{\text{deg}}}{180°} = \frac{\theta_{\text{rad}}}{\pi \text{ rad}} \quad \text{or} \quad \begin{aligned} \theta_{\text{deg}} &= \frac{180°}{\pi \text{ rad}} \theta_{\text{rad}} & \text{Radians to degrees} \\[2mm] \theta_{\text{rad}} &= \frac{\pi \text{ rad}}{180°} \theta_{\text{deg}} & \text{Degrees to radians} \end{aligned}$$

[*Note:* The proportion is usually easier to remember. Also, we will omit units in calculations until the final answer. Most calculators have a key labeled π that can be used with these formulas. If your calculator does not have such a key, use $\pi \approx 3.14159$. Some calculators convert decimal degrees directly into radians and vice versa.]

◆ EXAMPLE 1   **Radian-Degree Conversion**

Find each in exact form and to four significant digits.

(A) The degree measure of 1.5 rad
(B) The radian measure of 50°

SOLUTIONS   (A) $\dfrac{\theta_{\text{deg}}}{180°} = \dfrac{\theta_{\text{rad}}}{\pi \text{ rad}}$

$\theta_{\text{deg}} = \dfrac{180°}{\pi \text{ rad}} \theta_{\text{rad}}$

$= \dfrac{180}{\pi}(1.5)$

$= \dfrac{270°}{\pi}$ Exact form

$= 85.94°$ To four significant digits

(B) $\dfrac{\theta_{\text{deg}}}{180°} = \dfrac{\theta_{\text{rad}}}{\pi \text{ rad}}$

$\theta_{\text{rad}} = \dfrac{\pi \text{ rad}}{180°} \theta_{\text{deg}}$

$= \dfrac{\pi}{180}(50)$

$= \dfrac{5}{18} \pi \text{ rad}$ Exact form

$= 0.8727 \text{ rad}$ To four significant digits   ◆

MATCHED PROBLEM 1   Find each in exact form and to three significant digits.

(A) The radian measure of an angle of 20°
(B) The degree measure of an angle of 1 rad

◆ EXAMPLE 2   Radian-Degree Conversion

(A) Find the degree measure of -12 rad to two significant digits.
(B) Find the radian measure of $642°$ to three significant digits.

SOLUTIONS   (A) $\dfrac{\theta_{deg}}{180°} = \dfrac{\theta_{rad}}{\pi \text{ rad}}$

$$\theta_{deg} = \dfrac{180°}{\pi \text{ rad}} \theta_{rad}$$

$$= \dfrac{180}{\pi}(-12)$$

$$= -690° \qquad \textit{To two significant digits}$$

(B) $\dfrac{\theta_{deg}}{180°} = \dfrac{\theta_{rad}}{\pi \text{ rad}}$

$$\theta_{rad} = \dfrac{\pi \text{ rad}}{180°} \theta_{deg}$$

$$= \dfrac{\pi}{180}(642)$$

$$= 11.2 \text{ rad} \qquad \textit{To three significant digits}$$ ◆

MATCHED PROBLEM 2   Find each to three significant digits.

(A) The degree measure of 17.5 rad
(B) The radian measure of $-897°$

◆ Angle in Standard Position

To generalize the concept of trigonometric ratios, we first locate an angle in **standard position** in a rectangular coordinate system. To do this, we place the vertex at the origin and the initial side along the positive x axis. Recall that when the rotation is counterclockwise, the angle is positive and when the rotation is clockwise, the angle is negative. Figure 2 illustrates several angles in standard position.

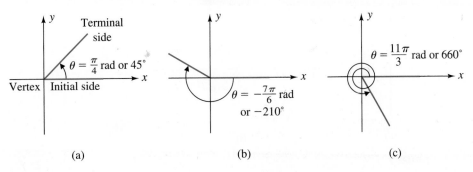

(a)   (b)   (c)

FIGURE 2
Angles in standard position

◆ EXAMPLE 3   Sketching Angles in Standard Position

Sketch the following angles in their standard positions:

(A) −60°   (B) 3π/2 rad   (C) −3π rad   (D) 405°

SOLUTIONS   (A)

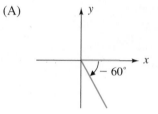

FIGURE 3

(B)

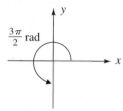

FIGURE 4

(C)

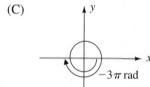

FIGURE 5

(D)

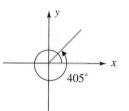

FIGURE 6   ◆

MATCHED PROBLEM 3   Sketch the following angles in their standard positions.

(A) 120°   (B) −π/6 rad   (C) 7π/2 rad   (D) −495°

Two angles are said to be **coterminal** if their terminal sides coincide when both angles are placed in their standard positions in the same rectangular coordinate system. Figure 7 shows two pairs of coterminal angles.

coterminal in the they both end in the same place

does not mater which way they go (+ or −) as long as they end up at the same point

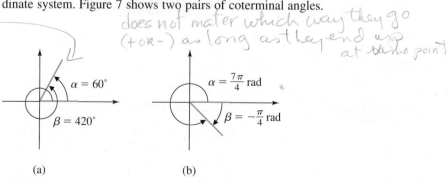

(a)   (b)

FIGURE 7
Coterminal angles

REMARK   1. The degree measures of two coterminal angles differ by an integer* multiple of 360°.

2. The radian measures of two coterminal angles differ by an integer multiple of 2π.   ◇

♦ EXAMPLE 4   Recognizing Coterminal Angles

Which of the following pairs of angles are coterminal?

(A) $\alpha = -135°$ (B) $\alpha = 120°$
    $\beta = 225°$ $\beta = -420°$
(C) $\alpha = -\pi/3$ rad   (D) $\alpha = \pi/3$ rad
    $\beta = 2\pi/3$ rad $\beta = 7\pi/3$ rad

SOLUTIONS   (A) The angles are coterminal if $\alpha - \beta$ is an integer multiple of 360°.

$$\alpha - \beta = (-135°) - 225° = -360° = -1(360°)$$

Thus, α and β are coterminal.

(B) $\alpha - \beta = 120° - (-420°) = 540°$

The angles are not coterminal, since 540° is not an integer multiple of 360°.

(C) The angles are coterminal if $\alpha - \beta$ is an integer multiple of 2π.

$$\alpha - \beta = (-\pi/3) - 2\pi/3 = -3\pi/3 = -\pi$$

The angles are not coterminal, since $-\pi$ is not an integer multiple of 2π.

(D) $\alpha - \beta = \pi/3 - 7\pi/3 = -6\pi/3 = -2\pi = (-1)(2\pi)$

Thus, α and β are coterminal.   ♦

MATCHED PROBLEM 4   Which of the following pairs of angles are coterminal?

(A) $\alpha = 90°$ (B) $\alpha = 750°$
    $\beta = -90°$ $\beta = 30°$
(C) $\alpha = -\pi/6$ rad   (D) $\alpha = 3\pi/4$ rad
    $\beta = -25\pi/6$ rad $\beta = 7\pi/4$ rad

♦ Arc Length and Sector Area

At first it may appear that radian measure of angles is more complicated and less useful than degree measure. However, just the opposite is true. The computation of arc length and the area of a circular sector, which we now discuss, should begin to convince you of some of the advantages of radian measure over degree measure. Refer to Figure 8 in the following discussion.

---

* An integer is a positive or negative whole number or 0; that is, the set of integers is {. . . , −4, −3, −2, −1, 0, 1, 2, 3, 4, . . .}.

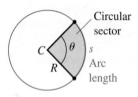

FIGURE 8
Circular sector

From the definition of radian measure of an angle,

$$\theta = \frac{s}{R} \text{ rad}$$

Solving for s, we obtain a formula for arc length

$$s = R\theta \qquad \theta \text{ in radian measure} \tag{1}$$

If θ is in degree measure, we must multiply by $\pi/180$ first (to convert to radians); then formula (1) becomes

$$s = \frac{\pi}{180} R\theta \qquad \theta \text{ in degree measure} \tag{2}$$

We see that the formula for arc length is much simpler when θ is in radian measure.

EXAMPLE 5   Arc Length

In a circle of radius 4.00 cm, find the arc length subtended by a central angle of

(A) 3.40 rad   (B) 10.0°

SOLUTIONS   (A) $s = R\theta$ (B) $s = \frac{\pi}{180} R\theta$

   $= 4.00(3.40) = 13.6$ cm

   $= \frac{\pi}{180}(4.0)(10.0) = 0.698$ cm

MATCHED PROBLEM 5   In a circle of radius 6.00 ft, find the arc length subtended by a central angle of

(A) 1.70 rad   (B) 40.0°

The formula for the area A of a circular sector in a circle with radius R and central angle θ in radian measure (see Figure 8) can be found by starting with the following proportion.

$$\frac{A}{\pi R^2} = \frac{\theta}{2\pi}$$

$$A = \frac{1}{2} R^2 \theta \qquad \theta \text{ in radian measure} \tag{3}$$

If θ is in degree measure, we must multiply by $\pi/180$ first (to convert to radians); then formula (3) becomes

$$A = \frac{\pi}{360} R^2 \theta \qquad \theta \text{ in degree measure} \tag{4}$$

Again we see that the formula for sector area is much simpler when θ is in radian measure.

◆ **EXAMPLE 6**   Area of a Sector

In a circle of radius 3 m, find the area (to three significant digits) of the circular sector with central angle

(A) 0.4732 rad   (B) 25°

SOLUTIONS   (A) $A = \dfrac{1}{2} R^2 \theta$ (B) $A = \dfrac{\pi}{360} R^2 \theta$

$\quad\quad\quad\quad = \dfrac{1}{2}(3)^2(0.4732) = 2.13 \text{ m}^2$ $\quad\quad = \dfrac{\pi}{360}(3)^2(25) = 1.96 \text{ m}^2$ ◆

MATCHED PROBLEM 6   In a circle of radius 7 in., find the area (to four significant digits) of the circular sector with central angle

(A) 0.1332 rad   (B) 110°

Angular velocity is another significant application of radian measure in engineering and physics problems. A detailed discussion of angular velocity is presented in the next section.

It is important to gain experience in the use of radian measure, since the concept will be used extensively in many developments that follow. By the time you finish this book you should feel as comfortable with radian measure as you now do with degree measure.

Answers to Matched Problems   **1.** (A) $\pi/9$ rad $= 0.349$ rad   (B) $(180/\pi)° = 57.3°$
2. (A) $(1.00 \times 10^3)°$ (B) -15.7 rad

3. (A)   (B)

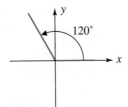   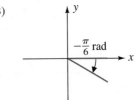

(C)   (D)

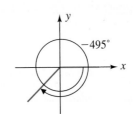

4. (A) Not coterminal   (B) Coterminal   (C) Coterminal   (D) Not coterminal
5. (A) 10.2 ft   (B) 4.19 ft
6. (A) 3.263 in.^{2} (B) 47.04 in.^{2}

EXERCISE 2.1

A *Remember that 180° corresponds to π rad; mentally convert each degree measure to radian measure in terms of π.*

1. 90°   **2.** 45°   **3.** 60°

4. 30°   **5.** 120°   **6.** 150°

Remember that π rad corresponds to 180°; mentally convert each radian measure to degree measure.

7. π/4 rad   **8.** π/2 rad   **9.** π/6 rad

10. π/3 rad   **11.** 5π/6 rad   **12.** 2π/3 rad

13. Convert to radian measure: 30°, 60°, 90°, 120°, 150°, 180°. (Start with 30° and take multiples.)

14. Convert to radian measure: 45°, 90°, 135°, 180°. (Start with 45° and take multiples.)

Sketch each angle in its standard position and find the degree measure of the two nearest angles (one negative and one positive) that are coterminal to the given angle.

15. 60°   **16.** 45°   **17.** −30°

18. −45°   **19.** 240°   **20.** −135°

B **21.** If the radius of a circle is 3.0 cm, find the radian measure of an angle subtended by an arc of length
(A) 6 cm   (B) 4.5 cm

22. If the radius of a circle is 4.0 m, find the radian measure of an angle subtended by an arc of length
(A) 12 m   (B) 18 m

Find the radian measure for each angle. Express the answer in exact form and in approximate form to four significant digits.

23. 18°   **24.** 9°   **25.** 27°

26. 36°   **27.** 130°   **28.** 140°

Find the degree measure for each angle. Express the answer in exact form and in approximate form with decimal degrees to four significant digits.

29. 1.6 rad   **30.** 0.5 rad   **31.** π/12 rad

32. π/36 rad   **33.** π/60 rad   **34.** π/180 rad

Sketch each angle in its standard position.

35. −π/6 rad   **36.** −π/3 rad   **37.** 300°

38. 390°   **39.** −7π/3 rad   **40.** −11π/4 rad

41. (A) Find the degree measures of 8.30 rad and −11.6 rad.
(B) Find the radian measures of 563° and −1,230°.

42. (A) Find the degree measures of 12.04 rad and −7.10 rad.
(B) Find the radian measures of 2,672° and −431.8°.

43. In a circle of radius 25.0 m, find the length of the arc subtended by a central angle of
(A) 2.33 rad   (B) 19.0°
(C) 0.821 rad   (D) 108°

44. In a circle of radius 14.0 in., find the length of the arc subtended by a central angle of
(A) 0.447 rad   (B) 68.0°
(C) 2.68 rad   (D) 212°

45. In a circle of radius 14.0 cm, find the area of the circular sector with central angle:
(A) 0.473 rad   (B) 25.0°
(C) 1.02 rad   (D) 112°

46. In a circle of radius 115 ft, find the area of the circular sector with central angle:
(A) 2.49 rad   (B) 33.0°
(C) 0.382 rad   (D) 204°

C *In which quadrant* does the terminal side of each angle lie?*

47. 432°   **48.** 821°   **49.** −14π/3 rad

50. −17π/4 rad   **51.** 1,243°   **52.** −942°

Find each value to four decimal places. Convert radians to decimal degrees.

53. 57.3421° = ? rad   **54.** 103.2187° = ? rad

55. 0.3184 rad = ?°   **56.** 1.0394 rad = ?°

57. 26°23′14″ = ? rad   **58.** 179°3′43″ = ? rad

Applications

59. **Radian Measure**   What is the radian measure of the smaller angle made by the hands of a clock at 2:30 (see the figure)? Express the answer in terms of π and as a decimal fraction to two decimal places.

---

* Recall that a rectangular coordinate system divides a plane into four parts called quadrants. These quadrants are numbered in a counterclockwise direction starting in the upper right-hand corner.

Figure for 59

60. Radian Measure Repeat Problem 59 for 4:30.

61. Pendulum A clock has a pendulum 22 cm long. If it swings through an angle of 32°, how far does the bottom of the bob travel in one swing?

Figure for 61

62. Pendulum If the bob on the bottom of the 22 cm pendulum in Problem 61 traces a 9.5 cm arc on each swing, through what angle in degrees does the pendulum rotate on each swing?

63. Engineering Oil is pumped from some wells using a donkey pump as shown in the figure. Through how many

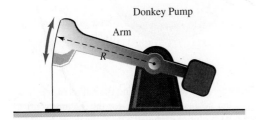

Figure for 63

degrees must an arm with a 72 in. radius rotate to produce a 24 in. vertical stroke at the pump down in the ground? Note that a point at the end of the arm must travel through a 24 in. arc to produce a 24 in. vertical stroke at the pump.

64. Engineering In Problem 63, find the arm length R that would produce an 18 in. vertical stroke while rotating through 21°.

65. Astronomy The sun is about 1.5×10^8 km from the earth. If the angle subtended by the diameter of the sun on the surface of the earth is 9.3×10^{-3} rad, approximately what is the diameter of the sun? [*Hint:* Use the intercepted arc to approximate the diameter.]

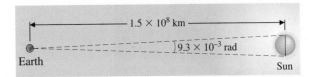

Figure for 65

∗66. Surveying If a natural gas tank 5.000 km away subtends an angle of 2.44°, approximate its height to the nearest meter. (See Problem 65.)

67. Photography The angle of view for a 300 mm telephoto lens is 8°. At 1,250 ft, what is the approximate width of the field of view? Use an arc length to approximate the chord length to the nearest foot.

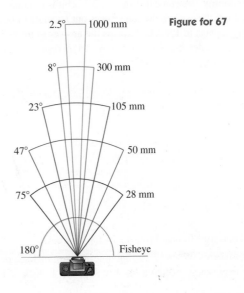

Figure for 67

68. Photography   The angle of view for a 1,000 mm tele-photo lens is 2.5°. At 865 ft, what is the approximate width of the field of view? Use an arc length to approximate the chord length to the nearest foot.

Geography   *In Problems 69–74 find the distance to the nearest mile between each pair of cities, given their latitudes. (Each pair of cities has the same longitude, and the radius of the earth is 3,964 mi.)*

Figure for 69–74

69. Columbus, Ohio (40°0′N);   Detroit, Michigan (42°20′N)

70. Cheyenne, Wyoming (41°9′N);   Denver, Colorado (39°45′N)

71. Wichita, Kansas (37°43′N);   Winnipeg, Canada (49°54′N)

72. Atlanta, Georgia (33°45′N);   Cincinnati, Ohio (39°8′N)

73. Havana, Cuba (23°8′N);   Tampa, Florida (27°51′N)

74. Mexico City (12°26′N);   San Antonio, Texas (29°23′N)

75. Astronomy   Assume that the earth's orbit is circular. A line from the earth to the sun sweeps out an angle of how many radians in 1 week? Express the answer in terms of π and as a decimal fraction to two decimal places. (Assume exactly 52 weeks in a year.)

Figure for 75

76. Astronomy   Repeat Problem 75 for 13 weeks.

77. **Astronomy**   When measuring time, an error of 1 sec per day may not seem like a lot. But suppose a clock is in

error by at most 1 sec per day. Then in 1 year the accumulated error could be as much as 365 sec. If we assume the earth's orbit about the sun is circular, with a radius of 9.3×10^7 mi, what would be the maximum error (in miles) in computing the distance the earth travels in its orbit after 1 year?

78. **Astronomy**   Using the clock described in Problem 77, what would be the maximum error (in miles) in computing the distance that Venus travels in a "Venus year"? Assume Venus' orbit around the sun is circular, with a radius of 6.7×10^7 mi, and that Venus completes one orbit (a "Venus year") in 224 earth days.

79. **Geometry**   A circular sector has an area of 52.39 ft^2 and a radius of 10.5 ft. Calculate the perimeter of the sector to the nearest foot.

80. **Geometry**   A circular sector has an area of 145.7 cm^2 and a radius of 8.4 cm. Calculate the perimeter of the sector to the nearest centimeter.

81. **Revolutions and Radians**   How many radians are in 5 revolutions? In 3.6 revolutions? In n revolutions? (Give answers in exact form in terms of π.)

82. **Revolutions and Radians**   Through how many radians does a pulley with a 10 cm diameter turn when 5.00 m of rope have been pulled through it without slippage? How many revolutions result? (Give answers to the nearest whole number.)

83. **Engineering**   Rotation of a drive wheel causes a shaft to rotate (see the figure). If the drive wheel completes three revolutions, how many revolutions will the shaft complete? Through how many radians will the shaft turn? Compute answers to one decimal place.

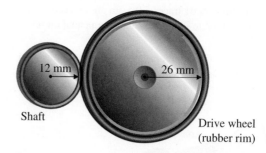

Figure for 83

84. **Engineering**   In Problem 83, find the radius of the drive wheel, to the nearest millimeter, for the 12 mm shaft to make seven revolutions when the drive wheel makes three revolutions.

85. Radians and Arc Length A bicycle wheel of diameter 32 in. travels a distance of 20 ft. Find the angle, to the nearest degree, swept out by one of the spokes.

86. Radians and Arc Length A bicycle has a front wheel with a diameter of 24 cm and a back wheel with a diameter of 60 cm. Through what angle (in radians) does the front wheel turn if the back wheel turns through 12 rad?

Figure for 85

☆**2.2** LINEAR AND ANGULAR VELOCITY

We will put our notion of radian measure to use in defining two special kinds of velocity that involve rotating motion. These velocities, called **angular velocity** and **linear velocity on a circle,** are used extensively in engineering and physics.

Starting with θ in radian measure (see Figure 1), recall (Section 2.1) that

$$\theta = \frac{s}{R}$$

FIGURE 1

We rewrite the formula in the equivalent form

$$s = R\theta \tag{1}$$

If we think of a point moving on the circumference of the circle with uniform speed, then a radial line from the center of the circle to this point will sweep out an angle θ at a uniform rate. Thus, if we divide both sides of equation (1) by time t, we obtain

$$\frac{s}{t} = R\frac{\theta}{t} \tag{2}$$

Now,

$\frac{s}{t}$ = Change in arc length per unit change in time

    = Linear velocity of point on the circle

    = V

and

$\frac{\theta}{t}$ = Change in angle in radian measure per unit change in time

    = Angular velocity

    = ω

☆ Sections marked with a star may be omitted without loss of continuity.

Making these substitutions in equation (2), we obtain the following additional important formulas.

LINEAR AND ANGULAR VELOCITY ON A CIRCLE

$V = R\omega$ *V is linear velocity of a point on a circle*

$\omega = \dfrac{V}{R}$ *ω is angular velocity (radians per unit time)*

◆ EXAMPLE 1   Electrical Wind Generator

An electrical wind generator (see Figure 2) has propeller blades that are 5.00 m long. If the blades are rotating at 8π rad/sec, what is the linear velocity (to the nearest meter per second) of a point on the tip of one of the blades?

FIGURE 2
Electrical wind generator

SOLUTION $V = R\omega$

$= 5.00(8\pi)$

$= 126$ m/sec   ◆

MATCHED PROBLEM 1   If a 3.0 ft diameter wheel turns at 12 rad/min, what is the velocity of a point on the wheel in feet per minute?

◆ EXAMPLE 2   Angular Velocity

A point on the rim of a 6.0 in. diameter wheel is traveling at 75 ft/sec. What is the angular velocity of the wheel in radians per second?

SOLUTION $\omega = \dfrac{V}{R}$

$= \dfrac{75}{0.25} = 300$ rad/sec   [*Note:*   3.0 in. = 0.25 ft]   ◆

MATCHED PROBLEM 2   A point on the rim of a 4.00 in. diameter wheel is traveling at 88.0 ft/sec. What is the angular velocity of the wheel in radians per second?

◆ **EXAMPLE 3**   Linear Velocity

If a 6 cm shaft is rotating at 4,000 rpm (revolutions per minute), what is the speed (to two significant digits) of a particle on its surface in centimeters per minute?

SOLUTION   Since 1 revolution is equivalent to 2π rad, we multiply 4,000 by 2π to obtain the angular velocity of the shaft in radians per minute.

$$\omega = 8,000\pi \text{ rad/min}$$

Now we use $V = R\omega$ to complete the solution.

$$V = 3(8,000\pi) = 75,000 \text{ cm/min}$$ ◆

MATCHED PROBLEM 3   If an 8 cm diameter drive shaft in a boat is rotating at 350 rpm, what is the speed (to three significant digits) of a particle on its surface in centimeters per second?

◆ **EXAMPLE 4**   Hubble Space Telescope

The 25,000 lb Hubble space telescope (see Figure 3) was launched April 1990 and placed in a 380 mi circular orbit above the earth's surface. It completes one orbit every 97 min, going from a dawn-to-dusk cycle nearly 15 times a day. If the radius of the earth is 3,964 mi, what is the linear velocity of the space telescope in miles per hour (mph)?

SOLUTION   The telescope completes 1 revolution (2π rad) in

$$\frac{97}{60} \text{ hr} = 1.6 \text{ hr}$$

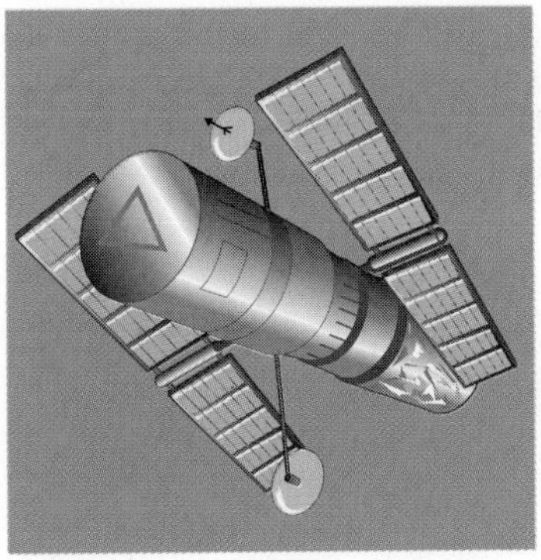

The angular velocity generated by the space telescope relative to the center of the earth is

$$\omega = \frac{\theta}{t} = \frac{2\pi}{1.6} = 3.9 \text{ rad/hr}$$

The linear velocity of the telescope is

$$V = R\omega$$
$$= (3,964 + 380)(3.9)$$
$$= 17,000 \text{ mph}$$ ◆

FIGURE 3
Hubble space telescope

MATCHED PROBLEM 4   The space shuttle *Columbia* was placed in a circular orbit 250 mi above the earth's surface. One orbit is completed in 1.51 hr. If the radius of the earth is 3,964 mi, what is the linear velocity (to three significant digits) of the shuttle in miles per hour?

Answers to   **1.** 18 ft/min   **2.** 528 rad/sec   **3.** 147 cm/sec
Matched Problems   **4.** 17,500 mph

EXERCISE 2.2

A *Find the velocity V of a point on the rim of a wheel, given the indicated information.*

1. $R = 6$ mm, $\omega = 0.5$ rad/sec

2. $R = 4,000$ cm, $\omega = 0.05$ rad/hr

Find the angular velocity ω given the following information.

3. $R = 6.0$ cm, $V = 102$ cm/sec

4. $R = 250$ km, $V = 500$ km/hr

B *Find the angular velocity of a wheel turning through θ radians in time t:*

5. $\theta = 2\pi$ rad, $t = 1.7$ hr

6. $\theta = 2\pi$ rad, $t = 3.04$ sec

7. $\theta = 8.07$ rad, $t = 13.6$ sec

8. $\theta = 13.67$ rad, $t = 21.03$ sec

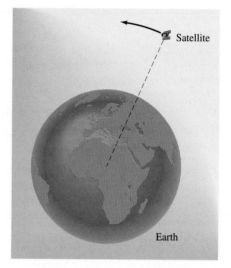
Figure for 11

Satellite

Earth

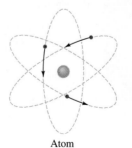

Applications

9. Engineering   A 16 mm diameter shaft rotates at 1,500 rps (revolutions per second). What is the speed (to the nearest meter) of a particle on its surface in meters per second?

10. Engineering   A 6 cm diameter shaft rotates at 500 rps. What is the speed (to the nearest meter) of a particle on its surface in meters per second?

11. Space Science   An earth satellite travels in a circular orbit at 20,000 mph. If the radius of the orbit is 4,300 mi, what angular velocity (in radians per hour, to three significant digits) is generated?

12. Engineering   A bicycle is ridden at a speed of 7.0 m/sec. If the wheel diameter is 64 cm, what is the angular velocity in radians per second?

13. Physics   The velocity of sound in air is approx. 335.3 m/sec. If an airplane has a 3.000 m diameter propeller, at what angular velocity will its tip pass through the sound barrier?

14. Physics   If an electron in an atom travels around the nucleus in a circular orbit at 8.11×10^6 cm/sec (see the figure), what angular velocity (in radians per second)

Figure for 14

Atom

does it generate, assuming the radius of the orbit is 5.00 $\times 10^{-9}$ cm?

15. **Astronomy** The earth revolves about the sun in an orbit that is approximately circular with a radius of 9.3×10^7 mi (see the figure). The radius of the orbit sweeps out an angle with what exact angular velocity in radians per hour? How fast (to the nearest hundred miles per hour) is the earth traveling along its orbit?

Figure for 15

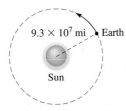

Velocity of the earth

16. **Astronomy** Take into consideration only the daily rotation of the earth to find out how fast (in miles per hour) a person would be moving on the equator. The radius of the earth is approx. 3,964 mi, and a daily rotation takes 23.93 hr (remember leap year). [*Hint:* Find the angular velocity first.]

17. **Astronomy** Jupiter makes one full revolution about its axis every 9 hr 55 min. If Jupiter's equatorial diameter is 88,700 mi,
   (A) What is its angular velocity relative to its axis of rotation?
   (B) What is the linear velocity of a point on Jupiter's equator?

18. **Astronomy** The sun makes one full revolution about its axis every 27.0 days. Assume 1 day = 24 hr. If its equatorial diameter is 865,400 mi,
   (A) What is the sun's angular velocity relative to its axis of rotation in radians per hour?
   (B) What is the linear velocity of a point on the sun's equator?

19. **Space Science** For an earth satellite to stay in orbit over a given stationary spot on earth, it must be placed in orbit 22,300 mi above the earth's surface. It will then take the satellite the same time to complete one orbit as the earth, 23.93 hr (remember leap year). Such satellites are called **geostationary satellites** and are used for communications and tracking space shuttles. If the radius of the earth is 3,964 mi, what is the linear velocity of a geostationary satellite?

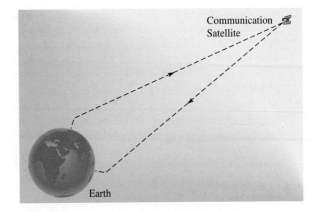

Figure for 19

20. **Astronomy** Until 1999 the planet Neptune is the planet farthest from the sun, 2.795×10^9 mi. (After 1999 and for the next 228 years Pluto will have this honor.) If Neptune takes 164 years to complete one orbit, what is its linear velocity in miles per hour?

*21. **Space Science** The earth rotates on its axis once every 23.93 hr, and a space shuttle revolves around the earth in the plane of the earth's equator once every 1.51 hr. Both are rotating in the same direction (see the figure). What is the length of time between consecutive passages of the shuttle over a particular point P on the equator? [*Hint:* The shuttle will make one complete revolution (2π rad) and a little more to be over the same point P again. Thus, the central angle generated by point P on earth must equal the central angle generated by the space shuttle minus 2π.]

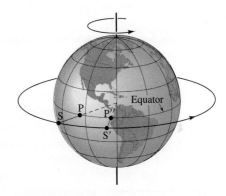

Figure for 21

*22. **Astronomy**   One of the moons of Jupiter rotates around the planet in its equatorial plane once every 42 hr 30 min. Jupiter rotates around its axis once every 9 hr 55 min. What is the length of time between consecutive passages of the moon over the same point on Jupiter's equator? [See the hint for Problem 21.]

*23. **Precalculus: Rotating Beacon**   A beacon light 15 ft from a wall rotates clockwise at the rate of exactly 1 rps (see the figure). If we start counting time in seconds when the light spot is at C, express the distance a the light spot travels along the wall in terms of time t.

*24. **Precalculus: Rotating Beacon**   Refer to Problem 23, and express the length of the light beam c in terms of t.

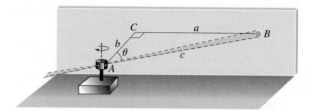

Figure for 23

2.3   TRIGONOMETRIC FUNCTIONS*

♦   **Trigonometric Functions with Angle Domains**
♦   **Trigonometric Functions with Real Number Domains**
♦   **Calculator Evaluation**
♦   **Summary of Sign Properties**

In Chapter 1 we introduced the concept of trigonometric ratios and tied this idea to right triangles. We were able to use these ratios to define six trigonometric functions with angle domains restricted to 0°–90°. In this section we introduce more general definitions that will apply to angle domains of arbitrary size, positive, negative, or zero, in degree or radian measure. We then move one (giant) step further and define these functions for arbitrary real numbers. With these new functions we will be able to do everything we did with the trigonometric ratios in the first chapter, plus a great deal more. In Section 2.6 we will approach the subject from a more modern point of view, where angles are not a necessary part of the definition. Each approach has its advantages for certain applications and uses.

♦   Trigonometric Functions with Angle Domains

We start with an arbitrary angle θ located in a rectangular coordinate system in a standard position. We then choose an arbitrary point $P(a, b)$ on the terminal side of θ, but away from the origin. If R is the distance of $P(a, b)$ from the origin, we can form six ratios involving R and the coordinates of P. We will use these six ratios to define six trigonometric functions, which are direct generalizations of the six trigonometric ratios given in Section 1.3. It is important to understand the definitions of the six trigonometric functions—a great deal depends on them.

---

* A brief review of Appendix B.1 on functions might prove helpful before starting this section.

TRIGONOMETRIC FUNCTIONS WITH ANGLE DOMAINS

For an arbitrary angle θ,

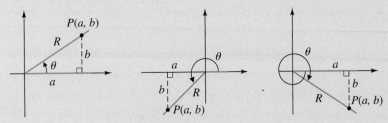

$$\sin \theta = \frac{b}{R} \qquad\qquad \csc \theta = \frac{R}{b} \quad b \neq 0 \qquad\qquad R = \sqrt{a^2 + b^2} > 0$$

$$\cos \theta = \frac{a}{R} \qquad\qquad \sec \theta = \frac{R}{a} \quad a \neq 0 \qquad\qquad P(a, b) \text{ is an}$$
arbitrary point on
the terminal
$$\tan \theta = \frac{b}{a} \quad a \neq 0 \qquad \cot \theta = \frac{a}{b} \quad b \neq 0 \qquad\qquad \text{side of } \theta,$$
$$(a, b) \neq (0, 0)$$

Domains: Sets of all possible angles for which the ratios are defined
Ranges: Subsets of the set of real numbers

[*Note:*   A trigonometric function has the same value at coterminal angles. Also, more precise statements about domain and ranges will be given in Chapter 3.]

REMARKS

1. We state the preceding definition of the six trigonometric functions in terms of an *ab* coordinate system instead of an *xy* coordinate system because when we generalize the definition of trigonometric functions still further (later in this section and in Section 2.6), we want to reserve *x* for an independent variable and *y* for a dependent variable. In addition, using an *ab* coordinate system shows more clearly that the preceding definition of the six trigonometric functions is a direct generalization of the ratio definition given in Section 1.3.

2. Because of properties of similar triangles, the preceding definition of the six trigonometric functions is independent of the choice of $P(a, b)$ on the terminal side of θ. (See Problems 1–4 in Exercise 2.3.)

3. The right triangle formed by dropping a perpendicular from $P(a, b)$ to the horizontal axis is called the **reference triangle** associated with the angle θ. We will often refer to this triangle. (A more complete discussion of the reference triangle is given in Section 2.5.)

4. The ratios in the definition may be negative as well as positive, depending on the quadrant in which the terminal side of θ lies.   ◇

◆   **EXAMPLE 1**   Evaluating Trigonometric Functions, Given a Point on the
Terminal Side of θ

Find the exact value of each of the six trigonometric functions for the angle θ
with terminal side containing $P(-4, -3)$ (see Figure 1).

SOLUTION

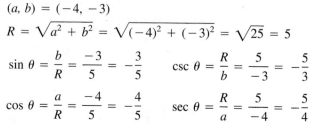

$$(a, b) = (-4, -3)$$
$$R = \sqrt{a^2 + b^2} = \sqrt{(-4)^2 + (-3)^2} = \sqrt{25} = 5$$

$$\sin \theta = \frac{b}{R} = \frac{-3}{5} = -\frac{3}{5} \qquad \csc \theta = \frac{R}{b} = \frac{5}{-3} = -\frac{5}{3}$$

$$\cos \theta = \frac{a}{R} = \frac{-4}{5} = -\frac{4}{5} \qquad \sec \theta = \frac{R}{a} = \frac{5}{-4} = -\frac{5}{4}$$

$$\tan \theta = \frac{b}{a} = \frac{-3}{-4} = \frac{3}{4} \qquad \cot \theta = \frac{a}{b} = \frac{-4}{-3} = \frac{4}{3}$$

◆

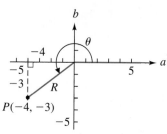

FIGURE 1

MATCHED PROBLEM 1   Find the exact value of each of the six trigonometric functions if the terminal
side of θ contains the point $(-8, -6)$. [*Note:* This point lies on the terminal
side of the same angle as in Example 1.]

◆   **EXAMPLE 2**   Using Given Information to Evaluate Trigonometric
Functions

Find the exact value of each of the other five trigonometric functions for the
given angle θ—without finding θ—given that the terminal side of θ is in quadrant
III and

$$\cos \theta = \frac{-3}{5}$$

SOLUTION   The information given is sufficient for us to locate a reference triangle in quadrant
III for θ even though we do not know θ (see Figure 2). We sketch the reference
triangle, label what we know, and then complete the problem as indicated.

Since $\cos \theta = \dfrac{a}{R} = \dfrac{-3}{5}$, we know that $a = -3$ and $R = 5$ (R is never

negative). If we can find b, we can determine the values of the other five functions
using their definitions.

We use the Pythagorean theorem to find b.

$$(-3)^2 + b^2 = 5^2$$
$$b^2 = 25 - 9 = 16 \qquad \text{\textit{b is negative since P(a, b) is in}}$$
$$\text{\textit{quadrant III.}}$$
$$b = -4$$

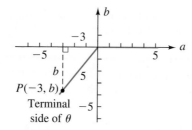

FIGURE 2

Thus,

$$(a, b) = (-3, -4) \qquad \text{and} \qquad R = 5$$

We can now find the other five functions using their definitions.

$$\sin \theta = \frac{b}{R} = \frac{-4}{5} = -\frac{4}{5} \qquad \csc \theta = \frac{R}{b} = \frac{5}{-4} = -\frac{5}{4}$$

$$\tan \theta = \frac{b}{a} = \frac{-4}{-3} = \frac{4}{3} \qquad \cot \theta = \frac{a}{b} = \frac{-3}{-4} = \frac{3}{4}$$

$$\sec \theta = \frac{R}{a} = \frac{5}{-3} = -\frac{5}{3}$$

◆

MATCHED PROBLEM 2   Repeat Example 2 for $\tan \theta = -\frac{3}{4}$ and the terminal side of θ in quadrant II.

◆ **Trigonometric Functions with Real Number Domains**

We now turn to the problem of defining trigonometric functions for real number domains. First note that to each real number x there corresponds an angle of x radians, and to each angle of x radians there corresponds the real number x. We define trigonometric functions with real number domains in terms of trigonometric functions with angle domains.

TRIGONOMETRIC FUNCTIONS WITH REAL NUMBER DOMAINS

For x any real number,

$$\sin x = \sin(x \text{ rad}) \qquad \csc x = \csc(x \text{ rad})$$
$$\cos x = \cos(x \text{ rad}) \qquad \sec x = \sec(x \text{ rad})$$
$$\tan x = \tan(x \text{ rad}) \qquad \cot x = \cot(x \text{ rad})$$

Domains: Subsets of the set of real numbers
Ranges: Subsets of the set of real numbers

Thus, for example, sin 3 = sin(3 rad), cos 1.23 = cos(1.23 rad), tan(−9) = tan(−9 rad), and so on.

REMARK   Because of this definition, we will often omit "rad" after x and interpret x as a real number or an angle with radian measure x, whichever fits the context in which x appears.   ◇

At first glance, the definition of trigonometric functions with real number domains appears artificial, but we will see that it frees the trigonometric functions from angles and opens them up to a large variety of significant applications not directly connected to angles.

◆ Calculator Evaluation

We used a calculator in Section 1.3 to approximate trigonometric ratios for acute angles in degree measure. These same calculators are internally programmed to approximate (to eight or ten significant digits) trigonometric functions for *any* angle (however large or small, positive or negative) in degree or in radian measure, or for *any* real number. (Remember, most graphing calculators use different sequences of steps than scientific calculators. Consult your owner's manual for your calculator.) In Section 2.5 we will show how to obtain exact values for certain special angles (integer multiples of 30° and 45° or integer multiples of $\pi/6$ and $\pi/4$) without the use of a calculator.

⚠ Caution   1. Set the calculator in **degree mode** when evaluating trigonometric functions of angles in degree measure (degree measure must be in decimal degrees).

2. Set the calculator in **radian mode** when evaluating trigonometric functions of angles in radian measure or trigonometric functions of real numbers.   ◇

We generalize the reciprocal relationships stated in Section 1.3 to evaluate secant, cosecant, and cotangent.

RECIPROCAL RELATIONSHIPS

For x any real number or angle in degree or radian measure,

$$\csc x = \frac{1}{\sin x} \qquad \sin x \neq 0$$

$$\sec x = \frac{1}{\cos x} \qquad \cos x \neq 0$$

$$\cot x = \frac{1}{\tan x} \qquad \tan x \neq 0$$

◆ **EXAMPLE 3**   Calculator Evaluation of Trigonometric Functions

With a calculator, evaluate to four significant digits:

(A) sin 286.38°   (B) tan(3.472 rad)   (C) cot 5.063
(D) cos(−107°35′)   (E) sec(−4.799)   (F) csc 192°47′22″

SOLUTIONS   (A) sin 286.38° = −0.9594   *Degree mode*
                                                  Use [sin] *key*

              (B) tan(3.472 rad) = 0.3430   *Radian mode*
                                                  Use [tan] *key*

              (C) cot 5.063 = −0.3657   *Radian mode*
                                                  Use [tan] *and* [1/x] *or* [x⁻¹] *keys*

              (D) cos(−107°35′) = cos(−107.5833. . .)   *Change to decimal degrees.*
                                                         Degree mode
                          = −0.3021   *Use* [cos] *key*

              (E) sec(−4.799) = 11.56   *Radian mode*
                                                  Use [cos] *and* [1/x] *or* [x⁻¹] *keys*

              (F) csc 192°47′22″ = csc(192.7894. . .)   *Change to decimal degrees.*
                                                         Degree mode
                          = −4.517   *Use* [sin] *and* [1/x] *or* [x⁻¹] *keys*

                                                                                  ◆

MATCHED PROBLEM 3   With a calculator, evaluate to four significant digits:

(A) cos 303.73°   (B) sec(−2.805)   (C) tan(−83°29′)
(D) sin(12 rad)   (E) csc 100°52′43″   (F) cot 9

◆ **Summary of Sign Properties**

To close this important section, we summarize the sign properties of the six
trigonometric functions in Table 1. There is no need to memorize the table since

TABLE 1
(*R* is always positive)

	Quadrant I			Quadrant II			Quadrant III			Quadrant IV		
	a	*b*	*R*	*a*	*b*	*R*	*a*	*b*	*R*	*a*	*b*	*R*
	+	+	+	−	+	+	−	−	+	+	−	+
$\sin x = b/R$ $\csc x = R/b$	+			+			−			−		
$\cos x = a/R$ $\sec x = R/a$	+			−			−			+		
$\tan x = b/a$ $\cot x = a/b$	+			−			+			−		

you can readily determine particular entries from the definitions of the functions involved and the signs of a and b (see Figure 3). In the table, x is associated with an angle that terminates in the respective quadrant; $P(a, b)$ is a point on the terminal side of the angle; and $R = \sqrt{a^2 + b^2} > 0$.

FIGURE 3

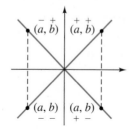

Answers to Matched Problems

1. $\sin \theta = -\frac{3}{5},$ $\cos \theta = -\frac{4}{5},$ $\tan \theta = \frac{3}{4},$ $\csc \theta = -\frac{5}{3},$ $\sec \theta = -\frac{5}{4},$ $\cot \theta = \frac{4}{3}$
[What do you think the values of the six trigonometric functions would be if we picked another point, say $(-12, -9)$, on the terminal side of the same angle θ as in Example 1? *Answer:* The same. (Why?)]

2. $\sin \theta = \frac{3}{5}, \csc \theta = \frac{5}{3}, \cot \theta = -\frac{4}{3}, \cos \theta = -\frac{4}{5}, \sec \theta = -\frac{5}{4}$

3. (A) 0.5553   (B) -1.059 (C) -8.754 (D) -0.5366
     (E) 1.018   (F) -2.211

EXERCISE 2.3

A *Find the exact value of each of the six trigonometric functions if the terminal side of θ contains the point $P(a, b)$. Do the same for $Q(a, b)$.*

1. $P(3, 4); Q(6, 8)$ **2.** $P(-3, -4); Q(-9, -12)$
3. $P(4, -3); Q(12, -9)$ **4.** $P(-3, 4); Q(-6, 8)$

Find the exact value of each of the other five trigonometric functions for the angle θ (without finding θ) given the indicated information. It would be helpful to sketch a reference triangle.

5. $\cos \theta = \dfrac{3}{5}$
   θ is a quadrant I angle

6. $\sin \theta = \dfrac{3}{5}$
   θ is a quadrant I angle

7. $\cos \theta = \dfrac{3}{5}$
   θ is a quadrant IV angle

8. $\sin \theta = \dfrac{3}{5}$
   θ is a quadrant II angle

9. $\csc \theta = -\dfrac{5}{4}$
   θ is a quadrant III angle

10. $\tan \theta = -\dfrac{4}{3}$
   θ is a quadrant II angle

11. $\csc \theta = -\dfrac{5}{4}$
   θ is a quadrant IV angle

12. $\tan \theta = -\dfrac{4}{3}$
   θ is a quadrant IV angle

Use a calculator to find Problems 13–32 to four significant digits. Make sure the calculator is in the correct mode (degree or radian) for each problem.

13. $\tan 89°$ **14.** $\sin 37°$ **15.** $\cos (3 \text{ rad})$
16. $\sin (4 \text{ rad})$ **17.** $\csc 162°$ **18.** $\sec 283°$
19. $\cot 341°$ **20.** $\csc 269°$ **21.** $\sin 13$
22. $\cos 7$ **23.** $\cot 2$ **24.** $\tan 1$
25. $\sec 74$ **26.** $\cot 108$ **27.** $\sin 428°$
28. $\tan 269°$ **29.** $\cos(-12)$ **30.** $\csc(-72)$
31. $\cot(-167°)$ **32.** $\sec(-273°)$

B *Find the exact value of each of the six trigonometric functions for an angle θ that has a terminal side containing the indicated point.*

33. $(\sqrt{3}, 1)$ **34.** $(1, 1)$ **35.** $(1, -\sqrt{3})$
36. $(-1, \sqrt{3})$ **37.** $(\sqrt{2}, -\sqrt{2})$ **38.** $(-2, -2)$

In which quadrants must the terminal side of an angle θ lie in order for

39. cos θ > 0   **40.** sin θ > 0   **41.** tan θ > 0

42. cot θ > 0   **43.** sec θ > 0   **44.** csc θ > 0

45. sin θ < 0   **46.** cos θ < 0   **47.** cot θ < 0

48. tan θ < 0   **49.** csc θ < 0   **50.** sec θ < 0

Find the exact value of each of the other five trigonometric functions for an angle θ (without finding θ) given the indicated information. It would be helpful to sketch a reference triangle.

51. $\sin \theta = -\dfrac{2}{3}$; cot θ > 0   **52.** $\cos \theta = -\dfrac{3}{5}$; tan θ < 0

53. $\sin \theta = -\dfrac{2}{3}$; tan θ < 0   **54.** $\cos \theta = -\dfrac{3}{5}$; sin θ < 0

55. $\sec \theta = \sqrt{3}$; sin θ < 0   **56.** tan θ = −2; csc θ > 0

Use a calculator to find Problems 57–74 to four significant digits.

57. cos 308.25°   **58.** sin 170.23°

59. tan 1.371   **60.** sin 3.519

61. cot(−265.33°)   **62.** csc(−45.27°)

63. sec(−4.013)   **64.** cot(−0.1578)

65. cos 208°12′55″   **66.** cos 192°45′13″

67. csc 112°5′38″   **68.** cot 321°18′5″

69. sec(−1,000)   **70.** cot(−3,000)

71. sin 405.33°   **72.** tan 623.05°

73. cos(−168°32′5″)   **74.** csc(−263°6′12″)

C 75. Which trigonometric functions are not defined when the terminal side of an angle lies along the positive or negative vertical axis?

76. Which trigonometric functions are not defined when the terminal side of an angle lies along the positive or negative horizontal axis?

For Problems 77–80, refer to the figure.

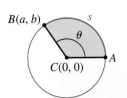

Figure for 77–80

77. In the figure, the coordinates of the center of the circle are (0, 0). If the coordinates of A are (5, 0) and arc length

s is exactly 6 units, find
  (A) The exact radian measure of θ
  (B) The coordinates of B to three significant digits

78. In the figure, the coordinates of the center of the circle are (0, 0). If the coordinates of A are (4, 0) and arc length s is exactly 10 units, find
  (A) The exact radian measure of θ
  (B) The coordinates of B to three significant digits

79. In the figure, the coordinates of the center of the circle are (0, 0). If the coordinates of A are (1, 0) and the arc length s is exactly 2 units, find
  (A) The exact radian measure of θ
  (B) The coordinates of B to three significant digits

80. In the figure, the coordinates of the center of the circle are (0, 0). If the coordinates of A are (1, 0) and the arc length s is exactly 4 units, find
  (A) The exact radian measure of θ
  (B) The coordinates of B to three significant digits

81. A circle with its center at the origin in a rectangular coordinate system passes through the point (4, 3). What is the length of the arc on the circle in the first quadrant between the positive horizontal axis and the point (4, 3)? Compute the answer to two decimal places.

82. Repeat Problem 81 with the circle passing through (3, 4).

Applications

83. Solar Energy   Light intensity on a solar cell changes with the angle of the sun and is given by the indicated formula (see the figure below). Find the intensity in terms of the constant k for θ = 0°, θ = 20°, θ = 40°, θ = 60°,

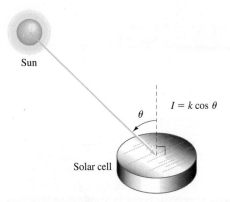

$I = k \cos \theta$

Figure for 83

and $\theta = 80°$. Compute each answer to two decimal places.

84. Solar Energy In Problem 83, at what angle will the light intensity be 50% of the vertical intensity?

The reason we have summers and winters is because the earth's axis of rotation tilts 23.5° away from the perpendicular, as indicated in the figure. A formula in Problems 85 and 86 quantifies this phenomenon.

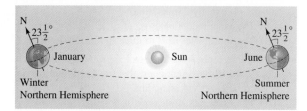

85. Sun's Energy and Seasons The amount of heat energy E from the sun received per square meter per unit of time in a given region on the surface of the earth is approximately proportional to the cosine of the angle θ that the sun makes with the vertical. Thus,

$$E = k \cos \theta$$

where k is the constant of proportionality for a given region. For a region with a latitude of 40°N, compare the energy received at the summer solstice ($\theta = 15°$) with the energy received at the winter solstice ($\theta = 63°$). Express answers in terms of k to two significant digits.

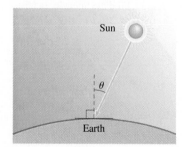

Figure for 85, 86

86. Sun's Energy and Seasons For a region with a latitude of 32°N, compare the energy received at the summer solstice ($\theta = 8°$) with the energy received at the winter solstice ($\theta = 55°$). Refer to Problem 85, and express answers in terms of k to two significant digits.

For Problems 87 and 88, refer to the figure.

Figure for 87, 88

87. Precalculus: Calculator Experiment It can be shown that the area of a polygon of n equal sides inscribed in a circle of radius 1 is given by

$$\frac{n}{2} \sin \left(\frac{360}{n} \right)^{\circ}$$

(A) Find the area of inscribed polygons for $n = 6$, $n = 10$, $n = 100$, $n = 1{,}000$, and $n = 10{,}000$. Compute each answer to five decimal places.

(B) As n gets larger and larger, what number does the area seem to approach? [*Hint:* What is the area of a circle with radius 1?]

(C) Will an inscribed polygon ever be a circle for any n?

88. Precalculus: Calculator Experiment It can be shown that the area of a polygon of n equal sides circumscribed around a circle of radius 1 is given by

$$n \tan \left(\frac{180}{n} \right)^{\circ}$$

(A) Find the area of circumscribed polygons for $n = 6$, $n = 10$, $n = 100$, $n = 1{,}000$, and $n = 10{,}000$. Compute each answer to five decimal places.

(B) As n gets larger and larger, what number does the area seem to approach? [*Hint:* What is the area of a circle with radius 1?]

(C) Will a circumscribed polygon ever be a circle for any n?

89. Engineering The figure shows a piston connected to a wheel that turns at 10 revolutions per second (rps). If P

Figure for 89

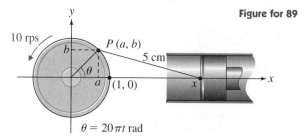

$\theta = 20\pi t$ rad

is at $(1, 0)$ when $t = 0$, then $\theta = 20\pi t$, where t is time in seconds. Show that

$$x = a + \sqrt{5^2 - b^2} = \cos 20\pi t + \sqrt{25 - (\sin 20\pi t)^2}$$

*90. **Engineering** In Problem 89, find the position (to two decimal places) of the piston (the value of x) for $t = 0$ and $t = 0.01$ sec.

91. **Alternating Current** An alternating current generator produces an electric current (measured in amperes) that is described by the equation

$$I = 35 \sin(48\pi t - 12\pi)$$

where t is time in seconds. (Section 3.4 has a detailed discussion of this subject.) What is the current I when $t = 0.13$ sec?

Figure for 91

92. **Alternating Current** What is the current I in Problem 91 when $t = 0.310$ sec?

93. **Precalculus: Logistics** A log of length L floats down a canal that has a right angle turn, as indicated in the figure. Express the length of the log in terms of θ and

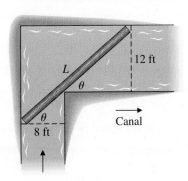

Figure for 93

the two widths of the canal (neglect the width of the log). The log is to touch the sides of the canal as shown. (In calculus, this equation is used to find the longest log that will float around the corner.)

94. **Precalculus: Logistics** Refer to Problem 93. What are the lengths (to three significant digits) of the logs associated with $\theta = 40°, 45°, 50°, 55°, 60°$?

95. **Precalculus: Angle of Inclination** The slope of a nonvertical line passing through points $P_1(x_1, y_1)$ and $P_2(x_2, y_2)$ is given by the formula

$$\text{Slope} = m = \frac{y_2 - y_1}{x_2 - x_1}$$

The angle θ that the line L makes with the x axis, $0° \le \theta < 180°$, is called the **angle of inclination** of the line L. Thus,

$$\text{Slope} = m = \tan\theta, \qquad 0° \le \theta < 180°$$

(A) Compute the slopes, to two decimal places, of the lines with angles of inclination 63.5° and 172°.

(B) Find the equation of a line passing through $(-3, 6)$ with an angle of inclination 143°. [*Hint:* Recall, $y - y_1 = m(x - x_1)$.] Write the answer in the form $y = mx + b$, with m and b to two decimal places.

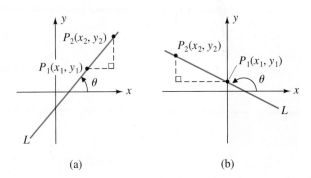

(a)   (b)

Figure for 95

96. **Precalculus: Angle of Inclination** Refer to Problem 95.

(A) Compute the slopes, to two decimal places, of the lines with angles of inclination 89.2° and 179°.

(B) Find the equation of a line passing through $(7, -4)$ with an angle of inclination 101°. Write the answer in the form $y = mx + b$, with m and b to two decimal places.

☆2.4 ADDITIONAL APPLICATIONS

- ◆ **Modeling Light Waves and Refraction**
- ◆ **Modeling Bow Waves**
- ◆ **Modeling Sonic Booms**
- ◆ **High-Energy Physics: Modeling Particle Energy**
- ◆ **Psychology: Modeling Perception**

If time permits, the material in this section will provide additional understanding of the use of trigonometry relative to several interesting applications. If the material must be omitted, at least look over the next few pages to gain a better appreciation of some additional applications of trigonometric functions.

◆ Modeling Light Waves and Refraction

Did you ever look at a pencil in a glass of water or poke a straight pole into a clear pool of water? The objects appear to bend at the surface (Figure 1). This bending phenomenon is caused by refracted light: When light waves pass from one medium to another of less or greater density, they bend. The reason for this is that light waves behave according to Fermat's least-time principle; that is, they follow the path from A to B that requires the least amount of time. In these cases, it can be shown that the least-time path is a bent path, as shown in Figure 2.

FIGURE 1
One or two pencils? There is actually only one.

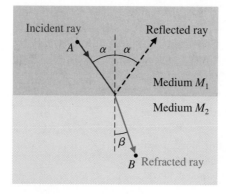

FIGURE 2
Refraction

In physics it is shown that

$$\frac{c_1}{c_2} = \frac{\sin \alpha}{\sin \beta} \tag{1}$$

where c_1 is the speed of light in medium M_1, c_2 is the speed of light in medium M_2, and α and β are as indicated in Figure 2.

---

* Sections marked with a star may be omitted without loss of continuity.

A more convenient form of equation (1) uses the notion of the **index of refraction,** which is the ratio of the speed of light in a vacuum to the speed of light in a given substance:

$$\text{Index of refraction, } n = \frac{\text{Speed of light in a vacuum}}{\text{Speed of light in a substance}}$$

If we let c represent the speed of light in a vacuum, then

$$\frac{c_1}{c_2} = \frac{c_1/c}{c_2/c} = \frac{1/n_1}{1/n_2} = \frac{n_2}{n_1} \tag{2}$$

where n_1 is the index of refraction in medium M_1, and n_2 is the index of refraction in medium M_2. (The index of refraction of a substance, rather than the velocity of light in that substance, is the property that is generally tabulated.) Substituting (2) into (1), we obtain **Snell's law:**

$$\frac{n_2}{n_1} = \frac{\sin \alpha}{\sin \beta} \tag{3}$$

[*Note:* n_2 is on the top and n_1 is on the bottom in (3), while c_1 is on the top and c_2 is on the bottom in (1).]

For any two given substances, the ratio n_2/n_1 is a constant. Thus, equation (3) is equivalent to

$$\frac{\sin \alpha}{\sin \beta} = \text{Constant}$$

The discovery that the sines of the angles of incidence and refraction are in a constant ratio to each other has been attributed to Willegrord Snell, a Dutch astronomer and mathematician (1591–1626), though there is now some doubt that he actually made the discovery. In any case, we will yield to common usage and continue to refer to the law as "Snell's law."

◆ EXAMPLE 1   Refracted Light

A spotlight shining on a pond strikes the water so that the angle of incidence α in Figure 3 is 23.5°. Find the refracted angle β. Use Snell's law and the fact that $n = 1.33$ for water and $n = 1.00$ for air.

SOLUTION   Use

$$\frac{n_2}{n_1} = \frac{\sin \alpha}{\sin \beta}$$

where $n_2 = 1.33$, $n_1 = 1.00$, and $\alpha = 23.5°$, and solve for β:

$$\frac{1.33}{1.00} = \frac{\sin 23.5°}{\sin \beta}$$

$$\sin \beta = \frac{\sin 23.5°}{1.33}$$

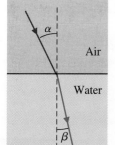

FIGURE 3
Refraction

$$\beta = \sin^{-1}\left(\frac{\sin 23.5°}{1.33}\right) \approx 17.4°$$

◆

MATCHED PROBLEM 1   Repeat Example 1 with $\alpha = 18.4°$.

The fact that light bends when passing from one medium into another is what makes telescopes, microscopes, and cameras possible. It is the carefully controlled bending of light rays that produces the useful results in optical instruments. Figures 4–6 illustrate a variety of phenomena connected with refracted light.

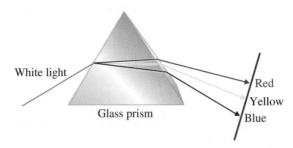

White light

Glass prism

Red
Yellow
Blue

FIGURE 4
Light spectrum—different wavelengths, different indexes
of refraction, different colors

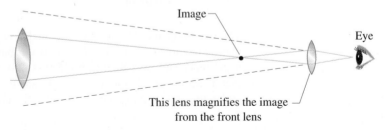

Image

Eye

This lens magnifies the image
from the front lens

FIGURE 5
Telescope optics

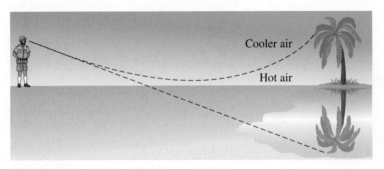

Cooler air

Hot air

FIGURE 6
Mirage—light is refracted and bent to appear as if reflected off water

Table 1 lists refractive indexes for several common materials.

TABLE 1
Refractive indexes

Material	Refractive index
Air	1.0003
Crown glass	1.52
Diamond	2.42
Flint glass	1.66
Ice	1.31
Water	1.33

◆ EXAMPLE 2   Reflected Light

If an underwater flashlight is directed toward the surface of a swimming pool, at what angle of incidence α will the light beam be totally reflected?

SOLUTION   The index of refraction for water is $n_1 = 1.33$ and that for air is $n_2 = 1.00$. Find the angle of incidence α in Figure 7 such that the angle of refraction β is 90°.

$$\frac{\sin \alpha}{\sin \beta} = \frac{n_2}{n_1}$$

$$\sin \alpha = \frac{1.00}{1.33} \sin 90°$$

$$\sin \alpha = \frac{1.00}{1.33} \quad (1)$$

$$\alpha = \sin^{-1} \frac{1.00}{1.33} = 48.8°$$

Thus, the light will be totally reflected if $\alpha \geq 48.8°$ (no light will be transmitted through the surface).

FIGURE 7
Underwater light refraction

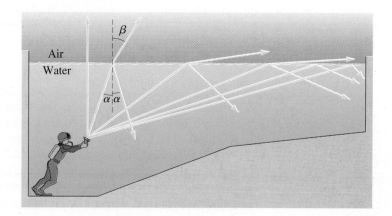

◆

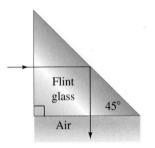

FIGURE 8
Flint Glass Prism

Show that the light beam passing through the flint glass prism (shown in Figure 8) is totally reflected off the slanted surface.

In general, the angle of incidence α such that the angle of refraction β is 90° is called the **critical angle**. For any angle of incidence larger than the critical angle, a light ray will be totally reflected. This critical angle (for total reflection) is important in the design of many optical instruments (such as binoculars) and is at the heart of the science of **fiber optics**. A small-diameter glass fiber bent in a curve (shown in Figure 9) will ''trap'' a light ray entering one end and the ray will be totally reflected and emerge out the other end.

Important uses of fiber optics are found in medicine and communications. Physicians use fiber optic instruments to see inside functioning organs. Surgeons use fiber optics to perform surgery involving only small incisions and outpatient facilities. A fiber optics communication cable carries information using high-speed pulses of laser light that is essentially distortion free. Using underwater armored cable (Figure 10), fiber optics communication networks are now in place worldwide. One fiber optic communication cable can carry 40,000 simultaneous telephone conversations or transmit the entire contents of the Encyclopedia Britannica in less than one minute.

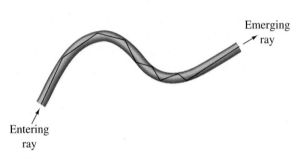

FIGURE 9
Fiber optics

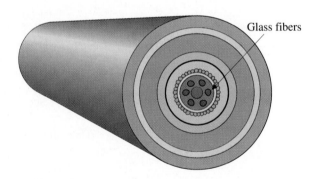

FIGURE 10
Transoceanic fiber optic armored cable
(about 1 in. in diameter)

◆ Modeling Bow Waves

A boat moving at a constant rate, faster than the water waves it produces, generates a **bow wave** that extends back from the bow of the boat at a given angle (Figure 11). If we know the speed of the boat and the speed of the waves produced by the boat, then we can determine the angle of the bow wave. Actually, if we know any two of these quantities, we can always find the third. Surprisingly, the solution to this problem also can be applied to sonic booms and high-energy particle physics, as we will see later in this section.

FIGURE 11
Bow waves of boats, sonic booms, and high-energy physics are related in a curious way; this section explains how

FIGURE 11
Bow waves of boats, sonic booms, and high-energy physics are related in a curious way; this section explains how

Refer to Figure 12; we reason as follows: When the boat is at P_1, the water wave it produces will radiate out in a circle, and by the time the boat reaches P_2, the wave will have moved a distance of r_1, which is less than the distance between P_1 and P_2 since the boat is assumed to be traveling faster than the wave. By the time the boat reaches the apex position B in Figure 12, the wave motion at P_2 will have moved r_2 units, and the wave motion at P_1 will have continued on out to r_3. Because of the constant speed of the boat and the constant speed of the wave motion, these circles of wave radiation will all have a common tangent that passes through the boat. Of course, the motion of the boat is continuous, and what we have said about P_1 and P_2 applies to all points along the path of the boat. The result of this phenomenon is the clearly visible wave fronts produced by the bow of a boat. Refer again to Figure 12, and you will see that the boat travels from P_1 to B in the same time t that the bow wave travels from P_1 to A; hence, if S_b is the speed of the boat and S_w is the speed of the bow wave, then (using $d = rt$),

Distance from P_1 to $A = S_w t$ Distance from P_1 to $B = S_b t$

FIGURE 12

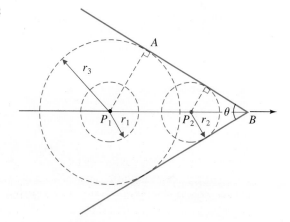

and, since triangle P_1BA is a right triangle,

$$\sin\frac{\theta}{2} = \frac{S_w t}{S_b t} = \frac{S_w}{S_b}$$

Thus,

$$\sin\frac{\theta}{2} = \frac{S_w}{S_b}$$

where S_w is the speed of the bow wave, S_b is the speed of the boat, and $S_b > S_w$.

◆ **EXAMPLE 3**   Bow Wave Speed

If a speedboat travels at 45 km/hr and the angle between the bow waves is 72°, how fast is the bow wave traveling?

SOLUTION   We use

$$\sin\frac{\theta}{2} = \frac{S_w}{S_b}$$

where $\theta = 72°$ and $S_b = 45$ km/hr. Then we solve for S_w.

$$S_w = (45 \text{ km/hr})(\sin 36°) \approx 26 \text{ km/hr}$$ ◆

MATCHED PROBLEM 3   If a speedboat is traveling at 121 km/hr and the angle between the bow waves is 74.5°, how fast is the bow wave moving?

◆ **Modeling Sonic Booms**

If we follow exactly the same line of reasoning as in the discussion of bow waves, we see that an aircraft flying faster than the speed of sound produces sound waves that pile up behind the aircraft in the form of a cone (see Figure 13). The cone intersects the ground in the form of a hyperbola, and along this curve we experience a phenomenon called a **sonic boom**. As in the bow wave analysis, we have

$$\sin\frac{\theta}{2} = \frac{S_s}{S_a}$$

where S_s is the speed of the sound wave, S_a is the speed of the aircraft, and $S_a > S_s$.

FIGURE 13
Sound cones and sonic booms

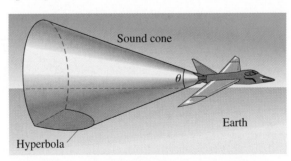

◆ High-Energy Physics: Modeling Particle Energy

Nuclear particles can be made to move faster than the velocity of light in certain materials, such as glass. In 1958, three physicists (Cerenkov, Frank, and Tamm) jointly received a Nobel prize for the work they did based on this fact. Interestingly, the bow wave analysis for boats applies equally well here. Instead of a sound cone, as in Figure 13, they obtained a light cone, as shown in Figure 14.

FIGURE 14
Particle energy

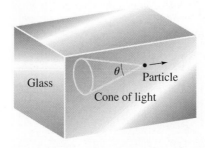

By measuring the cone angle θ, Cerenkov, Frank, and Tamm were able to determine the speed of the particle because the speed of light in glass is readily determined. They used the formula

$$\sin \frac{\theta}{2} = \frac{S_\ell}{S_p}$$

where S_ℓ is the speed of light in glass, S_p is the speed of the particle, and $S_p > S_\ell$. By determining the speed of the particle, they were then able to determine its energy by routine procedures.

◆ Psychology: Modeling Perception

An important field of study in psychology concerns sensory perception—hearing, seeing, smelling, feeling, and tasting. It is well known that individuals see certain objects differently in different surroundings. Lines that appear to be parallel in one setting may appear to be curved in another. Lines of the same length may appear to have different lengths in two different settings. Is a square always a square? (See Figure 15.)

FIGURE 15
Illusions

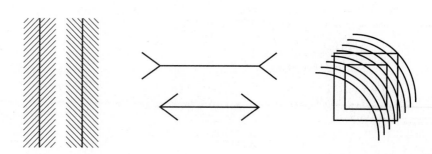

Figure 16 illustrates a perspective illusion in which the people farthest away appear to be larger than those closest—they are actually all the same size.

FIGURE 16
Which person is tallest?

An interesting experiment in visual perception was conducted by psychologists Berliner and Berliner. A tilted field of parallel lines was presented to several subjects who were then asked to estimate the position of a horizontal line in the field. Berliner and Berliner (*American Journal of Psychology,* vol. 65, pp. 271–277, 1952) reported that most subjects were consistently off, and that the difference in degrees d between their estimates and the actual horizontal could be approximated by the equation

$$d = a + b \sin 4\theta$$

where a and b were constants associated with a particular individual and θ was the angle of tilt of the visual field in degrees (see Figure 17).

FIGURE 17
Visual perception

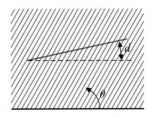

Answers to
Matched Problems

1. $13.7°$

2. $\dfrac{\sin 45°}{\sin \beta} = \dfrac{1.00}{1.66}$

   $\sin \beta = 1.66 \sin 45°$

   $\sin \beta = 1.17$

3. 73.2 km/hr

Since $\sin \beta$ cannot exceed 1 (see the definition of trigonometric functions on p. 61), the condition $\sin \beta = 1.17$ cannot physically happen! In other words, the light must be totally reflected, as indicated in Figure 8.

EXERCISE 2.4

Problems 1–8 refer to the following figure and table (repeated from the text for convenience):

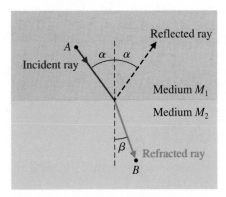

Figure for 1–8

Refractive indexes

Material	Refractive index
Air	1.0003
Crown glass	1.52
Diamond	2.42
Flint glass	1.66
Ice	1.31
Water	1.33

Applications

1. **Light Waves and Refraction** A light ray passing through air strikes the surface of a pool of water so that the angle of incidence $\alpha = 40.6°$. Find the angle of refraction β.

2. **Light Waves and Refraction** Repeat Problem 1 with $\alpha = 34.2°$.

3. **Light Waves and Refraction** A light ray from an underwater spotlight passes through a porthole (of flint glass) of a sunken ocean liner. If the angle of incidence α is 32.0°, what is the angle of refraction β?

4. **Light Waves and Refraction** Repeat Problem 3 with $\alpha = 45.0°$.

5. **Light Waves and Refraction** If light inside a diamond strikes one of its facets (flat surfaces), what is the critical angle of incidence α for total reflection? (The diamond is surrounded by air.) Compute your answer in decimal degrees to three significant digits.

Figure for 5

6. **Light Waves and Refraction** If light inside a triangular flint glass prism strikes one of the flat surfaces, what is the critical angle of incidence α for total reflection? (The prism is surrounded by air.)

7. **Light Waves and Refraction** A golfer hits a ball into a pond. From the side of the pond the ball is spotted on the bottom, five to eight ft from the edge. Is the ball actually closer or farther away than the observed position? Study the figure to determine where the ball appears to be relative to its actual position.

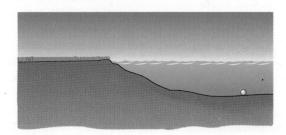

Figure for 7

*8. **Light Waves and Refraction** In an attempt to determine the speed of light in a sample of glass to be used for a lens, we send a light ray from a vacuum into the glass and measure angles α and β to be 30.00° and 18.22°, respectively. If light travels 3.00×10^8 m/sec in a vacuum, how fast will it travel in the glass?

9. **Bow Waves** If the bow waves of a boat travel at 20 km/hr and create an angle of 60°, how fast is the boat traveling?

10. Bow Waves   A boat traveling at 55 km/hr produces bow waves that separate at an angle of 54°. How fast is a bow wave traveling?

11. Bow Waves   If a supersonic aircraft flies at twice the speed of sound in air, what will the cone angle be?

12. Bow Waves   If a supersonic aircraft flies at three times the speed of sound in air, what will the cone angle be (to the nearest degree)?

13. Bow Waves   In crown glass, light travels at approx. 2×10^{10} cm/sec. If a high-energy particle passing through this glass creates a light cone of 90°, how fast is it traveling?

14. Bow Waves   Repeat Problem 13 using a light cone angle of 60°.

Using the empirical formula from Berliner and Berliner's study of perception, $d = a + b \sin 4\theta$, determine d to the nearest degree for the given values of a, b, and θ.

15. Psychology: Perception $a = -2.2$, $b = -4.5$, $\theta = 30°$

16. Psychology: Perception $a = -1.8$, $b = -4.2$, $\theta = 40°$

2.5   EXACT VALUE FOR SPECIAL ANGLES AND REAL NUMBERS

   ◆   **Introduction**
   ◆   **Evaluation of Trigonometric Functions for Quadrantal Angles**
   ◆   **Reference Triangles and Angles**
   ◆   **Special 30°–60° and 45° Right Triangles**
   ◆   **Evaluation of Trigonometric Functions for Angles or Real Numbers with 30°–60° and 45° Reference Triangles**

   ◆   Introduction

If an angle is an integer multiple of 30°, 45°, π/6 rad, or π/4 rad, and if a real number is an integer multiple of π/6 or π/4 (see Figure 1), then, for those values for which each trigonometric function is defined, the function can be evaluated exactly without the use of any calculator (which is different from finding approximate values using a calculator). With a little practice, you will—mentally—be able to determine these exact values. In many situations, working with exact values has substantial advantages over working with approximate values from calculators.

FIGURE 1
Some multiples of 30°, 45°, π/6, and π/4

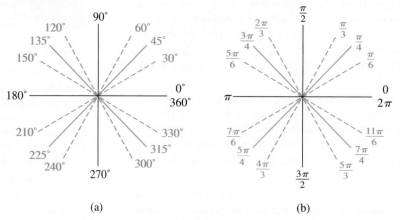

(a)   (b)

There are many significant applications of trigonometric functions. Some require angle domains, and others require real number domains. Our definitions of the trigonometric functions enable us to shift from angle domains to real number domains, and vice versa, with relative ease.

Before we start evaluating trigonometric functions for special values, it is useful to visualize the definition of the trigonometric functions for both angle and real number domains in terms of a "function machine." Figure 2 illustrates a "cosine machine" with an angle domain and a "sine machine" with a real number domain.

We are now ready to evaluate trigonometric functions for special angles and special real numbers. For these angles and real numbers, it will be relatively easy to find exact coordinates of a point on the terminal side of the indicated angle. Once coordinates are found, R can be found, and then exact values for all six trigonometric functions can be determined (when defined).

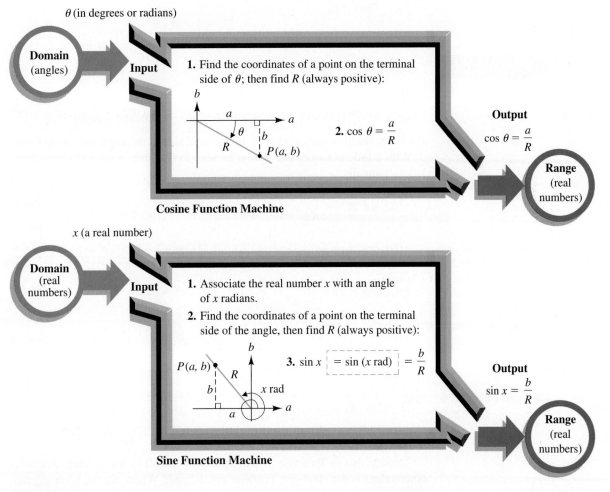

FIGURE 2
Cosine and sine "function machines"

◆   Evaluation of Trigonometric Functions for
Quadrantal Angles

The easiest angles to deal with are **quadrantal angles**—that is, angles with their
terminal side lying along a coordinate axis. These angles are integer multiples of
90° or $\pi/2$. It is easy to find coordinates of a point on a coordinate axis. Since
any nonorigin point will do, we shall, for convenience, choose points 1 unit from
the origin (Figure 3).

FIGURE 3
Quadrantal angle points

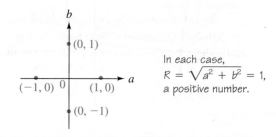

In each case,
$R = \sqrt{a^2 + b^2} = 1$,
a positive number.

◆   EXAMPLE 1   Evaluation Involving Quadrantal Angles

Find

(A) $\sin 90°$ (B) $\cos \pi$ (C) $\tan(-2\pi)$ (D) $\cot(-180°)$

SOLUTIONS   For each, visualize the location of the terminal side of the angle relative to Figure
3. With a little practice, you should be able to do most of the following calculations mentally.

(A) $\sin 90° = \dfrac{b}{R} = \dfrac{1}{1} = 1$ $(a, b) = (0, 1),\quad R = 1$

(B) $\cos \pi = \dfrac{a}{R} = \dfrac{-1}{1} = -1$ $(a, b) = (-1, 0),\quad R = 1$

(C) $\tan(-2\pi) = \dfrac{b}{a} = \dfrac{0}{1} = 0$ $(a, b) = (1, 0),\quad R = 1$

(D) $\cot(-180°) = \dfrac{a}{b} = \dfrac{-1}{0}$ $(a, b) = (-1, 0),\quad R = 1$

Not defined   ◇

MATCHED PROBLEM 1   Find

(A) $\sin(3\pi/2)$ (B) $\sec(-\pi)$ (C) $\tan 90°$ (D) $\cot(-270°)$

Notice that in Example 1, part (D), $\cot(-180°)$ is not defined. For what
other values is the cotangent function not defined? Where are the other trigonometric functions not defined? These important questions are considered in Exercise 2.5.

◆ Reference Triangle and Angle

Because the reference triangle is going to play a very important role in the work that follows, we now restate its definition as well as that of a reference angle.

REFERENCE TRIANGLE AND ANGLE

For a nonquadrantal angle θ:

1. To form a **reference triangle** for θ, drop a perpendicular from a point $P(a, b)$ on the terminal side of θ to the horizontal axis.
2. The **reference angle** α is the acute angle (always taken positive) between the terminal side of θ and the horizontal axis.

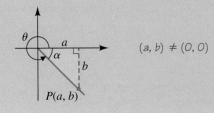

$(a, b) \neq (0, 0)$

◆ **EXAMPLE 2**   Reference Triangles and Angles

Sketch the reference triangle and find the reference angle α for each of the following angles.

(A) $\theta = 330°$ (B) $\theta = -315°$ (C) $\theta = -\pi/4$ (D) $\theta = 4\pi/3$

SOLUTIONS   (A)

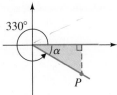

$\alpha = 360° - 330° = 30°$

FIGURE 4

(B)

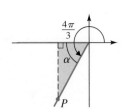

$\alpha = 360° - 315° = 45°$

FIGURE 5

(C)

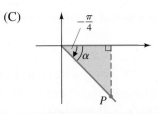

$\theta = |-\pi/4| = \pi/4$

FIGURE 6

(D)

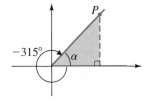

$\alpha = 4\pi/3 - \pi = \pi/3$

FIGURE 7   ◇

MATCHED PROBLEM 2   Sketch the reference triangle and find the reference angle α for each of the following angles.

(A) $\theta = -225°$ (B) $\theta = 420°$
(C) $\theta = -5\pi/6$ (D) $\theta = 2\pi/3$

◆ Special 30°–60° and 45° Right Triangles

If a reference triangle of a given angle is a 30°–60° right triangle or a 45° right triangle, then we will be able to find exact nonorigin coordinates on the terminal side of the given angle.

    If we take a 30°–60° right triangle, we note that it is one-half of an equilateral triangle, as indicated in Figure 8. Since all sides are equal in an equilateral triangle, we can apply the Pythagorean theorem to obtain a useful relationship among the three sides of the original triangle.

FIGURE 8

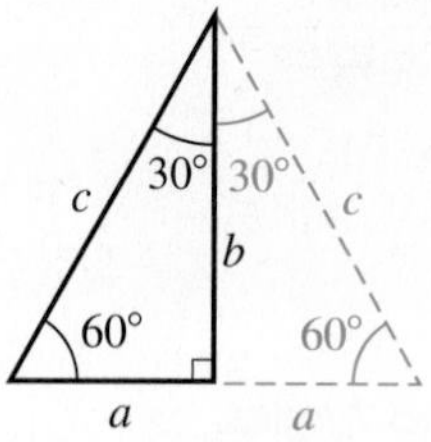

$$b = \sqrt{c^2 - a^2}$$
$$= \sqrt{(2a)^2 - a^2} \qquad \text{Since } c = 2a$$
$$= \sqrt{3a^2}$$
$$= a\sqrt{3}$$

Similarly, using the Pythagorean theorem on a 45° right triangle, we obtain the following (see Figure 9).

FIGURE 9

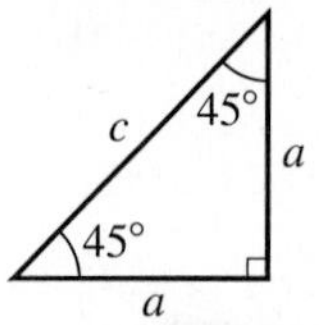

$$c = \sqrt{a^2 + a^2}$$
$$= \sqrt{2a^2}$$
$$= a\sqrt{2}$$

We summarize these results in the figures in the box at the top of page 85, along with some frequently used special cases. The ratios of these special triangles should be learned, since they will be used often in this and subsequent sections. The two triangles in color are the easiest to remember. The others can be obtained from these by multiplying or dividing the length of each side by the same nonzero quantity.

◆ Evaluation of Trigonometric Functions for
    Angles or Real Numbers with 30°–60° and
    45° Reference Triangles

If an angle θ has a 30°–60° or 45° reference triangle, then it is easy to find exact coordinates of a point P on the terminal side of θ and the exact distance of P

30°–60° AND 45° SPECIAL TRIANGLES

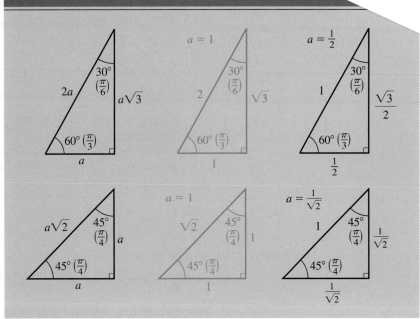

from the origin. Then, using the definitions of the six trigonometric functions given in Section 2.3, we can find the exact value of any of the six functions for the given θ. Several examples will illustrate the process.

◆ EXAMPLE 3   Exact Evaluation for Special Angles and Real Numbers

Evaluate exactly:

(A) $\sin 30°$, $\cos(\pi/6)$, $\cot(\pi/6)$ (B) $\cos 45°$, $\tan(\pi/4)$, $\csc(\pi/4)$

SOLUTIONS   (A) Use the special 30°–60° triangle (Figure 10) as the reference triangle for $\theta = 30°$ and $\theta = \pi/6$. Use the sides of the reference triangle to determine $P(a, b)$ and R; then use Definition 1.

FIGURE 10

$$\sin 30° = \frac{b}{R} = \frac{1}{2}$$

$$\cos \frac{\pi}{6} = \frac{a}{R} = \frac{\sqrt{3}}{2}$$

$$\cot \frac{\pi}{6} = \frac{a}{b} = \frac{\sqrt{3}}{1} = \sqrt{3}$$

$(\sqrt{3}, 1)$

2

$30° \left(\frac{\pi}{6}\right)$ 1

$\sqrt{3}$

$(a, b) = (\sqrt{3}, 1)$
$R = 2$

(B) Use the special 45° triangle (Figure 11) as the reference triangle for $\theta = 45°$ and $\theta = \pi/4$. Use the sides of the reference triangle to determine $P(a, b)$ and R; then use the appropriate definition from Section 2.3.

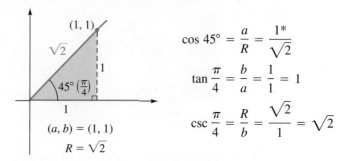

$$\cos 45° = \frac{a}{R} = \frac{1^*}{\sqrt{2}}$$

$$\tan \frac{\pi}{4} = \frac{b}{a} = \frac{1}{1} = 1$$

$$\csc \frac{\pi}{4} = \frac{R}{b} = \frac{\sqrt{2}}{1} = \sqrt{2}$$

FIGURE 11   ◇

MATCHED PROBLEM 3   Evaluate exactly:

(A) $\cos 60°$, $\sin(\pi/3)$, $\tan(\pi/3)$ (B) $\sin 45°$, $\cot(\pi/4)$, $\sec(\pi/4)$

Before proceeding with examples of reference triangles in quadrants other than the first quadrant, it is useful to recall the multiples of $\pi/3$ (60°), $\pi/6$ (30°), and $\pi/4$ (45°). See Figure 12.

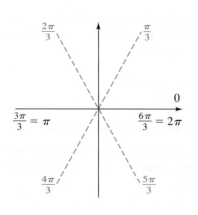

(a) Multiples of $\frac{\pi}{3}$ (60°)

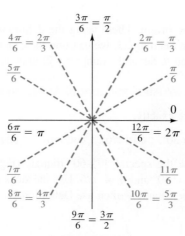

(b) Multiples of $\frac{\pi}{6}$ (30°)

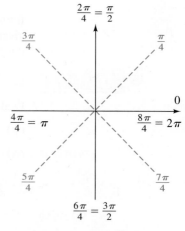

(c) Multiples of $\frac{\pi}{4}$ (45°)

FIGURE 12
Multiples of special angles

_____

* Whether we rationalize a denominator or not depends entirely on what we want to do with the answer. Leave answers in unrationalized form unless directed otherwise.

◆ EXAMPLE 4   Exact Evaluation for Special Angles and Real Numbers

Evaluate exactly:

(A) $\sin 210°$ (B) $\cos(2\pi/3)$ (C) $\tan(7\pi/4)$
(D) $\sec 300°$ (E) $\cot(-5\pi/6)$ (F) $\csc(-240°)$

SOLUTIONS   Each angle has a 30°–60° or 45° reference triangle. Locate it, determine $P(a, b)$ and R, and then evaluate.

(A) $\sin 210° = \dfrac{-1}{2} = -\dfrac{1}{2}$

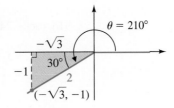

FIGURE 13

(B) $\cos \dfrac{2\pi}{3} = \dfrac{-1}{2} = -\dfrac{1}{2}$

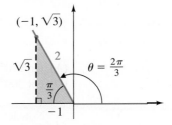

FIGURE 14

(C) $\tan \dfrac{7\pi}{4} = \dfrac{-1}{1} = -1$

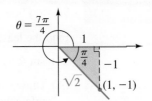

FIGURE 15

(D) $\sec 300° = \dfrac{2}{1} = 2$

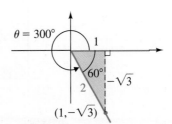

FIGURE 16

(E) $\cot\left(-\dfrac{5\pi}{6}\right) = \dfrac{-\sqrt{3}}{-1} = \sqrt{3}$

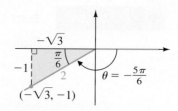

FIGURE 17

(F) $\csc(-240°) = \dfrac{2}{\sqrt{3}}$

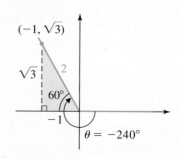

FIGURE 18

MATCHED PROBLEM 4   Evaluate exactly:

(A) $\tan 210°$ (B) $\sin(2\pi/3)$ (C) $\cos(7\pi/4)$
(D) $\cot 300°$ (E) $\csc(-5\pi/6)$ (F) $\sec(-240°)$

Now let us reverse the problem. That is, we are given the exact value of one of the six trigonometric functions that corresponds to one of the special reference triangles, and we must find, θ.

◆   EXAMPLE 5   Finding Special Angles θ

Find the least positive θ in degree and radian measure for which each is true.

(A) $\sin \theta = \sqrt{3}/2$ (B) $\cos \theta = -1/\sqrt{2}$

SOLUTIONS   (A) Draw a reference triangle (Figure 19) in the first quadrant with side opposite reference angle $\sqrt{3}$ and hypotenuse 2. Observe that this is a special 30°–60° triangle.

FIGURE 19

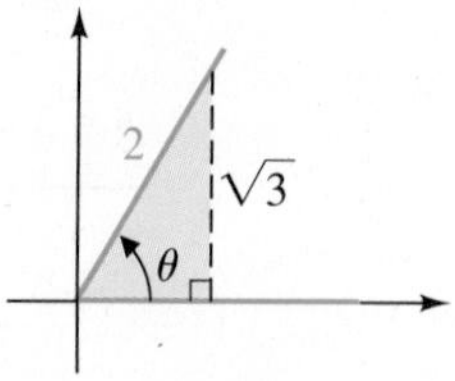

$\theta = 60°$ or $\dfrac{\pi}{3}$

(B) Draw a reference triangle (Figure 20) in the second quadrant with side adjacent reference angle -1 and hypotenuse $\sqrt{2}$. Observe that this is a special 45° triangle.

FIGURE 20

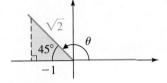

$$\theta = 180° - 45° = 135° \quad \text{or} \quad \frac{3\pi}{4}$$

◇

MATCHED PROBLEM 5 Repeat Example 5 for

(A) $\tan \theta = 1/\sqrt{3}$ (B) $\sec \theta = -\sqrt{2}$

We conclude this section with a summary of special values in Table 1. Some people memorize this table; others use the definition in Section 2.3 and special triangles.

TABLE 1
Special values

θ	$\sin \theta$	$\csc \theta$	$\cos \theta$	$\sec \theta$	$\tan \theta$	$\cot \theta$
0° or 0	0	N.D.	1	1	0	N.D.
30° or $\pi/6$	1/2	2	$\sqrt{3}/2$	$2/\sqrt{3}$	$1/\sqrt{3}$	$\sqrt{3}$
45° or $\pi/4$	$1/\sqrt{2}$	$\sqrt{2}$	$1/\sqrt{2}$	$\sqrt{2}$	1	1
60° or $\pi/3$	$\sqrt{3}/2$	$2/\sqrt{3}$	1/2	2	$\sqrt{3}$	$1/\sqrt{3}$
90° or $\pi/2$	1	1	0	N.D.	N.D.	0

N.D. = Not defined

Answers to
Matched Problems

1. (A) -1 (B) -1 (C) Not defined   (D) 0

2. (A)

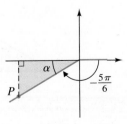

$\alpha = 225° - 180° = 45°$

(B)

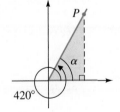

$\alpha = 420° - 360° = 60°$

(C)

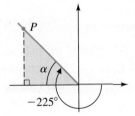

$\alpha = \pi - 5\pi/6 = \pi/6$

(D)

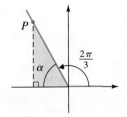

$\alpha = \pi - 2\pi/3 = \pi/3$

3. (A) $1/2$, $\sqrt{3}/2$, $\sqrt{3}$ (B) $1/\sqrt{2}$, 1, $\sqrt{2}$
4. (A) $1/\sqrt{3}$ (B) $\sqrt{3}/2$ (C) $1/\sqrt{2}$ (D) $-1/\sqrt{3}$
(E) -2 (F) -2
5. (A) $30°$ or $\pi/6$ (B) $135°$ or $3\pi/4$

EXERCISE 2.5

Do not use a calculator for any of the problems in this exercise.

A *Sketch the reference triangle and find the reference angle α for each of the following angles.*

1. $\theta = 60°$ **2.** $\theta = 45°$ **3.** $\theta = -60°$

4. $\theta = -45°$ **5.** $\theta = \dfrac{-\pi}{3}$ **6.** $\theta = \dfrac{-\pi}{4}$

7. $\theta = \dfrac{3\pi}{4}$ **8.** $\theta = \dfrac{5\pi}{6}$ **9.** $\theta = -210°$

10. $\theta = -150°$ **11.** $\theta = \dfrac{-5\pi}{4}$ **12.** $\theta = \dfrac{-5\pi}{3}$

Find the exact value of each of the following.

13. $\cos 0°$ **14.** $\sin 0°$ **15.** $\sin 30°$

16. $\cos 45°$ **17.** $\sin \dfrac{\pi}{2}$ **18.** $\cot \dfrac{\pi}{4}$

19. $\tan 45°$ **20.** $\tan 0$ **21.** $\tan \dfrac{\pi}{6}$

22. $\cot 0$ **23.** $\cos 90°$ **24.** $\cot \dfrac{\pi}{3}$

25. $\sin(-30°)$ **26.** $\cos(-60°)$ **27.** $\cos \dfrac{-\pi}{2}$

28. $\tan \pi$ **29.** $\tan 120°$ **30.** $\sin(-45°)$

31. $\cos \dfrac{-\pi}{6}$ **32.** $\tan \dfrac{-\pi}{4}$ **33.** $\cot 2\pi$

34. $\cos \dfrac{2\pi}{3}$ **35.** $\cot 150°$ **36.** $\cot(-60°)$

B *In Problems 37–54, find the exact value of each:*

37. $\sin \dfrac{7\pi}{6}$ **38.** $\sin \dfrac{3\pi}{4}$ **39.** $\sin \dfrac{3\pi}{2}$

40. $\cos \dfrac{11\pi}{6}$ **41.** $\sin 225°$ **42.** $\cos 300°$

43. $\cot \dfrac{-5\pi}{4}$ **44.** $\tan \dfrac{-4\pi}{3}$ **45.** $\cos \dfrac{-5\pi}{6}$

46. $\sin \dfrac{-5\pi}{3}$ **47.** $\cot(-390°)$ **48.** $\tan(-405°)$

49. $\sin \dfrac{-15\pi}{4}$ **50.** $\sin \dfrac{-7\pi}{2}$ **51.** $\csc 420°$

52. $\sec 390°$ **53.** $\cos \dfrac{5\pi}{2}$ **54.** $\cos \dfrac{8\pi}{3}$

Find all angles θ, where
(A) $0° \le \theta < 360°$ (B) $0 \le \theta < 2\pi$
for which the following functions are not defined.

55. tangent   **56.** cotangent   **57.** cosecant   **58.** secant

In Problems 59–64, find the least positive θ in
(A) degree measure   (B) radian measure
for which each is true.

59. $\sin \theta = \dfrac{1}{2}$ **60.** $\cos \theta = \dfrac{1}{\sqrt{2}}$ **61.** $\cos \theta = \dfrac{-1}{2}$

62. $\sin \theta = \dfrac{-1}{2}$ **63.** $\tan \theta = -\sqrt{3}$ **64.** $\cot \theta = -1$

C **65.** Find the exact value of all the angles between $0°$ and $360°$ for which $\sin \theta = -\sqrt{3}/2$.

66. Find the exact value of all the angles between $0°$ and $360°$ for which $\tan \theta = -\sqrt{3}$.

67. Find the exact value of all the angles between 0 rad and 2π rad for which $\cot \theta = -\sqrt{3}$.

68. Find the exact value of all the angles between 0 rad and 2π rad for which $\cos \theta = -\sqrt{3}/2$.

Find the exact values of x and y in Problems 69 and 70.

69. (A)   (B)   (C)

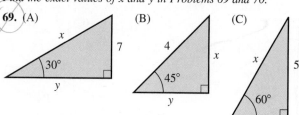

70. (A)

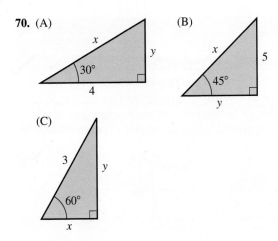

(C)

Later we will show that the area of an n-sided regular polygon inscribed in a circle of radius R (such as in the figure) is

given by

$$A = \frac{nR^2}{2} \sin \frac{2\pi}{n}$$

FIGURE for 71–74
Inscribed regular polygon

Use this formula to find the exact areas of the polygons defined in Problems 71–74. Assume each radius is exact.

71. $n = 3$, $R = 2$ cm   **72.** $n = 4$, $R = 5$ ft

73. $n = 6$, $R = 10$ in.   **74.** $n = 8$, $R = 4$ mm

2.6 CIRCULAR FUNCTIONS

- ◆ **Definition of Circular Functions**
- ◆ **Domain and Range for Sine and Cosine Functions**
- ◆ **Periodic Properties**
- ◆ **Fundamental Identities**
- ◆ **Circular Functions and Trigonometric Functions**
- ◆ **Evaluating Circular Functions**

Our treatment of trigonometric functions has progressed from the concrete to the abstract, which follows the historical development of the subject. We first defined trigonometric ratios relative to right triangles—a procedure used by the ancient Greeks. We then expanded the meaning of these functions by using generalized angles in standard positions in a rectangular coordinate system.

We now turn to the modern, more abstract definition of these functions, a definition involving real number domains, not angles or triangles. This is the preferred and most useful definition for advanced mathematics and the sciences, including calculus. We will be able to quickly observe some very useful trigonometric properties and relationships that were not as apparent using the angle approach.

◆ Definition of Circular Functions

If we graph the equation $a^2 + b^2 = 1$ in a rectangular coordinate system, we obtain a circle with center at the origin and radius 1 called the **unit circle.** Using this circle, we define the **circular functions** with real number domains, as follows.

CIRCULAR FUNCTIONS

Let x be an arbitrary real number and let U be the unit circle with equation $a^2 + b^2 = 1$. Start at $(1, 0)$ and proceed counterclockwise if x is positive and clockwise if x is negative around the unit circle until an arc length of $|x|$ has been covered. Let $P(a, b)$ be the point at the terminal end of the arc.

$c = 1$

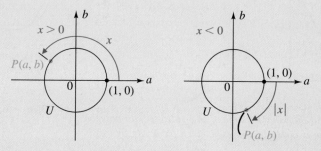

The following six circular functions* are defined in terms of the coordinates of P.

$$y = \sin x = \frac{b}{c=1} \qquad y = \cos x = \frac{a}{c=1} \qquad y = \tan x = \frac{b}{a}$$

$$y = \csc x = \frac{1}{b} \qquad y = \sec x = \frac{1}{a} \qquad y = \cot x = \frac{a}{b}$$

The independent variable is x and the dependent variable is y.

REMARKS
1. The circular function definition does not involve any angles.
2. The circular function definition uses standard function notation

$$y = f(x)$$

with f replaced by the name of a particular circular function; for example, $y = \sin x$ actually means $y = \sin(x)$. (See Appendix B.1 for a brief discussion of function and function notation.)
3. When we write $y = f(x) = x^2 - 1$, the expression on the right has a recipe for evaluating f (square x and subtract 1). When we write $y = f(x) = \sin x$, the expression on the right only identifies the function; for its evaluation we must refer to the circular function definition.   ◇

We will now investigate a few important properties of the circular functions that are easily observed from their definitions. In the next chapter we will graph the circular functions and discuss many additional properties.

---

* Following common usage, we also refer to circular functions as trigonometric functions.

◆ Domain and Range for Sine and Cosine Functions

Since each real number x can be associated with an arc $|x|$ units in length (counterclockwise if x is positive and clockwise if x is negative), the domain of each circular function is the set of real numbers for which the function is defined.

We now investigate the domains and ranges of the sine and cosine functions. (The domains and ranges of the other four circular functions will be covered briefly in Exercise 2.6 and in detail in the next chapter.) Referring to the definition of circular functions, observe that

$$a = \cos x \quad \text{and} \quad b = \sin x$$

Thus, as in Figure 1, the coordinates of a point P at the end of an arc of length $|x|$, starting at $(1, 0)$, are

$$P(\cos x, \sin x)$$

FIGURE 1

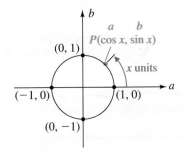

It is clear that $\cos x$ and $\sin x$ exist for each real number x, since $(\cos x, \sin x)$ are the coordinates of the point P on the unit circle corresponding to the arc length $|x|$. Thus, the domain of each function is the set of all real numbers.

What about the range of each function? As $P(\cos x, \sin x)$ moves around the unit circle, the abscissa of the point, $\cos x = a$, and the ordinate of the point, $\sin x = b$, both vary between -1 and 1. Thus, we conclude that the range of each function is the set of all real numbers y such that $-1 \le y \le 1$.

The above results are summarized in the box below for each reference.

DOMAIN AND RANGE FOR SINE AND FOR COSINE

Domain: All real numbers R
Range: $-1 \le y \le 1$, y a real number

◆ EXAMPLE 1   Finding Domain Values That Correspond to a Given Range Value

Find the domain values of the sine function, $-2\pi \le x \le 2\pi$, that have a range value -1. That is, find x, $-2\pi \le x \le 2\pi$, such that $\sin x = -1$.

SOLUTION   Refer to Figure 1; see that $\sin x = -1$ at the point $(0, -1)$. Thus, for any domain value x associated with an arc starting at $(1, 0)$ and terminating at $(0, -1)$, we have $\sin x = -1$. For the interval $-2\pi \le x \le 2\pi$, $\sin x = -1$ for $x = -\pi/2$ and $x = 3\pi/2$, as indicated in Figure 2.

FIGURE 2

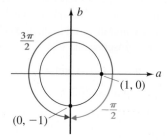

◇

MATCHED PROBLEM 1   Find the domain values of the cosine function, $-2\pi \le x \le 2\pi$, that have a range value -1. That is, find x, $-2\pi \le x \le 2\pi$, such that $\cos x = -1$.

♦   **Periodic Properties**

The circumference of the unit circle is

$$C = 2\pi R = 2\pi(1) = 2\pi$$

Thus, for a given value x (see Figure 1), if we add or subtract 2π, we will return to exactly the same point P with the same coordinates. And we conclude that

$$\sin(x + 2\pi) = \sin x \quad \text{and} \quad \cos(x + 2\pi) = \cos x$$
$$\sin(x - 2\pi) = \sin x \quad \text{and} \quad \cos(x - 2\pi) = \cos x$$

In general, if we add any integer multiple of 2π to x, we will return to exactly the same point P. Thus, for k any integer $(\ldots, -3, -2, -1, 0, 1, 2, 3, \ldots)$,

$$\sin(x + 2k\pi) = \sin x \quad \text{and} \quad \cos(x + 2k\pi) = \cos x$$

Functions with this kind of repetitive behavior are called **periodic functions.** In general:

PERIODIC FUNCTIONS

A function f is **periodic** if there is a positive real number p such that

$$f(x + p) = f(x)$$

for all x in the domain of f. The smallest such positive p, if it exists, is called **the period of f.**

From the definition of a periodic function, we conclude that

Both the sine function and cosine function have a period of 2π.

The other four circular functions also have periodic properties, which we discuss in detail in the next chapter.

◆ EXAMPLE 2 **Finding Domain Values That Correspond to a Given Range Value**

If $\cos x = -0.0315$, what is the value of each of the following?

(A) $\cos(x + 2\pi)$ (B) $\cos(x - 2\pi)$
(C) $\cos(x + 18\pi)$ (D) $\cos(x - 34\pi)$

SOLUTIONS All are equal to -0.0315, because the cosine function is periodic with period 2π. That is, $\cos(x + 2k\pi) = \cos x$ for *all* integers k. In part (A), $k = 1$; in part (B), $k = -1$; in part (C), $k = 9$; and in part (D), $k = -17$. ◇

MATCHED PROBLEM 2 If $\sin x = 0.7714$, what is the value of each of the following?

(A) $\sin(x + 2\pi)$ (B) $\sin(x - 2\pi)$
(C) $\sin(x + 14\pi)$ (D) $\sin(x - 26\pi)$

The periodic properties of the circular functions are of paramount importance in the development of further mathematics as well as the applications of mathematics. As we will see in the next chapter, the circular functions are made to order for the analysis of real-world periodic phenomena: light, sound, electrical, and water waves; motion in buildings during earthquakes; motion in suspension systems in automobiles; planetary motion; business cycles; and so on.

◆ **Fundamental Identities**

Returning to the definition of the circular functions and noting that

$$\sin x = b \qquad \text{and} \qquad \cos x = a$$

we can obtain the following useful relationships among the six functions.

$$\csc x = \frac{1}{b} = \frac{1}{\sin x} \tag{1}$$

$$\sec x = \frac{1}{a} = \frac{1}{\cos x} \tag{2}$$

$$\cot x = \frac{a}{b} = \frac{1}{b/a} = \frac{1}{\tan x} \tag{3}$$

$$\tan x = \frac{b}{a} = \frac{\sin x}{\cos x} \tag{4}$$

$$\cot x = \frac{a}{b} = \frac{\cos x}{\sin x} \tag{5}$$

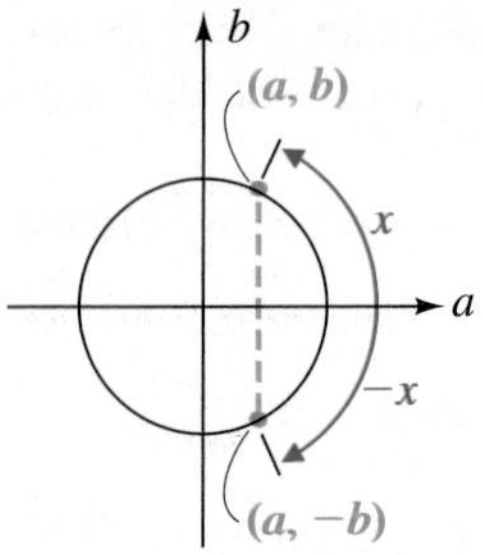

2 TRIGONOMETRIC FUNCTIONS

$(a, b) = (\cos x, \sin x)$

Because the terminal points of x and $-x$ are symmetric with respect to the horizontal axis (see Figure 3), we have the following sign properties.

$$\sin(-x) = -b = -\sin x \qquad (6)$$

$b = \sin x$, $\qquad a = \cos x$

$-b = -\sin x$

$$\cos(-x) = a = \cos x \qquad (7)$$

$$\tan(-x) = \frac{-b}{a} = -\frac{b}{a} = -\tan x \qquad (8)$$

Finally, because $(a, b) = (\cos x, \sin x)$ is on the unit circle $a^2 + b^2 = 1$, it follows that

$$(\cos x)^2 + (\sin x)^2 = 1$$

which is usually written in the form

$$\sin^2 x + \cos^2 x = 1 \qquad (9)$$

where $\sin^2 x$ and $\cos^2 x$ are concise ways of writing $(\sin x)^2$ and $(\cos x)^2$, respectively.

⚠️ **Caution**   Note that $(\cos x)^2 \neq \cos x^2$ and $(\sin x)^2 \neq \sin x^2$.   ◇

Equations (1)–(9) are called **fundamental identities.** They hold true for all replacements of x by real numbers (or angles in degree or radian measure, as we will see before the conclusion of this section) for which both sides of an equation are defined.

◆ **EXAMPLE 3**   **Use of Identities**

Simplify each expression using the fundamental identities.

(A) $\dfrac{\sin^2 x + \cos^2 x}{\tan x}$ (B) $\dfrac{\sin(-x)}{\cos(-x)}$

SOLUTIONS   (A) $\dfrac{\sin^2 x + \cos^2 x}{\tan x}$ Use identity (9).

$\qquad\quad = \dfrac{1}{\tan x}$ Use identity (3).

$\qquad\quad = \cot x$

(B) $\dfrac{\sin(-x)}{\cos(-x)}$ Use identity (4).

$\qquad\quad = \tan(-x)$ Use identity (8).

$\qquad\quad = -\tan x$ ◇

MATCHED PROBLEM 3   Simplify each expression using the fundamental identities.

(A) $\dfrac{1 - \cos^2 x}{\sin^3 x}$ (B) $\tan(-x)\cos(-x)$

FIGURE 3

Some of the fundamental identities will be used in Chapter [...]
graphing some of the trigonometric functions. A detailed discussion o[...]
is found in Chapter 4.

◆ Circular Functions and Trigonometric Functions

We now show how the earlier definitions of the trigonometric functions (involving angle domains) can be related to the circular functions (involving real number domains and the unit circle). To this end, let us look at the radian measure of an angle θ subtended by an arc of x units on the unit circle (Figure 4).

We see that for the unit circle, the angle subtended by an arc of x units has a radian measure of x. Thus, every real number can be associated with an arc of x units on the unit circle or a central angle of x radians on the same circle (if x is positive, we go counterclockwise; and if x is negative, we go clockwise). Notice that the point on the terminal end of the arc of x units is also on the terminal side of the angle of x radians. This provides the following very useful relationships between the previously defined trigonometric functions with angle domains and the circular functions with real number domains:

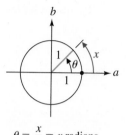

$$\theta = \frac{x}{1} = x \text{ radians}$$

FIGURE 4

CIRCULAR FUNCTIONS AND TRIGONOMETRIC FUNCTIONS

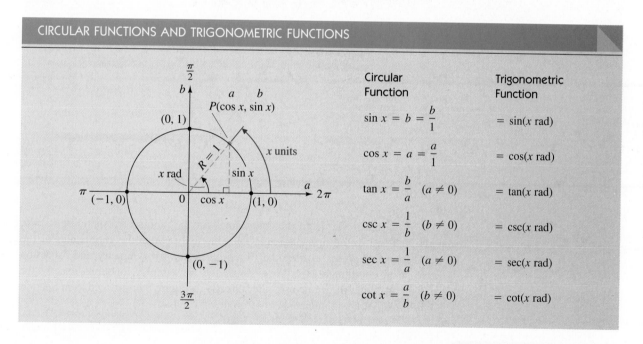

	Circular Function	Trigonometric Function
	$\sin x = b = \dfrac{b}{1}$	$= \sin(x \text{ rad})$
	$\cos x = a = \dfrac{a}{1}$	$= \cos(x \text{ rad})$
	$\tan x = \dfrac{b}{a} \quad (a \neq 0)$	$= \tan(x \text{ rad})$
	$\csc x = \dfrac{1}{b} \quad (b \neq 0)$	$= \csc(x \text{ rad})$
	$\sec x = \dfrac{1}{a} \quad (a \neq 0)$	$= \sec(x \text{ rad})$
	$\cot x = \dfrac{a}{b} \quad (b \neq 0)$	$= \cot(x \text{ rad})$

The information in the preceding box, which relates circular functions with trigonometric functions, is very useful and should be understood. We will use it to develop a number of properties that hold simultaneously for the circular functions and the trigonometric functions. (For example, the fundamental identities discussed above hold for all real numbers x or angles in degree or radian measure.)

◆ **Evaluating Circular Functions**

Because circular functions are related to trigonometric functions, we can evaluate circular functions by the procedures we used to evaluate trigonometric functions.

If x is an integer multiple of $\pi/4$ or $\pi/6$, we can find exact values of the six circular functions when they are defined. Corresponding coordinates of points on the unit circle can be found by using appropriate reference triangles and angles. (Recall the special 45° and 30°–60° triangles on page 85 in Section 2.5.) Using these relationships, we can determine coordinates of points that correspond to integer multiples of $\pi/4$ or $\pi/6$. Figure 5 shows these coordinates for multiples ranging from 0 to 2π.

FIGURE 5

Multiples of $\pi/6$ and $\pi/4$ on a unit circle

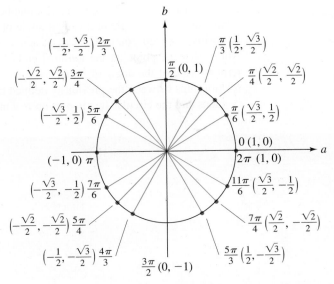

Unit circle $a^2 + b^2 = 1$

Note that if you learn the coordinates for the points corresponding to $\pi/6$, $\pi/4$, and $\pi/3$ in the first quadrant, you can easily obtain the coordinates of the corresponding points in the other three quadrants by symmetry and reflection across the coordinate axis. To evaluate the six circular functions exactly for integer multiples of $\pi/6$ or $\pi/4$, refer directly to Figure 5 or proceed as in Section 2.5 to relate the real number x to an angle of x radians.

◆ EXAMPLE 4   Evaluating Circular Functions Exactly

Find each exactly:

(A) $\sin \dfrac{8\pi}{3}$ (B) $\sec\left(-\dfrac{5\pi}{6}\right)$ (C) $\tan 7\pi$

SOLUTIONS   (A) Refer directly to Figure 5 or the fact that

$$\sin \frac{8\pi}{3} = \sin \left(\frac{8\pi}{3} \text{ rad} \right)$$

and proceed using reference triangles as in Section 2.5. In either case,

$$\sin \frac{8\pi}{3} = \frac{\sqrt{3}}{2}$$

(B) Refer directly to Figure 5 or the fact that

$$\sec \left(-\frac{5\pi}{6} \right) = \sec \left(-\frac{5\pi}{6} \text{ rad} \right)$$

and proceed using reference triangles as in Section 2.5. In either case,

$$\sec \left(-\frac{5\pi}{6} \right) = -\frac{2}{\sqrt{3}}$$

(C) 7π corresponds to the point $(-1, 0)$. Thus,

$$\tan 7\pi = \frac{0}{-1} = 0$$

◇

MATCHED PROBLEM 4   Find each exactly:

(A) $\cos \dfrac{13\pi}{6}$ (B) $\csc \left(-\dfrac{5\pi}{4} \right)$ (C) $\cot(-3\pi)$

To evaluate circular functions using a calculator, set the calculator in radian mode and proceed as in Section 2.5.

◆ EXAMPLE 5   Calculator Evaluation of Circular Functions

Evaluate to four significant digits:

(A) $\sin(-13.72)$ (B) $\sec 22.33$

SOLUTIONS   (A) $\sin(-13.72) = -0.9142$ Set in radian mode and use the $\boxed{\text{sin}}$ key

(B) $\sec 22.33 = -1.060$ Set in radian mode and use $\boxed{\text{cos}}$ and $\boxed{\text{1/x}}$ or $\boxed{\text{x}^{-1}}$ keys   ◇

MATCHED PROBLEM 5   Evaluate to four significant digits using a calculator:

(A) $\cos 505.3$ (B) $\cot(-0.003211)$

◆ EXAMPLE 6   Periodic Properties

Evaluate cos x to two significant digits for

(A) $x = 1.4$ (B) $x = 1.4 + 2\pi$ (C) $x = 1.4 - 2\pi$
(D) $x = 1.4 + 20\pi$ (E) $x = 1.4 - 8\pi$

SOLUTIONS   All have the same value because the cosine function is periodic with a period of 2π. That is, $\cos(x + 2k\pi) = \cos x$ for all integers k. Using a calculator (set in radian mode to evaluate real numbers), we obtain

(A) $\cos 1.4 = 0.17$ (B) $\cos(1.4 + 2\pi) = 0.17$
(C) $\cos(1.4 - 2\pi) = 0.17$ (D) $\cos(1.4 + 20\pi) = 0.17$
(E) $\cos(1.4 - 8\pi) = 0.17$ ◇

MATCHED PROBLEM 6   Evaluate sin x to two significant digits for

(A) $x = -3.2$ (B) $x = -3.2 + 2\pi$ (C) $x = -3.2 - 2\pi$
(D) $x = -3.2 + 22\pi$ (E) $x = -3.2 - 12\pi$

Answers to   **1.** $-\pi, \pi$ **2.** All 0.7714
Matched Problems **3.** (A) csc x (B) $-\sin x$
                  4. (A) $\sqrt{3}/2$ (B) $\sqrt{2}$ (C) Not defined
                  5. (A) -0.8793 (B) -311.4
                  6. All 0.058

EXERCISE 2.6

Figure 4 is repeated here for convenient reference.

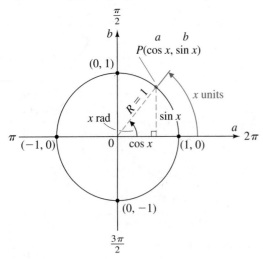

Figure for 1–24

A   **1.** Starting with the circumference of the unit circle, 2π, find:
  (A) $\frac{1}{2}$ circumference   (B) $\frac{3}{4}$ circumference

2. Starting with the circumference of the unit circle, 2π, find:
  (A) $\frac{1}{4}$ circumference   (B) $\frac{1}{8}$ circumference

3. Referring to the figure, state the coordinates of P for the indicated values of x:
  (A) $x = 0$ (B) $x = \pi/2$ (C) $x = -3\pi/2$
  (D) $x = \pi$ (E) $x = -2\pi$ (F) $x = 5\pi/2$

4. Referring to the figure, state the coordinates of P for the indicated values of x:
  (A) $x = 2\pi$ (B) $x = -\pi/2$ (C) $x = -\pi$
  (D) $x = 3\pi/2$ (E) $x = -5\pi/2$ (F) $x = -3\pi$

5. Given $y = \sin x$, how does y vary for the indicated variation in x?
  (A) x varies from 0 to $\pi/2$
  (B) x varies from $\pi/2$ to π
  (C) x varies from π to $3\pi/2$

(D) x varies from $3\pi/2$ to 2π
(E) x varies from 2π to $5\pi/2$

6. Given $y = \cos x$, how does y vary for the indicated variation in x?
(A) x varies from 0 to $\pi/2$
(B) x varies from $\pi/2$ to π
(C) x varies from π to $3\pi/2$
(D) x varies from $3\pi/2$ to 2π
(E) x varies from 2π to $5\pi/2$

7. Given $y = \cos x$, how does y vary for the indicated variation in x?
(A) x varies from 0 to $-\pi/2$
(B) x varies from $-\pi/2$ to $-\pi$
(C) x varies from $-\pi$ to $-3\pi/2$
(D) x varies from $-3\pi/2$ to -2π
(E) x varies from -2π to $-5\pi/2$

8. Given $y = \sin x$, how does y vary for the indicated variation in x?
(A) x varies from 0 to $-\pi/2$
(B) x varies from $-\pi/2$ to $-\pi$
(C) x varies from $-\pi$ to $-3\pi/2$
(D) x varies from $-3\pi/2$ to -2π
(E) x varies from -2π to $-5\pi/2$

Know!

B *Find the values of x from the indicated interval that satisfy the indicated equation or condition. (Refer to the figure at the beginning of the exercise.)*

9. $\sin x = 1$, $0 \le x \le 4\pi$
10. $\cos x = 1$, $0 \le x \le 4\pi$
11. $\sin x = 0$, $0 \le x \le 4\pi$
12. $\cos x = 0$, $0 \le x \le 4\pi$
13. $\tan x = 0$, $0 \le x \le 4\pi$
14. $\cot x = 0$, $0 \le x \le 4\pi$
15. $\sin x = -1$, $0 \le x \le 4\pi$
16. $\cos x = -1$, $0 \le x \le 4\pi$
17. $\cos x = 1$, $-2\pi \le x \le 2\pi$
18. $\sin x = 1$, $-2\pi \le x \le 2\pi$
19. $\cos x = 0$, $-2\pi \le x \le 2\pi$
20. $\sin x = 0$, $-2\pi \le x \le 2\pi$
21. $\tan x$ not defined, $0 \le x \le 4\pi$
22. $\cot x$ not defined, $0 \le x \le 4\pi$
23. $\csc x$ not defined, $0 \le x \le 4\pi$
24. $\sec x$ not defined, $0 \le x \le 4\pi$

In Problems 25–36, determine each to one significant digit. Use only the figure below, the definition of circular functions, and a calculator if necessary for multiplication and division.

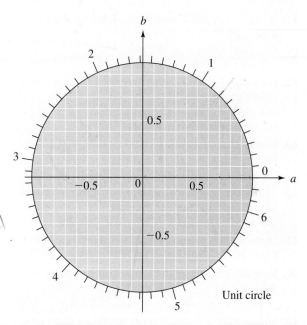

Unit circle

Figure for 25–36

25. $\sin 0.8$
26. $\cos 0.8$
27. $\cos 2.3$
28. $\sin 5.5$
29. $\sin(-0.9)$
30. $\cos(-2)$
31. $\sec 2.2$
32. $\csc 3.8$
33. $\tan 0.8$
34. $\cot 2.8$
35. $\cot(-0.4)$
36. $\tan(-4)$

Evaluate to four significant digits using a calculator.

37. $\sin(-0.2103)$
38. $\cos 14.78$
39. $\sec 1.432$
40. $\cot(-7.809)$
41. $\tan 4.704$
42. $\csc(-3.109)$
43. $\cos 105.2$
44. $\sin 23.04$
45. $\cot(-0.03333)$
46. $\tan(-4.7)$
47. $\csc 6.2$
48. $\sec(-1.5)$

Evaluate Problems 49–54 exactly.

49. $\cos \dfrac{3\pi}{4}$
50. $\sin \dfrac{2\pi}{3}$
51. $\csc\left(-\dfrac{\pi}{4}\right)$
52. $\tan\left(-\dfrac{7\pi}{6}\right)$
53. $\csc \dfrac{7\pi}{2}$
54. $\cos\left(-\dfrac{5\pi}{2}\right)$

55. $\cot\left(-\dfrac{\pi}{3}\right)$ **56.** $\tan\left(-\dfrac{2\pi}{3}\right)$ **57.** $\sec\dfrac{4\pi}{3}$

58. $\csc\dfrac{5\pi}{6}$ **59.** $\tan\left(-\dfrac{5\pi}{2}\right)$ **60.** $\cot(-3\pi)$

61. If $\sin x = 0.9525$, what is the value of each of the following?

    (A) $\sin(x + 2\pi)$ (B) $\sin(x - 2\pi)$
    (C) $\sin(x + 10\pi)$ (D) $\sin(x - 6\pi)$

62. If $\cos x = -0.0379$, what is the value of each of the following?

    (A) $\cos(x + 2\pi)$ (B) $\cos(x - 2\pi)$
    (C) $\cos(x + 8\pi)$ (D) $\cos(x - 12\pi)$

63. Evaluate $\tan x$ and $(\sin x)/(\cos x)$ to two significant digits for

    (A) $x = 1$ (B) $x = 5.3$ (C) $x = -2.376$

64. Evaluate $\cot x$ and $(\cos x)/(\sin x)$ to two significant digits for

    (A) $x = -1$ (B) $x = 8.7$ (C) $x = -12.64$

65. Evaluate $\sin(-x)$ and $-\sin x$ to two significant digits for
    (A) $x = 3$ (B) $x = -12.8$ (C) $x = 407$

66. Evaluate $\cos(-x)$ and $\cos x$ to two significant digits for

    (A) $x = 5$ (B) $x = -13.4$ (C) $x = -1,003$

67. Evaluate $\sin^2 x + \cos^2 x$ to two significant digits for
    (A) $x = 1$ (B) $x = -8.6$ (C) $x = 263$

68. Evaluate $1 - \sin^2 x$ and $\cos^2 x$ to two significant digits for

    (A) $x = 14$ (B) $x = -16.3$ (C) $x = 766$

Simplify each expression using the fundamental identities.

69. $\sin x \csc x$ **70.** $\cos x \sec x$

71. $\cot x \sec x$ **72.** $\tan x \csc x$

73. $\dfrac{\sin x}{1 - \cos^2 x}$ **74.** $\dfrac{\cos x}{1 - \sin^2 x}$

75. $\cot(-x)\sin(-x)$ **76.** $\tan(-x)\cos(-x)$

C *For Problems 77 and 78, fill the blanks in the* Reason *column with the appropriate identity, (1)–(9).*

77. Statement   **Reason**

$\tan^2 x + 1 = \left(\dfrac{\sin x}{\cos x}\right)^2 + 1$ (A) _____

$= \dfrac{\sin^2 x}{\cos^2 x} + 1$ Algebra

$= \dfrac{\sin^2 x + \cos^2 x}{\cos^2 x}$ Algebra

$= \dfrac{1}{\cos^2 x}$ (B) _____

$= \left(\dfrac{1}{\cos x}\right)^2$ Algebra

$= \sec^2 x$ (C) _____

78. Statement   **Reason**

$\cot^2 x + 1 = \left(\dfrac{\cos x}{\sin x}\right)^2 + 1$ (A) _____

$= \dfrac{\cos^2 x}{\sin^2 x} + 1$ Algebra

$= \dfrac{\cos^2 x + \sin^2 x}{\sin^2 x}$ Algebra

$= \dfrac{1}{\sin^2 x}$ (B) _____

$= \left(\dfrac{1}{\sin x}\right)^2$ Algebra

$= \csc^2 x$ (C) _____

79. What is the period of the cosecant function?

80. What is the period of the secant function?

Applications

81. Precalculus: Pi Estimate With s_n as shown in the figure, a sequence of numbers is formed as indicated. Compute the first five terms of the sequence to six decimal places, and compare the fifth term with the value of $\pi/2$.

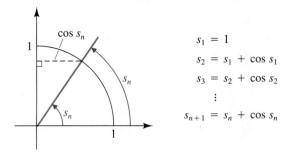

$s_1 = 1$

$s_2 = s_1 + \cos s_1$

$s_3 = s_2 + \cos s_2$

$\vdots$

$s_{n+1} = s_n + \cos s_n$

Figure for 81

82. Precalculus: Pi Estimate Repeat Problem 81 using $s_1 = 0.5$ as the first term of the sequence.

CHAPTER 2 SUMMARY

2.1
DEGREES AND RADIANS

An angle of **one radian** is a central angle of a circle subtended by an arc having ✳ the same length as the radius. The **radian measure** of a central angle subtending an arc of length s in a circle of radius R is $\theta = s/R$ radians (rad). **Radian and degree measure** are related by

$$\frac{\theta_{\text{deg}}}{180°} = \frac{\theta_{\text{rad}}}{\pi \text{ rad}}$$

An angle with its vertex at the origin and initial side along the positive x axis is in **standard position.** Two angles are **coterminal** if their terminal sides coincide when both angles are placed in their standard position in the same coordinate system. The arc length and the area of a **circular sector** (see Figure 1) are given by

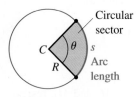

FIGURE 1

	Radian Measure	Degree Measure
Arc length	$s = R\theta$	$s = \dfrac{\pi}{180} R\theta$
Area	$A = \dfrac{1}{2} R^2\theta$	$A = \dfrac{\pi}{360} R^2\theta$

〜✐**2.2**
LINEAR AND ANGULAR
VELOCITY

The linear velocity V of a point moving on the circumference of a circle of radius R at a uniform rate, and the angular velocity ω (in radians per unit time) of the angle swept out by this point are related by $V = R\omega$ or, equivalently, $\omega = V/R$.

2.3
TRIGONOMETRIC FUNCTIONS

Trigonometric Functions with Angle Domains

For an arbitrary angle θ,

FIGURE 2

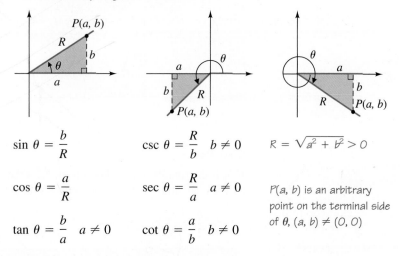

$$\sin\theta = \frac{b}{R} \qquad\qquad \csc\theta = \frac{R}{b} \quad b \neq 0 \qquad R = \sqrt{a^2 + b^2} > 0$$

$$\cos\theta = \frac{a}{R} \qquad\qquad \sec\theta = \frac{R}{a} \quad a \neq 0$$

$$\tan\theta = \frac{b}{a} \quad a \neq 0 \qquad \cot\theta = \frac{a}{b} \quad b \neq 0$$

$P(a, b)$ is an arbitrary point on the terminal side of θ, $(a, b) \neq (0, 0)$

Domains: Sets of all possible angles for which the ratios are defined
Ranges: Subsets of the set of real numbers

The right triangle formed by dropping a perpendicular from $P(a, b)$ to the horizontal axis is called the **reference triangle** associated with the angle θ.

Trigonometric Functions with Real Number Domains

For x any real number, we define

$$\sin x = \sin(x \text{ rad}) \qquad \csc x = \csc(x \text{ rad})$$
$$\cos x = \cos(x \text{ rad}) \qquad \sec x = \sec(x \text{ rad})$$
$$\tan x = \tan(x \text{ rad}) \qquad \cot x = \cot(x \text{ rad})$$

Domains: Subsets of the set of real numbers

Ranges: Subsets of the set of real numbers

If x is any real number or any angle in degree or radian measure, then the following **reciprocal relationships** hold (division by 0 excluded).

$$\csc x = \frac{1}{\sin x} \qquad \sec x = \frac{1}{\cos x} \qquad \cot x = \frac{1}{\tan x}$$

⋆**2.4**

ADDITIONAL APPLICATIONS

Refraction

If n_1 and n_2 represent the **index of refraction** for mediums M_1 and M_2, respectively, α is the **angle of incidence,** and β is the **angle of refraction** (see Figure 3), then according to **Snell's law,**

FIGURE 3

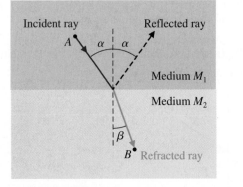

$$\frac{n_2}{n_1} = \frac{\sin \alpha}{\sin \beta}$$

The angle α for which $\beta = 90°$ is the **critical angle.**

Bow Waves

If S_b is the (uniform) speed of a boat, S_w is the speed of the bow waves generated by the boat, and θ is the angle between the bow waves, then

$$\sin \frac{\theta}{2} = \frac{S_w}{S_b}, \qquad S_b > S_w$$

This relationship also applies to **sound waves** when planes travel faster than the speed of sound and to **nuclear particles** that travel faster than the speed of light.

Perception

When individuals try to estimate the position of a horizontal line in a field of parallel lines tilted at an angle θ (in degrees), the difference between their estimate and the actual horizontal in degrees d can be approximated by

$$d = a + b \sin 4\theta$$

where a and b are constants associated with a particular individual.

2.5
EXACT VALUE FOR SPECIAL ANGLES AND REAL NUMBERS

Reference Triangle and Angle

For a nonquadrantal angle θ (see Figure 4):

1. To form a **reference triangle** for θ, drop a perpendicular from a point $P(a, b)$ on the terminal side of θ to the horizontal axis.
2. The **reference angle** α is the acute angle (always taken positive) between the terminal side of θ and the horizontal axis.

If the reference triangle for an angle or a real number θ is a 30°–60° right triangle or a 45° right triangle, then the relationships in Figure 5 can be used to find exact values of the trigonometric functions of θ.

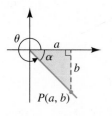

$(a, b) \neq (0, 0)$

FIGURE 4

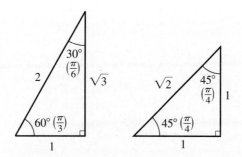

FIGURE 5

These exact values are summarized in Table 1.

TABLE 1
Special values

θ	$\sin \theta$	$\csc \theta$	$\cos \theta$	$\sec \theta$	$\tan \theta$	$\cot \theta$
0° or 0	0	N.D.	1	1	0	N.D.
30° or $\pi/6$	1/2	2	$\sqrt{3}/2$	$2/\sqrt{3}$	$1/\sqrt{3}$	$\sqrt{3}$
45° or $\pi/4$	$1/\sqrt{2}$	$\sqrt{2}$	$1/\sqrt{2}$	$\sqrt{2}$	1	1
60° or $\pi/3$	$\sqrt{3}/2$	$2/\sqrt{3}$	1/2	2	$\sqrt{3}$	$1/\sqrt{3}$
90° or $\pi/2$	1	1	0	N.D.	N.D.	0

N.D. = Not defined

2.6
CIRCULAR FUNCTIONS

The circular functions for a real number x are defined in terms of the coordinates of a point on a unit circle; they are related to the trigonometric functions of an angle of x radians as follows (see Figure 6):

Circular Functions and Trigonometric Functions

Circular Function		Trigonometric Function
$\sin x = b = \dfrac{b}{1}$		$= \sin(x \text{ rad})$
$\cos x = a = \dfrac{a}{1}$		$= \cos(x \text{ rad})$
$\tan x = \dfrac{b}{a}$	$(a \neq 0)$	$= \tan(x \text{ rad})$
$\csc x = \dfrac{1}{b}$	$(b \neq 0)$	$= \csc(x \text{ rad})$
$\sec x = \dfrac{1}{a}$	$(a \neq 0)$	$= \sec(x \text{ rad})$
$\cot x = \dfrac{a}{b}$	$(b \neq 0)$	$= \cot(x \text{ rad})$

FIGURE 6

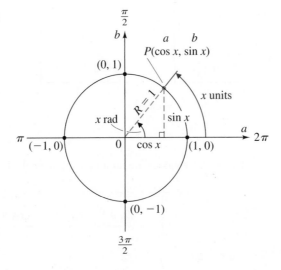

A function f is **periodic** if there is a positive real number p such that $f(x + p) = f(x)$ for all x in the domain of f. The smallest such positive p, if it exists, is called **the period of f**. Both the sine function and the cosine function have a period of 2π.

The properties of circular functions can be used to establish the following **fundamental identities:**

1. $\csc x = \dfrac{1}{\sin x}$

2. $\sec x = \dfrac{1}{\cos x}$

3. $\cot x = \dfrac{1}{\tan x}$

4. $\tan x = \dfrac{\sin x}{\cos x}$

5. $\cot x = \dfrac{\cos x}{\sin x}$

6. $\sin(-x) = -\sin x$

7. $\cos(-x) = \cos x$

8. $\tan(-x) = -\tan x$

9. $\sin^2 x + \cos^2 x = 1$

Figure 7 is useful for finding exact values of the circular functions at multiples of $\pi/6$ and $\pi/4$.

FIGURE 7

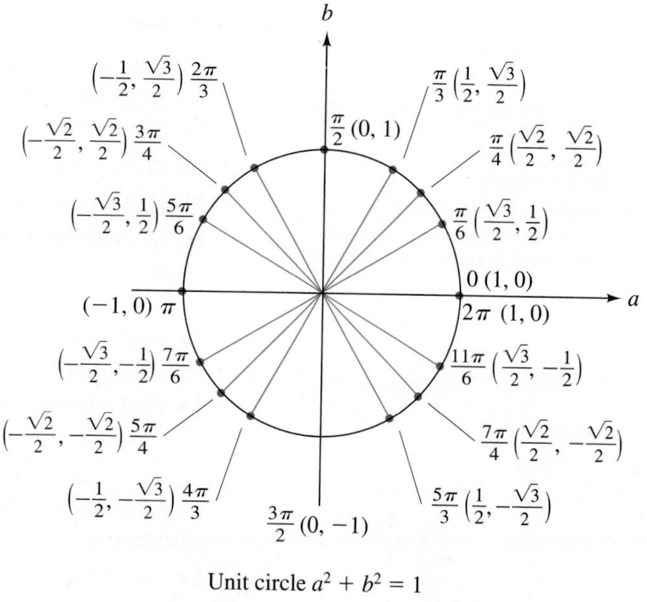

Unit circle $a^2 + b^2 = 1$

CHAPTER 2 REVIEW EXERCISE

Work through all the problems in this chapter review and check the answers. Answers to all review problems appear in the back of the book; following each answer is an italic number that indicates the section in which that type of problem is discussed. Where weaknesses show up, review the appropriate sections in the text. Review problems flagged with a star (☆) are from optional sections.

A **1.** Convert to radian measure in terms of π.
   (A) 60°   (B) 45°   (C) 90°

2. Convert to degree measure.
   (A) $\pi/6$ (B) $\pi/2$ (C) $\pi/4$

3. (A) Find the degree measure of 15.26 rad.
   (B) Find the radian measure of $-389.2°$.

☆4. Find the velocity V of a point on the rim of a wheel if $R = 25$ ft and $\omega = 7.4$ rad/min.

☆5. Find the angular velocity ω of a point on the rim of a wheel if $R = 5.2$ m and $V = 415$ m/hr.

6. Find the value of $\sin \theta$ and $\tan \theta$ if the terminal side of θ contains $P(-4, 3)$.

Evaluate Problems 7–9 to four significant digits using a calculator.

7. (A) cot 53°40′   (B) csc 67°10′

8. (A) cos 23.5°   (B) tan 42.3°

9. (A) cos 0.35   (B) tan 1.38

10. Sketch the reference triangle and find the reference angle α for
   (A) $\theta = 120°$ (B) $\theta = -\dfrac{7\pi}{4}$

11. Evaluate exactly without a calculator.
   (A) sin 60°   (B) cos($\pi/4$)   (C) tan 0°

Figure 1 from the text is repeated here for use in Problems 12–14.

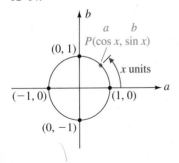

Figure for 12–14

12. Refer to the figure and state the coordinates of P for the indicated values of x.
   (A) $x = -2\pi$ (B) $x = \pi$ (C) $x = -3\pi/2$
   (D) $x = \pi/2$ (E) $x = -5\pi$ (F) $x = 7\pi/2$

13. Refer to the figure: Given $y = \sin x$, how does y vary for the indicated variations in x?
   (A) x varies from 0 to $\pi/2$
   (B) x varies from $\pi/2$ to π
   (C) x varies from π to $3\pi/2$
   (D) x varies from $3\pi/2$ to 2π
   (E) x varies from 2π to $5\pi/2$
   (F) x varies from $5\pi/2$ to 3π

14. Refer to the figure: Given $y = \cos x$, how does y vary for the indicated variations in x?
   (A) x varies from 0 to $-\pi/2$
   (B) x varies from $-\pi/2$ to $-\pi$
   (C) x varies from $-\pi$ to $-3\pi/2$
   (D) x varies from $-3\pi/2$ to -2π
   (E) x varies from -2π to $-5\pi/2$
   (F) x varies from $-5\pi/2$ to -3π

B **15.** What is the degree measure of a central angle subtended by an arc exactly $\frac{7}{60}$ of the circumference of a circle?

16. What is the radian measure of a central angle subtended by an arc of length 24 cm if the radius of the circle is 8 cm?

17. If the radius of a circle is 4 cm, find the length of an arc intercepted by an angle of 1.5 rad.

18. Convert 212° to radian measure in terms of π.

19. Convert $\pi/12$ rad to degree measure.

20. Find the tangent of 0, $\pi/2$, π, and $3\pi/2$.

21. In which quadrant does the terminal side of each angle lie?
   (A) 732°   (B) -7 rad

22. Find reference angles corresponding to:
   (A) 187.4°   (B) 103°20′   (C) $-37°40′$

23. Find reference angles corresponding to:
   (A) 2.39 rad   (B) 5.00 rad   (C) -4.0 rad
   (Use $\pi \approx 3.14$.)

Evaluate Problems 24–35 to three significant digits using a calculator.

24. cos 187.4°   **25.** tan 187.4°   **26.** sin 103°20′

27. sec 103°20′   28. cot(−37°40′)   29. sec(−37°40′)

30. sin 2.39   31. cot 2.39   32. cos 5

33. tan 5   34. sin(−4)   35. cot(−4)

36. Find reference angles in exact form in terms of π for
(A) $5\pi/6$ (B) $7\pi/4$ (C) $-4\pi/3$

In Problems 37–48 find the exact value of each without using a calculator.

37. $\cos \dfrac{5\pi}{6}$ 38. $\tan \dfrac{5\pi}{6}$ 39. $\sin \dfrac{7\pi}{4}$

40. $\cot \dfrac{7\pi}{4}$ 41. $\sin \dfrac{3\pi}{2}$ 42. $\cos \dfrac{3\pi}{2}$

43. $\sin \dfrac{-4\pi}{3}$ 44. $\sec \dfrac{-4\pi}{3}$ 45. $\cos 3\pi$

46. $\cot 3\pi$ 47. $\cos \dfrac{-11\pi}{6}$ 48. $\sin \dfrac{-11\pi}{6}$

In Problems 49–52 use a calculator to evaluate each to five decimal places.

49. sin 384.0314°   50. tan(−198°43′6″)

51. cos 26   52. cot(−68.005)

53. If $\sin \theta = -\frac{4}{5}$ and the terminal side of θ does not lie in the third quadrant, find the exact values of $\cos \theta$ and $\tan \theta$ without finding θ.

54. Find the least positive exact value of θ in radian measure such that $\sin \theta = -\frac{1}{2}$.

55. Find the exact value of each of the other five trigonometric functions if

$$\sin \theta = -\frac{2}{5} \quad \text{and} \quad \tan \theta < 0$$

56. Find all the angles exactly between 0° and 360° for which $\tan \theta = -1$.

57. Find all the angles exactly between 0 and 2π for which $\cos \theta = -\sqrt{3}/2$.

58. In a circle of radius 12.0 cm, find the length of an arc subtended by a central angle of
(A) 1.69 rad   (B) 22.5°

59. In a circle with diameter 80 ft, find the area (to three significant digits) of the circular sector with central angle.
(A) 0.773 rad   (B) 135°

60. Find the distance between Charleston, West Virginia (38°21′N latitude) and Cleveland, Ohio (41°28′N latitude). Both cities have the same longitude and the radius of the earth is 3,964 mi.

61. Find the angular velocity of a wheel turning through 6.43 radians in 15.24 seconds.

62. Evaluate cos x to three significant digits for
(A) $x = 7$ (B) $x = 7 + 2\pi$ (C) $x = 7 - 2\pi$
(D) $x = 7 + 4\pi$ (E) $x = 7 - 30\pi$

63. Evaluate $\tan(-x)$ and $-\tan x$ to three significant digits for
(A) $x = 7$ (B) $x = -17.9$ (C) $x = -2{,}135$

64. One of the following is not an identity. Indicate which one.
(A) $\csc x = \dfrac{1}{\sin x}$ (B) $\cot x = \dfrac{1}{\tan x}$
(C) $\tan x = \dfrac{\sin x}{\cos x}$ (D) $\sec x = \dfrac{1}{\sin x}$
(E) $\sin^2 x + \cos^2 x = 1$ (F) $\cot x = \dfrac{\cos x}{\sin x}$

65. Simplify: $(\csc x)(\cot x)(1 - \cos^2 x)$

66. Simplify: $\cot(-x) \sin(-x)$

67. A point on the unit circle with its center at the origin starts at (1, 0) and moves counterclockwise around the circle until it travels a distance of 1.4 units on the circle. What are the coordinates of the point in its terminal position? Compute the answer to three decimal places.

68. A circular sector has an area of 342.5 m² and a radius of 12 m. Calculate the arc length of the sector to the nearest meter.

69. A circle with its center at the origin in a rectangular coordinate system passes through the point (4, 5). What is the length of the arc on the circle in the first quadrant between the positive horizontal axis and the point (4, 5)? Compute the answer to two decimal places.

Applications

70. **Engineering** Through how many radians does a pulley with 10 cm diameter turn when 10 m of rope has been pulled through it without slippage? How many revolutions result? (Give answers to one decimal place.)

71. **Engineering** If the large gear in the figure at the top of page 110 completes five revolutions, how many revolutions will the middle gear complete? How many revolutions will the small gear complete?

Figure for 71

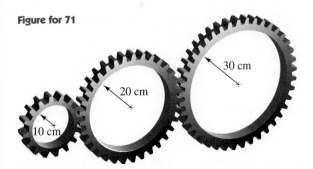

72. Engineering   An automobile is traveling at 70 ft/sec. If the wheel diameter is 27 in., what is the angular velocity in rad/sec?

73. Space Science   A satellite is placed in a circular orbit 1,000 miles above the earth's surface. If the satellite completes one orbit every 114 min and the radius of the earth is 3,964 mi, what is the linear velocity (to three significant digits) of the satellite in miles per hour?

74. Electric Current   An alternating current generator produces an electrical current (measured in amperes) that is described by the equation

$$I = 30 \sin(120\pi t - 60\pi)$$

where t is time in seconds. What is the current I when $t = 0.015$ sec? (Give answers to one decimal place.)

75. Precalculus   A ladder of length L leaning against a building just touches a 10 ft high fence located 2 ft from the building (see the figure).
(A) Express the length of the ladder in terms of θ.
(B) Find the length of the ladder to three significant digits for $\theta = 0.9, 1.0, 1.1, 1.2,$ and 1.3.
(In calculus this equation is used to find the shortest ladder that will clear the fence.)

Figure for 75

76. Light Waves   A light wave passing through air strikes the surface of a pool of water so that the angle of incidence is $\alpha = 31.7°$. Find the angle of refraction. (Water has a refractive index of 1.33 and air has a refractive index of 1.00.)

77. Light Waves   A triangular crown glass prism is surrounded by air. If light inside the prism strikes one of its facets, what is the critical angle of incidence α for total reflection? (The refractive index for crown glass is 1.52, and the refractive index for air is 1.00.)

78. Bow Waves   If a boat traveling at 25 mph produces bow waves that separate at an angle of 51°. how fast are the bow waves traveling?

GRAPHING TRIGONOMETRIC FUNCTIONS

☆ Sections marked with a star may be
omitted without loss of continuity.

ith the trigonometric functions defined, we are in a position to consider a substantially expanded list of applications and properties. As a brief preview, look at Figure 1.

What feature seems to be shared by the different illustrations? All appear to be repetitive—that is, periodic. The trigonometric functions, as we will see shortly, can be used to describe such phenomena with remarkable precision. Glance through Section 3.4, "Additional Applications," to preview an even greater variety of applications.

In this chapter you will learn how to sketch graphs of the trigonometric functions quickly and easily. You will also learn how to recognize certain fundamental and useful properties of these functions.

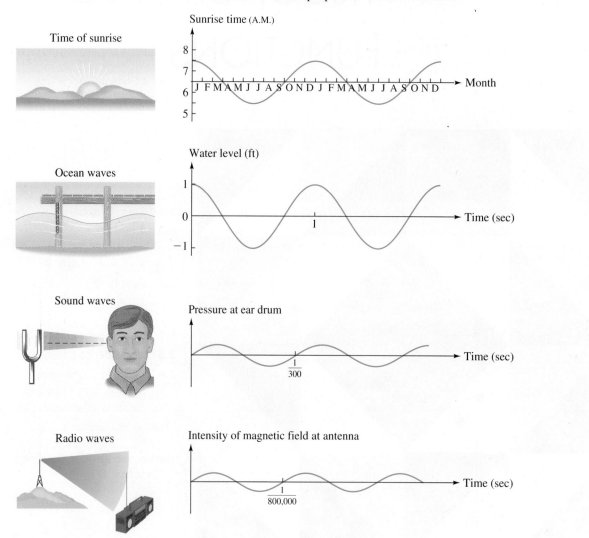

FIGURE 1

3.1 BASIC GRAPHS

- ◆ **Graphs of $y = \sin x$ and $y = \cos x$**
- ◆ **Graphs of $y = \tan x$ and $y = \cot x$**
- ◆ **Graphs of $y = \csc x$ and $y = \sec x$**
- ◆ **Graphing with a Graphing Calculator**

In this section we will discuss the graphs of the six trigonometric functions introduced in Chapter 2. We will also discuss the domains, ranges, and periodic properties of these functions. Section 2.6 on circular functions will prove particularly important in our work.

Although it appears that there is a lot to remember in this section, you only need to be familiar with the graphs and properties of the sine, cosine, and tangent functions. The reciprocal relationships we discussed in Section 2.6 will enable you to determine the graphs and properties of the other three trigonometric functions from the sine, cosine, and tangent functions.

◆ Graphs of $y = \sin x$ and $y = \cos x$

First, we consider

$$y = \sin x, \qquad x \text{ a real number} \tag{1}$$

The graph of the sine function is the graph of the set of all ordered pairs of real numbers (x, y) that satisfy equation (1). How do we find these pairs of numbers? To help make the process clear, we refer again to a function machine with the unit circle definition inside (Figure 1).

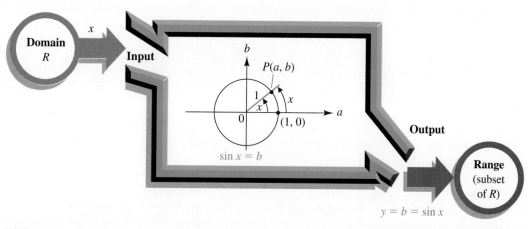

FIGURE 1
Sine function machine

We are interested in graphing, in an xy coordinate system, all ordered pairs of real numbers (x, y) produced by the function machine. We could resort to point-by-point plotting using a calculator, which becomes tedious and tends to obscure some important properties. Instead, we choose to speed up the process by using some of the properties discussed in Section 2.6 and by observing how $y = \sin x = b$ varies as $P(a, b)$ moves around the unit circle. Recall that in Section 2.6 we found that the domain of the sine function is the set of all real numbers R, and its range is the set of all real numbers y such that $-1 \leq y \leq 1$. In addition, we found that the sine function is periodic with period 2π.

Because the sine function is periodic with period 2π, we will concentrate on the graph over one period, from 0 to 2π. Once we have the graph for one period, we can complete as much of the rest of the graph as we wish by repeating the graph to the left and to the right.

Figure 2 illustrates how $y = \sin x = b$ varies as x increases from 0 to 2π and $P(a, b)$ moves around the unit circle.

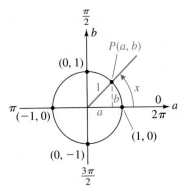

As x increases	$y = \sin x = b$
from 0 to $\pi/2$	increases from 0 to 1
from $\pi/2$ to π	decreases from 1 to 0
from π to $3\pi/2$	decreases from 0 to -1
from $3\pi/2$ to 2π	increases from -1 to 0

FIGURE 2
$y = \sin x = b$

The information in Figure 2 can be translated into a graph of $y = \sin x$, $0 \leq x \leq 2\pi$, as shown in Figure 3. (Where the graph is uncertain, fill in with calculator values.)

FIGURE 3
$y = \sin x,$
$0 \leq x \leq 2\pi$

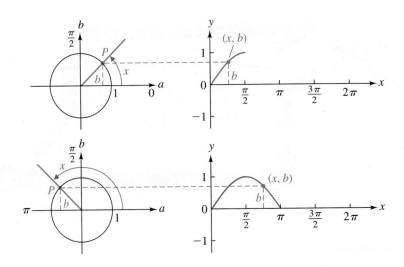

FIGURE 3 (*continued*)
$y = \sin x,$
$0 \le x \le 2\pi$

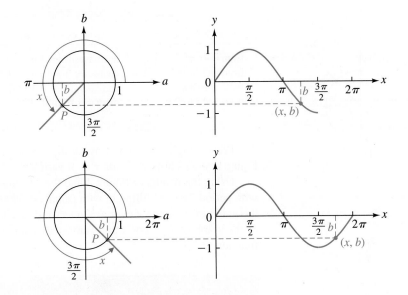

To complete the graph of $y = \sin x$ over any interval desired, we need only to repeat the final graph in Figure 3 to the left and to the right as far as we wish. The next box summarizes what we now know about the graph of $y = \sin x$ and its basic properties.

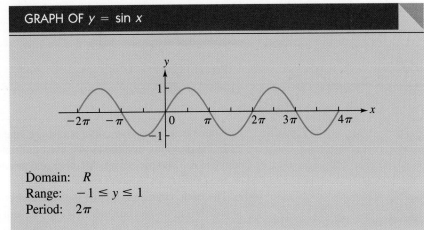

GRAPH OF $y = \sin x$

Domain: R
Range: $-1 \le y \le 1$
Period: 2π

Both the x and y axes are real number lines (see Appendix A.1). Because the domain of the sine function is all real numbers, the graph of $y = \sin x$ extends without limit in both horizontal directions. Also, because the range is $-1 \le y \le 1$, no point on the graph can have a y coordinate greater than 1 or less than -1.

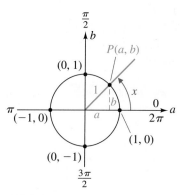

FIGURE 4

$y = \cos x = a$

As x increases	$y = \cos x = a$
from 0 to $\pi/2$	decreases from 1 to 0
from $\pi/2$ to π	decreases from 0 to -1
from π to $3\pi/2$	increases from -1 to 0
from $3\pi/2$ to 2π	increases from 0 to 1

Proceeding in the same way for the cosine function, we can obtain its graph. Figure 4 shows how $y = \cos x = a$ varies as x increases from 0 to 2π.

Using the results of Figure 4 and the fact that the cosine function is periodic with period 2π (and filling in with calculator values where necessary), we obtain the graph of $y = \cos x$ over any interval desired. The next box shows a portion of the graph of $y = \cos x$ and summarizes some of the properties of the cosine function.

GRAPH OF $y = \cos x$

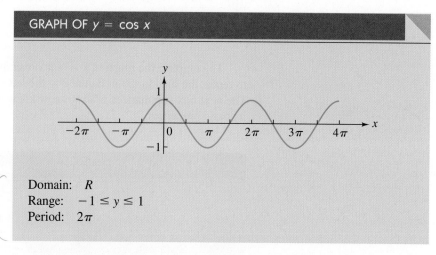

Domain: R

Range: $-1 \le y \le 1$

Period: 2π

REMARKS   1. Notice that the cosine curve is just the sine curve shifted $\pi/2$ units to the left.

2. In the previous two boxes, we used multiples of π as labels on the x axis for convenience and for clarity. All other real numbers are also associated with points on the axis. (Both axes are real number lines.)

3. The basic characteristics of the sine and cosine graphs should be understood so that the curves can be quickly sketched from memory. In particular, you should be able to answer the following questions:

(A) What is the period of each function (how often does the graph repeat)?

(B) What are the x intercepts of each function (where does the graph cross the x axis)?

(C) What is the y intercept of each function (where does the graph cross the y axis)?

(D) How far does the graph deviate from the x axis?

(E) Where do the high and low points occur?

◆ Graphs of $y = \tan x$ and $y = \cot x$

We first discuss the graph of $y = \tan x$. Later, because $\cot x = 1/(\tan x)$, we will be able to get the graph of $y = \cot x$ from the graph of $y = \tan x$ using reciprocals of ordinates.

From Figure 5 we can see that whenever $P(a, b)$ is on the horizontal axis of the unit circle (that is, whenever $x = k\pi$, k an integer), then $(a, b) = (\pm 1, 0)$ and $\tan x = b/a = 0/(\pm 1) = 0$. These values of x are the x intercepts of the graph of $y = \tan x$; that is, where the graph crosses the x axis.

x intercepts: $x = k\pi$, k an integer

Thus, as a first step in graphing $y = \tan x$, we locate the x intercepts with solid dots along the x axis, as indicated in Figure 6.

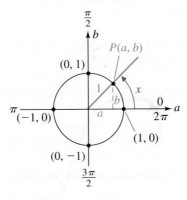

FIGURE 5

$y = \tan x = \dfrac{b}{a}$

FIGURE 6

x intercepts and vertical asymptotes for $y = \tan x$

Also, from Figure 5 we can see that whenever $P(a, b)$ is on the vertical axis of the unit circle (that is, whenever $x = \pi/2 + k\pi$, k an integer), then $(a, b) = (0, \pm 1)$ and $\tan x = b/a = \pm 1/0$, which is not defined. There can be no points plotted for these values of x. Thus, as a second step in graphing $y = \tan x$, we draw dashed vertical lines through each of these points on the x axis where $\tan x$ is not defined; the graph cannot touch these lines (see Figure 6). These dashed lines will become guidelines, called *vertical asymptotes,* which are very helpful for sketching the graph of $y = \tan x$.

Vertical asymptotes: $x = \dfrac{\pi}{2} + k\pi$, k an integer

We next investigate the behavior of the graph of $y = \tan x$ over the interval $0 \leq x < \pi/2$. Two points are easy to plot: $\tan 0 = 0$ and $\tan(\pi/4) = 1$. What happens to $\tan x$ as x approaches $\pi/2$ from the left? [Remember that $\tan(\pi/2)$ is not defined.] When x approaches $\pi/2$ from the left, $P(a, b)$ approaches $(0, 1)$ and stays in the first quadrant. Thus, a approaches 0 through positive values and b approaches 1. What happens to $y = \tan x$ in this process? In Example 1, we perform a calculator experiment to determine an answer.

◆ **EXAMPLE 1**   Calculator Experiment

Form a table of values for $y = \tan x$ with x approaching $\pi/2 \approx 1.570\ 796\ 3$ from the left (through values less than $\pi/2$). Any conclusions?

SOLUTION   We create a table as follows.

x	0	0.5	1	1.5	1.57	1.5707	1.570 796 3
tan x	0	0.5	1.6	14.1	1,256	10,381	37,320,540

Conclusion: As x approaches $\pi/2$ from the left, $y = \tan x$ appears to increase without bound.   ◆

MATCHED PROBLEM 1   Repeat Example 1, but with x approaching $-\pi/2 \approx -1.570\ 796\ 3$ from the right. Any conclusions?

Figure 7a shows the results of the analysis in Example 1: $y = \tan x$ increases without bound when x approaches $\pi/2$ from the left.

Now we examine the behavior of the graph of $y = \tan x$ over the interval $-\pi/2 < x \le 0$. Because of the identity $\tan(-x) = -\tan x$ (see Section 2.6), we can reflect the graph of $y = \tan x$ for $0 \le x < \pi/2$ through the origin to obtain the full graph over the interval $-\pi/2 < x < \pi/2$ (Figure 7b).

FIGURE 7
$y = \tan x$

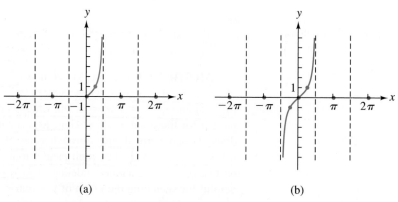

(a)   (b)

Proceeding in the same way for the other intervals between the asymptotes (the dashed vertical lines), it appears that the tangent function is periodic with period π. We confirm this as follows: If (a, b) are the coordinates of P associated with x (see Figure 8), then using unit circle symmetry and congruent reference triangles, $(-a, -b)$ are the coordinates of the point Q associated with $x + \pi$. Consequently,

$$\tan(x + \pi) = \frac{-b}{-a} = \frac{b}{a} = \tan x$$

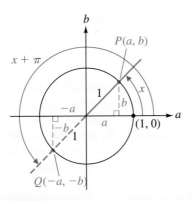

FIGURE 8
tan(x + π) = tan x

We conclude that the tangent function is periodic with period π. In general,

$$\tan(x + k\pi) = \tan x, \quad k \text{ an integer}$$

for all values of x for which both sides of the equation are defined.

Now, to complete as much of the general graph of $y = \tan x$ as we wish, all we need to do is to repeat the graph in Figure 7b to the left and to the right over intervals of π units. The main characteristics of the graph of $y = \tan x$ should be learned so that the graph can be sketched quickly. The figure in the next box summarizes the above discussion.

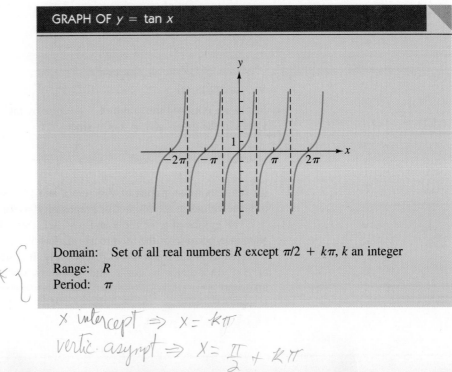

GRAPH OF $y = \tan x$

Domain: Set of all real numbers R except $\pi/2 + k\pi$, k an integer
Range: R
Period: π

x intercept ⟹ x = kπ
vertic. asympt ⟹ x = $\frac{\pi}{2}$ + kπ

To graph $y = \cot x$, we recall that

$$\cot x = \frac{1}{\tan x}$$

and proceed by taking reciprocals of ordinate values in the graph of $y = \tan x$. Note that the x intercepts for the graph of $y = \tan x$ become vertical asymptotes for the graph of $y = \cot x$, and the vertical asymptotes for the graph of $y = \tan x$ become x intercepts for the graph of $y = \cot x$. The graph of $y = \cot x$ is shown in the following box. Again, you should learn the main characteristics of this function so that its graph can be sketched readily.

GRAPH OF $y = \cot x$

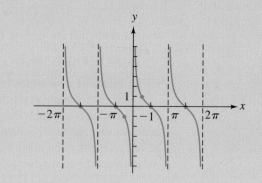

Domain: Set of all real numbers R except $k\pi$, k an integer
Range: R
Period: π

x intercept ⇒ x = π/2 + kπ
Vertic asym ⇒ x = kπ

◆ **Graphs of** $y = \csc x$ **and** $y = \sec x$

Just as we obtained the graph of $y = \cot x$ by taking reciprocals of the ordinate values in the graph of $y = \tan x$, since

$$\csc x = \frac{1}{\sin x} \qquad \text{and} \qquad \sec x = \frac{1}{\cos x}$$

we can obtain the graphs of $y = \csc x$ and $y = \sec x$ by taking reciprocals of ordinate values in the respective graphs of $y = \sin x$ and $y = \cos x$.

The graphs of $y = \csc x$ and $y = \sec x$ are shown in the next two boxes. Note that, because of the reciprocal relationship, vertical asymptotes occur at the x intercepts of $\sin x$ and $\cos x$, respectively. It is helpful to first draw the graphs of $y = \sin x$ and $y = \cos x$, and then draw vertical asymptotes through the x intercepts.

GRAPH OF $y = \csc x$

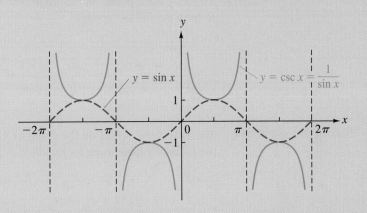

Domain: All real numbers x, except $x = k\pi$, k an integer
Range: All real numbers y such that $y \leq -1$ or $y \geq 1$
Period: 2π

GRAPH OF $y = \sec x$

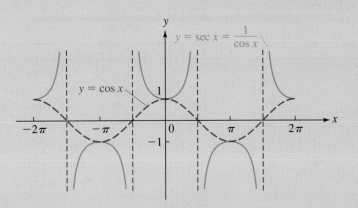

Domain: All real numbers x, except $x = \pi/2 + k\pi$, k an integer
Range: All real numbers y such that $y \leq -1$ or $y \geq 1$
Period: 2π

◆ EXAMPLE 2   Calculator Experiment

To verify some points on the graph of $y = \csc x$, $0 < x < \pi$, we use a calculator to form the following table:

x	0.001	0.01	0.1	1.57	3.00	3.13	3.14
csc x	1,000	100	10	1	7.1	86.3	628

1/sin .001 = 1,000

◆

MATCHED PROBLEM 2   To verify some points on the graph of $y = \sec x$, $-\pi/2 < x < \pi/2$, complete the following table using a calculator:

x	−1.57	−1.56	−1.4	0	1.4	1.56	1.57
sec x	1257	92.6	5.9	1	5.9	92.6	1257

en la calculadora :
1/cos −1.57 = 1255.7

◆ Graphing with a Graphing Calculator*

We determined the graphs of the six basic trigonometric functions by analyzing the behavior of the functions, exploiting relationships between the functions, and plotting only a few points. We refer to this process as curve sketching; one of the major objectives of this course is that you master this technique.

Graphing calculators also can be used to sketch graphs of functions; their accuracy depends on the screen resolution of the calculator. The smallest darkened rectangular area on the screen that the calculator can display is called a pixel. Most graphing calculators have a resolution of about 50 pixels per in., which results in rough, but useful sketches. Note that the graphs shown earlier in this section were created using sophisticated computer software and printed at a resolution of about 1,000 pixels per in.

The portion of the *xy* coordinate plane displayed on the screen of a graphing calculator is called the **viewing rectangle** and is determined by the **range** and **scale** for *x* and for *y*. Figure 9 illustrates a standard viewing rectangle using the following range and scale:

xmin = −10   xmax = 10   xscl = 1
ymin = −10   ymax = 10   yscl = 1

---

* Graphing calculator material is included in the text and exercise sets in this and subsequent chapters. This material is clearly identified with either the icon [icon] or [C]. Any or all of the graphing calculator material may be omitted without loss of continuity. Treatments are generic in nature. For those who need help with specific calculators, refer to the user's manual for your calculator or to the graphing calculator supplement that accompanies this text. All references to graphing calculators also apply to the wide variety of graphing software available for personal computers.

FIGURE 9

Standard viewing rectangle

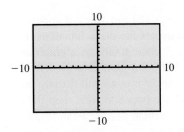

Most graphing calculators do not display labels on the axes. We have added numeric labels at the end of each axis to make the graphs easier to read. Now we want to see what the graphs of the trigonometric functions will look like on a graphing calculator.

♦ EXAMPLE 3   Trigonometric Graphs on a Graphing Calculator

Use a graphing calculator to graph the functions

$$y = \sin x, \qquad y = \tan x, \qquad \text{and} \qquad y = \sec x$$

for $-2\pi \le x \le 2\pi$, $-5 \le y \le 5$. Display each graph in a separate viewing rectangle.

SOLUTION   First, set the calculator to the radian mode. Most calculators remember this setting, so you should have to do this only once. Next, enter the following values; use 6.28 to approximate 2π:

xmin = −6.28	xmax = 6.28	xscl = 1
ymin = −5	ymax = 5	yscl = 1

This defines a viewing rectangle ranging from −6.28 to 6.28 on the horizontal axis and from −5 to 5 on the vertical axis with tick marks one unit apart on each axis. Now enter the function $y = \sin x$ and draw the graph (see Figure 10(a)). Repeat for $y = \tan x$ and $y = \sec x = 1/\cos x$ to obtain the graphs in Figures 10(b) and 10(c).

(handwritten notes in margin:)
① hit window
② enter values for x min
   x max ....etc..
③ hit (Y=) enter value
   y = sin x
     on
   y = tan x
     on
   y = sec x
④ hit graph
⑤ to find values hit trace and move arrows left & right

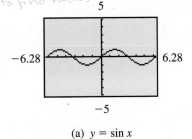

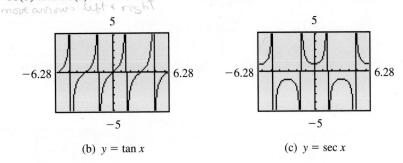

(a) $y = \sin x$ (b) $y = \tan x$ (c) $y = \sec x$

FIGURE 10

Graphing calculator graphs of trigonometric functions

♦

MATCHED PROBLEM 3   Repeat Example 3 for (A) $y = \cos x$, (B) $y = \cot x$, and (C) $y = \csc x$.

In Figures 10(b) and 10(c), it appears that the calculator has also drawn the vertical asymptotes for these functions, but this is not the case. Most graphing calculators calculate points on a graph and connect these points with line segments. The last point plotted to the left of the asymptote and the first plotted to the right of the asymptote will usually have very large y coordinates. If these y coordinates have opposite sign, then the calculator will connect the two points with a nearly vertical line segment, which gives the appearance of an asymptote. There is no harm in this as long as you understand that the calculator is not performing any analysis to identify asymptotes; it is simply connecting points with line segments. If you wish, you can set the calculator to plot the points without the connecting line segments, as illustrated in Figure 11.

FIGURE 11

Graph of $y = \sec x$ in the plot-points-only mode

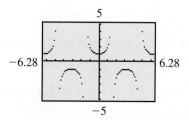

Answers to
Matched Problems

1.

x	0	-0.5	-1	-1.5	-1.57	-1.5707	$-1.570\,796\,3$
$\tan x$	0	-0.5	-1.6	-14.1	$-1,256$	$-10,381$	$-37,320,540$

Conclusion: As x approaches $-\pi/2$ from the right, $y = \tan x$ appears to decrease without bound.

2.

x	-1.57	-1.56	-1.4	0	1.4	1.56	1.57
$\sec x$	1,256	92.6	5.9	1	5.9	92.6	1,256

3. (A) $y = \cos x$

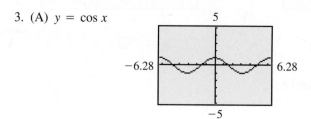

(B) $y = \cot x$

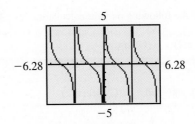

(C) $y = \csc x$

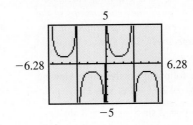

EXERCISE 3.1

A

1. What are the periods of the cosine, secant, and tangent functions?

2. What are the periods of the sine, cosecant, and cotangent functions?

3. How far does the graph of each of the following functions deviate from the x axis?
(A) $\sin x$ (B) $\cot x$ (C) $\sec x$

4. How far does the graph of each of the following functions deviate from the x axis?
(A) $\cos x$ (B) $\tan x$ (C) $\csc x$

In Problems 5–10 what are the x intercepts for the graph of each function over the interval $-2\pi \le x \le 2\pi$?

5. $\cos x$ **6.** $\sin x$ **7.** $\tan x$
8. $\cot x$ **9.** $\sec x$ **10.** $\csc x$

B

11. For what values of x, $-2\pi \le x \le 2\pi$, are the following not defined?
(A) $\sin x$ (B) $\cot x$ (C) $\sec x$

12. For what values of x, $-2\pi \le x \le 2\pi$, are the following not defined?
(A) $\cos x$ (B) $\tan x$ (C) $\csc x$

13. Use a calculator to produce an accurate graph of $y = \cos x$, $0 \le x \le 1.6$, using domain values 0, 0.1, 0.2, . . . , 1.5, 1.6.

14. Use a calculator to produce an accurate graph of $y = \sin x$, $0 \le x \le 1.6$, using domain values 0, 0.1, 0.2, . . . , 1.5, 1.6.

15. Use a calculator to produce an accurate graph of $y = \tan x$, $-1.4 \le x \le 1.4$, using domain values -1.4, -1.2, -1.0, . . . , 1.2, 1.4.

16. Use a calculator to produce an accurate graph of $y = \cot x$, $0.2 \le x \le 3.0$, using domain values 0.2, 0.4, 0.6, . . . , 2.8, 3.0.

17. Use a calculator to produce an accurate graph of $y = \csc x$, $0.2 \le x \le 3.0$, using domain values 0.2, 0.4, 0.6, . . . , 2.8, 3.0.

18. Use a calculator to produce an accurate graph of $y = \sec x$, $-1.4 \le x \le 1.4$, using domain values -1.4, -1.2, -1.0, . . . , 1.2, 1.4.

In Problems 19–24 make a sketch of each trigonometric function without looking at the text or using a calculator. Label each point where the graph crosses the x axis in terms of π.

19. $y = \sin x$, $-2\pi \le x \le 2\pi$
20. $y = \cos x$, $-2\pi \le x \le 2\pi$
21. $y = \tan x$, $0 \le x \le 2\pi$
22. $y = \cot x$, $0 < x < 2\pi$
23. $y = \csc x$, $-\pi < x < \pi$
24. $y = \sec x$, $-\pi \le x \le \pi$

C 25. Try to calculate each of the following on your calculator. Explain the problem.

(A) cot 0   (B) tan(π/2)   (C) csc π

26. Try to calculate each of the following on your calculator. Explain the problem.

(A) tan($-\pi$/2)   (B) cot($-\pi$)   (C) sec(π/2)

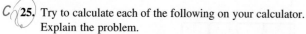

Problems 27–36 require the use of a graphing calculator. Problems 27–34 graphically explore the relationships of the graphs of $y = \sin x$ *and* $y = \cos x$ *with the graphs of* $y = A \sin x$ *and* $y = A \cos x$, *respectively. This topic is discussed in detail in the next section.*

27. (A) Graph each of the following in the same viewing rectangle ($-2\pi \le x \le 2\pi$, $-3 \le y \le 3$).

$$y = \sin x \qquad y = 2 \sin x \qquad y = 3 \sin x$$

(B) What is the y coordinate of the highest point on each graph? Of the lowest point?

(C) Based on the observations in part (B), what is the y coordinate of the highest point on the graph of $y = p \sin x$, $p > 0$? Of the lowest point?

28. (A) Graph each of the following in the same viewing rectangle ($-2\pi \le x \le 2\pi$, $-3 \le y \le 3$).

$$y = \cos x \qquad y = 2 \cos x \qquad y = 3 \cos x$$

(B) What is the y coordinate of the highest point on each graph? Of the lowest point?

(C) Based on the observations in part (B), what is the y coordinate of the highest point on the graph of $y = p \cos x$, $p > 0$? Of the lowest point?

29. (A) Graph each of the following in the same viewing rectangle ($-2\pi \le x \le 2\pi$, $-3 \le y \le 3$).

$$y = -\cos x \qquad y = -2 \cos x \qquad y = -3 \cos x$$

(B) What is the y coordinate of the highest point on each graph? Of the lowest point?

(C) Based on the observations in part (B), what is the y coordinate of the highest point on the graph of $y = p \cos x$, $p < 0$? Of the lowest point?

30. (A) Graph each of the following in the same viewing rectangle ($-2\pi \le x \le 2\pi$, $-3 \le y \le 3$).

$$y = -\sin x \qquad y = -2 \sin x \qquad y = -3 \sin x$$

(B) What is the y coordinate of the highest point on each graph? Of the lowest point?

(C) Based on the observations in part (B), what is the y coordinate of the highest point on the graph of $y = p \sin x$, $p < 0$? Of the lowest point?

31. (A) Graph each of the following in the same viewing rectangle ($-\pi \le x \le \pi$, $-2 \le y \le 2$).

$$y = \cos x \qquad y = \cos 2x \qquad y = \cos 3x$$

(B) How many periods of each graph appear in this viewing rectangle?

(C) Based on the observations in part (B), how many periods of the graph of $y = \cos nx$, n a positive integer, would appear in this viewing rectangle?

32. (A) Graph each of the following in the same viewing rectangle ($0 \le x \le 2\pi$, -2, $\le y \le 2$).

$$y = \sin x \qquad y = \sin 2x \qquad y = \sin 3x$$

(B) How many periods of each graph appear in this viewing rectangle?

(C) Based on the observations in part (B), how many periods of the graph of $y = \sin nx$, n a positive integer, would appear in this viewing rectangle?

33. (A) Graph each of the following in the viewing rectangle ($-\pi/2 \le x \le \pi/2$, $-5 \le y \le 5$), one viewing rectangle for each.

$$y = \tan x \qquad y = \tan 2x \qquad y = \tan 3x$$

(B) How many periods of each graph appear in this viewing rectangle?

(C) Based on the observations in part (B), how many periods of the graph of $y = \tan nx$, n a positive integer, would appear in this viewing rectangle?

34. (A) Graph each of the following in the viewing rectangle ($0 \le x \le \pi$, $-5 \le y \le 5$), one viewing rectangle for each.

$$y = \cot x \qquad y = \cot 2x \qquad y = \cot 3x$$

(B) How many periods of each graph appear in this viewing rectangle?

(C) Based on the observations in part (B), how many periods of the graph of $y = \cot nx$, n a positive integer, would appear in this viewing rectangle?

35. Graph $y = \sin x$ and $y = x$ in the same viewing rectangle ($-1 \le x \le 1$ and $-1 \le y \le 1$). Notice that for $-0.5 \le x \le 0.5$, the two graphs are almost indistinguishable. In many applied problems, $\sin x$ is replaced by x for small $|x|$.

36. Graph $y = \tan x$ and $y = x$ in the same viewing rectangle ($-1 \le x \le 1$ and $-1 \le y \le 1$). Notice that for $-0.5 \le x \le 0.5$, the two graphs are very close together. In many applied problems, $\tan x$ is replaced by x for small $|x|$.

3.2 GRAPHING $y = k + A \sin Bx$ AND $y = k + A \cos Bx$

◆ **Graphs of $y = A \sin x$ and $y = A \cos x$**
◆ **Graphs of $y = \sin Bx$ and $y = \cos Bx$**
◆ **Graphs of $y = A \sin Bx$ and $y = A \cos Bx$**
◆ **Graphs of $y = k + A \sin Bx$ and $y = k + A \cos Bx$**
◆ **Application: Sound Frequency**
◆ **Application: Floating Objects**

Graphing the equations $y = k + A \sin Bx$ and $y = k + A \cos Bx$ is not difficult if you have a clear understanding of the graphs of the basic equations $y = \sin x$ and $y = \cos x$ studied in Section 3.1. Being able to graph trigonometric forms involving the constants k, A, and B significantly increases the variety of applications that we will be able to consider.

◆ Graphs of $y = A \sin x$ and $y = A \cos x$

To start, we investigate the effect of the constant A by comparing

$$y = \sin x \quad \text{and} \quad y = A \sin x$$

We can obtain the graph of $y = A \sin x$ from the graph of $y = \sin x$ by multiplying each y value of $y = \sin x$ by the constant A. The graph of $y = A \sin x$ will cross the x axis everywhere the graph of $y \sin x$ crosses the x axis, because $A \cdot 0 = 0$. Thus, the graphs of $y = A \sin x$ and $y = \sin x$ have the same x intercepts.

We know that the maximum deviation of the graph of $y = \sin x$ from the x axis is 1. Thus, the maximum deviation of the graph of $y = A \sin x$ from the x axis is $|A| \cdot 1 = |A|$. The constant $|A|$ is called the **amplitude** of the graph of $y = A \sin x$ and represents the maximum deviation of the graph from the x axis.

◆ **EXAMPLE 1**   Comparing Amplitudes

Compare the graphs of $y = \frac{1}{3} \sin x$ and $y = -3 \sin x$ with the graph of $y = \sin x$ by graphing each on the same coordinate system for $0 \le x \le 2\pi$.

SOLUTION   We see that the graph of $y = \frac{1}{3} \sin x$ has an amplitude of $\left|\frac{1}{3}\right| = \frac{1}{3}$, the graph of $y = -3 \sin x$ has an amplitude of $|-3| = 3$, and the graph of $y = \sin x$ has an amplitude of $|1| = 1$. The negative sign in $y = -3 \sin x$ turns the graph of $y = 3 \sin x$ upside down. That is, the graph of $y = -3 \sin x$ is the same as the graph of $y = 3 \sin x$ reflected across the x axis. The graphs of all three equations, $y = \frac{1}{3} \sin x$, $y = -3 \sin x$, and $y = \sin x$, are shown in Figure 1.

FIGURE 1
Comparing amplitudes

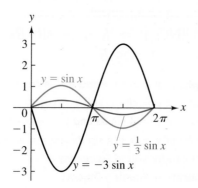

In summary, the results of Example 1 show that the effect of A in $y = A \sin x$ is to increase or decrease the y values of $y = \sin x$ without affecting the x values. A similar analysis applies to $y = A \cos x$; this function also has an amplitude of $|A|$ and a period of 2π.

MATCHED PROBLEM 1   Compare the graphs of $y = \frac{1}{2} \cos x$ and $y = -2 \cos x$ with the graph of $y = \cos x$ by graphing each on the same coordinate system for $0 \le x \le 2\pi$.

◆ **Graphs of $y = \sin Bx$ and $y = \cos Bx$**

We now examine the effect of B by comparing

$$y = \sin x \qquad \text{and} \qquad y = \sin Bx, \qquad B > 0$$

Both have the same amplitude, 1, but how do their periods compare? Since $\sin x$ has a period of 2π, it follows that $\sin Bx$ completes one cycle as Bx varies from

$$Bx = 0 \qquad \text{to} \qquad Bx = 2\pi$$

or as x varies from

$$x = \frac{0}{B} = 0 \qquad \text{to} \qquad x = \frac{2\pi}{B}$$

Thus, the period of $\sin Bx$ is $2\pi/B$. We check this result as follows: If $f(x) = \sin Bx$, then

$$f\left(x + \frac{2\pi}{B}\right) = \sin\left[B\left(x + \frac{2\pi}{B}\right)\right] = \sin(Bx + 2\pi) = \sin Bx = f(x)$$

◆ EXAMPLE 2   Comparing Periods

Compare the graphs of $y = \sin 2x$ and $y = \sin(x/2)$ with the graph of $y = \sin x$ by graphing each on the same coordinate system for one period starting at the origin.

SOLUTION   Period for $\sin 2x$: $\sin 2x$ completes one cycle as $2x$ varies from

$$2x = 0 \qquad \text{to} \qquad 2x = 2\pi$$

or as x varies from

$$x = 0 \quad \text{to} \quad x = \pi$$

Thus, the period for sin $2x$ is π.

Period for sin $(x/2)$:   sin $(x/2)$ completes one cycle as $x/2$ varies from

$$\frac{x}{2} = 0 \quad \text{to} \quad \frac{x}{2} = 2\pi$$

or as x varies from

$$x = 0 \quad \text{to} \quad x = 4\pi$$

Thus, the period for sin$(x/2)$ is 4π. The graphs of all three equations, $y = \sin 2x$, $y = \sin(x/2)$, and $y = \sin x$, are shown in Figure 2.

FIGURE 2
Comparing periods

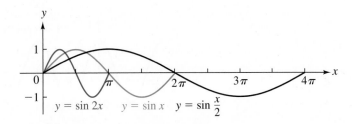

We see from Example 2 that the effect of B is to compress or stretch the basic sine curve by changing the period of sin x. A similar analysis applies to $y \cos Bx$, where $B > 0$; its period is $2\pi/B$.

MATCHED PROBLEM 2   Compare the graphs of $y = \cos 2x$ and $y = \cos (x/2)$ with the graph of $y = \cos x$ by graphing each on the same coordinate system for one period starting at the origin.

◆ **Graphs of $y = A \sin Bx$ and $y = A \cos Bx$**

We summarize the results of the discussions of amplitude and period in the box below.

For $y = A \sin Bx$ or $y = A \cos Bx$, $B > 0$:

$$\text{Amplitude} = |A| \quad \text{Period} = \frac{2\pi}{B}$$

If $B > 1$, the basic sine or cosine curve is compressed.
If $0 < B < 1$, the basic sine or cosine curve is stretched.

We now consider several examples where we show how graphs of $y = A \sin Bx$ and $y = A \cos Bx$ can be sketched rather quickly.

◆ **EXAMPLE 3**   Graphing the Form $y = A \cos Bx$

State the amplitude and period for $y = 3 \cos 2x$, and graph the equation for $-\pi \le x \le 2\pi$.

SOLUTION $\text{Amplitude} = |A| = |3| = 3 \qquad \text{Period} = \dfrac{2\pi}{B} = \dfrac{2\pi}{2} = \pi$

To sketch the graph, divide the interval of one period, from 0 to π, into four equal parts, locate x intercepts, and locate high and low points (see Figure 3(a)). Sketch the graph for one period; then extend this graph to fill out the desired interval (Figure 3(b)).

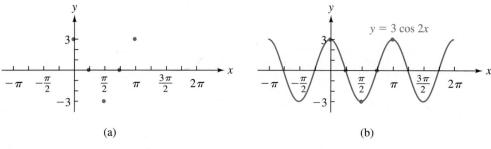

(a)   (b)

FIGURE 3   ◆

Notice that we scaled the x axis using the period divided by 4; that is, the basic unit on the x axis is $\pi/4$. Also, we adjusted the scale on the y axis to accommodate the amplitude 3.

The scales on both axes do not have to be the same.

MATCHED PROBLEM 3   State the amplitude and period for $y = \frac{1}{3} \sin(x/2)$, and graph the equation for $-2\pi \le x \le 6\pi$.

◆ **EXAMPLE 4**   Graphing the Form $y = A \sin Bx$

State the amplitude and period for $y = -\frac{1}{2} \sin(\pi x/2)$, and graph the equation for $-5 \le x \le 5$.

SOLUTION $\text{Amplitude} = |A| = \left| -\dfrac{1}{2} \right| = \dfrac{1}{2} \qquad \text{Period} = \dfrac{2\pi}{B} = \dfrac{2\pi}{\pi/2} = 4$

Because of the $-\frac{1}{2}$, the graph of $y = -\frac{1}{2} \sin(\pi x/2)$ is the graph of $y = \frac{1}{2} \sin(\pi x/2)$ reflected across the x axis (turned upside down). As before, we divide one period, from 0 to 4, into four equal parts, locate x intercepts, and locate high and low points (see Figure 4(a)). Then we graph the equation for one period, and extend the graph over the desired interval (see Figure 4(b)).

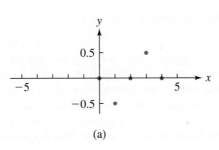

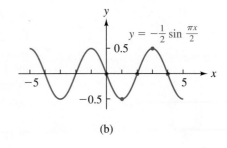

(a)   (b)

FIGURE 4

MATCHED PROBLEM 4   State the amplitude and period for $y = -2 \cos 2\pi x$, and graph the equation for $-2 \le x \le 2$.

◆   **Graphs of $y = k + A \sin Bx$ and $y = k + A \cos Bx$**

By adding a constant k to either $A \sin Bx$ or $A \cos Bx$, we are simply adding k to the ordinates of the points on their graphs. That is, we are **vertically translating** the graphs of $y = A \sin Bx$ and $y = A \cos Bx$ up k units if $k > 0$ or down $|k|$ units if $k < 0$.

◆   EXAMPLE 5   Graphing the Form $y = k + A \cos Bx$

Graph

$$y = -2 + 3 \cos 2x, \qquad -\pi \le x \le 2\pi$$

SOLUTION   We first graph $y = 3 \cos 2x$ (as we did in Example 3), and then move the graph down $|k| = |-2| = 2$ units (since $k = -2 < 0$).

FIGURE 5

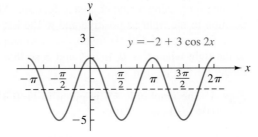

You may find it helpful to first draw the horizontal dashed line shown in Figure 5. In this case, the line is 2 units below the x axis, which represents a vertical translation of -2. Then the graph of $y = 3 \cos 2x$ is drawn relative to the dashed line and the original y axis.

MATCHED PROBLEM 5   Graph

$$y = 3 - 2 \cos 2\pi x, \qquad -2 \le x \le 2$$

FIGURE 6
Tuning fork

◆   Application: Sound Frequency

In many applications involving periodic phenomena and time (sound waves, pendulum motion, water waves, electromagnetic waves, and so on), we speak of the **period** as the length of time taken for one complete cycle of motion. For example, since each complete vibration of an A440 tuning fork (see Figure 6) lasts $\frac{1}{440}$ sec, we say that its period is $\frac{1}{440}$ sec.

Closely related to the period is the concept of **frequency,** which is the number of periods or cycles per second. The frequency of the A440 tuning fork is 440 cycles/sec, which is the reciprocal of the period. Instead of "cycles per second," we usually write "Hz," where Hz (read "hertz") is the standard unit for frequency (cycles per unit of time), and is named after the German physicist Heinrich Rudolph Hertz (1857–1894), who discovered and produced radio waves.

PERIOD AND FREQUENCY

For any periodic phenomenon, if P is the period and f is the frequency, then:

$$P = \frac{1}{f}$$ *Period is the time for one complete cycle.*

$$f = \frac{1}{P}$$ *Frequency is the number of cycles per unit of time.*

◆   EXAMPLE 6   An Equation of a Sound

Referring to the tuning fork in Figure 6, we want to model the motion of the tip of one prong using a sine or cosine function. We choose deviation from the rest position to the right as positive and to the left as negative. If the prong motion has a frequency of 440 Hz (cycles/sec), an amplitude of 0.04 cm, and is 0.04 cm to the right when $t = 0$, find A and B so that $y = A \cos Bt$ is an approximate model for this motion.

SOLUTION   *Find A.* The amplitude $|A|$ is given to be 0.04. Since $y = 0.04$ when $t = 0$, $A = 0.04$ (and not -0.04).

Find B. We are given that the frequency, f, is 440 Hz; hence, the period is found using the reciprocal formula:

$$P = \frac{1}{f} = \frac{1}{400} \text{ sec}$$

But from the earlier discussion, $P = 2\pi/B$. Thus,

$$B = \frac{2\pi}{P} = \frac{2\pi}{\frac{1}{440}} = 880\pi$$

Write the equation

$$y = A \cos Bt = 0.04 \cos 880\pi t$$

✔Check   Amplitude $= |A| = |0.04| = 0.04$ Period $= \dfrac{2\pi}{B} = \dfrac{2\pi}{880\pi} = \dfrac{1}{440}$ sec

Frequency $= \dfrac{1}{\text{Period}} = \dfrac{1}{\frac{1}{440}} = 440$ Hz

And when $t = 0$,

$$y = 0.04 \cos 880\pi(0)$$
$$= 0.04(1) = 0.04$$ ◆

MATCHED PROBLEM 6   Repeat Example 6 if the prong motion has a frequency of 262 Hz (cycles/sec), an amplitude of 0.065 cm, and is 0.065 cm to the left of the rest position when $t = 0$.

◆ Application: Floating Objects

Did you know that it is possible to determine the mass of a floating object (Figure 7) simply by making it bob up and down in the water and timing its period of oscillation?

FIGURE 7
Floating object

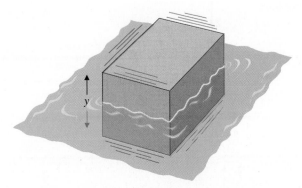

If the rest position of the floating object is taken to be 0 and we start counting time as the object passes up through 0 when it is made to oscillate, then its equation of motion can be shown to be (neglecting water and air resistance)

$$y = D \sin \sqrt{\frac{1{,}000gA}{M}} \, t$$

where y and D are in meters, t is time in seconds, $g = 9.75$ m/sec^2, A is the horizontal cross-sectional area in square meters, and M is mass in kilograms. The amplitude and period for the motion are given by

$$\text{Amplitude} = |D| \qquad \text{Period} = \frac{2\pi}{\sqrt{1{,}000gA/M}}$$

◆ **EXAMPLE 7**   Mass of a Buoy

A cylindrical buoy with cross-sectional area 1.25 m² is observed (after being pushed) to bob up and down with a period of 0.50 sec. Approximately what is the mass of the buoy in kilograms?

SOLUTION   We use the formula

$$\text{Period} = \frac{2\pi}{\sqrt{1{,}000gA/M}}$$

with $\pi \approx 3.14$, $g = 9.75$ m/sec², $A = 1.25$ m², and period $= 0.50$ sec, and solve for M. Thus,

$$0.50 = \frac{2(3.14)}{\sqrt{(1{,}000)(9.75)(1.25)/M}}$$

$$M \approx 77 \text{ kg} \qquad \qquad \blacklozenge$$

MATCHED PROBLEM 7   Write the equation of motion of the buoy in Example 7 in the form $y = D \sin Bt$, assuming its amplitude is 0.4 m.

Answers to   **1.**
Matched Problems

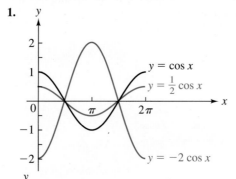

2.

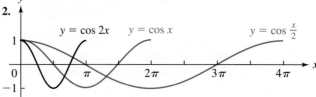

3. Amplitude $= 1/3$, period $= 4\pi$

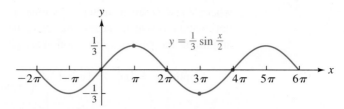

4. Amplitude = 2, period = 1

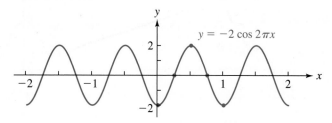

5.

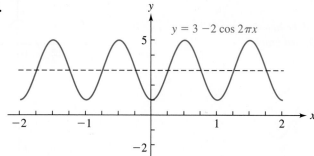

6. $y = -0.065 \cos 524\pi t$ **7.** $y = 0.4 \sin 4\pi t$

EXERCISE 3.2

A *Make a sketch of each trigonometric function without looking at the text or using a calculator. Label each point where the graph crosses the x axis.*

1. $y = \sin x$, $-2\pi \leq x \leq 2\pi$
2. $y = \cos x$, $-2\pi \leq x \leq 2\pi$

State the amplitude and period for each equation, and graph it over the indicated interval.

3. $y = 3 \cos x$, $-2\pi \leq x \leq 2\pi$
4. $y = 5 \sin x$, $-2\pi \leq x \leq 2\pi$
5. $y = -2 \sin x$, $0 \leq x \leq 4\pi$
6. $y = -3 \cos x$, $0 \leq x \leq 4\pi$
7. $y = \dfrac{1}{2} \sin x$, $0 \leq x \leq 2\pi$
8. $y = \dfrac{1}{3} \cos x$, $0 \leq x \leq 2\pi$
9. $y = \cos 2x$, $-\pi \leq x \leq \pi$
10. $y = \sin 4x$, $-\pi \leq x \leq \pi$

11. $y = \sin 2\pi x$, $-2 \leq x \leq 2$
12. $y = \cos 4\pi x$, $-1 \leq x \leq 1$
13. $y = \cos \dfrac{x}{4}$, $0 \leq x \leq 8\pi$
14. $y = \sin \dfrac{x}{2}$, $0 \leq x \leq 4\pi$

B *State the amplitude and period for each equation, and graph it over the indicated interval.*

15. $y = 2 \sin 4x$, $-\pi \leq x \leq \pi$
16. $y = 3 \cos 2x$, $-\pi \leq x \leq \pi$
17. $y = \dfrac{1}{3} \cos 2\pi x$, $-2 \leq x \leq 2$
18. $y = \dfrac{1}{2} \sin 2\pi x$, $-2 \leq x \leq 2$
19. $y = -\dfrac{1}{4} \sin \dfrac{x}{2}$, $-4\pi \leq x \leq 4\pi$
20. $y = -3 \cos \dfrac{x}{2}$, $-4\pi \leq x \leq 4\pi$

Graph each equation over the indicated interval.

21. $y = -1 + \dfrac{1}{3} \cos 2\pi x, \quad -2 \le x \le 2$

22. $y = -\dfrac{1}{2} + \dfrac{1}{2} \sin 2\pi x, \quad -2 \le x \le 2$

23. $y = 2 - \dfrac{1}{4} \sin \dfrac{x}{2}, \quad -4\pi \le x \le 4\pi$

24. $y = 3 - 3 \cos \dfrac{x}{2}, \quad -4\pi \le x \le 4\pi$

In Problems 25–28 find the equation of the form $y = A \sin Bx$ that produces the indicated graph.

25.

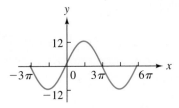

26.

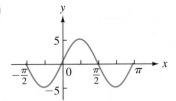

27.

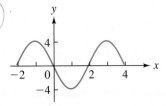

28.

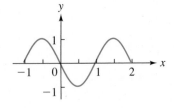

In Problems 29–32 find the equation of the form $y = A \cos Bx$ that produces the indicated graph.

29.

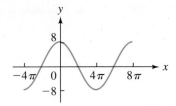

30.

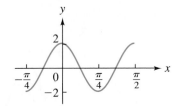

31.

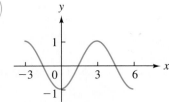

32.

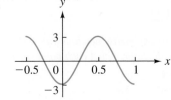

 Problems 33–44 require the use of a graphing calculator. Adjust the ranges in the viewing rectangles so that you see at least two periods of a particular function.

In Problems 33–38, graph the given equation. Find an equation of the form $y = k + A \sin Bx$ or $y = k + A \cos Bx$ that has the same graph. These problems suggest the existence of further indentities in addition to the basic identities discussed in Section 2.6. Further identities are discussed in detail in Chapter 4.

33. $y = \sin x \cos x$ **34.** $y = \cos^2 x - \sin^2 x$

35. $y = 2 \cos^2 x$ **36.** $y = 2 \sin^2 x$

37. $y = 2 - 4 \sin^2 2x$ **38.** $y = 6 \cos^2 \dfrac{x}{2} - 3$

Problems 39–44 graphically explore the relationship of the graphs of $y = \sin Bx$ and $y = \cos Bx$ with the graphs of $y = \sin(Bx + C)$ and $y = \cos(Bx + C)$, respectively. This topic is discussed in detail in the next section.

39. (A) Graph each of the following functions in the same viewing rectangle.

$$y = \sin x$$

$$y = \sin (x - 1)$$

$$y = \sin (x - 2)$$

(B) Based on the graphs in part (A), describe the relationship between the graphs of $y = \sin x$ and $y = \sin (x - a)$ where a is a positive real number.

40. (A) Graph each of the following functions in the same viewing rectangle.

$$y = \cos x$$

$$y = \cos (x - 1)$$

$$y = \cos (x - 2)$$

(B) Based on the graphs in part (A), describe the relationship between the graphs of $y = \cos x$ and $y = \cos (x - a)$ where a is a positive real number.

41. (A) Graph each of the following functions in the same viewing rectangle.

$$y = \cos x$$

$$y = \cos (x + 1)$$

$$y = \cos (x + 2)$$

(B) Based on the graphs in part (A), describe the relationship between the graphs of $y = \cos x$ and $y = \cos (x + a)$ where a is a positive real number.

42. (A) Graph each of the following functions in the same viewing rectangle.

$$y = \sin x$$

$$y = \sin (x + 1)$$

$$y = \sin (x + 2)$$

(B) Based on the graphs in part (A), describe the relationship between the graphs of $y = \sin x$ and $y = \sin (x + a)$ where a is a positive real number.

43. (A) Graph each of the following functions in the same viewing rectangle.

$$y = \sin 2x$$

$$y = \sin (2x - 1)$$

$$y = \sin (2x - 2)$$

(B) Based on the graphs in part (A), describe the relationship between the graphs of $y = \sin 2x$ and $y = \sin (2x - a)$ where a is a positive real number.

44. (A) Graph each of the following functions in the same viewing rectangle.

$$y = \cos 3x$$

$$y = \cos (3x - 1)$$

$$y = \cos (3x - 2)$$

(B) Based on the graphs in part (A), describe the relationship between the graphs of $y = \cos 3x$ and $y = \cos (3x - a)$ where a is a positive real number.

 Applications

In these applications, assume all given values are exact unless indicated otherwise.

45. **Electrical Circuits** The voltage E in an electrical circuit is given by $E = 110 \sin 120\pi t$, where t is time in seconds. What are the amplitude and period of the function? What is the frequency of the function? Graph the function for $0 \le t \le \frac{3}{60}$.

46. **Spring-Mass System** The equation $y = -4 \cos 8t$, where t is time in seconds, represents the motion of a weight hanging on a spring after it has been pulled 4 cm below its equilibrium point and released (see the figure). What are the amplitude, period, and frequency of the function? [Air resistance and friction (damping forces) are neglected.] Graph the function for $0 \le t \le 3\pi/4$.

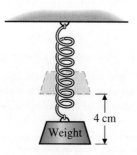

Figure for 46

*47. **Electrical Circuits**   If the voltage E in an electrical circuit has an amplitude of 12 V and a frequency of 40 Hz, and if $E = 12$ V when $t = 0$ sec, find an equation of the form $E = A \cos Bt$ that gives the voltage at any time t.

*48. **Spring-Mass System**   If the motion of the weight in Problem 46 has an amplitude of 6 in. and a frequency of 2 Hz, and if its position when $t = 0$ sec is 6 in. above its position at rest (above the rest position is positive and below is negative), find an equation of the form $y = A \cos Bt$ that describes the motion at any time t. (Neglect any damping forces—that is, air resistance and friction.)

*49. **Floating Objects**
  (A) A 3 m $\times$ 3 m $\times$ 1 m float in the shape of a rectangular solid is observed to bob up and down with a period of 1 sec. What is the mass of the float in kilograms (to three significant digits)?
  (B) Write an equation of motion for the float in part (A) in the form $y = D \sin Bt$, assuming the amplitude of the motion is 0.2 m.
  (C) Graph the equation found in part (B) for $0 \le t \le 2$.

*50. **Floating Objects**
  (A) A cylindrical buoy with diameter 0.6 m is observed (after being pushed) to bob up and down with a period of 0.4 sec. What is the mass of the buoy (to the nearest kilogram)?
  (B) Write an equation of motion for the buoy in the form $y = D \sin Bt$, assuming the amplitude of the motion is 0.1 m.
  (C) Graph the equation for $0 \le t \le 1.2$.

Figure for 50

51. **Physiology**   A normal seated adult breathes in and exhales about 0.80 liter of air every 4.00 sec. The volume of air $V(t)$ in the lungs t seconds after exhaling is given approximately by

$$V(t) = 0.45 - 0.35 \cos \frac{\pi t}{2}, \qquad 0 \le t \le 8$$

Graph the function over the indicated interval.

52. **Pollution**   In a large city, the amount of sulfur dioxide pollutant released into the atmosphere due to the burning of coal and oil for heating purposes varies seasonally. Suppose the number of tons of pollutant released into the atmosphere during the nth week after January 1 is given approximately by

$$P(n) = 1 + \cos \frac{n\pi}{26}, \qquad 0 \le n \le 104$$

Graph the function over the indicated interval.

*53. **Rotary and Linear Motion**   A Ferris wheel with a diameter of 40 m rotates counterclockwise at 4 rpm (revolutions per minute), as indicated in the figure. It starts at $\theta = 0$ when time t is 0. At the end of t min, $\theta = 8\pi t$. Why? Convince yourself that the position of the person's shadow (from the sun directly overhead) on the x axis is given by

$$x = 20 \sin 8\pi t$$

Graph this equation for $0 \le t \le 1$.

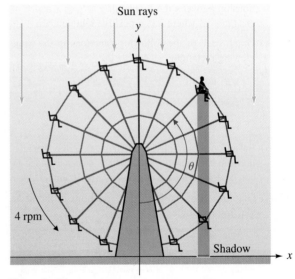

Figure for 53

*54. **Rotary and Linear Motion**   Repeat Problem 53 for a Ferris wheel with a diameter of 30 m rotating at 6 rpm.

3.3 GRAPHING $y = k + A \sin(Bx + C)$ AND $y = k + A \cos(Bx + C)$

◆   **Graphing $y = A \sin(Bx + C)$ and $y = A \cos(Bx + C)$**
◆   **Graphing $y = k + A \sin(Bx + C)$ and $y = k + A \cos(Bx + C)$**
◆   **Finding Phase Shift Using a Graphing Calculator**

We are now ready to consider sine and cosine functions of a more general form. These functions are used extensively to model real-world phenomena. See Problems 35–42 in Exercise 3.3 and the applications discussed in Section 3.4.

◆   **Graphing $y = A \sin(Bx + C)$ and $y = A \cos(Bx + C)$**

We are interested in graphing equations of the form

$$y = A \sin(Bx + C) \quad \text{and} \quad y = A \cos(Bx + C)$$

We will find that the graphs of these equations are simply the graphs of

$$y = A \sin Bx \quad \text{or} \quad y = A \cos Bx$$

translated horizontally to the left or to the right, which can be seen as follows: Since $A \sin x$ has a period of 2π, it follows that $A \sin(Bx + C)$ completes one cycle as $Bx + C$ varies from

$$Bx + C = 0 \quad \text{to} \quad Bx + C = 2\pi$$

or, solving for x, as x varies from

$$x = -\frac{C}{B} \quad \text{to} \quad x = -\frac{C}{B} + \frac{2\pi}{B}$$

with $-\frac{C}{B}$ to $-\frac{C}{B} + \frac{2\pi}{B}$ labeled Phase shift and Period respectively.

Thus, $y = A \sin(Bx + C)$ has a period of $2\pi/B$, and its graph is the graph of $y = A \sin Bx$ translated horizontally $|-C/B|$ units to the right if $-C/B > 0$ and $|-C/B|$ units to the left if $-C/B < 0$. The horizontal translation, determined by the number $-C/B$, is often referred to as the **phase shift.**

What are the period and phase shift for $y = \sin(x + \pi/2)$? To answer this question, you can use the formulas above for period and phase shift or you can go through the process used to derive the formulas. It is probably easier for most people to remember and use the process. We set $x + \pi/2$ equal to the end points of one complete cycle of the sine function, 0 and 2π, then solve for x.

$$x + \frac{\pi}{2} = 0 \qquad x + \frac{\pi}{2} = 2\pi$$

$$x = -\frac{\pi}{2} \qquad x = -\frac{\pi}{2} + 2\pi$$

Thus, $-\pi/2$ is the phase shift, and the graph of $y = \sin x$ is horizontally translated $|-\pi/2| = \pi/2$ units to the left (since $-\pi/2 < 0$). The period is 2π. Figure 1(a)

FIGURE 1
Phase shift

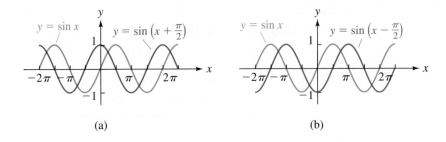

(a)   (b)

shows the graphs of $y = \sin x$ and $y = \sin(x + \pi/2)$. Going through a similar process, Figure 1(b) shows the graphs of $y = \sin x$ and $y = \sin(x - \pi/2)$. Here, the phase shift is $\pi/2$, and the graph of $y = \sin x$ is horizontally translated $\pi/2$ units to the right (since $\pi/2 > 0$).

A similar analysis applies to $y = A \cos(Bx + C)$, and the results are summarized in the following box.

PROPERTIES OF $y = A \sin(Bx + C)$ AND $y = A \cos(Bx + C)$

For $B > 0$,

$$\text{Amplitude} = |A| \qquad \text{Period} = \frac{2\pi}{B} \qquad \text{Phase shift} = -\frac{C}{B}$$

As we have already indicated, it is not necessary to memorize the formulas for period and phase shift unless you wish to. The period and phase shift are easily found in the following steps for graphing.

STEPS FOR GRAPHING $y = A \sin(Bx + C)$ AND $y = A \cos(Bx + C)$

Step 1   Find the amplitude: $|A|$

Step 2   Solve $Bx + C = 0$ and $Bx + C = 2\pi$:

$$Bx + C = 0 \qquad \text{and} \qquad Bx + C = 2\pi$$

$$x = -\frac{C}{B} \qquad\qquad\qquad x = -\frac{C}{B} + \frac{2\pi}{B}$$

$$\underbrace{\text{Phase shift}}_{} \quad \underset{\text{Period}}{\uparrow}$$

The graph completes one full cycle as $Bx + C$ varies from 0 to 2π; that is, as x varies from $-C/B$ to $(-C/B) + (2\pi/B)$.

Step 3   Graph one cycle over the interval from $-C/B$ to $(-C/B) + (2\pi/B)$.

Step 4   Extend the graph in step 3 to the left or to the right as desired.

◆   EXAMPLE 1   Graphing the Form $y = A \cos(Bx + C)$

Graph

$$y = 20 \cos\left(\pi x - \frac{\pi}{2}\right), \qquad -1 \le x \le 3$$

SOLUTION   **Step 1**   Find the amplitude.

Amplitude $= |A| = |20| = 20$

Step 2   Solve $Bx + C = 0$ and $Bx + C = 2\pi$.

$$\pi x - \frac{\pi}{2} = 0 \qquad\qquad \pi x - \frac{\pi}{2} = 2\pi$$

$$x = \frac{1}{2} \qquad\qquad x = \frac{1}{2} + 2$$

Phase shift   Period

$$\text{Phase shift} = \frac{1}{2} \qquad \text{Period} = 2$$

Step 3   Graph one cycle over the interval from $\frac{1}{2}$ to $(\frac{1}{2} + 2) = \frac{5}{2}$ (Figure 2).

FIGURE 2

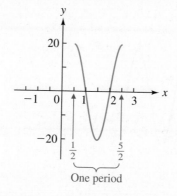

One period

Step 4   Extend the graph from -1 to 3 (Figure 3).

FIGURE 3

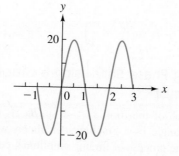

MATCHED PROBLEM 1   State the amplitude, period, and phase shift for $y = -5 \sin(x/2 + \pi/2)$. Graph the equation for $-3\pi \le x \le 5\pi$.

◆ Graphing $y = k + A \sin(Bx + C)$
 and $y = k + A \cos(Bx + C)$

In order to graph an equation of the form $y = k + A \sin(Bx + C)$ or $y = k + A \cos(Bx + C)$, we graph $y = A \sin(Bx + C)$ or $y = A \cos(Bx + C)$, as outlined previously, and then vertically translate the graph up k units if $k > 0$ or down $|k|$ units if $k < 0$. Thus, graphing equations of the form $y = k + A \sin(Bx + C)$ or $y = k + A \cos(Bx + C)$, where $k \neq 0$ and $C \neq 0$, involves both a horizontal translation (phase shift) and a vertical translation of the basic equation $y = A \sin Bx$ or $y = A \cos Bx$.

◆ EXAMPLE 2   Graphing the Form $y = k + A \cos(Bx + C)$

Graph

$$y = 10 + 20 \cos\left(\pi x - \frac{\pi}{2}\right), \qquad -1 \le x \le 3$$

SOLUTION   We graph $y = 20 \cos(\pi x - \pi/2)$, $-1 \le x \le 3$, as in Example 1, and then vertically translate the graph up 10 units (Figure 4).

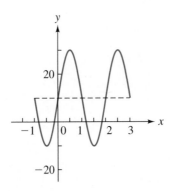

FIGURE 4   ◆

MATCHED PROBLEM 2   Graph

$$y = -2 - 5 \sin\left(\frac{x}{2} + \frac{\pi}{2}\right), \qquad -3\pi \le x \le 5\pi$$

◆ Finding Phase Shift Using a Graphing
  Calculator

Given a graph of the form $y = A \sin(Bx + C)$ or $y = A \cos(Bx + C)$ in a viewing rectangle of a graphing calculator, with A, B, and C not known, we investigate the process of finding amplitude, period, and phase shift.

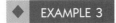

◆   EXAMPLE 3   Using a Graphing Calculator to Find a Phase Shift

Graph

$$y = 4 \sin x - 3 \cos x \qquad (1)$$

and find an equation of the form $y = A \sin(Bx + C)$, which has the same graph.

SOLUTION   The graph of equation (1) is shown in Figure 5. This graph appears to be a sine wave with amplitude 5 and period 2π that has been shifted to the right. Thus, we conclude that $A = 5$ and $B = 2\pi/p = 1$. To determine C, we will first find the phase shift by examining the graph. The phase shift of the graph in Figure 5 is the same as the smallest positive x intercept of the graph, or equivalently, as the smallest positive root of equation (1). Some graphing calculators have a built-in routine that will approximate the roots of an equation, others require you to zoom in on a graph to approximate its intercepts. Using either of these approximation techniques, we find that the smallest positive intercept of this graph to three decimal places is $x = 0.644$. This is also the phase shift for the graph. To find C, we substitute $B = 1$ and $x = 0.644$ in the phase-shift equation $x = -C/B$.

$$x = -\frac{C}{B}$$

$$0.644 = -\frac{C}{1}$$

$$C = -0.644$$

Thus, graphing

$$y = 5 \sin(x - 0.644) \qquad (2)$$

produces a translated sine wave that agrees with the graph in Figure 5.   ◆

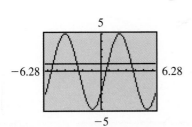

FIGURE 5

MATCHED PROBLEM 3   Graph $y = 3 \sin x + 4 \cos x$, find the x intercept closest to the origin, correct to three decimal places, and find an equation of the form $y = A \sin(Bx + C)$ that has the same graph.

REMARKS   1. We did not include a graph of equation (2) in Example 3 because it would look just like the graph in Figure 5. Graphing equivalent forms of the same equation on a calculator is not very instructive. After the first equation is graphed, it appears that nothing is happening, when in fact the calculator is drawing the second graph over the first one. One way to check that two different equations have the same graph is to graph the difference of the two equations with a vertical shift to separate the graph from the x axis. Figure 6 was obtained by graphing

$$y1 = 4 \sin x - 3 \cos x$$
$$y2 = 5 \sin(x - 0.644)$$

and

$$y3 = y1 - y2 + 1$$

FIGURE 6
Graphs of $y1$, $y2$, and $y3$

The horizontal line through (0, 1) shows that the graphs of *y*1 and *y*2 are identical in this viewing rectangle.

2. Since we approximated the value of *C* in Example 3, the graph of *y*2 is actually an approximation to the graph of *y*1. However, the difference will not be apparent unless the graphs are greatly magnified. If more accuracy is required, we simply approximate *C* to more decimal places.

3. Using graphs to discover relationships between trigonometric forms is an important concept. However, it is also important to be able to verify these relationships analytically. Techniques for this analysis will be discussed in the next chapter.   ◇

Answers to
Matched Problems

1. Amplitude = 5,   Period = 4π,   Phase shift = $-\pi$.

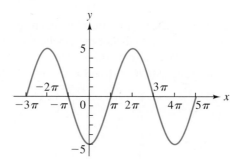

2.

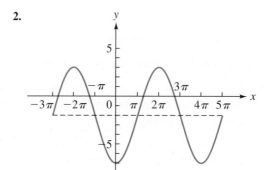

3. *x* intercept: -0.927; $y = 5 \sin(x + 0.927)$

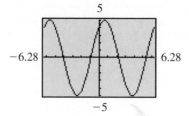

EXERCISE 3.3

A *Indicate the phase shift for each equation, and graph it over the stated interval.*

1. $y = \cos\left(x + \dfrac{\pi}{2}\right)$, $\dfrac{-\pi}{2} \le x \le \dfrac{3\pi}{2}$

2. $y = \cos\left(x - \dfrac{\pi}{2}\right)$, $\dfrac{\pi}{2} \le x \le \dfrac{5\pi}{2}$

3. $y = \sin\left(x - \dfrac{\pi}{4}\right)$, $-\pi \le x \le 2\pi$

4. $y = \cos\left(x + \dfrac{\pi}{4}\right)$, $-\pi \le x \le 2\pi$

5. $y = 3 \sin\left(x - \dfrac{\pi}{2}\right)$, $\dfrac{-\pi}{2} \le x \le \dfrac{5\pi}{2}$

6. $y = \dfrac{1}{4}\cos\left(x + \dfrac{\pi}{2}\right)$, $-\pi \le x \le 2\pi$

B *State the amplitude, period, and phase shift for each equation, and graph it over the indicated interval.*

7. $y = \sin(2\pi x - \pi)$, $-1 \le x \le 2$

8. $y = \cos(\pi x - \pi)$, $-2 \le x \le 3$

9. $y = 4 \cos\left(\pi x + \dfrac{\pi}{4}\right)$, $-1 \le x \le 3$

10. $y = 2 \sin\left(\pi x - \dfrac{\pi}{2}\right)$, $-2 \le x \le 2$

11. $y = -2 \cos(2x + \pi)$, $-\pi \le x \le 3\pi$

12. $y = -3 \sin(4x - \pi)$, $-\pi \le x \le \pi$

Graph each equation over the indicated interval.

13. $y = -2 + 4 \cos\left(\pi x + \dfrac{\pi}{4}\right)$, $-1 \le x \le 3$

14. $y = -3 + 2 \sin\left(\pi x - \dfrac{\pi}{2}\right)$, $-2 \le x \le 2$

15. $y = 3 - 2 \cos(2x + \pi)$, $-\pi \le x \le 3\pi$

16. $y = 4 - 3 \sin(4x - \pi)$, $-\pi \le x \le \pi$

C *In Problems 17 and 18 state the amplitude, period, and phase shift for each equation, and graph it over the indicated interval.*

17. $y = 2 \sin\left(3x - \dfrac{\pi}{2}\right)$, $\dfrac{-2\pi}{3} \le x \le \dfrac{5\pi}{3}$

18. $y = -4 \cos\left(4x + \dfrac{\pi}{2}\right)$, $\dfrac{-\pi}{2} \le x \le \pi$

In Problems 19 and 20 graph each equation over the indicated interval.

19. $y = 4 + 2 \sin\left(3x - \dfrac{\pi}{2}\right)$, $\dfrac{-2\pi}{3} \le x \le \dfrac{5\pi}{3}$

20. $y = 6 - 4 \cos\left(4x + \dfrac{\pi}{2}\right)$, $\dfrac{-\pi}{2} \le x \le \pi$

21. Graph

$$y = \cos\left(x - \dfrac{\pi}{2}\right) \quad \text{and} \quad y = \sin x$$

in the same coordinate system. Conclusion?

22. Graph

$$y = \sin\left(x + \dfrac{\pi}{2}\right) \quad \text{and} \quad y = \cos x$$

in the same coordinate system. Conclusion?

For Problems 23 and 24, refer to the graph.

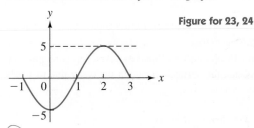

Figure for 23, 24

23. If the graph is a graph of an equation of the form $y = A \sin(Bx + C)$, $0 < -C/B < 2$, find the equation.

24. If the graph is a graph of an equation of the form $y = A \sin(Bx + C)$, $-2 < -C/B < 0$, find the equation.

For Problems 25 and 26, refer to the graph.

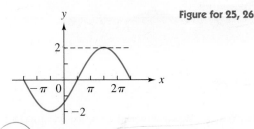

Figure for 25, 26

25. If the graph is a graph of an equation of the form $y = A \cos(Bx + C)$, $-2\pi < -C/B < 0$, find the equation.

26. If the graph is a graph of an equation of the form $y = A \cos(Bx + C)$, $0 < -C/B < 2\pi$, find the equation.

Problems 27–34 require the use of a graphing calculator. Graph the given equation and find the x intercept closest to the origin, correct to three decimal places. Use this intercept to find an equation of the form $y = A \sin(Bx + C)$ that has the same graph as the given equation.

27. $y = \sin x + \sqrt{3} \cos x$

28. $y = \sqrt{3} \sin x - \cos x$

29. $y = \sqrt{2} \sin x - \sqrt{2} \cos x$

30. $y = \sqrt{2} \sin x + \sqrt{2} \cos x$

31. $y = 1.4 \sin 2x + 4.8 \cos 2x$

32. $y = 4.8 \sin 2x - 1.4 \cos 2x$

33. $y = 2 \sin \dfrac{x}{2} - \sqrt{5} \cos \dfrac{x}{2}$

34. $y = \sqrt{5} \sin \dfrac{x}{2} + 2 \cos \dfrac{x}{2}$

Applications

In these applications, assume all given values are exact unless indicated otherwise.

35. Water Waves (see Section 3.4)   At a particular point in the ocean, the vertical change in the water due to wave action is given by

$$y = 5 \sin \frac{\pi}{6}(t + 3)$$

where y is in meters and t is time in seconds. What are the amplitude, period, and phase shift? Graph the equation for $0 \le t \le 39$.

Figure for 35

36. Water Waves   Repeat Problem 35 if the wave equation is

$$y = 8 \cos \frac{\pi}{12}(t - 6), \qquad 0 \le t \le 72$$

37. Electrical Circuit (see Section 3.4)   The current I (in amperes) in an electrical circuit is given by $I = 30 \sin(120\pi t - \pi)$, where t is time in seconds. State the amplitude, period, frequency (cycles per second), and phase shift. Graph the equation for the interval $0 \le t \le \frac{3}{60}$.

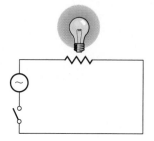

Figure for 37

38. Electrical Circuit   Repeat Problem 37 for

$$I = 110 \cos\left(120\pi t + \frac{\pi}{2}\right), \qquad 0 \le t \le \frac{2}{60}$$

∗39. Rotary and Linear Motion   If, in Problem 53, Exercise 3.2 (the Ferris wheel problem), we start θ at $\pi/2$ when $t = 0$, find the equation of motion of the shadow of a person. State the amplitude, period, and phase shift. Graph the equation for $0 \le t \le 1$.

∗40. Rotary and Linear Motion   Repeat Problem 39 with $\theta = -\pi/2$ when $t = 0$.

∗41. Modeling Temperature Variation   The 30 year average monthly temperature, F°, for each month of the year for San Antonio, TX, is given in Table 1 (*World Almanac*).

(A) Enter the data for a two-year period in your graphing calculator and produce a scatter plot in the viewing rectangle: $1 \le x \le 24$, $50 \le y \le 85$. Use 1 month as the basic time unit.

(B) It appears a sine curve of the form

$$y = k + A \sin(Bx + C) \qquad (3)$$

will closely model this data. The constants k, A, and B are easily determined from Table 1 as follows:

$$A = (\text{Max } y - \text{Min } y)/2$$
$$B = 2\pi/\text{Period}$$
$$k = \text{Min } y + A$$

TABLE 1

x (month)	1	2	3	4	5	6	7	8	9	10	11	12
y (temp.)	50	54	62	70	76	82	85	84	79	70	60	53

To estimate C, visually estimate to one decimal place the smallest positive phase shift from the plot in part (A). After determining A, B, k, and C, write the resulting equation. (Your value of C may differ slightly from the answer in the book.)

(C) Plot the results of parts (A) and (B) in the same viewing rectangle. (An improved fit may result by adjusting your value of C up or down.)

 *42. Modeling Sunrise Times   Sunrise times for the fifth of each month over a one year period were taken from a tide booklet for San Francisco Bay to form Table 2. Daylight savings time was ignored. Repeat Problem 41 for this data, but replace equation (3) with

$$y = k + A \cos(Bx + C)$$

and change the y values of the viewing rectangle to $4 \le y \le 8$. (Before entering Table 2 data into your graphing calculator, convert sunrise times from hours and minutes to decimal hours rounded to two decimal places.)

TABLE 2

x (months)	1	2	3	4	5	6	7	8	9	10	11	12
y (sun rise)	7:26	7:11	6:36	5:50	5:10	4:48	4:53	5:16	5:43	6:09	6:39	7:10

*3.4 ADDITIONAL APPLICATIONS

♦   **Modeling Electric Current**
♦   **Modeling Water Waves**
♦   **Modeling Light and Other Electromagnetic Waves**
♦   **Modeling Spring-Mass Systems**

Many types of applications of trigonometry have already been considered in this and the preceding chapters. This section provides a sampler of additional applications from several different fields. You are not expected to become an expert in any of the areas we discuss nor do you need any prior knowledge of any of the subjects to understand the discussion or work any of the problems. Several of the applications we consider use the following important properties of the sine and cosine functions:

---

* Sections marked with a star may be omitted without loss of continuity.

PROPERTIES OF SINE AND COSINE

For $y = A \sin(Bt + C)$ or $y = A \cos(Bt + C)$:

$$\text{Amplitude} = |A| \qquad \text{Period} = \frac{2\pi}{B} \qquad \text{Frequency} = \frac{1}{\text{Period}}$$

$$\text{Phase shift} = -\frac{C}{B} \begin{cases} \text{Right} & \text{if } -C/B > 0 \\ \text{Left} & \text{if } -C/B < 0 \end{cases}$$

As we indicated in the previous section, you do not need to memorize the formulas for period and phase shift. You can obtain the same results by solving the two equations

$$Bt + C = 0 \qquad \text{and} \qquad Bt + C = 2\pi$$

(We use the variable t, rather than x, because most of the applications here involve functions of time.)

Phenomena that can be described by either $y = A \sin(Bt + C)$ or $y = A \cos(Bt + C)$ are said to be **simple harmonic**.

◆ Modeling Electric Current

Michael Faraday (1791–1867), a research scientist at the Royal Institute in London, and Joseph Henry (1797–1878), a professor at the Albany Academy in New York, independently discovered that a flow of electricity could be created by moving a wire in a magnetic field. Suppose we bend a wire in the form of a rectangle, and we locate this wire between the south and north poles of magnets, as shown in Figure 1. We now rotate the wire at a constant counterclockwise speed, starting with BC in its lowest position. As BC turns toward the horizontal, an electrical current will flow from C to B (see Figure 1(a)). The strength of the current (measured in amperes) will be 0 at the lowest position, and will increase to a maximum value at the horizontal position. As BC continues to turn from the horizontal to the top position, the current flow decreases to 0. As BC starts down from the top position in a counterclockwise direction, the current flow reverses, going from B to C, and again reaches a maximum at the left horizontal position. When BC moves from this horizontal position back to the original bottom position, the current flow again decreases to 0. This pattern repeats itself for each revolution, and hence is periodic. Is it too much to expect that a trigonometric function can describe the relationship between current and time (as in Figure 1(b))? If we measure and graph the strength of the current I in the wire relative to time t, we can indeed find an equation of the form

$$I = A \sin(Bt + C)$$

FIGURE 1
Alternating current

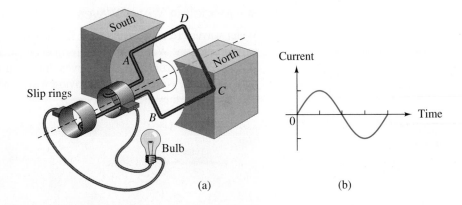

(a)   (b)

that will give us the relationship between current and time. For example,

$$I = 30 \sin 120\pi t$$

represents a 60 Hz alternating current flow with a maximum value of 30 amp.

◆   **EXAMPLE 1**   Alternating Current Generator

An alternating current generator produces an electrical current (measured in amperes) that is described by the equation

$$I = 35 \sin(40\pi t - 10\pi)$$

where t is time in seconds. What are the amplitude, frequency, and phase shift for the current?

SOLUTION
$$\overset{A}{\ } \quad \overset{B}{\ } \quad \overset{C}{\ }$$
$$y = 35 \sin(40\pi t - 10\pi)$$

Amplitude $= |35| = 35$ amp

To find the period and phase shift, solve $Bt + C = 0$ and $Bt + C = 2\pi$:

$$40\pi t - 10\pi = 0 \qquad\qquad 40\pi t - 10\pi = 2\pi$$
$$40\pi t = 10\pi \qquad\qquad 40\pi t = 10\pi + 2\pi$$
$$t = \frac{1}{4} \qquad\qquad t = \frac{1}{4} + \frac{1}{20}$$

Phase shift $= \dfrac{1}{4}$ sec (right) $\qquad$ Period $= \dfrac{1}{20}$ sec

$$\text{Frequency} = \frac{1}{\text{Period}} = 20 \text{ Hz}$$

◆

MATCHED PROBLEM 1   Repeat Example 1 for $I = -25 \sin(30\pi t + 5\pi)$.

◆ Modeling Water Waves

Water waves are perhaps the most familiar wave form. You can actually see the wave in motion! Water waves are formed by particles of water rotating in circles (Figure 2). A particle actually moves only a short distance as the wave passes through. These wave forms are moving waves, and it can be shown that in their simplest form, they are sine curves.

FIGURE 2
Water waves

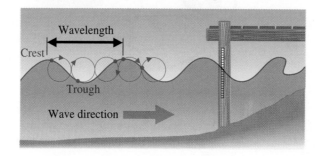

The waves can be represented by an equation of the form

$$y = A \sin 2\pi\left(ft - \frac{r}{\lambda}\right)$$ *Moving wave equation**   (1)

where f is frequency, t is time, r is distance from the source, and λ is wavelength. Thus, for a wave of a given frequency and wavelength, y is a function of the two variables, t and r.

Equation (1) is typical for moving waves. If a wave passes a pier piling with a vertical scale attached (Figure 2), the graph of the water level on the scale relative to time would look something like Figure 3(a), since r would be fixed. On the other hand, if we actually photograph a wave, freezing the motion in time, then the profile of the wave would look something like Figure 3(b).

FIGURE 3
Water wave

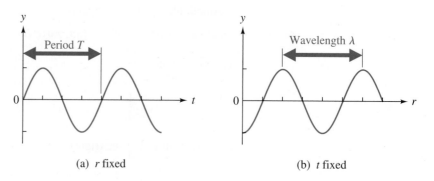

(a) r fixed   (b) t fixed

---

* Actually, this general wave equation can be used to describe any longitudinal (compressional) or transverse wave motion of one frequency in a nondispersive medium. A sound wave is an example of a longitudinal wave; water waves and electromagnetic waves are examples of transverse waves.

Experiments have shown that the **wavelength** λ for water waves is given approximately by

$$\lambda = 5.12T^2 \qquad \text{In feet}$$

where T is the period of the wave in seconds (see Figure 3(a)), and the **speed S of the wave** is given approximately by

$$S = \sqrt{\frac{g\lambda}{2\pi}} \qquad \text{In feet per second}$$

where $g = 32$ ft/sec^2. Thus, a wave with a period of exactly 8 sec and an amplitude of exactly 3 ft would have an equation of the form (holding r fixed)

$$y = 3 \sin 2\pi\left(\frac{1}{8}t - \frac{r}{\lambda}\right)$$

$$= 3 \sin\left(\frac{\pi}{4}t + c\right)$$

where c is a constant. Its wavelength would be

$$\lambda = 5.12(8^2) \approx 328 \text{ ft}$$

and it would move at a speed of approximately

$$S \approx \sqrt{\frac{32(328)}{2(3.14)}} \approx 41 \text{ ft/sec}$$

or, in miles per hour (mph),

$$(41 \text{ ft/sec})\frac{3{,}600 \text{ sec/hr}}{5{,}280 \text{ ft/mi}} \approx 28 \text{ mph}$$

◆ Modeling Light and Other Electromagnetic Waves

Visible light is a transverse wave form with a frequency range between 4×10^{14} Hz (red) and 7×10^{14} Hz (violet). The retina of the eye responds to these vibrations, and through a complicated chemical process, the vibrations are eventually perceived by the brain as light in various colors. Light is actually a small part of a continuous spectrum of electromagnetic wave forms—most of which are not visible (see Figure 4). Included in the spectrum are radio waves (AM and FM), microwaves, x rays, and gamma rays. All these waves travel at the speed of light, approx. 3×10^{10} cm/sec (186,000 mph), and many of them are either partially or totally adsorbed in the atmosphere of the earth, as indicated in Figure 4. Their distinguishing characteristics are wavelength and frequency, which are related by the formula

$$\lambda v = c \qquad\qquad (2)$$

where c is the speed of light, λ is the wavelength, and v is the frequency.

FIGURE 4
Electromagnetic wave spectrum

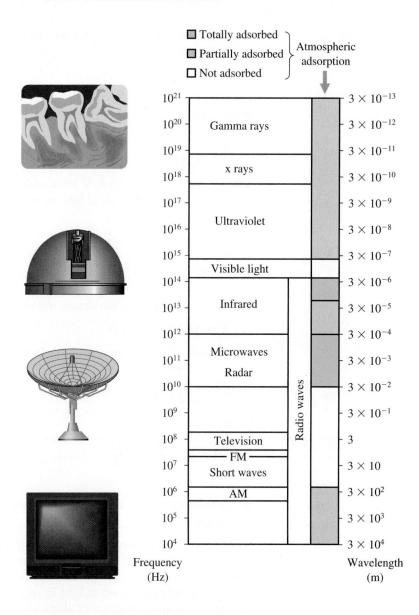

Electromagnetic waves are traveling waves that can be described by an equation of the form

$$E = A \sin 2\pi\left(vt - \frac{r}{\lambda}\right)$$ *Traveling wave equation*   (3)

where t is time and r is the distance from the source. This equation is a function of two variables, t and r. If we freeze time, then the graph of (3) looks something like Figure 5(a). If we look at the electromagnetic field at a single point in space—that is, if we hold r fixed—then the graph of (3) looks something like Figure 5(b).

FIGURE 5
Electromagnetic wave

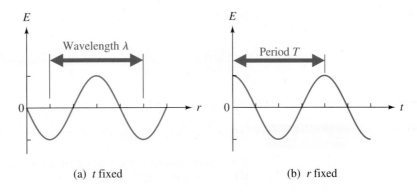

(a) t fixed

(b) r fixed

Electromagnetic waves are produced by electrons that have been excited into oscillatory motion. This motion creates a combination of electric and magnetic fields that move through space at the speed of light. The frequency of the oscillation of the electron determines the nature of the wave (see Figure 4), and a receiver responds to the wave through induced oscillation of the same frequency (see Figure 6).

FIGURE 6
Electromagnetic field

Electron

Electron

Figures 7 and 8 illustrate FM, AM, and radar waves at fixed points in space.

FIGURE 7
FM and AM electromagnetic waves

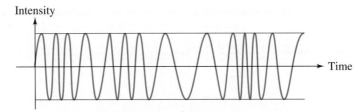

(a) FM–a frequency-modulated carrier wave
with constant amplitude and variable frequency

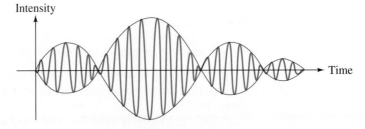

(b) AM–an amplitude-modulated carrier wave
with constant frequency and variable amplitude

FIGURE 8
Radar pulses

Intensity

Time

◆ EXAMPLE 2	Electromagnetic Waves

If an electromagnetic wave has a frequency of $v = 10^{12}$ Hz, what is its period? What is its wavelength in meters?

SOLUTION $\text{Period} = \dfrac{1}{v} = \dfrac{1}{10^{12}} = 10^{-12} \text{ sec}$

To find the wavelength λ, we use the formula

$$\lambda v = c$$

with the speed of light $c \approx 3 \times 10^8$ m/sec:

$$\lambda = \frac{3 \times 10^8 \text{ m/sec}}{10^{12} \text{ Hz}} = 3 \times 10^{-4} \text{ m}$$ ◆

MATCHED PROBLEM 2   Repeat Example 2 for $v = 10^6$ Hz.

◆ EXAMPLE 3	Ultraviolet Waves

An ultraviolet wave has an equation of the form

$$y = A \sin Bt$$

Find B if the wavelength is $\lambda = 3 \times 10^{-9}$ m.

SOLUTION   We first use $\lambda v = c$ ($c \approx 3 \times 10^8$ m/sec) to find the frequency, and then use $v = B/2\pi$ to find B.

$$v = \frac{c}{\lambda} = \frac{3 \times 10^8 \text{ m/sec}}{3 \times 10^{-9} \text{ m}} = 10^{17} \text{ Hz}$$

$$B = 2\pi v = 2\pi \times 10^{17}$$ ◆

MATCHED PROBLEM 3   A gamma ray has an equation of the form $y = A \sin Bt$. Find B if the wavelength is $\lambda = 3 \times 10^{-12}$ m.

◆ Modeling Spring-Mass Systems

An object of mass M hanging on a spring will produce simple harmonic motion when pulled down and released (if we neglect friction and air resistance; see Figure 9).

Figure 10 illustrates simple harmonic motion, damped harmonic motion, and resonance. **Damped harmonic motion** occurs when amplitude decreases to 0 as time increases. **Resonance** occurs when amplitude increases as time increases.

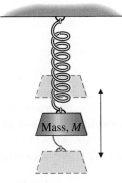

FIGURE 9

Damped harmonic motion is essential in the design of suspension systems for cars, buses, trains, and motorcycles, as well as in the design of buildings, bridges, and aircraft. Resonance is useful in some electric circuits and in some mechanical systems, but it can be disastrous in bridges, buildings, and aircraft. Commercial jets have lost wings in flight, and large bridges have collapsed because of resonance. Marching soldiers must break step while crossing a bridge in order to avoid the creation of resonance—and the collapse of the bridge!

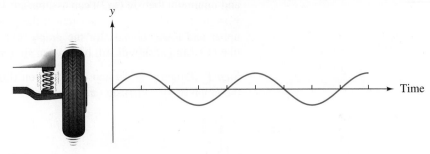

Spring only
Simple harmonic motion (neglect friction and air resistance)

(a)

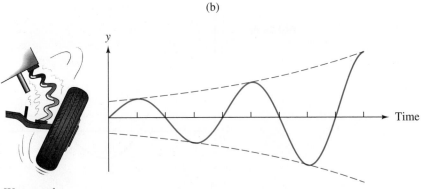

Spring and shock absorber
Damped harmonic motion

(b)

Wavy road
Resonance—disaster!

FIGURE 10   (c)

EXAMPLE 4   Damped Harmonic Motion

Graph

$$y = \frac{1}{t} \sin \frac{\pi}{2}t, \qquad 1 \le t \le 8$$

SOLUTION   The $1/t$ factor in front of $\sin(\pi/2)t$ affects the amplitude. Since the maximum and minimum that $\sin(\pi/2)t$ can assume are 1 and -1, respectively, if we graph $y = 1/t$ first ($1 \le t \le 8$) and reflect the graph across the t axis, we will have upper and lower bounds for the graph of $y = (1/t) \sin(\pi/2)t$. We also note that $(1/t) \sin(\pi/2)t$ will still be 0 when $\sin(\pi/2)t$ is 0.

Step 1   Graph $y = 1/t$ and its reflection (Figure 11)—called the **envelope** for the graph of $y = (1/t) \sin(\pi/2)t$.

FIGURE 11

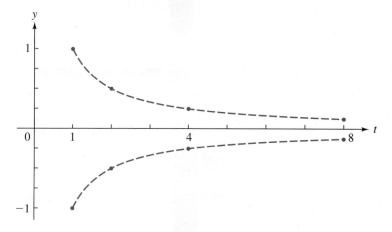

Step 2   Now sketch the graph of $y = \sin(\pi/2)t$, but keep high and low points within the envelope (Figure 12).

FIGURE 12

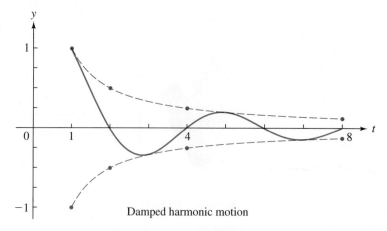

Damped harmonic motion

MATCHED PROBLEM 4   Graph

$$y = t \sin \frac{\pi}{2}t, \quad 1 \le t \le 8$$

Answers to
Matched Problems

1. Amplitude = 25 amp,   Frequency = 15 Hz,   Phase shift = $-\frac{1}{6}$ sec (left)
2. Period = 10^{-6} sec, λ = 300 m
3. $B = 2\pi \times 10^{20}$
4.

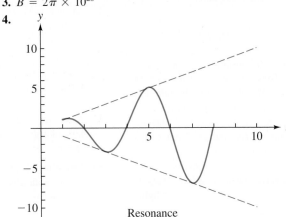

Resonance

EXERCISE 3.4

In these applications assume all given values are exact unless indicated otherwise.

 Applications

1. **Modeling Electric Current**   An alternating current generator produces a current given by

$$I = 10 \sin(120\pi t - 60\pi)$$

where t is time in seconds and I is in amperes. What are the amplitude, frequency, and phase shift for this current?

2. **Modeling Electric Current**   An alternating current generator produces a current given by

$$I = -60 \cos\left(80\pi t + \frac{2\pi}{3}\right)$$

What are the amplitude, frequency, and phase shift for the current?

3. **Modeling Electric Current**   An alternating current generator produces a 30 Hz current flow with a maximum value of 20 amp. Write an equation in the form $I = A \cos Bt$, $A > 0$, for this current.

4. **Modeling Electric Current**   An alternating current generator produces a 60 Hz current flow with a maximum value of 10 amp. Write an equation of the form $I = A \sin Bt$ for this current.

5. **Modeling Water Waves**   A water wave at a fixed position has an equation of the form

$$y = 15 \sin \frac{\pi}{8}t$$

where t is time in seconds and y is in feet. How high is the wave from trough to crest? (See Figure 2.) What is its wavelength in feet? How fast is it traveling in feet per second? Calculate your answers to the nearest foot.

6. **Modeling Water Waves**   A water wave has an amplitude of 30 ft and a period of 14 sec. If its equation is

given by

$$y = A \sin Bt$$

at a fixed position, find A and B. What is the wavelength in feet? How fast is it traveling in feet per second? How high would the wave be from trough to crest? Calculate your answers to the nearest foot.

7. Modeling Water Waves   Graph the equation in Problem 5 for $0 \leq t \leq 32$.

8. Modeling Water Waves   Graph the equation in Problem 6 for $0 \leq t \leq 28$.

9. Modeling Water Waves   A **tsunami** is a sea wave caused by an earthquake. These are very long waves with wave lengths of 150 mi that travel at nearly the rate of a jet airliner (around 470 mi/hr). At sea, these waves have an amplitude of only 1 or 2 ft, so a ship would be unaware of the wave passing underneath. As a tsunami approaches a shore, however, the waves are slowed down and the water piles up to a virtual wall of water, sometimes over 100 ft high. When such a wave crashes into land, considerable destruction can result—a three story concrete lighthouse 50 ft above sea level was completely washed away when a tsunami hit Scotch Cap, Alaska, in 1946.

(A) If at a particular moment in time, a tsunami has an equation of the form

$$y = A \sin Br$$

where y is in ft and r is the distance from the source in mi, find the equation if the amplitude is 2 ft and the wavelength is 150 mi.

(B) Find the period of the tsunami in (A) in seconds.

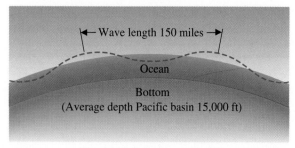

Tsunami (vertical exaggerated)

Figure for 9

10. Modeling Water Waves   Because of increased friction from the ocean bottom, a wave will begin to break when

the ocean depth becomes less than one-half the wavelength. If an ocean wave at a fixed position has an equation of the form

$$y = 8 \sin \frac{\pi}{5} t$$

at what depth (to the nearest foot) will the wave start to break?

11. Modeling Water Waves   A water wave has an equation of the form

$$y = 25 \sin 2\pi \left(\frac{t}{10} - \frac{r}{512} \right)$$

Graph the equation for a fixed position, say $r = 1{,}024$ ft, for $0 \leq t \leq 20$.

12. Modeling Water Waves   Graph the equation in Problem 11 for a fixed time, say $t = 0$, for $0 \leq r \leq 1{,}024$. [*Hint:*   Use $\sin(-x) = -\sin x$.]

13. Modeling Light and Other Electromagnetic Waves Suppose an electron oscillates at 10^8 Hz ($v = 10^8$ Hz) creating an electromagnetic wave. What is its period? What is its wavelength in meters?

14. Modeling Light and Other Electromagnetic Waves Repeat Problem 13 with $v = 10^{18}$ Hz.

15. Modeling Light and Other Electromagnetic Waves An x ray has an equation of the form

$$y = A \sin Bt$$

Find B if the wavelength of the x ray, λ, is 3×10^{-10} m. [*Note:* $c \approx 3 \times 10^8$ m/sec]

16. Modeling Light and Other Electromagnetic Waves A microwave has an equation of the form

$$y = A \sin Bt$$

Find B if the wavelength is $\lambda = 0.003$ m.

17. Modeling Light and Other Electromagnetic Waves An AM radio wave for a given station has an equation of the form

$$y = A(1 + 0.02 \sin 2\pi \cdot 1{,}200t) \sin 2\pi \cdot 10^5 t$$

for a given 1,200 Hz tone. The expression in parentheses modulates the amplitude A of the carrier wave

$$y = A \sin 2\pi \cdot 10^5 t$$

(see Figure 7(b)). What are the period and frequency of the carrier wave for time t in seconds? Can a wave with this frequency pass through the atmosphere? (See Figure 4.)

18. Modeling Light and Other Electromagnetic Waves
An FM radio wave for a given station has an equation of the form

$$y = A \sin(2\pi \cdot 10^8 t + 0.02 \sin 2\pi \cdot 1{,}200t)$$

for a given tone of 1,200 Hz. The second term within the parentheses modulates the frequency of the carrier wave $y = A \sin 2\pi \cdot 10^8 t$ (see Figure 7(a)). What are the period and frequency of the carrier wave? Can a wave with this frequency pass through the atmosphere? (See Figure 4.)

19. Spring-Mass Systems   Graph
$y = \sin 2\pi t, \quad 0 \le t \le 2$

20. Spring-Mass Systems   Graph
$$y = \frac{1}{t} \sin 2\pi t, \quad \frac{1}{4} \le t \le 2$$

21. Spring-Mass Systems   Graph
$y = t \sin 2\pi t, \quad 0 \le t \le 2$

22. Spring-Mass Systems   Which form of oscillatory motion (simple harmonic motion, damped harmonic motion, or resonance) is illustrated by each equation in Problems 19, 20, and 21?

3.5   ADDITION OF ORDINATES

◆   **Graphing Using Addition of Ordinates**
◆   **Sound Waves**
◆   **Fourier Series: A Brief Look**

After having considered the trigonometric functions individually, we now consider them in combination with each other and with other functions. Graphing these combinations is best illustrated through examples.

◆   Graphing Using Addition of Ordinates

 EXAMPLE 1   Addition of Ordinates

Graph

$$y = \frac{x}{2} + \sin x, \quad 0 \le x \le 2\pi$$

SOLUTION   A simple and fast method of hand graphing equations involving two or more terms is to graph each term separately on the same coordinate system, and then add ordinates. In this case, we form

$$y_1 = \frac{x}{2} \quad \text{and} \quad y_2 = \sin x$$

We sketch the graph of each equation in the same coordinate system, then use a compass, dividers, ruler, or eye to add the ordinates $y_1 + y_2$ (see Figure 1). The final graph of $y = (x/2) + \sin x$ is shown in Figure 2. Use a calculator to determine specific points on the graph if greater accuracy is desired.   ◆

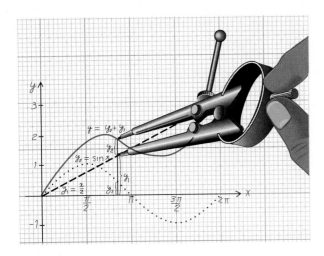

FIGURE 1
Addition of ordinates

FIGURE 2
Addition of ordinates

MATCHED PROBLEM 1   Graph $y = (x/2) + \cos x$, $0 \leq x \leq 2\pi$, using addition of ordinates.

◆   **EXAMPLE 2**   Addition of Ordinates

Graph

$$y = 3 \sin x + \cos 2x, \qquad 0 \leq x \leq 3\pi$$

SOLUTION   We use the method of addition of ordinates as in Example 1, letting $y_1 = 3 \sin x$ and $y_2 = \cos 2x$ (see Figure 3).   ◆

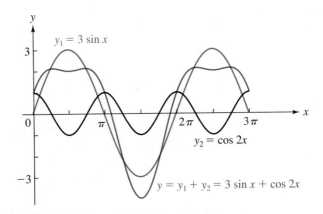

FIGURE 3
Addition of ordinates

◆

MATCHED PROBLEM 2   Graph $y = 3 \sin x + \sin 3x$, $0 \leq x \leq 2\pi$, using addition of ordinates.

◆ Sound Waves

Sound is produced by a vibrating object that, in turn, excites air molecules into motion. The vibrating air molecules cause a periodic change in air pressure that travels through air at about 1,100 ft/sec. (Sound is not transmitted in a vacuum.) When this periodic change in air pressure reaches your eardrum, the drum vibrates at the same frequency as the source, and the vibration is transmitted to the brain as sound. The range of audible frequencies covers about ten octaves, extending from 20 Hz up to 20,000 Hz. Low frequencies are associated with low pitch and high frequencies with high pitch.

If in place of an eardrum we use a microphone, then the air disturbance can be changed into a pulsating electrical signal that can be visually displayed on an oscilloscope (Figure 4). A pure tone from a tuning fork will look like a sine curve (also called a **sine wave**). The sound from a tuning fork can be accurately described by either the sine function or the cosine function. For example, a tuning fork vibrating at 264 Hz ($f = 264$) with an amplitude of 0.002 in., produces C on the musical scale, and the wave on an oscilloscope can be described by an equation of the form

$$y = A \sin 2\pi ft$$
$$= 0.002 \sin 2\pi(264)t$$

FIGURE 4
A simple sound wave

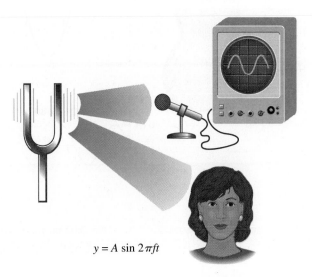

$$y = A \sin 2\pi ft$$

Most sounds are more complex than that produced by a tuning fork. Figure 5 illustrates a note produced by a guitar. You may be surprised to learn that even these more complex sound forms can be described in terms of simple sine waves by means of a *Fourier series* (which we discuss below). Theoretically, any sound can be reproduced by an appropriate combination of pure tones from tuning forks. The separate components of a complex sound are called **partial tones**. The partial tone with the smallest frequency is the **fundamental tone**. The other partial tones are called **overtones** (see Figure 5).

FIGURE 5

A complex sound wave—the guitar note shown at the top can be approximated very closely by adding the three sine waves (pure tones) below

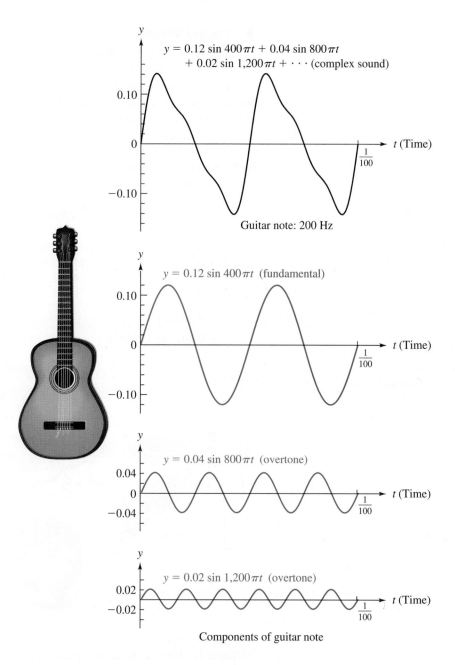

A complex sound wave—the guitar note shown at the top can be approximated very closely by adding the three sine waves (pure tones) below

$y = 0.12 \sin 400\pi t + 0.04 \sin 800\pi t + 0.02 \sin 1{,}200\pi t + \cdots$ (complex sound)

Guitar note: 200 Hz

$y = 0.12 \sin 400\pi t$ (fundamental)

$y = 0.04 \sin 800\pi t$ (overtone)

$y = 0.02 \sin 1{,}200\pi t$ (overtone)

Components of guitar note

◆ Fourier Series: A Brief Look

We now point out a significant use of combinations of trigonometric functions called **Fourier series** (named after the French mathematician Joseph Fourier, 1768–1830). These combinations may be encountered in advanced applied mathematics in the study of sound, heat flow, electrical fields and circuits, and spring-mass systems. The following discussion is included only for illustrative purposes—the reader is not expected to become proficient in this area at this time.

The following are examples of Fourier series.

$$y = \sin x + \frac{\sin 3x}{3} + \frac{\sin 5x}{5} + \cdots \tag{1}$$

$$y = \sin \pi x + \frac{\sin 2\pi x}{2} + \frac{\sin 3\pi x}{3} + \cdots \tag{2}$$

The three dots at the end of each series indicate that the pattern established in the first three terms continues indefinitely. If we graph the first term of each series, then the sum of the first two terms, and so on, we will obtain a sequence of graphs that will get closer and closer to a **square wave** for (1) and a **sawtooth wave** for (2). The greater the number of terms we take in the series, the more the graph will look like the indicated wave form. Figures 6 and 7 illustrate these phenomena.

FIGURE 6
Square wave

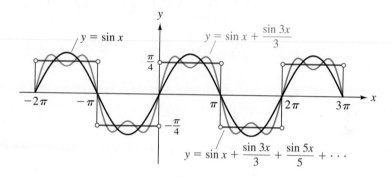

FIGURE 7
Sawtooth wave

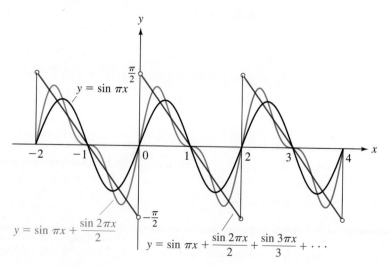

Answers to   **1.**
Matched Problems

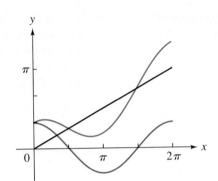

2.

EXERCISE 3.5

A *Graph each function over the indicated interval.*

1. $y = 2 + \sin x, \quad 0 \le x \le 2\pi$

2. $y = -1 + \cos x, \quad -\pi/2 \le x \le 3\pi/2$

3. $y = x + \cos x, \quad 0 \le x \le 5\pi/2$

4. $y = x + \sin x, \quad 0 \le x \le 2\pi$

B **5.** $y = \dfrac{x}{2} + \cos \pi x, \quad 0 \le x \le 3$

6. $y = \dfrac{x}{2} - \sin 2\pi x, \quad 0 \le x \le 2$

7. $y = 3 \cos x + \sin 2x, \quad 0 \le x \le 3\pi$

8. $y = 3 \cos x + \cos 3x, \quad 0 \le x \le 2\pi$

9. $y = \sin x + 2 \cos 2x, \quad 0 \le x \le 3\pi$

10. $y = \cos x + 3 \sin 2x, \quad 0 \le x \le 3\pi$

C **11.** $y = \sin x + \dfrac{\sin 3x}{3}, \quad 0 \le x \le 2\pi$

(Compare to the square wave in Figure 6.)

12. $y = \sin \pi x + \dfrac{\sin 2\pi x}{2}, \quad -2 \le x \le 2$

(Compare to the sawtooth wave in Figure 7.)

13. $y = 0.06 \sin 400\pi t + 0.03 \sin 800\pi t, \quad 0 \le t \le \dfrac{1}{200}$

(A sound wave)

14. $y = 0.04 \sin 800\pi t + 0.02 \sin 1{,}200\pi t,$

$0 \le t \le \dfrac{1}{100}$ (A sound wave)

Problems 15–20 require the use of a graphing calculator.

15. Graph the following equation for $-2\pi \le x \le 2\pi$ and $-1 \le y \le 1$ (compare to the square wave in Figure 6).

$$y = \sin x + \frac{\sin 3x}{3} + \frac{\sin 5x}{5} + \frac{\sin 7x}{7} + \frac{\sin 9x}{9}$$

16. Graph the following equation for $-4 \le x \le 4$ and $-2 \le y \le 2$ (compare to the sawtooth wave in Figure 7).

$$y = \sin \pi x + \frac{\sin 2\pi x}{2} + \frac{\sin 3\pi x}{3} + \frac{\sin 4\pi x}{4}$$
$$+ \frac{\sin 5\pi x}{5}$$

17. The sum of the first three terms of the Fourier series for a wave form are

$$y = \cos \frac{x}{2} + \frac{1}{2} \cos x + \frac{1}{9} \cos \frac{3x}{2}$$

Graph this equation for $-4\pi \le x \le 4\pi$ and $-2 \le y \le 2$, and sketch by hand the corresponding wave form. (Assume that the graph of the wave form is comprised entirely of straight line segments.)

18. Repeat Problem 17 for

$$y = \cos \frac{x}{2} + \frac{1}{9} \cos \frac{3x}{2} + \frac{1}{25} \cos \frac{5x}{2}$$

19. Graph the sound wave given by

$$y = 0.08 \sin 400\pi t + 0.04 \sin 800\pi t$$
$$+ 0.02 \sin 1200\pi t, \quad 0 \le t \le 0.01$$

20. Graph the sound wave given by

$$y = 0.06 \sin 200\pi t + 0.05 \sin 400\pi t$$
$$+ 0.03 \sin 600\pi t, \quad 0 \le t \le 0.02$$

Applications

21. Physiology A normal seated adult breathes in and exhales about 0.80 liter of air every 4.00 sec. See the graph.

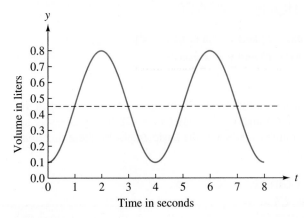

Figure for 21

The volume V of air in the lungs t sec after exhaling can be approximated by an equation of the form $V = k + A \cos Bt$, $0 \le t \le 8$. Find the equation.

22. Earth Science During the autumnal equinox, the surface temperature (°C) of the water of a lake x hours after sunrise was recorded over a 24-hr period, and the results were recorded on the following graph. The surface temperature can be approximated by an equation of the form $T = k + A \cos Bx$. Find the equation.

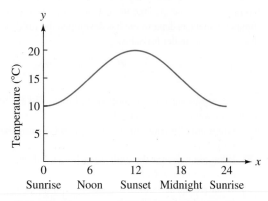

Figure for 22

23. Wildlife Management In a wilderness region north of the Grand Canyon (Arizona), the deer and mountain lion populations are interrelated, since the mountain lions rely on the deer as a food source. Their populations go up and down in cycles, but out of phase with each other. It was found that the deer population in the area changed approximately according to the formula

$$y = 1,200 + 300 \sin \frac{\pi}{4} t$$

where t is time in years and $t = 0$ at the beginning of 1980.
(A) Graph the equation for a 16 year period starting at the beginning of 1980.
(B) What is the maximum expected population over the time period?
(C) What is the minimum expected population over the time period?

24. Wildlife Management Referring to Problem 23, it was found that the mountain lion population in the area changed approximately according to the formula

$$y = 100 + 25 \sin \left(\frac{\pi}{4} t - \frac{\pi}{2} \right)$$

where t is time in years and $t = 0$ at the beginning of 1980.

(A) Graph the equation for a 16 year period starting at the beginning of 1980.

(B) What is the maximum expected population over the time period?

(C) What is the minimum expected population over the time period?

25. Modeling a Seasonal Business Cycle   A large soft drink company's sales vary seasonally, but, overall, the company is also experiencing a steady growth rate. The financial analysis department has developed the following mathematical model for sales.

$$S = 5 + \frac{t}{52} - 4 \cos \frac{\pi t}{26}$$

where S represents sales in millions of dollars for a week of sales t weeks after January 1.

(A) Graph this function for a 3 year period starting January 1.

(B) What are sales for the 26th week in the 3rd year (to three significant digits)?

(C) What are the sales for the 52nd week in the 3rd year (to one significant digit)?

26. Modeling a Seasonal Business Cycle   Repeat Problem 25 for the following mathematical model.

$$S = 4 + \frac{t}{52} - 2 \cos \frac{\pi t}{26}$$

C 27. Sound   A pure tone is transmitted through a speaker that is also emitting high frequency, low volume static. The resulting sound is given by

$$y = 0.1 \sin 500\pi t + 0.003 \sin 24{,}000\pi t$$

(A) Graph this equation, using a graphing calculator, for $0 \le t \le 0.004$ and $-0.3 \le y \le 0.3$. Notice that the result looks like a pure sine wave.

(B) Use the zoom feature (or its equivalent) to zoom in on the graph in the vicinity of $(0.001, 0.1)$. You should now *see* the static element of the sound.

C 28. Sound   Another pure tone is transmitted through the same speaker as in Problem 27. The resulting sound is given by

$$y = 0.08 \cos 500\pi t + 0.004 \sin 26{,}000\pi t$$

(A) Graph this equation, using a graphing calculator, for $0 \le t \le 0.004$ and $-0.3 \le y \le 0.3$. Notice that the sound appears to be a pure tone.

(B) Use the zoom feature (or its equivalent) to zoom in on the graph in the vicinity of $(0.002, -0.08)$. Now you should be able to *see* the static.

3.6 TANGENT, COTANGENT, SECANT, AND COSECANT FUNCTIONS REVISITED

- ◆ **Graphing $y = A \tan(Bx + C)$ and $y = A \cot(Bx + C)$**
- ◆ **Graphing $y = A \sec(Bx + C)$ and $y = A \csc(Bx + C)$**

We now graph the more general forms of the tangent, cotangent, secant, and cosecant functions following essentially the same process we developed for graphing $y = A \sin(Bx + C)$ and $y = A \cos(Bx + C)$. The process is not difficult if you have a clear understanding of the basic graphs for these functions, including periodic properties.

◆ Graphing $y = A \tan(Bx + C)$ and
  $y = A \cot(Bx + C)$

We repeat the basic graphs for $y = \tan x$ and $y = \cot x$, that we developed in Section 3.1.

GRAPH OF $y = \tan x$

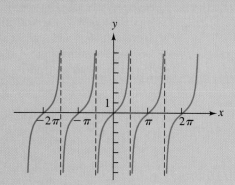

Domain: Set of all real numbers R except $\pi/2 + k\pi$, k an integer

Range: R

Period: π

GRAPH OF $y = \cot x$

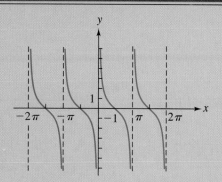

Domain: Set of all real numbers R except $k\pi$, k an integer

Range: R

Period: π

   Given $y = A \tan(Bx + C)$ or $y = A \cot(Bx + C)$, we now investigate how the constants A, B, and C affect the basic graphs of $y = \tan x$ and $y = \cot x$, respectively.

   We first note that **amplitude is not defined for the tangent and cotangent functions,** since the deviation from the x axis for both functions is indefinitely far in both directions. The effect of A is to make the <u>graph steeper if $|A| > 1$ or</u> <u>to make the curve less steep if $|A| < 1$. If A is negative, the graph is reflected</u> <u>across the x axis (turned upside down).</u>

Just as with the sine and cosine functions, the constants B and C involve a period change and a phase shift. Since both $A \tan x$ and $A \cot x$ have a period of π, it follows that both $A \tan(Bx + C)$ and $A \cot(Bx + C)$ complete one cycle as $Bx + C$ varies from

$$Bx + C = 0 \qquad \text{to} \qquad Bx + C = \pi$$

or, solving for x, as x varies from

$$x = -\frac{C}{B} \qquad \text{to} \qquad x = -\frac{C}{B} + \frac{\pi}{B}$$

Phase shift ———┐   Period

Thus, $y = A \tan(Bx + C)$ and $y = A \cot(Bx + C)$ each have a period of π/B, and their <u>graphs are translated horizontally $-C/B$ units to the right if $-C/B$</u> > 0 and $|-C/B|$ units to the left if $-C/B < 0$. As before, the horizontal translation determined by the number $-C/B$ is referred to as the phase shift.

Again, you do not need to memorize the formulas for period and phase shift. You need only remember the process used above.

◆ EXAMPLE 1   Graphing $y = A \tan Bx$

Find the period and phase shift for $y = 3 \tan(\pi x/2)$. Then sketch its graph for $-3 < x < 3$.

SOLUTION   One cycle of $y = 3 \tan(\pi x/2)$ is completed as $\pi x/2$ varies from 0 to π. Solve for x.

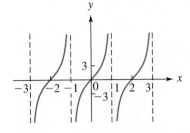

$$\frac{\pi x}{2} = 0 \qquad \frac{\pi x}{2} = \pi$$

$$x = 0 \qquad x = 0 + 2$$

$$\text{Period} = 2 \qquad \text{Phase shift} = 0$$

[*Note:* In general, if $C = 0$, then there is no phase shift.] Graph the equation for one period, and extend this graph over the interval from -3 to 3 (Figure 1).   ◆

FIGURE 1

MATCHED PROBLEM 1   Find the period and phase shift for $y = 2 \cot(x/2)$. Then sketch its graph for $-2\pi < x < 2\pi$.

◆ EXAMPLE 2   Graphing $y = A \tan (Bx + C)$

Find the period and phase shift for

$$y = \tan\left(\frac{\pi}{2}x + \frac{\pi}{4}\right)$$

Then sketch the graph for $-1.5 < x < 2.5$.

SOLUTION   ***Step 1***   Find the period and phase shift by solving $Bx + C = 0$ and $Bx + C = \pi$ for x.

$$\frac{\pi}{2}x + \frac{\pi}{4} = 0 \qquad\qquad \frac{\pi}{2}x + \frac{\pi}{4} = \pi$$

$$\frac{\pi}{2}x = -\frac{\pi}{4} \qquad\qquad \frac{\pi}{2}x = -\frac{\pi}{4} + \pi$$

$$x = -\frac{1}{2} \qquad\qquad x = -\frac{1}{2} + 2$$

$$\text{Period} = 2 \qquad\qquad \text{Phase shift} = -\frac{1}{2}$$

Step 2   Sketch one period of the graph starting at $x = -\frac{1}{2}$ (the phase shift) and ending at $x = (-\frac{1}{2}) + 2 = \frac{3}{2}$ (the phase shift plus one period) (Figure 2). Note that a vertical asymptote is at $x = \frac{1}{2}$.

FIGURE 2

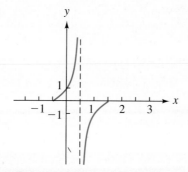

Step 3   Extend the graph from -1.5 to 2.5 (Figure 3).

FIGURE 3

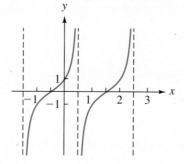

MATCHED PROBLEM 2   Find the period and phase shift for

$$y = \cot\left(2x + \frac{\pi}{2}\right)$$

Then sketch the graph for $-\pi/2 \le x \le \pi$.

◆   Graphing $y = A \sec(Bx + C)$ and
     $y = A \csc(Bx + C)$

For convenient reference, we repeat the basic graphs for $y = \sec x$ and $y = \csc x$ that we developed in Section 3.1.

GRAPH OF $y = \csc x$

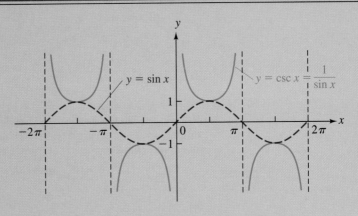

Domain: All real numbers x, except $x = k\pi$, k an integer

Range: All real numbers y such that $y \le -1$ or $y \ge 1$

Period: 2π

GRAPH OF $y = \sec x$

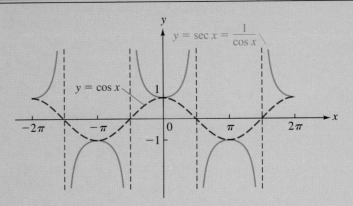

Domain: All real numbers x, except $x = \pi/2 + k\pi$, k an integer

Range: All real numbers y such that $y \le -1$ or $y \ge 1$

Period: 2π

As with the tangent and cotangent functions, **amplitude is not defined for either the secant or the cosecant** functions. Since the period of each function is the same as that for the sine and cosine, 2π, we find the period and phase shift by solving $Bx + C = 0$ and $Bx + C = 2\pi$. To graph either $y = A \sec(Bx + C)$ or $y = A \csc(Bx + C)$, you may find it easier to graph

$$y = \frac{1}{A}\cos(Bx + C) \qquad \text{or} \qquad y = \frac{1}{A}\sin(Bx + C)$$

with a dashed curve and then take reciprocals. An example should make the process clear.

◆ **EXAMPLE 3** Graphing $y = A \csc(Bx + C)$

Find the period and phase shift for

$$y = 2 \csc\left(\frac{\pi}{2}x - \pi\right)$$

Then sketch the graph for $-2 < x < 10$.

SOLUTION **Step 1** Find the period and phase shift by solving $Bx + C = 0$ and $Bx + C = 2\pi$ for x.

$$\frac{\pi}{2}x - \pi = 0 \qquad\qquad \frac{\pi}{2}x - \pi = 2\pi$$

$$\frac{\pi}{2}x = \pi \qquad\qquad \frac{\pi}{2}x = \pi + 2\pi$$

$$x = 2 \qquad\qquad\qquad x = 2 + 4$$

$$\text{Period} = 4 \qquad \text{Phase shift} = 2$$

Step 2 Since

$$2 \csc\left(\frac{\pi}{2}x - \pi\right) = \frac{1}{\dfrac{1}{2}\sin\left(\dfrac{\pi}{2}x - \pi\right)}$$

we graph

$$y = \frac{1}{2}\sin\left(\frac{\pi}{2}x - \pi\right)$$

for one cycle from 2 to $2 + 4 = 6$ with a dashed curve, and then take reciprocals. From our work in Section 3.3, we know that this graph is the graph of

$$y = \frac{1}{2}\sin\frac{\pi}{2}x$$

shifted to the right 2 units, as shown in Figure 4. Notice that we also place vertical asymptotes through the x intercepts of the sine graph to guide us when we sketch the cosecant function.

FIGURE 4

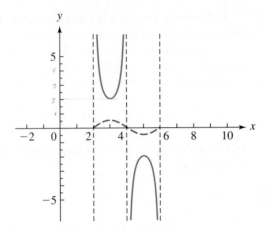

Step 3 Extend the one cycle from step 2 over the required interval from -2 to 10 (see Figure 5).

FIGURE 5

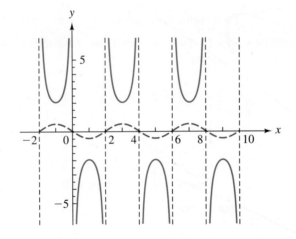

MATCHED PROBLEM 3   Find the period and phase shift for $y = \frac{1}{2} \sec(2x + \pi)$. Then sketch the graph for $-3\pi/4 < x < 3\pi/4$.

Answers to
Matched Problems

1. Period $= 2\pi$
     Phase shift $= 0$

2. Period $= \pi/2$
     Phase shift $= -\pi/4$

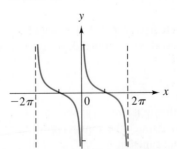

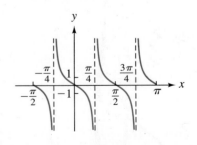

3. Period $= \pi$

   Phase shift $= -\pi/2$

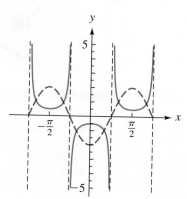

EXERCISE 3.6

A *Sketch a graph of each of the following without looking at the text or using a calculator.*

1. $y = \tan x, \quad 0 \le x \le 2\pi$

2. $y = \cot x, \quad 0 < x < 2\pi$

3. $y = \csc x, \quad -\pi < x < \pi$

4. $y = \sec x, \quad -\pi \le x \le \pi$

B *Indicate the period of each function, and graph the function over the indicated interval.*

5. $y = 3 \tan 2x, \quad -\pi \le x \le \pi$

6. $y = 2 \cot 4x, \quad 0 < x < \pi/2$

7. $y = -\frac{1}{2} \cot 2\pi x, \quad 0 < x < 1$

8. $y = -\frac{1}{4} \tan 8\pi x, \quad 0 \le x \le \frac{1}{2}$

9. $y = \sec \pi x, \quad -1.5 < x < 3.5$

10. $y = \csc(x/2), \quad -3\pi \le x \le 3\pi$

11. $y = \frac{1}{2} \tan(x/2), \quad -\pi < x < 3\pi$

12. $y = \frac{1}{2} \cot(x/2), \quad 0 < x < 4\pi$

13. $y = 2 \csc(x/2), \quad 0 < x < 8\pi$

14. $y = 2 \sec \pi x, \quad -1 < x < 3$

Indicate the period and phase shift, and graph each function.

15. $y = \tan\left(x - \frac{\pi}{2}\right), \quad -\pi < x < \pi$

16. $y = \cot\left(x + \frac{\pi}{2}\right), \quad -\frac{\pi}{2} < x < \frac{3\pi}{2}$

17. $y = \cot(2x - \pi), \quad -\frac{\pi}{2} < x < \frac{\pi}{2}$

18. $y = \tan(2x + \pi), \quad -\frac{3\pi}{4} < x < \frac{3\pi}{4}$

19. $y = \csc\left(\pi x - \frac{\pi}{2}\right), \quad -\frac{1}{2} < x < \frac{5}{2}$

20. $y = \sec\left(\pi x + \frac{\pi}{2}\right), \quad -1 < x < 1$

C *Indicate the period and phase shift, and graph each function.*

21. $y = 4 \tan(2x + \pi), \quad -\pi \le x \le \pi$

22. $y = 2 \tan\left(\frac{x}{4} - \frac{\pi}{4}\right), \quad -4\pi \le x \le 4\pi$

23. $y = -3 \cot(\pi x - \pi), \quad -2 < x < 2$

24. $y = -2 \tan\left(\frac{\pi}{4}x - \frac{\pi}{4}\right), \quad -1 < x < 7$

25. $y = 2 \sec\left(\pi x - \frac{\pi}{2}\right), \quad -1 < x < 3$

26. $y = 3 \csc\left(\frac{\pi}{2}x + \frac{\pi}{2}\right), \quad -1 < x < 3$

Problems 27–34 require the use of a graphing calculator. Graph the given equation and find an equation of the form $y = A \tan Bx$, $y = A \cot Bx$, $y = A \sec Bx$, or $y = A \csc Bx$ that has the same graph. These problems suggest additional identities beyond the fundamental ones we discussed in Section 2.6; we will discuss additional important identities in detail in Chapter 4.

27. $y = \csc x - \cot x$

28. $y = \csc x + \cot x$

29. $y = \cot x + \tan x$

30. $y = \cot x - \tan x$

31. $y = \cos 2x + \sin 2x \tan 2x$

32. $y = \sin 3x + \cos 3x \cot 3x$

33. $y = \dfrac{\sin 6x}{1 - \cos 6x}$ **34.** $y = \dfrac{\sin 4x}{1 + \cos 4x}$

 Applications

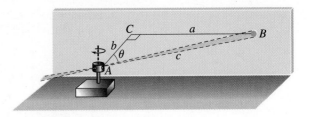

Figure for 35

*35. **Precalculus**   A beacon light 15 ft from a wall rotates clockwise at the rate of exactly 1 rps (see the figure); thus, $\theta = 2\pi t$.
   (A) If we start counting time in seconds when the light spot is at C, write an equation for the distance a the light spot travels along the wall in terms of time t.
   (B) Graph the equation found in part (A) for the time interval $0 \le t < 0.25$.

*36. **Precalculus**   Refer to Problem 35.
   (A) Write an equation for the length of the light beam c in terms of t.
   (B) Graph the equation found in part (A) for the time interval $0 \le t < 0.25$.

CHAPTER 3 SUMMARY

3.1
BASIC GRAPHS

FIGURE 1
Graph of $y = \sin x$

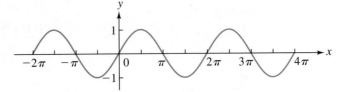

Domain: R
Range: $-1 \le y \le 1$
Period: 2π

FIGURE 2
Graph of $y = \cos x$

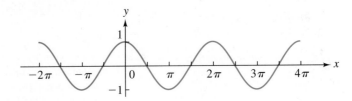

Domain: R
Range: $-1 \le y \le 1$
Period: 2π

FIGURE 3
Graph of $y = \tan x$

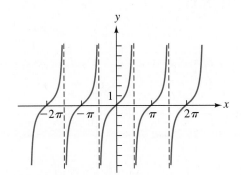

Domain: Set of all real numbers R except $\pi/2 + k\pi$, k an integer
Range: R
Period: π

FIGURE 4
Graph of $y = \cot x$

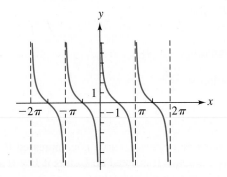

Domain: Set of all real numbers R except $k\pi$, k an integer
Range: R
Period: π

FIGURE 5
Graph of $y = \csc x$

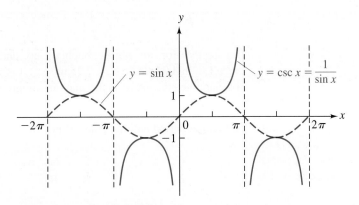

Domain: All real numbers x, except $x = k\pi$, k an integer
Range: All real numbers y such that $y \le -1$ or $y \ge 1$
Period: 2π

FIGURE 6
Graph of $y = \sec x$

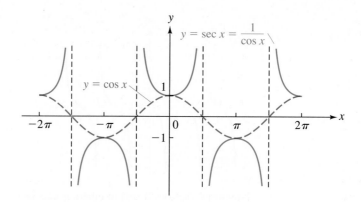

Domain: All real numbers x, except $x = \pi/2 + k\pi$, k an integer
Range: All real numbers y such that $y \le -1$ or $y \ge 1$
Period: 2π

3.2

GRAPHING $y = k + A \sin Bx$
and $y = k + A \cos Bx$

Amplitude $= |A|$

Period $= \dfrac{2\pi}{B}$

Vertical translation $= k$

The basic sine or cosine curve is compressed if $B > 1$ and stretched if $0 < B < 1$. The graph is translated up k units if $k > 0$ and down $|k|$ units if $k < 0$.
    If P is the period and f the **frequency** of a periodic phenomenon, then

$$P = \frac{1}{f} \qquad \textit{Period is the time for one complete cycle}$$

$$f = \frac{1}{P} \qquad \textit{Frequency is the number of cycles per unit of time}$$

The period for an object bobbing up and down in water is given by $2\pi/\sqrt{1{,}000gA/M}$ where $g = 9.75$ m/sec², A is the horizontal cross-sectional area in square meters, and M is the mass in kilograms.

3.3
GRAPHING
$y = k + A \sin(Bx + C)$ and
$y = k + A \cos(Bx + C)$

To find the period and the phase shift, solve $Bx + C = 0$ and $Bx + C = 2\pi$.

$$x = -\frac{C}{B} \qquad\qquad x = -\frac{C}{B} + \frac{2\pi}{B}$$

$$\underset{\textit{Phase Shift}}{\underbrace{\qquad\qquad\qquad\qquad}} \quad \underset{\textit{Period}}{\uparrow}$$

The graph completes one full cycle as x varies from $-C/B$ to $(-C/B) + (2\pi/B)$.

*3.4
ADDITIONAL APPLICATIONS

Modeling Electric Current

Alternating current flows are represented by equations of the form $I = A \sin(Bt + C)$ where I is the current in amperes and t is time in seconds.

Modeling Water Waves

Water waves can be represented by equations of the form $y = A \sin 2\pi(ft - r/\lambda)$ where f is frequency, t is time, r is distance from the source, and λ is the wavelength.

Modeling Light and Other Electromagnetic Waves

The wavelength λ and the frequency v of an electromagnetic wave are related by $\lambda v = c$, where c is the speed of light. These waves can be represented by an equation of the form $E = A \sin 2\pi(vt - r/\lambda)$ where t is time and r is distance from the source.

Spring-Mass Systems

An object hanging on a spring that is set in motion is said to exhibit **simple harmonic motion** if its amplitude remains constant; **damped harmonic motion** if its amplitude decreases to 0 as time increases; and **resonance** if its amplitude increases as time increases.

3.5
ADDITION OF ORDINATES

Combinations of trigonometric and other functions often can be graphed by first graphing separate components and then adding the ordinates. This technique can be applied to sound waves where separate components are called **partial tones**. The partial tone with the smallest frequency is called the **fundamental tone** and the other partial tones are called **overtones**. **Fourier series** use sums of trigonometric functions to approximate periodic wave forms such as a **square wave** or a **sawtooth wave**.

3.6
TANGENT, COTANGENT,
SECANT, AND COSECANT
FUNCTIONS REVISITED

Amplitude is not defined for the tangent, cotangent, secant, or cosecant functions.

To find the period and the phase shift for the graph of $y = A \tan(Bx + C)$ or $y = A \cot(Bx + C)$, solve $Bx + C = 0$ and $Bx + C = \pi$.

$$x = -\frac{C}{B} \qquad\qquad x = -\frac{C}{B} + \frac{\pi}{B}$$

$$\underbrace{\qquad\text{Phase Shift}\qquad}\quad \underset{\text{Period}}{\uparrow}$$

The graph completes one full cycle as x varies from $-C/B$ to $(-C/B) + (\pi/B)$.

Find the period and the phase shift for the graph of $y = A \sec(Bx + C)$ or $y = A \csc(Bx + C)$ by solving $Bx + C = 0$ and $Bx + C = 2\pi$. It is helpful to first graph

$$y = \frac{1}{A} \cos(Bx + C) \qquad \text{or} \qquad y = \frac{1}{A} \sin(Bx + C)$$

with a dashed curve and then take reciprocals.

CHAPTER 3 REVIEW EXERCISE

Work through all the problems in this chapter review and check the answers. Answers to all review problems appear in the back of the book; following each answer is an italic number that indicates the section in which that type of problem is discussed. Where weaknesses show up, review the appropriate sections in the text. Review problems flagged with a star (☆) are from optional sections.

A Sketch a graph of each function for $-2\pi \le x \le 2\pi$.

1. $y = \sin x$ **2.** $y = \cos x$ **3.** $y = \tan x$

4. $y = \cot x$ **5.** $y = \sec x$ **6.** $y = \csc x$

In Problems 7–9 sketch a graph of each function for the indicated interval.

7. $y = 3 \cos \dfrac{x}{2}, \quad -4\pi \le x \le 4\pi$

8. $y = \frac{1}{2} \sin 2x, \quad -\pi \le x \le \pi$

9. $y = 4 + \cos x, \quad 0 \le x \le 2\pi$

B In Problems 10–26 sketch a graph of each function for the indicated interval.

10. $y = -2 \sin \pi x, \quad -2 \le x \le 2$

11. $y = -\frac{1}{3} \cos 2\pi x, \quad -2 \le x \le 2$

12. $y = -1 + \frac{1}{2} \sin 2x, \quad -\pi \le x \le \pi$

13. $y = 3 - 2 \cos \dfrac{\pi}{2} x, \quad -4 \le x \le 4$

14. $y = \sin\!\left(x - \dfrac{\pi}{2}\right), \quad 0 \le x \le 2\pi$

15. $y = \cos(x + \pi), \quad 0 \le x \le 2\pi$

16. $y = -2 \sin(\pi x - \pi), \quad 0 \le x \le 2$

17. $y = -\frac{1}{4} \cos(2x + \pi), \quad 0 \le x \le 2\pi$

18. $y = 4 - 2 \sin(\pi x - \pi), \quad 0 \le x \le 2$

19. $y = \tan 2x, \quad -\pi \le x \le \pi$

20. $y = \cot \pi x, \quad -2 < x < 2$

21. $y = 3 \csc \pi x, \quad -1 < x < 2$

22. $y = 2 \sec \dfrac{x}{2}, \quad -\pi < x < 3\pi$

23. $y = \tan\!\left(x + \dfrac{\pi}{2}\right), \quad -\pi < x < \pi$

24. $y = \cot\!\left(x - \dfrac{\pi}{2}\right), \quad -\dfrac{\pi}{2} < x < \dfrac{3\pi}{2}$

25. $y = x + \sin \pi x, \quad 0 \le x \le 2$

26. $y = 2 \sin x + \cos 2x, \quad 0 \le x \le 4\pi$

27. What are the period and amplitude of the function in Problem 11?

28. What is the period of the function in Problem 20?

29. What is the phase shift for the function in Problem 14?

30. Find the amplitude, period, and phase shift for

$$y = -3 \cos(\pi x + \pi)$$

31. Find the period and phase shift for

$$y = -2 \tan\!\left(\dfrac{\pi}{2} x + \dfrac{\pi}{2}\right)$$

32. Find the period and phase shift for

$$y = 2 \sec(2x - \pi)$$

33. Find the equation of the form $y = A \sin Bx$ whose graph is

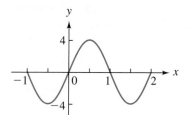

34. Find the equation of the form $y = A \cos Bx$ whose graph is

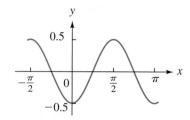

C In Problems 35 and 36 sketch a graph of each function for the indicated interval.

35. $y = 2 \tan\!\left(\pi x + \dfrac{\pi}{2}\right), \quad -1 < x < 1$

36. $y = 2 \sec(2x - \pi), \quad 0 \le x < \dfrac{5\pi}{4}$

37. If the following graph is a graph of an equation of the form $y = A \sin(Bx + C)$, $0 < -C/B < 1$, find the equation.

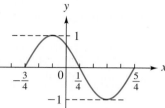

 Problems 38–41 require the use of a graphing calculator.

38. Graph $y = 1/(1 + \tan^2 x)$ in a viewing rectangle that displays at least two full periods of the graph. Find an equation of the form $y = k + A \sin Bx$ or $y = k + A \cos Bx$ that has the same graph.

39. Graph $y = 1.2 \sin 2x + 1.6 \cos 2x$ and approximate the x intercept closest to the origin, correct to three decimal places. Use this intercept to find an equation of the form $y = A \sin (Bx + C)$ that has the same graph as the given equation.

40. The sum of the first five terms of the Fourier series for a wave form are

$$y = -\frac{\pi}{4} + \frac{4}{\pi} \cos \frac{x}{2} + \frac{2}{\pi} \cos x$$
$$+ \frac{4}{9\pi} \cos \frac{3x}{2} + \frac{4}{25\pi} \cos \frac{5x}{2}$$

Graph this equation for $-4\pi \le x \le 4\pi$ and $-3 \le y \le 3$, and sketch by hand the corresponding wave form. (Assume that the graph of the wave form is comprised entirely of straight line segments.)

41. Graph each of the following equations and find an equation of the form $y = A \tan Bx$, $y = A \cot Bx$, $y = A \sec Bx$, or $y = A \csc Bx$ that has the same graph as the given equation.

(A) $y = \dfrac{2 \sin x}{\sin 2x}$ (B) $y = \dfrac{2 \cos x}{\sin 2x}$

(C) $y = \dfrac{2 \cos^2 x}{\sin 2x}$ (D) $y = \dfrac{2 \sin^2 x}{\sin 2x}$

 Applications

42. Spring-Mass System If the motion of a weight hung on a spring has an amplitude of 4 cm and a frequency of

8 Hz, and if its position when $t = 0$ sec is 4 cm below its position at rest (above the rest position is positive and below is negative), find an equation of the form $y = A \cos Bt$ that describes the motion at any time t. (Neglect any damping forces such as air resistance and friction.)

43. Pollution In a large city the amount of sulfur dioxide pollutant released into the atmosphere due to the burning of coal and oil for heating purposes varies seasonally. If measurements over a 2 year period produced the following graph, find an equation of the form $P = K + A \cos Bn$, $0 \le n \le 104$, where P is the number of tons of pollutants released into the atmosphere during the nth week after January 31.

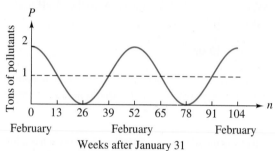

Figure for 43

44. Floating Objects A cylindrical buoy with diameter 1.2 m is observed (after being pushed down) to bob up and down with a period of 0.8 sec and an amplitude of 0.6 m.
(A) Find the mass of the buoy to the nearest kilogram.
(B) Write an equation of motion for the buoy in the form $y = D \sin Bt$.
(C) Graph the equation in part (B) for $0 \le t \le 1.6$.

45. Sound Waves The motion of one tip of a tuning fork is given by an equation of the form $y = 0.05 \cos Bt$.
(A) If the frequency of the fork is 280 Hz, what is the period? What is the value of B?
(B) If the period of the motion of the fork is 0.0025 sec, what is the frequency? What is the value of B?
(C) If $B = 700\pi$, what is the period? What is the frequency?

46. Electrical Circuits If the voltage E in an electrical circuit has amplitude 18 and frequency 30 Hz, and $E = 18$ V when $t = 0$ sec, find an equation of the form $y = A \cos Bt$ that gives the voltage at any time t.

47. Water Waves At a particular point in the ocean, the vertical change in the water due to wave action is given by

$$y = 6 \cos \frac{\pi}{10}(t - 5)$$

where y is in meters and t is time in seconds. Find the amplitude, period, and phase shift. Graph the equation for $0 \le t \le 80$.

48. Electrical Circuits   The voltage E and current I in a circuit are given by

$$E = 20 \sin 100\pi t \quad \text{and} \quad I = 15 \sin\left(100\pi t - \frac{\pi}{2}\right)$$

where t is time in seconds. Find the amplitude and period for both equations and the phase shift for I. Graph both equations on the same set of axes for $0 \le t \le \frac{3}{50}$.

☆ **49. Water Waves**   A water wave at a fixed position has an equation of the form

$$y = 12 \sin \frac{\pi}{3} t$$

where t is time in seconds and y is in feet. How high is the wave from trough to crest? What is its wavelength in feet? How fast is it traveling in feet per second? Compute answers to the nearest foot.

☆ **50. Electromagnetic Waves**   An ultraviolet wave has a frequency of $v = 10^{15}$ Hz. What is its period? What is its wavelength in meters? (Recall the speed of light $c \approx 3 \times 10^8$ m/sec.)

☆ **51. Spring-Mass Systems**   Graph each of the following equations for $0 \le t \le 2$ and identify each as an example of simple harmonic motion, damped harmonic motion, or resonance.
   (A) $y = \sin 2\pi t$ (B) $y = (1 + t) \sin 2\pi t$
   (C) $y = \dfrac{1}{1 + t} \sin 2\pi t$

52. Modeling a Seasonal Business Cycle   The sales for a national chain of ice cream shops are growing steadily but are subject to seasonal variations. The following mathematical model has been developed to project the monthly sales for the next two years:

$$S = 5 + t + 5 \sin \frac{\pi t}{6}$$

where S represents the sales in millions of dollars and t is time in months. Graph this equation for $0 \le t \le 24$.

53. Rocket Flight   A camera recording the launch of a rocket is located 1,000 m from the launching pad (see the figure).
   (A) Write an equation for the rocket's altitude h in terms of the camera's angle of elevation θ.
   (B) Graph this equation for $0 \le \theta < \pi/2$.

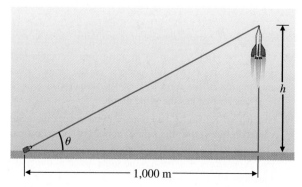

Figure for 53

C ***54. Modeling Sunset Times**   Sunset times for the fifth of each month over a one year period were taken from a tide booklet for San Francisco Bay to form Table 1. Daylight savings time was ignored.
   (A) Convert the data in Table 1 from hours and minutes to two-place decimal hours. Enter this data for a two year period in your graphing calculator and produce a scatter plot in the following viewing rectangle: $1 \le x \le 24$, $16 \le y \le 20$.
   (B) A function of the form $y = k + A \sin(Bx + C)$ can be used to model this data. Use the converted data from Table 1 to determine k and A (to two decimal places), and B (exact value). Use the graph in part A to visually estimate C (to one decimal place).
   (C) Plot the data from part (A) and the equation from part (B) in the same viewing rectangle. If necessary, adjust your value of C to produce a better fit.

TABLE 1

x (months)	1	2	3	4	5	6	7	8	9	10	11	12
y (sunset)	17:05	17:38	18:07	18:36	19:04	19:29	19:35	19:15	18:34	17:47	17:07	16:51

Work through all the problems in this cumulative review and check the answers. Answers to all review problems are in the back of the book; following each answer is an italic number that indicates the section in which that type of problem is discussed. Where weaknesses show up, review the appropriate sections in the text. Review problems flagged with a star (☆) are from optional sections.

A **1.** An arc of $\frac{1}{4}$ the circumference of a circle subtends a central angle of how many degrees? Of how many radians?

2. Change 21°47′ to decimal degrees to two decimal places.

3. Find the degree measure of 1.67 rad.

4. Find the radian measure of −715.3°.

5. The hypotenuse of a right triangle is 34 in. and one of the acute angles is 25°. Find the other acute angle and the other two sides.

6. Find the value of sin θ and tan θ if the terminal side of θ contains $P(8, -15)$.

7. Evaluate to four significant digits using a calculator.
(A) sin 23°12′   (B) sec 145.6°   (C) cot 0.88

8. Sketch the reference triangle and find the reference angle α for
(A) $\theta = \dfrac{11\pi}{6}$ (B) $\theta = -225°$

9. Sketch a graph of each function for $-2\pi \le x \le 2\pi$.
(A) $y = \sin x$ (B) $y = \tan x$ (C) $y = \sec x$

B **10.** Find θ to the nearest 10′ if tan $\theta = 0.9465$ and $-90° \le \theta \le 90°$.

11. If the second hand of a clock is 5.00 cm long, how far does the tip of the hand travel in 40 sec?

☆**12.** How fast (in cm/min) is the tip of the second hand in Problem 11 moving?

13. Find θ to the nearest second if $\theta = \sin^{-1} 0.4621$.

14. Find x in the following figure.

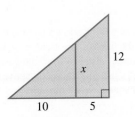

15. Convert 48° to radian measure in terms of π.

16. If the radius of a circle is 2.0 ft, find the length of an arc intercepted by a central angle of 145°.

17. Find reference angles corresponding to
(A) 237.6°   (B) −514°30′   (C) 9.00   (D) $-\dfrac{8\pi}{3}$

18. Find the exact value of each of the following without using a calculator.
(A) $\sin \dfrac{5\pi}{4}$ (B) $\cos \dfrac{7\pi}{6}$ (C) $\tan \dfrac{-5\pi}{3}$ (D) $\csc 3\pi$

19. Find the exact value of each of the other five trigonometric functions if $\cos \theta = -\frac{2}{3}$ and $\tan \theta < 0$.

20. The central angle of a circular sector subtends an arc of 7 cm in a circle with circumference 24 cm. Find the area of the sector (to the nearest tenth).

In Problems 21–26 sketch a graph of each function for the indicated interval. State the period and, if applicable, the amplitude and phase shift for each function.

21. $y = 1 - \dfrac{1}{2}\cos 2x, \quad -2\pi \le x \le 2\pi$

22. $y = 2 \sin\left(x - \dfrac{\pi}{4}\right), \quad -\pi \le x \le 3\pi$

23. $y = 5 \tan 4x, \quad 0 \le x \le \pi$

24. $y = \csc \dfrac{x}{2}, \quad -4\pi < x < 4\pi$

25. $y = 2 \sec \pi x, \quad -2 \le x \le 2$

26. $y = \cot\left(\pi x + \dfrac{\pi}{2}\right), \quad -1 \le x \le 3$

27. Sketch the graph of $y = \sin x + \sin 2x$ for $0 \le x \le 2\pi$.

28. Find the equation of the form $y = A \sin Bx$ whose graph is

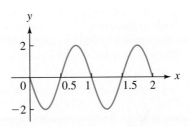

29. Simplify: $(\tan x)(\sin x) + \cos x$

30. Find the exact value of all angles between 0 and 360° for which $\sin \theta = -\frac{1}{2}$.

31. If the sides of a right triangle are 23.5 in. and 37.3 in., find the hypotenuse and find the acute angles to the nearest 0.1°.

32. Find the angles in Problem 27 to the nearest 10′.

C   33. A point on the unit circle with center at the origin starts at (1, 0) and moves counterclockwise around the circle until it travels a distance of 2.3 units on the circle. What are the coordinates of the point in its terminal position? Compute the answer to three decimal places.

34. A circle with its center at the origin in a rectangular coordinate system passes through the point (8, 15). What is the length of the arc on the circle in the first quadrant between the positive horizontal axis and the point (8, 15)? Compute the answer to two decimal places.

35. If θ is a first quadrant angle and $\tan \theta = a$, express the other five trigonometric functions of θ in terms of a.

36. Find the equation of the form $y = A \sin(Bx + C)$, $0 < -C/B < 1$, whose graph is

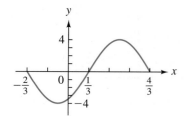

Problems 37–40 require the use of a graphing calculator.

37. Graph $y = \tan^2 x/(1 + \tan^2 x)$ in a viewing rectangle that displays at least two full periods of the graph. Find an equation of the form $y = k + A \sin Bx$ or $y = k + A \cos Bx$ that has the same graph as the given equation.

38. Graph $y = 2.4 \sin(x/2) - 1.8 \cos(x/2)$ and approximate the x intercept closest to the origin, correct to three decimal places. Use this intercept to find an equation of the form $y = A \sin (Bx + C)$ that has the same graph as the given equation.

39. The sum of the first four terms of the Fourier series for a wave form are

$$y = \frac{\pi}{2} - \frac{4}{\pi} \cos x - \frac{4}{9\pi} \cos 3x - \frac{4}{25\pi} \cos 5x$$

Graph this equation for $-2\pi \le x \le 2\pi$ and $-2 \le y \le 4$, and sketch by hand the corresponding wave form. (Assume that the graph of the wave form is comprised entirely of straight line segments.)

40. Graph each of the following equations and find an equation of the form $y = A \tan Bx$, $t = A \cot Bx$, $y = A \sec Bx$, or $y = A \csc Bx$ that has the same graph as the given equation.

(A) $y = \dfrac{\sin 2x}{1 + \cos 2x}$ (B) $y = \dfrac{2 \cos x}{1 + \cos 2x}$

(C) $y = \dfrac{2 \sin x}{1 - \cos 2x}$ (D) $y = \dfrac{\sin 2x}{1 - \cos 2x}$

Applications

∗41. Geography/Navigation   Find the distance (to the nearest mile) between Gary, IN with latitude 41°36′N and Pensacola, FL with latitude 30°25′N. (Both cities have approximately the same longitude.) Use $\pi \approx 3.14$ and $r \approx 3,960$ miles for the earth's radius.

42. Volume of a Cone   A paper drinking cup has the shape of a right circular cone with altitude 9 cm and radius 4 cm (see the figure). Find the volume of the water in the cup (to the nearest cm³) when the water is 6 cm deep. [*Recall:* $V = \frac{1}{3}\pi r^2 h$.]

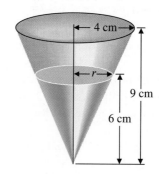

Figure for 42

43. Construction   The base of a 40 ft arm on a crane is 10 ft above the ground (see the figure). The crane is picking up an object whose horizontal distance from the base of the arm is 30 ft. If the tip of the arm is directly above the object, find the angle of elevation of the arm and the altitude of the tip.

Figure for 43

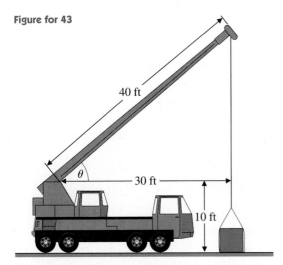

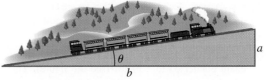

Angle of inclination: θ Grade: $\dfrac{a}{b}$

Figure for 45

46. Surveying   A 10 m tall house is located directly across the street from an office building (see the figure). The angle of elevation of the office building from the ground is 72° and from the top of the house is 68°. How tall is the office building? How wide is the street?

Figure for 46

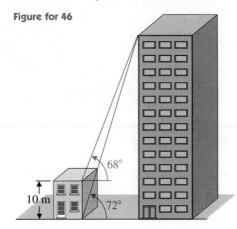

44. Billiards   Under ideal conditions, when a ball strikes a cushion on a pocket billiards table at an angle, it bounces off the cushion at the same angle (see the figure). Where should the ball in the center of the table strike the upper cushion so that it bounces off the cushion into the lower right corner pocket? Give the answer in terms of the distance from the center of the side pocket to the point of impact.

47. Forest Fires   Two fire towers are located 5 mi apart on a straight road. Observers at each tower spot a fire and report its location in terms of the angles in the figure.
(A) How far from Tower B is the point on the road closest to the fire?
(B) How close is the fire to the road?

Figure for 47

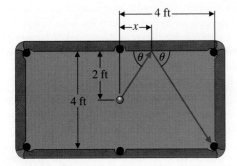

Figure for 44

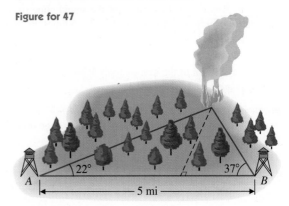

45. Railroad Grades   The amount of inclination of a railroad track is referred to as its grade and usually is given as a percentage (see the figure). Most major railroads limit grades to a maximum of 3%, although some logging and mining railroads have grades as high as 7%, requiring the use of specially geared locomotives. Find the angle of inclination in degrees (to the nearest 0.1°) for a 3% grade. Find the grade (to the nearest 1%) if the angle of inclination is 3°.

*48. **Geometry**   A circular sector has an area of 15.43 ft^2 and a radius of 4.2 ft. Calculate the perimeter of the sector to the nearest foot.

☆49. **Engineering**   A large saw in a sawmill is driven by a chain connected to a motor (see the figure). The drive wheel on the motor is rotating at 300 rpm.
(A) What is the angular velocity of the saw (in rad/min)?
(B) What is the linear velocity of a point on the rim of the saw (in in./min)?

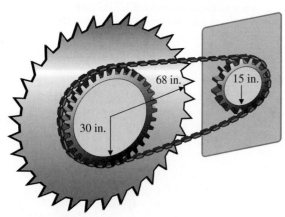

Figure for 49

☆50. **Light Waves**   A light ray passing through air strikes the surface of a pool of water so that the angle of incidence is $\alpha = 38.4°$. Find the angle of refraction β. (Water has a refractive index of 1.33 and air has a refractive index of 1.00.)

☆51. **Sonic Boom**   Find the angle of the sound cone (to the nearest degree) for a plane traveling at 1.5 times the speed of sound.

☆52. **Water Waves**   A wave is traveling at 25 ft/sec. What is its wavelength to the nearest foot? What is its period to the nearest tenth of a second?˙

Problems 53–57 refer to the figure. In Problems 53–55, express the indicated area in terms of x and/or trigonometric functions of x.

53. **Precalculus**   A1 = area of triangle *OAP*.
54. **Precalculus**   A2 = area of sector *OAP*.

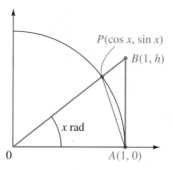

Figure for 53–57

55. **Precalculus**   A3 = area of triangle *OAB*.
56. **Precalculus**   Use the relationship A1 < A2 < A3 and the results of Problems 53–55 to show that

$$\cos x < \frac{\sin x}{x} < 1, \qquad x > 0$$

c 57. **Precalculus**   The result in Problem 56 is used in calculus to show that $(\sin x)/x$ approaches 1 as x approaches 0. To illustrate this behavior, graph $y1 = \cos x$, $y2 = (\sin x)/x$, and $y3 = 1$ in the same viewing rectangle: $-1 \leq x \leq 1$, $0.8 \leq y \leq 1.2$.

58. **Spring-Mass System**   The equation $y = -5 \cos 10t$, where t is time in seconds, represents the motion of a weight hanging on a spring after it has been pulled 5 cm below its equilibrium point and released. What are the amplitude, period, and frequency of the function? Graph the function for $0 \leq t \leq \pi$.

59. **Electrical Circuits**   The current I (in amperes) in an electrical circuit is given by $I = 12 \sin(60\pi t - \pi)$ where t is time in seconds. State the amplitude, period, frequency, and phase shift. Graph the equation for $0 \leq t \leq 0.1$.

☆60. **Electromagnetic Waves**   An infrared wave has a wavelength of 6×10^{-5} m. What is its period? (Recall the speed of light: $c \approx 3 \times 10^8$ m/sec)

61. **Precalculus: Surveying**   A university has just constructed a new building at the intersection of two perpendicular streets (see the figure). Now they want to connect the existing streets with a walkway that passes behind and just touches the building.
(A) Express the length of the new walkway in terms of θ.
(B) Graph this equation for $0 < \theta < \pi/2$. (In calculus, this equation is used to find the shortest walkway.)

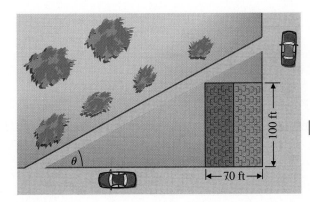

Figure for 61

*62. **Emergency Vehicles**   An emergency vehicle is parked 50 ft from a building (see the figure). A warning light on top of the vehicle is rotating at exactly 20 rpm.

Figure for 62

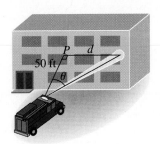

(A) Write an equation for the distance d in terms of the angle θ.

(B) If we start counting time t in minutes when the light is directed at P, write an equation for θ in terms of time t.

(C) Write an equation for d in terms of time t.

(D) Graph the equation in part (C) for $0 \le t < 0.0125$.

🄲 *63. **Modeling Temperature Variation**   The 30 year average monthly temperature, F°, for each month of the year for Milwaukee, WI, is given in Table 1 (*World Almanac*):

TABLE 1

x (months)	1	2	3	4	5	6	7	8	9	10	11	12
y (temperature)	19	23	32	45	55	65	71	69	62	51	37	25

(A) Enter the data in Table 1 for a two year period in your graphing calculator and produce a scatter plot in the viewing rectangle.

$$1 \le x \le 24, \qquad 15 \le y \le 75$$

(B) A function of the form $y = k + A \sin(Bx + C)$ can be used to model this data. Use the data in Table 1 to determine k, A, and B. Use the graph in part (A) to visually estimate C (to one decimal place).

(C) Plot the data from part (A) and the equation from part (B) in the same viewing rectangle. If necessary, adjust your value of C to produce a better fit.

IDENTITIES

☆ Sections marked with a star may be omitted without loss of continuity.

rigonometric functions have many uses. In addition to solving real-world problems, they are used in the development of mathematics—analytic geometry, calculus, and so on. Whatever their use, it is often of value to be able to change a trigonometric expression from one form to an equivalent form. This involves the use of identities. An equation in one or more variables is said to be an **identity** if the left side is equal to the right side for all replacements of the variables for which both sides are defined. The equation

$$x^2 - x - 6 = (x - 3)(x + 2)$$

is an identity, while

$$x^2 - x - 6 = 2x$$

is not. The latter is called a **conditional equation,** since it holds only for certain values of x and not for all values for which both sides are defined.

In this chapter we will develop a number of very useful classes of trigonometric identities, and you will get practice in using these identities to convert a variety of trigonometric expressions into equivalent forms. You will also get practice in using the identities directly to solve several other types of problems.

4.1 FUNDAMENTAL IDENTITIES AND THEIR USE

♦ **Fundamental Identities**
♦ **Evaluating Trigonometric Functions**
♦ **Converting to Equivalent Forms**

♦ Fundamental Identities

Our first encounter with trigonometric identities was in Section 2.6, where we established 11 fundamental forms. We restate and name these 11 identities in the box at the top of the next page for convenient reference. These fundamental identities will be used very frequently in the work that follows.

The second and third Pythagorean identities were established in Problems 77 and 78 in Exercise 2.6. An easy way to remember them is to note that the second can be obtained from the first by dividing both sides of the first by $\cos^2 x$, and the third can be obtained from the first by dividing both sides of the first by $\sin^2 x$, as follows:

$$\sin^2 x + \cos^2 x = 1 \qquad\qquad \sin^2 x + \cos^2 x = 1$$

$$\frac{\sin^2 x}{\cos^2 x} + \frac{\cos^2 x}{\cos^2 x} = \frac{1}{\cos^2 x} \qquad\qquad \frac{\sin^2 x}{\sin^2 x} + \frac{\cos^2 x}{\sin^2 x} = \frac{1}{\sin^2 x}$$

$$\tan^2 x + 1 = \sec^2 x \qquad\qquad 1 + \cot^2 x = \csc^2 x$$

FUNDAMENTAL TRIGONOMETRIC IDENTITIES

For x any real number or angle in degree or radian measure for which both sides are defined:

Reciprocal identities

$$\csc x = \frac{1}{\sin x} \qquad \sec x = \frac{1}{\cos x} \qquad \cot x = \frac{1}{\tan x}$$

Quotient identities

$$\tan x = \frac{\sin x}{\cos x} \qquad \cot x = \frac{\cos x}{\sin x}$$

Identities for negatives

$$\sin(-x) = -\sin x \qquad \cos(-x) = \cos x \qquad \tan(-x) = -\tan x$$

Pythagorean identities

$$\sin^2 x + \cos^2 x = 1 \qquad \tan^2 x + 1 = \sec^2 x \qquad 1 + \cot^2 x = \csc^2 x$$

◆ Evaluating Trigonometric Functions

Suppose we know that $\cos x = -\frac{4}{5}$ and $\tan x = \frac{3}{4}$. How can we find the exact value of the remaining trigonometric functions of x without finding x and without using reference triangles? We use fundamental identities.

◆ **EXAMPLE 1**   Using Fundamental Identities

If $\cos x = -\frac{4}{5}$ and $\tan x = \frac{3}{4}$, use the fundamental identities to find the exact values of the remaining four trigonometric functions at x.

SOLUTION   *Find sec x.*

$$\sec x = \frac{1}{\cos x} = \frac{1}{-\frac{4}{5}} = -\frac{5}{4}$$

Find cot x.

$$\cot x = \frac{1}{\tan x} = \frac{1}{\frac{3}{4}} = \frac{4}{3}$$

Find sin x: We can start with either a Pythagorean identity or a quotient identity. We choose the quotient identity $\tan x = (\sin x)/(\cos x)$ changed to the form

$$\sin x = (\cos x)(\tan x) = \left(-\frac{4}{5}\right)\left(\frac{3}{4}\right) = -\frac{3}{5}$$

Find csc x.

$$\csc x = \frac{1}{\sin x} = \frac{1}{-\frac{3}{5}} = -\frac{5}{3}$$

◆

MATCHED PROBLEM 1 If $\sin x = -\frac{4}{5}$ and $\cot x = -\frac{3}{4}$, use the fundamental identities to find the exact values of the remaining four trigonometric functions at x.

◆ **EXAMPLE 2** Using Fundamental Identities

Use the fundamental identities to find the exact values of the remaining trigonometric functions of x, given

$$\cos x = \frac{-4}{\sqrt{17}} \qquad \text{and} \qquad \tan x < 0$$

SOLUTION *Find sin x:* We start with the Pythagorean identity

$$\sin^2 x + \cos^2 x = 1$$

and solve for $\sin x$:

$$\sin x = \pm \sqrt{1 - \cos^2 x}$$

Since both $\cos x$ and $\tan x$ are negative, x is associated with the second quadrant, where $\sin x$ is positive; hence,

$$\sin x = \sqrt{1 - \cos^2 x}$$

$$= \sqrt{1 - \left(\frac{-4}{\sqrt{17}}\right)^2}$$

$$= \sqrt{\frac{1}{17}} = \frac{1^*}{\sqrt{17}}$$

Find sec x.

$$\sec x = \frac{1}{\cos x} = \frac{1}{-4/\sqrt{17}} = -\frac{\sqrt{17}}{4}$$

Find csc x.

$$\csc x = \frac{1}{\sin x} = \frac{1}{1/\sqrt{17}} = \sqrt{17}$$

---

* An equivalent answer is $1/\sqrt{17} = \sqrt{17}/(\sqrt{17}\sqrt{17}) = \sqrt{17}/17$, a form in which we have rationalized (eliminated radicals in) the denominator. Whether we rationalize the denominator or not depends entirely on what we want to do with the answer—sometimes an unrationalized form is more useful than a rationalized form. For the remainder of this book, you should leave answers to matched problems and exercises unrationalized, unless directed otherwise.

Find tan x.

$$\tan x = \frac{\sin x}{\cos x} = \frac{1/\sqrt{17}}{-4/\sqrt{17}} = -\frac{1}{4}$$

Find cot x.

$$\cot x = \frac{1}{\tan x} = \frac{1}{-\frac{1}{4}} = -4$$

◆

MATCHED PROBLEM 2 Use the fundamental identities to find the exact values of the remaining trigonometric functions of *x*, given:

$$\tan x = -\frac{\sqrt{21}}{2} \qquad \text{and} \qquad \cos x > 0$$

◆ Converting to Equivalent Forms

One of the most important and frequent uses of the fundamental identities is the conversion of trigonometric forms into equivalent simpler or more useful forms. A couple of examples will illustrate the process.

◆ **EXAMPLE 3** Simplifying Trigonometric Expressions

Use fundamental identities and appropriate algebraic operations to simplify the following expression.

$$\frac{1}{\cos^2 \alpha} - 1$$

SOLUTON We start by forming a single fraction.

$$\frac{1}{\cos^2 \alpha} - 1 = \frac{1 - \cos^2 \alpha}{\cos^2 \alpha} \qquad \text{Algebra}$$

$$= \frac{\sin^2 \alpha}{\cos^2 \alpha} \qquad \text{Pythagorean identity}$$

$$= \left(\frac{\sin \alpha}{\cos \alpha}\right)^2 \qquad \text{Algebra}$$

$$= \tan^2 \alpha \qquad \text{Quotient identity}$$

◆

Key Algebraic Steps in Example 3

$$\frac{1}{b^2} - 1 = \frac{1}{b^2} - \frac{b^2}{b^2} = \frac{1 - b^2}{b^2} \qquad \text{and} \qquad \frac{a^2}{b^2} = \left(\frac{a}{b}\right)^2$$

MATCHED PROBLEM 3 Use fundamental identities and appropriate algebraic operations to simplify the following expression.

$$\frac{\sin^2 \theta}{\cos^2 \theta} + 1$$

◆ **EXAMPLE 4**   Converting a Trigonometric Expression to an Equivalent Form

Using fundamental identities, write the following expression in terms of sines and cosines, and then simplify.

$$\frac{\tan x - \cot x}{\tan x + \cot x}$$

Write the final answer in terms of the cosine function.

SOLUTION

$$\frac{\tan x - \cot x}{\tan x + \cot x} = \frac{\dfrac{\sin x}{\cos x} - \dfrac{\cos x}{\sin x}}{\dfrac{\sin x}{\cos x} + \dfrac{\cos x}{\sin x}}$$ Change to sines and cosines.

$$= \frac{(\sin x \cos x)\left(\dfrac{\sin x}{\cos x} - \dfrac{\cos x}{\sin x}\right)}{(\sin x \cos x)\left(\dfrac{\sin x}{\cos x} + \dfrac{\cos x}{\sin x}\right)}$$ Multiply numerator and denominator by the least common denominator of all internal fractions.

$$= \frac{\sin^2 x - \cos^2 x}{\sin^2 x + \cos^2 x}$$ Algebra

$$= \frac{1 - \cos^2 x - \cos^2 x}{1}$$ Pythagorean identities

$$= 1 - 2\cos^2 x$$ Algebra   ◆

Key Algebraic Steps in Example 4

$$\frac{\dfrac{a}{b} - \dfrac{b}{a}}{\dfrac{a}{b} + \dfrac{b}{a}} = \frac{ab\left(\dfrac{a}{b} - \dfrac{b}{a}\right)}{ab\left(\dfrac{a}{b} + \dfrac{b}{a}\right)} = \frac{a^2 - b^2}{a^2 + b^2}$$

MATCHED PROBLEM 4   Using fundamental identities, write the following expression in terms of sines and cosines, and then simplify.

$$1 + \frac{\tan z}{\cot z}$$

Answers to Matched Problems

1. $\csc x = -\frac{5}{4}$, $\tan x = -\frac{4}{3}$, $\cos x = \frac{3}{5}$, $\sec x = \frac{5}{3}$
2. $\cot x = -2/\sqrt{21}$, $\sec x = \frac{5}{2}$, $\cos x = \frac{2}{5}$, $\sin x = -\sqrt{21}/5$, $\csc x = -5/\sqrt{21}$
3. $\sec^2 \theta$ **4.** $\sec^2 z$

EXERCISE 4.1

A *Use the fundamental identities to find the exact values of the remaining trigonometric functions of x, given the following:*

1. $\sin x = -\frac{2}{3}$ and $\cos x = \sqrt{5}/3$

2. $\sin x = 2/\sqrt{5}$ and $\cos x = -1/\sqrt{5}$

3. $\tan x = 2$ and $\sin x = -2/\sqrt{5}$

4. $\cot x = -\frac{2}{3}$ and $\sec x = \sqrt{13}/2$

Simplify each expression using the fundamental identities.

5. $\tan u \cot u$

6. $\sec x \cos x$

7. $\tan x \csc x$

8. $\sec \theta \cot \theta$

9. $\dfrac{\sec^2 x - 1}{\tan x}$

10. $\dfrac{\csc^2 v - 1}{\cot v}$

11. $\dfrac{\sin^2 \theta}{\cos \theta} + \cos \theta$

12. $\dfrac{1}{\csc^2 x} + \dfrac{1}{\sec^2 x}$

13. $\dfrac{1}{\sin^2 \beta} - 1$

14. $\dfrac{1 - \sin^2 u}{\cos u}$

15. $\dfrac{(1 - \cos x)^2 + \sin^2 x}{1 - \cos x}$

16. $\dfrac{\cos^2 x + (\sin x + 1)^2}{\sin x + 1}$

B *Use the fundamental identities to find the exact values of the remaining trigonometric functions of x, given the following.*

17. $\sin x = \frac{1}{4}$ and $\tan x < 0$

18. $\cos x = \frac{2}{3}$ and $\csc x < 0$

19. $\tan x = -2$ and $\sin x < 0$

20. $\cot x = -3$ and $\cos x > 0$

21. $\csc x = \frac{3}{2}$ and $\tan x < 0$

22. $\sec x = -\frac{5}{3}$ and $\cot x > 0$

Using fundamental identities, write the following expressions in terms of sines and cosines, and then simplify.

23. $\csc(-y) \cos(-y)$

24. $\sin(-\alpha) \sec(-\alpha)$

25. $\cot x \cos x + \sin x$

26. $\cos u + \sin u \tan u$

27. $\dfrac{\cot(-\theta)}{\csc \theta} + \cos \theta$

28. $\sin y - \dfrac{\tan(-y)}{\sec y}$

29. $\dfrac{\cot x}{\tan x} + 1$

30. $\dfrac{1 + \cot^2 y}{\cot^2 y}$

31. $\sec w \csc w - \sec w \sin w$

32. $\csc \theta \sec \theta - \csc \theta \cos \theta$

C 33. If $\sin x = \frac{2}{5}$, find:

(A) $\sin^2(x/2) + \cos^2(x/2)$

(B) $\csc^2(2x) - \cot^2(2x)$

34. If $\cos x = \frac{3}{7}$, find:

(A) $\sin^2(2x) + \cos^2(2x)$

(B) $\sec^2(x/2) - \tan^2(x/2)$

Each of the following is an identity in certain quadrants. Indicate which quadrants.

35. $\sqrt{1 - \cos^2 x} = \sin x$

36. $\sqrt{1 - \sin^2 x} = \cos x$

37. $\sqrt{1 - \sin^2 x} = -\cos x$

38. $\sqrt{1 - \cos^2 x} = -\sin x$

39. $\sqrt{1 - \sin^2 x} = |\cos x|$

40. $\sqrt{1 - \cos^2 x} = |\sin x|$

41. $\dfrac{\sin x}{\sqrt{1 - \sin^2 x}} = \tan x$

42. $\dfrac{\sin x}{\sqrt{1 - \sin^2 x}} = -\tan x$

 Applications

Precalculus: Trigonometric Substitution *In calculus, problems are frequently encountered that involve radicals of the forms $\sqrt{a^2 - u^2}$ and $\sqrt{a^2 + u^2}$. It is very useful to be able to make trigonometric substitutions and use fundamental identities to transform these expressions into nonradical forms. Problems 43–46 involve such transformations. (Recall from algebra that $\sqrt{N^2} = N$ if $N \geq 0$ and $\sqrt{N^2} = -N$ if $N < 0$.)*

43. In the expression $\sqrt{a^2 - u^2}$, $a > 0$, let $u = a \sin x$, $-\pi/2 < x < \pi/2$. After using an appropriate fundamental identity, write the given expression in a final form free of radicals.

44. In the expression $\sqrt{a^2 - u^2}$, $a > 0$, let $u = a \cos x$, $0 < x < \pi$. After using an appropriate fundamental identity, write the given expression in a final form free of radicals.

45. In the expression $\sqrt{a^2 + u^2}$, $a > 0$, let $u = a \tan x$, $0 < x < \pi/2$. After using an appropriate fundamental identity, write the given expression in a final form free of radicals.

46. In the expression $\sqrt{a^2 + u^2}$, $a > 0$, let $u = a \cot x$, $0 < x < \pi/2$. After using an appropriate fundamental identity, write the given expression in a final form free of radicals.

Precalculus: Parametric Equations   *Suppose we are given the parametric equations of a curve,*

$$\begin{cases} x = \cos t \\ y = \sin t \end{cases} \quad 0 \leq t \leq 2\pi$$

[The parameter t is assigned values and the corresponding points (cos t, sin t) are plotted in a rectangular coordinate system.] These parametric equations can be transformed into a standard rectangular form free of the parameter t by use of the fundamental identities as follows.

$$x^2 + y^2 = \cos^2 t + \sin^2 t = 1$$

Thus,

$$x^2 + y^2 = 1$$

is the nonparametric equation for the curve. The latter is the equation of a circle with radius 1 and center at the origin. Refer to this discussion for Problems 47 and 48.

47. Transform the parametric equations (by suitable use of a fundamental identity) into a nonparametric form.

$$\begin{cases} x = 5 \cos t \\ y = 2 \sin t \end{cases} \quad 0 \leq t \leq 2\pi$$

Hint: First write the parametric equations in the following form, then square and add.

$$\begin{cases} \dfrac{x}{5} = \cos t \\ \dfrac{y}{2} = \sin t \end{cases} \quad 0 \leq t \leq 2\pi$$

48. Transform the parametric equations (by suitable use of a fundamental identity) into a nonparametric form.

$$\begin{cases} x = 3 \cos t \\ y = 4 \sin t \end{cases} \quad 0 \leq t \leq 2\pi$$

4.2   VERIFYING TRIGONOMETRIC IDENTITIES

♦   **Comments on Algebra**
♦   **Verifying Identities**

We now use the experience gained in the last section to verify given identities. Before we begin, however, a few comments on algebraic operations on trigonometric functions will prove useful.

♦   Comments on Algebra

In the process of converting trigonometric expressions into equivalent forms, we often use basic algebraic operations such as multiplication, factoring, finding least common denominators (LCDs), combining and reducing fractions, and so on. If you are a little rusty in these areas, the following brief review may be helpful. We will limit our discussion to a couple of examples.

♦   EXAMPLE 1   Multiplication

(A) $(\sin x)(\sin x - 2) = \sin^2 x - 2 \sin x$

$a(a - 2) = a^2 - 2a$

(B) $(1 - \cos x)(1 + \cos x) = 1 - \cos^2 x$

$(1 - b)(1 + b) = 1 - b^2$

(C) $(\sin x + \cos x)^2 = \sin^2 x + 2 \sin x \cos x + \cos^2 x$

$\quad\quad (a + b)^2 = a^2 + 2ab + b^2$

(D) $(2 \tan x - 3)(\tan x + 1) = 2 \tan^2 x - \tan x - 3$

$\quad\quad (2c - 3)(c + 1) = 2c^2 - c - 3$

MATCHED PROBLEM 1   Multiply

(A) $(\sec u)(\sec u + \csc u)$ (B) $(\csc y - 1)(\csc y + 1)$

(C) $(\tan x - \cot x)^2$ (D) $(3 \sin x - 2)(\sin x + 1)$

◆ EXAMPLE 2   Factoring

(A) $\cos^2 t - \sin t \cos t = (\cos t)(\cos t - \sin t)$

$\quad\quad b^2 - ab = b(b - a)$

(B) $1 - \cos^2 x = (1 - \cos x)(1 + \cos x)$

$\quad\quad 1 - b^2 = (1 - b)(1 + b)$

(C) $2 \cos^2 y + 5 \cos y + 3 = (2 \cos y + 3)(\cos y + 1)$

$\quad\quad 2b^2 + 5b + 3 = (2b + 3)(b + 1)$

MATCHED PROBLEM 2   Factor

(A) $\sec^2 u - \sec u$ (B) $\tan^2 x - \cot^2 x$

(C) $4 \sin^2 x - 9 \sin x + 2$

◆ Verifying Identities

We will now verify (prove) some given trigonometric identities; this process will be helpful to you if you want to convert a trigonometric expression into a form that may be more useful. Verifying a trigonometric identity is different from solving an equation. When solving an equation you use properties of equality such as adding the same quantity to each side or multiplying both sides by a nonzero quantity. These operations are not valid in the process of verifying identities because, at the start, we do not know that the left and right expressions are equal.

VERIFYING AN IDENTITY

To verify an identity, start with the expression on one side and, through a sequence of valid steps involving the use of known identities or algebraic manipulation, convert that expression into the expression on the other side.

⚠ Caution   When verifying an identity, *do not* add the same quantity to each side, multiply both sides by the same nonzero quantity, or square (or take the square root of) both sides. ◇

The following examples illustrate some of the techniques used to establish certain identities. To become proficient in the process, it is important that you work many problems on your own.

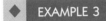 **EXAMPLE 3**   Identity Verification

Verify the identity

$$\csc(-x) = -\csc x$$

VERIFICATION $\csc(-x) = \dfrac{1}{\sin(-x)}$ *Reciprocal identity*

$$= \dfrac{1}{-\sin x}$$ *Identity for negatives*

$$= -\dfrac{1}{\sin x}$$ *Algebra*

$$= -\csc x$$ *Reciprocal identity*   ◆

MATCHED PROBLEM 3   Verify the identity

$$\sec(-x) = \sec x$$

 **EXAMPLE 4**   Identity Verification

Verify the identity

$$\tan x \sin x + \cos x = \sec x$$

VERIFICATION $\tan x \sin x + \cos x = \dfrac{\sin x}{\cos x}\, \sin x + \cos x$ *Quotient identity*

$$= \dfrac{\sin^2 x + \cos^2 x}{\cos x}$$ *Algebra*

$$= \dfrac{1}{\cos x}$$ *Pythagorean identity*

$$= \sec x$$ *Reciprocal identity*   ◆

Key Algebraic Steps in Example 4

$$\dfrac{a}{b}\,a + b = \dfrac{a^2}{b} + b = \dfrac{a^2 + b^2}{b}$$

MATCHED PROBLEM 4   Verify the identity

$$\cot x \cos x + \sin x = \csc x$$

There is no fixed procedure that works in the verifications for all identities. Nevertheless, certain steps can be taken that will help in many cases. Study the following suggestions.

SOME SUGGESTIONS FOR VERIFYING IDENTITIES	

1. Start with the more complicated side of the identity and transform it into the simpler side. In the process, keep the simpler side of the identity in mind; knowing where you are going often suggests steps that should be taken to get there.

2. Specifically, after you have taken any obvious steps, try to
   - Express each function in terms of sine and cosine; or
   - Use algebraic manipulation such as simplifying, multiplying, factoring, combining fractions, splitting single fractions, and so on.

◆ **EXAMPLE 5** Identity Verification

Verify the identity

$$\frac{\cot^2 x - 1}{1 + \cot^2 x} = 1 - 2 \sin^2 x$$

VERIFICATION

$$\frac{\cot^2 x - 1}{1 + \cot^2 x} = \frac{\dfrac{\cos^2 x}{\sin^2 x} - 1}{1 + \dfrac{\cos^2 x}{\sin^2 x}}$$ *Convert to sines and cosines.*

$$= \frac{(\sin^2 x)\left(\dfrac{\cos^2 x}{\sin^2 x} - 1\right)}{(\sin^2 x)\left(1 + \dfrac{\cos^2 x}{\sin^2 x}\right)}$$ *Multiply numerator and denominator by $\sin^2 x$, the LCD of all secondary fractions.*

$$= \frac{\cos^2 x - \sin^2 x}{\sin^2 x + \cos^2 x}$$ *Algebra*

$$= \frac{1 - \sin^2 x - \sin^2 x}{1}$$ *Pythagorean identities, twice*

$$= 1 - 2 \sin^2 x$$ *Algebra*   ◆

Key Algebraic Steps in Example 5

$$\frac{\dfrac{b^2}{a^2} - 1}{1 + \dfrac{b^2}{a^2}} = \frac{a^2\left(\dfrac{b^2}{a^2} - 1\right)}{a^2\left(1 + \dfrac{b^2}{a^2}\right)} = \frac{b^2 - a^2}{a^2 + b^2}$$

MATCHED PROBLEM 5   Verify the identity

$$\frac{\tan^2 x - 1}{1 + \tan^2 x} = 1 - 2\cos^2 x$$

◆ EXAMPLE 6   Identity Verification

Verify the identity

$$\frac{1 + \cos x}{\sin x} + \frac{\sin x}{1 + \cos x} = 2\csc x$$

VERIFICATION

$$\frac{1 + \cos x}{\sin x} + \frac{\sin x}{1 + \cos x} = \frac{(1 + \cos x)^2 + \sin^2 x}{(\sin x)(1 + \cos x)}$$ Algebra

$$= \frac{1 + 2\cos x + \cos^2 x + \sin^2 x}{(\sin x)(1 + \cos x)}$$ Algebra

$$= \frac{1 + 2\cos x + 1}{(\sin x)(1 + \cos x)}$$ Pythagorean identity

$$= \frac{2 + 2\cos x}{(\sin x)(1 + \cos x)}$$ Algebra

$$= \frac{2(1 + \cos x)}{(\sin x)(1 + \cos x)}$$ Algebra

$$= \frac{2}{\sin x}$$ Cancel common factor by division. Reciprocal identity ◆

$$= 2\csc x$$

Key Algebraic Steps in Example 6

$$\frac{1 + b}{a} + \frac{a}{1 + b} = \frac{(1 + b)^2 + a^2}{a(1 + b)} = \frac{1 + 2b + b^2 + a^2}{a(1 + b)}$$

and

$$\frac{2 + 2b}{a(1 + b)} = \frac{2(1 + b)}{a(1 + b)} = \frac{2}{a}$$

MATCHED PROBLEM 6   Verify the identity

$$\frac{1 + \sin x}{\cos x} + \frac{\cos x}{1 + \sin x} = 2\sec x$$

 EXAMPLE 7   Identity Verification

Verify the identity

$$\csc x + \cot x = \frac{\sin x}{1 - \cos x}$$

(A) Going from left to right   (B) Going from right to left

VERIFICATIONS   (A) Going from left to right:

$$\csc x + \cot x = \frac{1}{\sin x} + \frac{\cos x}{\sin x}$$ Convert to
sines and cosines.

$$= \frac{1 + \cos x}{\sin x}$$ Algebra

$$= \frac{(\sin x)(1 + \cos x)}{\sin^2 x}$$ We need a sin x on top, so we multiply numerator and denominator by sin x.

$$= \frac{(\sin x)(1 + \cos x)}{1 - \cos^2 x}$$ Pythagorean identity

$$= \frac{(\sin x)(1 + \cos x)}{(1 - \cos x)(1 + \cos x)}$$ Factor denominator.

$$= \frac{\sin x}{1 - \cos x}$$ Cancel common factor by division.

Key Algebraic Steps in Example 7(A)

$$\frac{1}{a} + \frac{b}{a} = \frac{1 + b}{a} = \frac{a(1 + b)}{a^2} \qquad \text{and} \qquad \frac{a(1 + b)}{1 - b^2} = \frac{a(1 + b)}{(1 - b)(1 + b)} = \frac{a}{1 - b}$$

(B) Going from right to left:

$$\frac{\sin x}{1 - \cos x} = \frac{(\sin x)(1 + \cos x)}{(1 - \cos x)(1 + \cos x)}$$ Multiply numerator and denominator by (1 + cos x) so that we can take advantage of the Pythagorean identity.

$$= \frac{(\sin x)(1 + \cos x)}{1 - \cos^2 x}$$ Algebra

$$= \frac{(\sin x)(1 + \cos x)}{\sin^2 x}$$ Pythagorean identity

$$= \frac{1 + \cos x}{\sin x}$$ Cancel common factor.

$$= \frac{1}{\sin x} + \frac{\cos x}{\sin x}$$ Algebra

$$= \csc x + \cot x$$ Fundamental identities   ◆

Key Algebraic Steps in Example 7(B)

$$\frac{a}{1 - b} = \frac{a(1 + b)}{(1 - b)(1 + b)} = \frac{a(1 + b)}{1 - b^2} \qquad \text{and} \qquad \frac{a(1 + b)}{a^2} = \frac{1 + b}{a} = \frac{1}{a} + \frac{b}{a}$$

MATCHED PROBLEM 7   Verify the identity

$$\sec m + \tan m = \frac{\cos m}{1 - \sin m}$$

(A) Going from left to right   (B) Going from right to left

1. (A) $\sec^2 u + \sec u \csc u$ (B) $\csc^2 y - 1$
   (C) $\tan^2 x - 2 \tan x \cot x + \cot^2 x$ (D) $3 \sin^2 x + \sin x - 2$

2. (A) $(\sec u)(\sec u - 1)$ (B) $(\tan x - \cot x)(\tan x + \cot x)$
   (C) $(4 \sin x - 1)(\sin x - 2)$

3. $\sec(-x) = \dfrac{1}{\cos(-x)} = \dfrac{1}{\cos x} = \sec x$

4. $\cot x \cos x + \sin x = \dfrac{\cos^2 x}{\sin x} + \sin x = \dfrac{\cos^2 x + \sin^2 x}{\sin x}$

$$= \dfrac{1}{\sin x} = \csc x$$

5. $\dfrac{\tan^2 x - 1}{1 + \tan^2 x} = \dfrac{\dfrac{\sin^2 x}{\cos^2 x} - 1}{1 + \dfrac{\sin^2 x}{\cos^2 x}} = \dfrac{(\cos^2 x)\left(\dfrac{\sin^2 x}{\cos^2 x} - 1\right)}{(\cos^2 x)\left(1 + \dfrac{\sin^2 x}{\cos^2 x}\right)}$

$$= \dfrac{\sin^2 x - \cos^2 x}{\cos^2 x + \sin^2 x} = \dfrac{1 - \cos^2 x - \cos^2 x}{1}$$

$$= 1 - 2 \cos^2 x$$

6. $\dfrac{1 + \sin x}{\cos x} + \dfrac{\cos x}{1 + \sin x} = \dfrac{(1 + \sin x)^2 + \cos^2 x}{(\cos x)(1 + \sin x)}$

$$= \dfrac{1 + 2 \sin x + \sin^2 x + \cos^2 x}{(\cos x)(1 + \sin x)}$$

$$= \dfrac{2 + 2 \sin x}{(\cos x)(1 + \sin x)} = \dfrac{2}{\cos x} = 2 \sec x$$

7. (A) Going from left to right:

$$\sec m + \tan m = \dfrac{1}{\cos m} + \dfrac{\sin m}{\cos m} = \dfrac{1 + \sin m}{\cos m}$$

$$= \dfrac{(\cos m)(1 + \sin m)}{\cos^2 m} = \dfrac{(\cos m)(1 + \sin m)}{1 - \sin^2 m}$$

$$= \dfrac{(\cos m)(1 + \sin m)}{(1 - \sin m)(1 + \sin m)} = \dfrac{\cos m}{1 - \sin m}$$

(B) Going from right to left:

$$\dfrac{\cos m}{1 - \sin m} = \dfrac{(\cos m)(1 + \sin m)}{(1 - \sin m)(1 + \sin m)} = \dfrac{(\cos m)(1 + \sin m)}{1 - \sin^2 m}$$

$$= \dfrac{(\cos m)(1 + \sin m)}{\cos^2 m} = \dfrac{1 + \sin m}{\cos m}$$

$$= \dfrac{1}{\cos m} + \dfrac{\sin m}{\cos m} = \sec m + \tan m$$

EXERCISE 4.2

A Match each function in Problems 1–12 with one of the functions in (A)–(L) to form a fundamental identity or a minor variation of one. Do not look at the list of 11 identities that was given in Section 4.1.

(A) $\dfrac{1}{\csc x}$

(B) $\cos x$

(C) 1

(D) $\tan x$

(E) $\dfrac{1}{\sec x}$

(F) $\sin^2 x$

(G) $\tan^2 x$

(H) $\cot x$

(I) $-\sin x$

(J) $1 - \sin^2 x$

(K) $1 - \cos^2 x$

(L) $\sec^2 x$

1. $\sin x$

2. $\cos(-x)$

3. $\cos x$

4. $\dfrac{\sin x}{\cos x}$

5. $\dfrac{\cos x}{\sin x}$

6. $\sin^2 x + \cos^2 x$

7. $\cos^2 x$

8. $\dfrac{1}{\cot x}$

9. $\tan^2 x + 1$

10. $\sin(-x)$

11. $\sin^2 x$

12. $\sec^2 x - 1$

Verify each identity.

13. $\cos x \sec x = 1$

14. $\sin x \csc x = 1$

15. $\tan x \cos x = \sin x$

16. $\cot x \sin x = \cos x$

17. $\tan x = \sin x \sec x$

18. $\cot x = \cos x \csc x$

19. $\csc(-x) = -\csc x$

20. $\sec(-x) = \sec x$

21. $\dfrac{\sin \alpha}{\cos \alpha \tan \alpha} = 1$

22. $\dfrac{\cos \alpha}{\sin \alpha \cot \alpha} = 1$

23. $\dfrac{\cos \beta \sec \beta}{\tan \beta} = \cot \beta$

24. $\dfrac{\tan \beta \cot \beta}{\sin \beta} = \csc \beta$

25. $(\sec \theta)(\sin \theta + \cos \theta) = \tan \theta + 1$

26. $(\csc \theta)(\cos \theta + \sin \theta) = \cot \theta + 1$

27. $\dfrac{\cos^2 t - \sin^2 t}{\sin t \cos t} = \cot t - \tan t$

28. $\dfrac{\cos \alpha - \sin \alpha}{\sin \alpha \cos \alpha} = \csc \alpha - \sec \alpha$

29. $\dfrac{\cos \beta}{\cot \beta} + \dfrac{\sin \beta}{\tan \beta} = \sin \beta + \cos \beta$

30. $\dfrac{\tan u}{\sin u} - \dfrac{\cot u}{\cos u} = \sec u - \csc u$

31. $\sec^2 \theta - \tan^2 \theta = 1$

32. $\csc^2 \theta - \cot^2 \theta = 1$

33. $(\sin^2 x)(1 + \cot^2 x) = 1$

34. $(\cos^2 x)(\tan^2 x + 1) = 1$

35. $(\csc \alpha + 1)(\csc \alpha - 1) = \cot^2 \alpha$

36. $(\sec \beta - 1)(\sec \beta + 1) = \tan^2 \beta$

37. $\dfrac{\sin t}{\csc t} + \dfrac{\cos t}{\sec t} = 1$

38. $\dfrac{1}{\sec^2 m} + \dfrac{1}{\csc^2 m} = 1$

B *Verify each identity.*

39. $\dfrac{\sin^2 x}{\cos x} + \cos x = \sec x$

40. $\dfrac{\cos^2 x}{\sin x} + \sin x = \csc x$

41. $\dfrac{1 - (\cos \theta - \sin \theta)^2}{\cos \theta} = 2 \sin \theta$

42. $\dfrac{1 - (\sin \theta - \cos \theta)^2}{\sin \theta} = 2 \cos \theta$

43. $\dfrac{\tan w + 1}{\sec w} = \sin w + \cos w$

44. $\dfrac{\cot y + 1}{\csc y} = \cos y + \sin y$

45. $\dfrac{\cos s}{\sin^2 s - 1} = -\sec s$

46. $\dfrac{\sin t}{\cos^2 t - 1} = -\csc t$

47. $\dfrac{1}{1 - \cos^2 \theta} = 1 + \cot^2 \theta$

48. $\dfrac{1}{1 - \sin^2 \theta} = 1 + \tan^2 \theta$

49. $\dfrac{\sin^2 \beta}{1 - \cos \beta} = 1 + \cos \beta$

50. $\dfrac{\cos^2 \beta}{1 + \sin \beta} = 1 - \sin \beta$

51. $\dfrac{2 - \cos^2 \theta}{\sin \theta} = \csc \theta + \sin \theta$

52. $\dfrac{2 - \sin^2 \theta}{\cos \theta} = \sec \theta + \cos \theta$

53. $\tan x + \cot x = \sec x \csc x$

54. $\dfrac{\csc x}{\cot x + \tan x} = \cos x$

55. $\dfrac{1 - \csc x}{1 + \csc x} = \dfrac{\sin x - 1}{\sin x + 1}$

56. $\dfrac{1 - \cos x}{1 + \cos x} = \dfrac{\sec x - 1}{\sec x + 1}$

57. $\csc^2 \alpha - \cos^2 \alpha - \sin^2 \alpha = \cot^2 \alpha$

58. $\sec^2 \alpha - \sin^2 \alpha - \cos^2 \alpha = \tan^2 \alpha$

59. $(\sin x + \cos x)^2 - 1 = 2 \sin x \cos x$

60. $\sec x - 2 \sin x = \dfrac{(\sin x - \cos x)^2}{\cos x}$

61. $(\sin u - \cos u)^2 + (\sin u + \cos u)^2 = 2$

62. $(\tan x - 1)^2 + (\tan x + 1)^2 = 2 \sec^2 x$

63. $\sin^4 x - \cos^4 x = 1 - 2 \cos^2 x$

64. $\sin^4 x + 2 \sin^2 x \cos^2 x + \cos^4 x = 1$

65. $\dfrac{1 + \cos^2 s}{1 - \cos^4 s} = \csc^2 s$ **66.** $\dfrac{1 + \sin^2 s}{1 - \sin^4 s} = \sec^2 s$

67. $\dfrac{\cos x}{1 - \sin x} + \dfrac{\cos x}{1 + \sin x} = 2 \sec x$

68. $\dfrac{\cos x}{\csc x + 1} + \dfrac{\cos x}{\csc x - 1} = 2 \tan x$

69. $\dfrac{\sin \alpha}{1 - \cos \alpha} - \dfrac{1 + \cos \alpha}{\sin \alpha} = 0$

70. $\dfrac{1 + \cos \alpha}{\sin \alpha} + \dfrac{\sin \alpha}{1 + \cos \alpha} = 2 \csc \alpha$

71. $\dfrac{1}{\csc \theta + \cot \theta} + \dfrac{1}{\csc \theta - \cot \theta} = 2 \csc \theta$

72. $\dfrac{1}{\sec \theta + \tan \theta} - \dfrac{1}{\sec \theta - \tan \theta} = -2 \tan \theta$

73. $\dfrac{\cos^2 n - 3 \cos n + 2}{\sin^2 n} = \dfrac{2 - \cos n}{1 + \cos n}$

74. $\dfrac{\sin^2 n + 4 \sin n + 3}{\cos^2 n} = \dfrac{3 + \sin n}{1 - \sin n}$

75. $\dfrac{1 - \cot^2 x}{\tan^2 x - 1} = \cot^2 x$

76. $\dfrac{\tan^2 x - 1}{1 - \cot^2 x} = \tan^2 x$

77. $\sec^2 x + \csc^2 x = \sec^2 x \csc^2 x$

78. $\tan^2 x - \sin^2 x = \tan^2 x \sin^2 x$

79. $(\sec x - \tan x)^2 = \dfrac{1 - \sin x}{1 + \sin x}$

80. $\dfrac{1 - \cos x}{1 + \cos x} = (\cot x - \csc x)^2$

81. $\dfrac{1 + \sin t}{\cos t} = \dfrac{\cos t}{1 - \sin t}$

82. $\dfrac{\sin t}{1 - \cos t} = \dfrac{1 + \cos t}{\sin t}$

83. $\dfrac{\cos \alpha}{\sec \alpha - 1} = \dfrac{\cos \alpha + 1}{\tan^2 \alpha}$

84. $\dfrac{\sin \alpha}{\csc \alpha - 1} = \dfrac{\sin \alpha + 1}{\cot^2 \alpha}$

Show that each of the following equations is not an identity by finding a value for which both sides are defined but are not equal to each other. For example, sin x = cos x is not an identity, since the two sides are not equal for x = 0.

85. $\tan x = \cot x$

86. $\sin^2 x + \sin x = 1$

87. $\sin^2 x - \cos^2 x = 1$

88. $\cot x = 2 \sin x \cos x$

89. $\cot^2 x + \cos x = \sin^2 x$

90. $\cos^2 x + 2 \cos x - 8 = 0$

C Verify each identity.

91. $\dfrac{\sin x}{1 - \cos x} - \cot x = \csc x$

92. $\dfrac{\cos x}{1 - \sin x} - \tan x = \sec x$

93. $\dfrac{\cot \beta}{\csc \beta + 1} = \dfrac{\csc \beta - 1}{\cot \beta}$

94. $\dfrac{\tan \beta}{\sec \beta - 1} = \dfrac{\sec \beta + 1}{\tan \beta}$

95. $\dfrac{3 \cos^2 m + 5 \sin m - 5}{\cos^2 m} = \dfrac{3 \sin m - 2}{1 + \sin m}$

96. $\dfrac{2 \sin^2 z + 3 \cos z - 3}{\sin^2 z} = \dfrac{2 \cos z - 1}{1 + \cos z}$

Problems 97 and 98 involve trigonometric functions with two variables. Be careful with the terms you combine and simplify.

97. $\dfrac{\sin x \cos y + \cos x \sin y}{\cos x \cos y - \sin x \sin y} = \dfrac{\tan x + \tan y}{1 - \tan x \tan y}$

98. $\dfrac{\tan \alpha + \tan \beta}{1 - \tan \alpha \tan \beta} = \dfrac{\cot \alpha + \cot \beta}{\cot \alpha \cot \beta - 1}$

 Problems 99–104 require the use of a graphing calculator. Graph f(x), find a simpler function g(x) which has the same graph as f(x), and verify the identity f(x) = g(x). [Assume g(x) = k + A t(x) where t(x) is one of the six trigonometric functions.]

99. $f(x) = \dfrac{1 + \sin x}{2 \cos x} - \dfrac{\cos x}{2 + 2 \sin x}$

100. $f(x) = \sin x \cos x + \dfrac{1 - \sin^2 x}{\tan x}$

101. $f(x) = \dfrac{\sin x \tan x}{1 - \cos x}$

102. $f(x) = \dfrac{\cos^2 x}{1 + \sin x - \cos^2 x}$

103. $f(x) = \dfrac{3 \sin x - 2 \sin x \cos x}{1 - \cos x} - \dfrac{1 + \cos x}{\sin x}$

104. $f(x) = \dfrac{1 + \cos x - 2 \cos^2 x}{1 - \cos x} - \dfrac{\sin^2 x}{1 + \cos x}$

4.3 SUM, DIFFERENCE, AND COFUNCTION IDENTITIES

♦ **Sum and Difference Identities for Cosine**
♦ **Cofunction Identities**
♦ **Sum and Difference Identities for Sine and Tangent**
♦ **Summary and Use**

♦ Sum and Difference Identities for Cosine

The fundamental identities we discussed in Section 4.1 involve only one variable. We now consider an important identity, called a **difference identity for cosine**, which involves two variables:

$$\cos(x - y) = \cos x \cos y + \sin x \sin y \qquad (1)$$

Many other useful identities can be readily established from this particular one. We will sketch a proof of identity (1) in which we assume that x and y are restricted as follows: $0 < y < x < 2\pi$. Identity (1) holds, however, for all real numbers and angles in radian or degree measure. In the proof we will make use of the **distance-between-two-points formula**,

$$d(P_1, P_2) = \sqrt{(x_2 - x_1)^2 + (y_2 - y_1)^2}$$

for points $P_1(x_1, y_1)$ and $P_2(x_2, y_2)$ in a rectangular coordinate system.

We associate x and y with arcs and angles on a unit circle, as indicated in Figure 1(a). Using the definitions of the circular functions in Section 2.6, the terminal points of x and y are labeled as indicated in Figure 1(a).

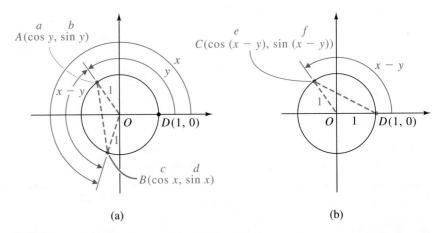

(a)   (b)

FIGURE 1

Now, if we rotate the triangle AOB clockwise about the origin until the terminal point A coincides with $D(1, 0)$, then terminal point B will be at C (see Figure 1(b)). Thus, since rotation preserves lengths, we have

$$d(A, B) = d(C, D)$$
$$\sqrt{(c - a)^2 + (d - b)^2} = \sqrt{(1 - e)^2 + (0 - f)^2}$$
$$(c - a)^2 + (d - b)^2 = (1 - e)^2 + f^2$$
$$c^2 - 2ac + a^2 + d^2 - 2db + b^2 = 1 - 2e + e^2 + f^2$$
$$(c^2 + d^2) + (a^2 + b^2) - 2ac - 2db = 1 - 2e + (e^2 + f^2) \qquad (2)$$

Since $c^2 + d^2 = 1$, $a^2 + b^2 = 1$, and $e^2 + f^2 = 1$ (Why?), equation (2) becomes

$$e = ac + bd \qquad (3)$$

Replacing e, a, c, b, and d with $\cos(x - y)$, $\cos y$, $\cos x$, $\sin y$, and $\sin x$, respectively (see Figure 1), we obtain

$$\cos(x - y) = \cos y \cos x + \sin y \sin x$$
$$= \cos x \cos y + \sin x \sin y \qquad (4)$$

If we replace y with $-y$ in (4) and use the identities for negatives, we obtain the **sum identity for cosine:**

$$\cos(x + y) = \cos x \cos y - \sin x \sin y \qquad (5)$$

◆ Cofunction Identities

To obtain sum and difference identities for the sine and tangent functions, we first derive **cofunction identities** directly from identity (1), the difference identity for cosine:

$$\cos(x - y) = \cos x \cos y + \sin x \sin y \qquad \text{Let } x = \pi/2.$$

$$\cos\left(\frac{\pi}{2} - y\right) = \cos\frac{\pi}{2}\cos y + \sin\frac{\pi}{2}\sin y$$

$$= 0\cos y + 1\sin y$$

$$= \sin y$$

Thus,

$$\cos\left(\frac{\pi}{2} - y\right) = \sin y \tag{6}$$

for y any real number or angle in radian measure. If y is in degree measure, replace $\pi/2$ with $90°$. Now, if in (6) we let $y = \pi/2 - x$, then we have

$$\cos\left[\frac{\pi}{2} - \left(\frac{\pi}{2} - x\right)\right] = \sin\left(\frac{\pi}{2} - x\right)$$

$$\cos x = \sin\left(\frac{\pi}{2} - x\right)$$

or

$$\sin\left(\frac{\pi}{2} - x\right) = \cos x \tag{7}$$

where x is any real number or angle in radian measure. If x is in degree measure, replace $\pi/2$ with $90°$.

Finally, we state the cofunction identities for tangent and secant (and leave the derivations to Problems 9 and 11 in Exercise 4.3).

$$\tan\left(\frac{\pi}{2} - x\right) = \cot x$$
$$\sec\left(\frac{\pi}{2} - x\right) = \csc x \tag{8}$$

for x any real number or angle in radian measure. If x is in degree measure, replace $\pi/2$ with $90°$.

REMARK   If $0 < x < 90°$, then x and $90° - x$ are complementary angles. Originally, *cosine, cotangent,* and *cosecant* meant, respectively, *complements sine, complements tangent,* and *complements secant.* Now we simply refer to cosine, cotangent, and cosecant as **cofunctions of** sine, tangent, and secant, respectively.   ◇

◆   Sum and Difference Identities for Sine
and Tangent

To derive a **difference identity for sine**, we use (7), (1), and (6) as follows:

$$\sin(x - y) = \cos\left[\frac{\pi}{2} - (x - y)\right] \qquad \text{Use (6).}$$

$$= \cos\left[\left(\frac{\pi}{2} - x\right) - (-y)\right] \qquad \text{Algebra}$$

$$= \cos\left(\frac{\pi}{2} - x\right)\cos(-y)$$

$$+ \sin\left(\frac{\pi}{2} - x\right)\sin(-y) \qquad \text{Use (1).}$$

$$= \sin x \cos y - \cos x \sin y \qquad \text{Use (6), (7), and identities for negatives.}$$

The same result is obtained by replacing $\pi/2$ with $90°$. Thus,

$$\sin(x - y) = \sin x \cos y - \cos x \sin y \tag{9}$$

Now, if we replace y with $-y$ (a good exercise to do), we obtain

$$\sin(x + y) = \sin x \cos y + \cos x \sin y \tag{10}$$

It is not difficult to derive sum and difference identities for the tangent function. See if you can supply the reason for each step.

$$\tan(x - y) = \frac{\sin(x - y)}{\cos(x - y)}$$

$$= \frac{\sin x \cos y - \cos x \sin y}{\cos x \cos y + \sin x \sin y}$$

$$= \frac{\dfrac{\sin x \cos y}{\cos x \cos y} - \dfrac{\cos x \sin y}{\cos x \cos y}}{\dfrac{\cos x \cos y}{\cos x \cos y} + \dfrac{\sin x \sin y}{\cos x \cos y}}$$

$$= \frac{\tan x - \tan y}{1 + \tan x \tan y}$$

Thus, for all angles or real numbers x and y,

$$\tan(x - y) = \frac{\tan x - \tan y}{1 + \tan x \tan y} \tag{11}$$

And if we replace y in (11) with $-y$ (another good exercise to do), we obtain

$$\tan(x + y) = \frac{\tan x + \tan y}{1 - \tan x \tan y} \tag{12}$$

◆   Summary and Use

Before proceeding with examples that illustrate the use of these new identities, we list them and the other cofunction identities for convenient reference.

SUMMARY OF IDENTITIES

For x and y any real numbers or angles in degree or radian measure for which both sides are defined:

Sum identities

$$\sin(x + y) = \sin x \cos y + \cos x \sin y$$
$$\cos(x + y) = \cos x \cos y - \sin x \sin y$$
$$\tan(x + y) = \frac{\tan x + \tan y}{1 - \tan x \tan y}$$

Difference identities

$$\sin(x - y) = \sin x \cos y - \cos x \sin y$$
$$\cos(x - y) = \cos x \cos y + \sin x \sin y$$
$$\tan(x - y) = \frac{\tan x - \tan y}{1 + \tan x \tan y}$$

Cofunction identities

(Replace $\pi/2$ with $90°$ if x is in degree measure.)

$$\sin\left(\frac{\pi}{2} - x\right) = \cos x \qquad \cos\left(\frac{\pi}{2} - x\right) = \sin x$$

$$\tan\left(\frac{\pi}{2} - x\right) = \cot x \qquad \cot\left(\frac{\pi}{2} - x\right) = \tan x$$

$$\sec\left(\frac{\pi}{2} - x\right) = \csc x \qquad \csc\left(\frac{\pi}{2} - x\right) = \sec x$$

◆ **EXAMPLE 1**   Using a Difference Identity

Simplify $\sin(x - \pi)$ using a difference identity.

SOLUTION   Use the difference identity for sine, replacing y with π.

$$\sin(x - y) = \sin x \cos y - \cos x \sin y$$
$$\sin(x - \pi) = \sin x \cos \pi - \cos x \sin \pi$$
$$= (\sin x)(-1) - (\cos x)(0)$$
$$= -\sin x$$

MATCHED PROBLEM 1   Simplify $\cos(x + 3\pi/2)$ using a sum identity.

◆ **EXAMPLE 2**   Using a Cofunction Identity

Write $\sin 75°$ in the form $\cos \theta, 0 \le \theta \le 90°$.

SOLUTION   Use $\sin x = \cos(90° - x)$. Thus,

$$\sin 75° = \cos(90° - 75°) = \cos 15°$$ ◆

MATCHED PROBLEM 2   Write $\cos 37°$ in the form $\sin \theta, 0 \le \theta \le 90°$.

◆ **EXAMPLE 3**   Using a Sum Identity

Find the value of $\tan 75°$ in exact radical form.

SOLUTION   Since we can write $75° = 45° + 30°$, the sum of two special angles, we can use the sum identity for tangents with $x = 45°$ and $y = 30°$.

$$\tan(x + y) = \frac{\tan x + \tan y}{1 - \tan x \tan y}$$

$$\tan(45° + 30°) = \frac{\tan 45° + \tan 30°}{1 - \tan 45° \tan 30°}$$

$$= \frac{1 + \dfrac{1}{\sqrt{3}}}{1 - 1 \cdot \dfrac{1}{\sqrt{3}}}$$ Multiply numerator and denominator by $\sqrt{3}$.

$$= \frac{\sqrt{3} + 1}{\sqrt{3} - 1} \cdot \frac{\sqrt{3} + 1}{\sqrt{3} + 1}$$ Rationalizing the denominator produces a simpler result.

$$= 2 + \sqrt{3}$$ ◆

MATCHED PROBLEM 3   Find the value of $\cos 15°$ in exact radical form.

◆ **EXAMPLE 4**   Using a Sum Identity

Find the exact value of $\cos(x + y)$, given $\sin x = \frac{3}{5}$, $\cos y = \frac{4}{5}$, x in quadrant II, and y in quadrant I. Do not use a calculator.

SOLUTION   We start with the sum identity for cosine:

$$\cos(x + y) = \cos x \cos y - \sin x \sin y$$

We know $\sin x$ and $\cos y$, but not $\sin y$ and $\cos x$. We find the latter two values by using reference triangles and the Pythagorean theorem. From Figure 2,

FIGURE 2

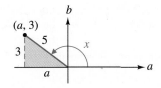

$$a = -\sqrt{5^2 - 3^2} = -4$$

$$\cos x = -\frac{4}{5}$$

From Figure 3,

FIGURE 3

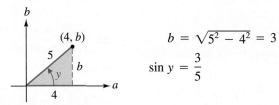

$$b = \sqrt{5^2 - 4^2} = 3$$

$$\sin y = \frac{3}{5}$$

Thus,

$$\cos(x + y) = \cos x \cos y - \sin x \sin y$$

$$= \left(-\frac{4}{5}\right)\left(\frac{4}{5}\right) - \left(\frac{3}{5}\right)\left(\frac{3}{5}\right) = \frac{-25}{25} = -1 \qquad \blacklozenge$$

MATCHED PROBLEM 4   Find the exact value of $\sin(x - y)$, given $\sin x = -\frac{2}{3}$, $\cos y = \sqrt{5}/3$, x in quadrant III, and y in quadrant IV.

◆ EXAMPLE 5   Verifying an Identity

Verify the identity:

$$\tan x + \cot y = \frac{\cos(x - y)}{\cos x \sin y}$$

VERIFICATION $\dfrac{\cos(x - y)}{\cos x \sin y} = \dfrac{\cos x \cos y + \sin x \sin y}{\cos x \sin y}$ *Difference identity for cosine*

$$= \frac{\cancel{\cos x} \cos y}{\cancel{\cos x} \sin y} + \frac{\sin x \cancel{\sin y}}{\cos x \cancel{\sin y}} \qquad \textit{Algebra}$$

$$= \cot y + \tan x \qquad \textit{Quotient identities}$$

$$= \tan x + \cot y \qquad \blacklozenge$$

MATCHED PROBLEM 5   Verify the identity:

$$\cot y - \cot x = \frac{\sin(x - y)}{\sin x \sin y}$$

Answers to   **1.** $\sin x$ **2.** $\sin 53°$ **3.** $\dfrac{1 + \sqrt{3}}{2\sqrt{2}}$ **4.** $\dfrac{-4\sqrt{5}}{9}$
Matched Problems

5. $\dfrac{\sin(x - y)}{\sin x \sin y} = \dfrac{\sin x \cos y - \cos x \sin y}{\sin x \sin y}$

$$= \dfrac{\sin x \cos y}{\sin x \sin y} - \dfrac{\cos x \sin y}{\sin x \sin y} = \cot y - \cot x$$

EXERCISE 4.3

A *We can use sum identities to verify periodic properties for the trigonometric functions. Verify the following identities using sum identities.*

1. $\cos(x + 2\pi) = \cos x$ **2.** $\sin(x + 2\pi) = \sin x$

3. $\cot(x + \pi) = \cot x$ **4.** $\tan(x + \pi) = \tan x$

5. $\sin(x + 2k\pi) = \sin x$,   **6.** $\cos(x + 2k\pi) = \cos x$,
     k an integer k an integer

7. $\tan(x + k\pi) = \tan x$,   **8.** $\cot(x + k\pi) = \cot x$,
     k an integer k an integer

Verify each identity using cofunction identities for sine and cosine and the fundamental identities discussed in Section 4.1.

9. $\tan\left(\dfrac{\pi}{2} - x\right) = \cot x$ **10.** $\cot\left(\dfrac{\pi}{2} - x\right) = \tan x$

11. $\sec\left(\dfrac{\pi}{2} - x\right) = \csc x$ **12.** $\csc\left(\dfrac{\pi}{2} - x\right) = \sec x$

Convert to forms involving sin x, cos x, and/or tan x using sum or difference identities.

13. $\sin(x - 45°)$ **14.** $\sin(30° - x)$

15. $\cos(x + 180°)$ **16.** $\sin(180° - x)$

17. $\tan\left(\dfrac{\pi}{4} - x\right)$ **18.** $\tan\left(x + \dfrac{\pi}{3}\right)$

B *Use appropriate identities to find exact values for each of the following. Do not use a calculator.*

19. $\sin 75°$ **20.** $\sec 75°$

21. $\cos\dfrac{\pi}{12}$ $\left[Hint:\ \dfrac{\pi}{12} = \dfrac{\pi}{4} - \dfrac{\pi}{6}\right]$

22. $\sin\dfrac{7\pi}{12}$ $\left[Hint:\ \dfrac{7\pi}{12} = \dfrac{\pi}{3} + \dfrac{\pi}{4}\right]$

23. $\sin 22° \cos 38° + \cos 22° \sin 38°$

24. $\cos 74° \cos 44° + \sin 74° \sin 44°$

25. $\dfrac{\tan 110° - \tan 50°}{1 + \tan 110° \tan 50°}$ **26.** $\dfrac{\tan 27° + \tan 18°}{1 - \tan 27° \tan 18°}$

Find sin(x − y) and tan(x + y) exactly without a calculator using the information given and appropriate identities.

27. $\sin x = \tfrac{2}{3}$, $\cos y = -\tfrac{1}{4}$, x in quadrant II, and y in quadrant III

28. $\sin x = -\tfrac{3}{5}$, $\sin y = \sqrt{8}/3$, x in quadrant IV, and y in quadrant I

29. $\cos x = -\tfrac{1}{3}$, $\tan y = \tfrac{1}{2}$, x in quadrant II, and y in quadrant III

30. $\tan x = \tfrac{3}{4}$, $\tan y = -\tfrac{1}{2}$, x in quadrant III, and y in quadrant IV

Verify each identity.

31. $\sin 2x = 2 \sin x \cos x$

32. $\cos 2x = \cos^2 x - \sin^2 x$

33. $\cot(x - y) = \dfrac{\cot x \cot y + 1}{\cot y - \cot x}$

34. $\cot(x + y) = \dfrac{\cot x \cot y - 1}{\cot x + \cot y}$

35. $\cot 2x = \dfrac{\cot^2 x - 1}{2 \cot x}$

36. $\tan 2x = \dfrac{2 \tan x}{1 - \tan^2 x}$

37. $\dfrac{\tan \alpha + \tan \beta}{\tan \alpha - \tan \beta} = \dfrac{\sin(\alpha + \beta)}{\sin(\alpha - \beta)}$

38. $\dfrac{\cot \alpha + \cot \beta}{\cot \alpha - \cot \beta} = \dfrac{\sin(\beta + \alpha)}{\sin(\beta - \alpha)}$

39. $\tan x - \tan y = \dfrac{\sin(x - y)}{\cos x \cos y}$

40. $\cot x - \tan y = \dfrac{\cos(x + y)}{\sin x \cos y}$

41. $\tan(x + y) = \dfrac{\cot x + \cot y}{\cot x \cot y - 1}$

42. $\tan(x - y) = \dfrac{\cot y - \cot x}{\cot x \cot y + 1}$

43. $\dfrac{\sin(x + h) - \sin x}{h} = (\sin x)\left(\dfrac{\cos h - 1}{h}\right)$
$$+ (\cos x)\left(\dfrac{\sin h}{h}\right)$$

44. $\dfrac{\cos(x + h) - \cos x}{h} = (\cos x)\left(\dfrac{\cos h - 1}{h}\right)$
$$- (\sin x)\left(\dfrac{\sin h}{h}\right)$$

Graph each equation in Problems 45–48. [Hint: First write each equation in terms of a single trigonometric function.]

45. $y = \sin 3x \cos x - \cos 3x \sin x, \quad 0 \le x \le 2\pi$

46. $y = \cos 3x \cos x - \sin 3x \sin x, \quad 0 \le x \le \pi$

47. $y = \sin \dfrac{\pi x}{4} \cos \dfrac{3\pi x}{4} + \cos \dfrac{\pi x}{4} \sin \dfrac{3\pi x}{4}, \quad 0 \le x \le 3$

48. $y = \cos \pi x \cos \dfrac{\pi}{2} x + \sin \pi x \sin \dfrac{\pi}{2} x, \quad -3 \le x \le 3$

C   *Verify the identities in Problems 49 and 50*
    [Hint: $\sin(x + y + z) = \sin[(x + y) + z].]$

49. $\sin(x + y + z) = \sin x \cos y \cos z$
$$+ \cos x \sin y \cos z + \cos x \cos y \sin z$$
$$- \sin x \sin y \sin z$$

50. $\cos(x + y + z) = \cos x \cos y \cos z$
$$- \sin x \sin y \cos z - \sin x \cos y \sin z$$
$$- \cos x \sin y \sin z$$

Problems 51–56 require the use of a graphing calculator. Use sum or difference identities to convert each equation to a form involving sin x, cos x, and/or tan x. Enter the original form in the calculator as y1, the converted form as y2, and graph y1, y2, and y3 = y1 − y2 + 1 in the same viewing rectangle.

51. $y = \sin\left(x + \dfrac{\pi}{3}\right)$ **52.** $\sin\left(x - \dfrac{\pi}{6}\right)$

53. $y = \cos\left(x - \dfrac{5\pi}{6}\right)$ **54.** $\cos\left(x + \dfrac{3\pi}{4}\right)$

55. $y = \tan\left(x + \dfrac{\pi}{4}\right)$ **56.** $\tan\left(x - \dfrac{2\pi}{3}\right)$

Applications

57. Precalculus: Angle of Intersection of Two Lines   Use the information in the figure to show that

$$\tan(\theta_2 - \theta_1) = \dfrac{m_2 - m_1}{1 + m_1 m_2}$$

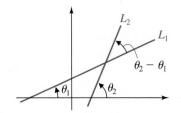

Figure for 57

$\tan \theta_1 = \text{Slope of } L_1 = m$
$\tan \theta_2 = \text{Slope of } L_2 = m_2$

58. Precalculus: Angle of Intersection of Two Lines   Find the acute angle of intersection between the two lines $y = 3x + 1$ and $y = \frac{1}{2}x - 1$. (Use the results of Problem 57.)

∗59. Surveying   A prominent geological feature of Yosemite National Park is the large monolithic granite peak called Half Dome. The dome rises straight up from the valley floor where Mirror Lake provides an early morning mirror image of the dome. How can the height of Half Dome, H, above the lake level be determined by using only a sextant h feet high to measure the angle of elevation, β, to the top of the dome, and the angle of depression, α, to the reflected dome top in the lake? (See the figure, which

Figure for 59

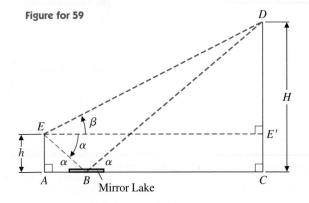

is not to scale.) [*Note:* *AB* and *BC* are not measured as in earlier problems of this type.]*

(A) Using right triangle relationships, show that

$$H = h\left(\frac{1 + \tan\beta\cot\alpha}{1 - \tan\beta\cot\alpha}\right)$$

(B) Using sum identities, show that the result in part (A) can be written in the form

$$H = h\left[\frac{\sin(\alpha + \beta)}{\sin(\alpha - \beta)}\right]$$

(C) If a sextant of height 5.50 ft measures α to be $45.00°$ and β to be $44.92°$, compute the height H of Half Dome above Mirror Lake to three significant digits.

∗∗60. Light Refraction   Light rays passing through a plate glass window are refracted when they enter the glass and again when they leave to continue on a path parallel to the entering rays (see the figure).

(A) If the plate glass is A inches thick, the parallel displacement of the light rays is B inches, the angle of incidence is α, and the angle of refraction is β, show that

$$\tan\beta = \tan\alpha - \frac{B}{A}\sec\alpha$$

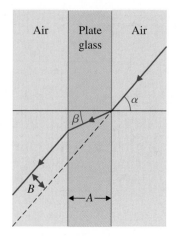

Figure for 60

Hint: First use geometric relationships to obtain

$$\frac{A}{\sin(90° - \beta)} = \frac{B}{\sin(\alpha - \beta)}$$

Then use sum identities and fundamental identities to complete the task.

(B) Using the results in part (A), find β to the nearest degree if $\alpha = 45°$, $A = 0.5$ in., and $B = 0.2$ in.

---

* The solution outlined in parts (A) and (B) is due to Richard J. Palmaccio of Fort Lauderdale, Florida.

4.4   DOUBLE-ANGLE AND HALF-ANGLE IDENTITIES

- ◆   **Double-Angle Identities**
- ◆   **Half-Angle Identities**

We now develop another important set of identities called **double-angle** and **half-angle identities**. We can obtain these identities directly from the sum and difference identities that were found in Section 4.3. In spite of names involving the word *angle,* the new identities hold for real numbers as well.

◆   Double-Angle Identities

If we start with the sum identity for sine,

$$\sin(x + y) = \sin x \cos y + \cos x \sin y$$

and let $y = x$, we obtain

$$\sin(x + x) = \sin x \cos x + \cos x \sin x$$

or

$$\sin 2x = 2 \sin x \cos x \tag{1}$$

Similarly, if we start with the sum identity for cosine,

$$\cos(x + y) = \cos x \cos y - \sin x \sin y$$

and let $y = x$, we obtain

$$\cos(x + x) = \cos x \cos x - \sin x \sin x$$

or

$$\cos 2x = \cos^2 x - \sin^2 x \tag{2}$$

Now, using the Pythagorean identities in the two forms

$$\cos^2 x = 1 - \sin^2 x \tag{3}$$
$$\sin^2 x = 1 - \cos^2 x \tag{4}$$

and substituting (3) into (2), we obtain

$$\cos 2x = 1 - \sin^2 x - \sin^2 x$$
$$\cos 2x = 1 - 2 \sin^2 x \tag{5}$$

Substituting (4) into (2), we obtain

$$\cos 2x = \cos^2 x - (1 - \cos^2 x)$$
$$\cos 2x = 2 \cos^2 x - 1 \tag{6}$$

A double-angle identity can be developed for the tangent function in the same way by starting with the sum identity for tangent. This is left as an exercise for you to do. We list these double-angle identities for convenient reference.

DOUBLE-ANGLE IDENTITIES

For x any real number or angle in degree or radian measure for which both sides are defined:

$$\sin 2x = 2 \sin x \cos x$$
$$\cos 2x = \cos^2 x - \sin^2 x$$
$$= 1 - 2 \sin^2 x$$
$$= 2 \cos^2 x - 1$$
$$\tan 2x = \frac{2 \tan x}{1 - \tan^2 x}$$

◆   **EXAMPLE 1**   Verifying an Identity

Verify the identity

$$\sin 2x = \frac{2 \tan x}{1 + \tan^2 x}$$

VERIFICATION   We start with the right side.

$$\frac{2 \tan x}{1 + \tan^2 x} = \frac{2\left(\dfrac{\sin x}{\cos x}\right)}{1 + \dfrac{\sin^2 x}{\cos^2 x}}$$ Quotient identity

$$= \frac{2 \sin x \cos x}{\cos^2 x + \sin^2 x}$$ Multiply numerator and denominator by $\cos^2 x$.

$$= \frac{\sin 2x}{1}$$ Double-angle and Pythagorean identities

$$= \sin 2x$$ ◆

MATCHED PROBLEM 1   Verify the identity

$$\cos 2x = \frac{1 - \tan^2 x}{1 + \tan^2 x}$$

◆   **EXAMPLE 2**   Using Double-Angle Identities

Find the exact value of $\cos 2x$ and $\tan 2x$ if $\sin x = \frac{4}{5}$, $\pi/2 < x < \pi$.

SOLUTION   First draw a reference triangle in the second quadrant, and find $\cos x$ and $\tan x$ (see Figure 1).

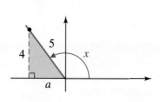

$$a = -\sqrt{5^2 - 4^2} = -3$$

$$\sin x = \frac{4}{5}$$

FIGURE 1

$$\cos x = -\frac{3}{5}$$

$$\tan x = -\frac{4}{3}$$

$$\cos 2x = 1 - 2 \sin^2 x$$ Use double-angle identity and the results above.

$$= 1 - 2\left(\frac{4}{5}\right)^2$$

$$= -\frac{7}{25}$$

$$\tan 2x = \frac{2 \tan x}{1 - \tan^2 x}$$ *Use double-angle identity and the*
preceding results.

$$= \frac{2(-\tfrac{4}{3})}{1 - (-\tfrac{4}{3})^2}$$

$$= \frac{24}{7}$$

◆

MATCHED PROBLEM 2   Find the exact value of $\sin 2x$ and $\cos 2x$ if $\tan x = -\frac{3}{4}$, $-\pi/2 < x < 0$.

◆   Half-Angle Identities

Half-angle identities are simply double-angle identities in an alternate form. We start with the double-angle identity for cosine in the form

$$\cos 2u = 1 - 2 \sin^2 u$$

and let $u = x/2$. Then

$$\cos x = 1 - 2 \sin^2 \frac{x}{2}$$

Now solve for $\sin(x/2)$ to obtain a half-angle formula for the sine function.

$$2 \sin^2 \frac{x}{2} = 1 - \cos x$$

$$\sin^2 \frac{x}{2} = \frac{1 - \cos x}{2}$$

$$\sin \frac{x}{2} = \pm \sqrt{\frac{1 - \cos x}{2}} \qquad (7)$$

In identity (7) the choice of the sign is determined by the quadrant in which $x/2$ lies.

Now we start with the double-angle identity for cosine in the form

$$\cos 2u = 2 \cos^2 u - 1$$

and let $u = x/2$. We then obtain a half-angle formula for the cosine function.

$$\cos x = 2 \cos^2 \frac{x}{2} - 1$$

$$2 \cos^2 \frac{x}{2} = 1 + \cos x$$

$$\cos^2 \frac{x}{2} = \frac{1 + \cos x}{2}$$

$$\cos \frac{x}{2} = \pm \sqrt{\frac{1 + \cos x}{2}} \qquad (8)$$

In identity (8) the choice of the sign is again determined by the quadrant in which $x/2$ lies.

To obtain a half-angle identity for the tangent function, we can use the quotient identity and the half-angle formulas for sine and cosine.

$$\tan \frac{x}{2} = \frac{\sin \dfrac{x}{2}}{\cos \dfrac{x}{2}} = \frac{\pm\sqrt{\dfrac{1 - \cos x}{2}}}{\pm\sqrt{\dfrac{1 + \cos x}{2}}} = \pm\sqrt{\frac{1 - \cos x}{1 + \cos x}} \tag{9}$$

where the sign is determined by the quadrant in which $x/2$ lies.

We now list all the half-angle identities for convenient reference. Two of the half-angle identities for tangent are left as Problems 27 and 28 in Exercise 4.4.

HALF-ANGLE IDENTITIES

For x any real number or angle in degree or radian measure for which both sides are defined:

$$\sin \frac{x}{2} = \pm\sqrt{\frac{1 - \cos x}{2}} \qquad\qquad \cos \frac{x}{2} = \pm\sqrt{\frac{1 + \cos x}{2}}$$

$$\tan \frac{x}{2} = \pm\sqrt{\frac{1 - \cos x}{1 + \cos x}} = \frac{\sin x}{1 + \cos x} = \frac{1 - \cos x}{\sin x}$$

where the sign is determined by the quadrant containing $x/2$.

◆ **EXAMPLE 3**   Using a Half-Angle Identity

Find $\cos 165°$ exactly by means of a half-angle identity.

SOLUTION $$\cos 165° = \cos \frac{330°}{2} = -\sqrt{\frac{1 + \cos 330°}{2}}$$

The negative square root is used since $165°$ is in the second quadrant and cosine is negative there. We complete the evaluation by noting that the reference triangle for $330°$ is a $30°$–$60°$ triangle in the fourth quadrant (see Figure 2).

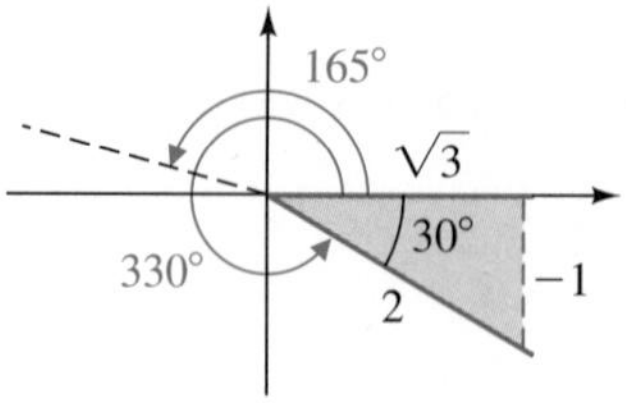

$$\cos 330° = \cos 30°$$

$$= \frac{\sqrt{3}}{2}$$

FIGURE 2

Thus,

$$\cos 165° = -\sqrt{\frac{1 + \sqrt{3}/2}{2}} = -\frac{\sqrt{2 + \sqrt{3}}}{2}$$

◆

MATCHED PROBLEM 3 Find the exact value of $\sin 165°$ using a half-angle identity.

◆ **EXAMPLE 4** Using Half-Angle Identities

Find the exact value of $\sin(x/2)$, $\cos(x/2)$, and $\tan(x/2)$ if $\sin x = -\frac{3}{5}$, $\pi < x < 3\pi/2$.

SOLUTION Draw a reference triangle in the third quadrant and find $\cos x$ (see Figure 3).

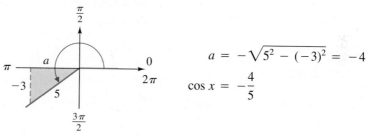

$$a = -\sqrt{5^2 - (-3)^2} = -4$$

$$\cos x = -\frac{4}{5}$$

FIGURE 3

If $\pi < x < 3\pi/2$, then

$$\pi/2 < x/2 < 3\pi/4 \qquad \text{Divide each member of } \pi < x < 3\pi/2 \text{ by 2.}$$

Thus, $x/2$ is in the second quadrant, where sine is positive and cosine and tangent are negative. Using half-angle identities, we obtain

$$\sin\frac{x}{2} = \sqrt{\frac{1 - \cos x}{2}} \qquad\qquad \cos\frac{x}{2} = -\sqrt{\frac{1 + \cos x}{2}}$$

$$= \sqrt{\frac{1 - (-\frac{4}{5})}{2}} \qquad\qquad = -\sqrt{\frac{1 + (-\frac{4}{5})}{2}}$$

$$= \sqrt{\frac{9}{10}} \quad \text{or} \quad \frac{3\sqrt{10}}{10} \qquad\qquad = -\sqrt{\frac{1}{10}} \quad \text{or} \quad \frac{-\sqrt{10}}{10}$$

$$\tan\frac{x}{2} = \frac{\sin(x/2)}{\cos(x/2)}$$

$$= \frac{3\sqrt{10}/10}{-\sqrt{10}/10} = -3$$

◆

MATCHED PROBLEM 4 Find the exact value for $\sin(x/2)$, $\cos(x/2)$, and $\tan(x/2)$ if $\cot x = -\frac{4}{3}$, $\pi/2 < x < \pi$.

◆ EXAMPLE 5   Verifying an Identity

Verify the identity

$$\cos^2 \frac{x}{2} = \frac{\tan x + \sin x}{2 \tan x}$$

VERIFICATION

$$\cos^2 \frac{x}{2} = \frac{1 + \cos x}{2}$$ *Square both sides of the half-angle identity for cosine.*

$$= \frac{\tan x}{\tan x} \cdot \frac{1 + \cos x}{2}$$ *Algebra*

$$= \frac{\tan x + \tan x \cos x}{2 \tan x}$$ *Algebra*

$$= \frac{\tan x + \sin x}{2 \tan x}$$ *Quotient identity and algebra*   ◆

MATCHED PROBLEM 5   Verify the identity

$$\sin^2 \frac{x}{2} = \frac{\tan x - \sin x}{2 \tan x}$$

Answers to
Matched Problems

1. $\dfrac{1 - \tan^2 x}{1 + \tan^2 x} = \dfrac{1 - \dfrac{\sin^2 x}{\cos^2 x}}{1 + \dfrac{\sin^2 x}{\cos^2 x}} = \cos^2 x - \sin^2 x = \cos 2x$

2. $\sin 2x = -\frac{24}{25}, \cos 2x = \frac{7}{25}$ **3.** $\dfrac{\sqrt{2 - \sqrt{3}}}{2}$

4. $\sin(x/2) = 3\sqrt{10}/10, \cos(x/2) = \sqrt{10}/10, \tan(x/2) = 3$

5. $\sin^2 \dfrac{x}{2} = \dfrac{1 - \cos x}{2} = \dfrac{\tan x}{\tan x} \cdot \dfrac{1 - \cos x}{2} = \dfrac{\tan x - \sin x}{2 \tan x}$

EXERCISE 4.4

A *Evaluate each side of the indicated identity for* $x = 60°$ *(thus verifying it for one particular case).*

 1. $\sin 2x = 2 \sin x \cos x$

2. $\cos 2x = \cos^2 x - \sin^2 x$

3. $\tan 2x = \dfrac{2 \tan x}{1 - \tan^2 x}$

4. $\sin \dfrac{x}{2} = \pm \sqrt{\dfrac{1 - \cos x}{2}}$

Use half-angle identities to find the exact value of Problems 5–8. Do not use a calculator.

5. $\sin 105°$ **6.** $\cos 105°$

7. $\tan 15°$ **8.** $\tan 75°$

B *Verify each of the following identities.*

9. $\dfrac{2 \tan x}{\sin 2x} = \sec^2 x$

10. $(\sin x + \cos x)^2 = 1 + \sin 2x$

11. $\sin 2x = (\tan x)(1 + \cos 2x)$

12. $1 - \cos 2x = \tan x \sin 2x$

13. $2 \sin^2 \dfrac{x}{2} = \dfrac{\sin^2 x}{1 + \cos x}$ 14. $2 \cos^2 \dfrac{x}{2} = \dfrac{\sin^2 x}{1 - \cos x}$

15. $(\sin \theta - \cos \theta)^2 = 1 - \sin 2\theta$

16. $\sin 2\theta = (\sin \theta + \cos \theta)^2 - 1$

17. $\cos^2 \dfrac{w}{2} = \dfrac{1 + \cos w}{2}$ 18. $\sin^2 \dfrac{w}{2} = \dfrac{1 - \cos w}{2}$

19. $\cot \dfrac{\alpha}{2} = \dfrac{1 + \cos \alpha}{\sin \alpha}$ 20. $\cot \dfrac{\alpha}{2} = \dfrac{\sin \alpha}{1 - \cos \alpha}$

21. $\tan 2\beta = \dfrac{2}{\cot \beta - \tan \beta}$

22. $\dfrac{\tan \alpha + \cot \alpha}{\cot \alpha - \tan \alpha} = \sec 2\alpha$

23. $\dfrac{\cos 2t}{1 - \sin 2t} = \dfrac{1 + \tan t}{1 - \tan t}$

24. $\cos 2t = \dfrac{1 - \tan^2 t}{1 + \tan^2 t}$

25. $\tan 2x = \dfrac{2 \tan x}{1 - \tan^2 x}$ 26. $\sin 2x = \dfrac{2 \tan x}{1 + \tan^2 x}$

27. $\tan \dfrac{x}{2} = \dfrac{\sin x}{1 + \cos x}$ 28. $\tan \dfrac{x}{2} = \dfrac{1 - \cos x}{\sin x}$

29. $\sec^2 x = (\sec 2x)(2 - \sec^2 x)$

30. $2 \csc 2x = \dfrac{1 + \tan^2 x}{\tan x}$

31. $\cos 2x = \dfrac{\cot x - \tan x}{\cot x + \tan x}$

32. $\cos x = \dfrac{1 - \tan^2(x/2)}{1 + \tan^2(x/2)}$

Graph each equation. [*Hint: Simplify first with an appropriate identity.*]

33. $y = \sin x \cos x, \quad -\pi \le x \le \pi$

34. $y = \cos^2 x - \sin^2 x, \quad -\pi \le x \le \pi$

35. $y = \dfrac{\sin x}{1 + \cos x}, \quad 0 \le x \le 2\pi$

36. $y = \dfrac{1 - \cos x}{\sin x}, \quad 0 < x < 2\pi$

37. $y = \sin^2 x, \quad -\pi \le x \le \pi$

38. $y = \cos^2 x, \quad -\pi \le x \le \pi$

Find the exact value of sin 2x, cos 2x, and tan 2x for the information given in Problems 39–44. Do not use a calculator.

39. $\sin x = \frac{3}{5}, \quad 0° < x < 90°$

40. $\cos x = \frac{4}{5}, \quad 0° < x < 90°$

41. $\cos x = -\frac{4}{5}, \quad \pi/2 < x < \pi$

42. $\sin x = \frac{3}{5}, \quad \pi/2 < x < \pi$

43. $\cot x = -\frac{5}{12}, \quad -\pi/2 < x < 0$

44. $\tan x = -\frac{5}{12}, \quad -\pi/2 < x < 0$

Find the exact value of sin(x/2) and cos(x/2) for the information given in Problems 45–50. Do not use a calculator.

45. $\cos x = \frac{1}{3}, \quad 0° < x < 90°$

46. $\sin x = \frac{4}{5}, \quad 0° < x < 90°$

47. $\sin x = -\frac{1}{3}, \quad \pi < x < 3\pi/2$

48. $\cos x = -\frac{1}{4}, \quad \pi < x < 3\pi/2$

49. $\cot x = \frac{3}{4}, \quad -\pi < x < -\pi/2$

50. $\tan x = \frac{3}{4}, \quad -\pi < x < -\pi/2$

C *Find the exact value of sin x, cos x, and tan x for the information given in Problems 51–56. Do not use a calculator.*

51. $\sin 2x = \frac{3}{5}, \quad 0 < x < \pi/4$

52. $\tan 2x = -\frac{4}{3}, \quad 0 < x < \pi/2$

53. $\sec 2x = -\frac{5}{3}, \quad -\pi/2 < x < 0$

54. $\csc 2x = -\frac{5}{3}, \quad -\pi/4 < x < 0$

55. $\tan 2x = -\frac{4}{3}, \quad -\pi < x < -\pi/2$

56. $\cot 2x = -\frac{4}{3}, \quad -\pi < x < -\pi/2$

Verify each identity.

57. $\sin 3x = 3 \sin x - 4 \sin^3 x$

58. $\cos 3x = 4 \cos^3 x - 3 \cos x$

59. $\sin 4x = (\cos x)(4 \sin x - 8 \sin^3 x)$

60. $\cos 4x = 8 \cos^4 x - 8 \cos^2 x + 1$

61. $\tan 3x = \dfrac{3 \tan x - \tan^3 x}{1 - 3 \tan^2 x}$

62. $4 \sin^4 x = 1 - 2 \cos 2x + \cos^2 2x$

63. $\dfrac{\sin x}{x} = \cos(x/2) \dfrac{\sin(x/2)}{x/2}$

64. $\dfrac{\sin x}{x} = \cos(x/2) \cos(x/4) \dfrac{\sin(x/4)}{x/4}$

Problems 65–78 require the use of a graphing calculator.

In Problems 65–68, graph y1, y2, and y3 = y1 − y2 + 1 in the same viewing rectangle.

65. $y1 = \sin 2x, \quad y2 = 2\sin x\cos x$

66. $y1 = \cos 2x, \quad y2 = 2\cos^2 x - 1$

67. $y1 = \cos 2x, \quad y2 = 1 - 2\sin^2 x$

68. $y1 = \tan 2x, \quad y2 = \dfrac{2\tan x}{1 - \tan^2 x}$

In Problems 69–72, graph y1, y2, and y3 = y1 − y2 + 1 in the same viewing rectangle for $-2\pi \le x \le 2\pi$ and state the interval(s) for which the graphs of y1 and y2 agree.

69. $y1 = \sin\dfrac{x}{2}, \quad y2 = \sqrt{\dfrac{1 - \cos x}{2}}$

70. $y1 = \sin\dfrac{x}{2}, \quad y2 = -\sqrt{\dfrac{1 - \cos x}{2}}$

71. $y1 = \cos\dfrac{x}{2}, \quad y2 = -\sqrt{\dfrac{1 + \cos x}{2}}$

72. $y1 = \cos\dfrac{x}{2}, \quad y2 = \sqrt{\dfrac{1 + \cos x}{2}}$

In Problems 73–78, graph f(x), find a simpler function g(x) that has the same graph as f(x), and verify the identity f(x) = g(x). [Assume g(x) = k + A t(Bx) where t(x) is one of the six trigonometric functions.]

73. $f(x) = \csc x + \cot x$ **74.** $f(x) = \csc x - \cot x$

75. $f(x) = \dfrac{\cot x}{1 + \cos 2x}$

76. $f(x) = \dfrac{1}{\cot x \sin 2x - 1}$

77. $f(x) = \dfrac{1 + 2\cos 2x}{1 + 2\cos x}$

78. $f(x) = \dfrac{1 - 2\cos 2x}{2\sin x - 1}$

Applications

79. **Projectile Distance**   In physics it can be shown that the distance *d* a javelin will travel (see the figure) is given

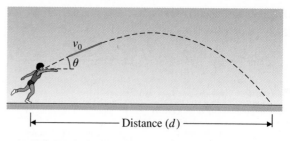

Figure for 79

approximately by

$$d = \frac{2v_0^2 \sin\theta\cos\theta}{32 \text{ ft/sec}^2}$$

where v_0 is the initial velocity of the javelin in feet per second. Write the formula in terms of sine only by using a suitable identity.

80. **Projectile Distance**   Using the resulting equation in Problem 79, determine the angle θ that will produce the maximum distance *d* for a given initial velocity v_0. [*Hint:*   For what value of θ, $0 < \theta < 90°$, will $\sin 2\theta$ be maximum?] This result is an important consideration for javelin, shot-put, and discus throwers.

81. **Engineering**   Find the exact value of *x* in the figure; then find *x* and θ to three decimal places. [*Hint:*   Use $\tan 2\theta = (2\tan\theta)/(1 - \tan^2\theta)$.]

Figure for 81

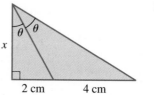

82. **Engineering**   Find the exact value of *x* in the figure; then find *x* and θ to three decimal places. [*Hint:*   Use $\cos 2\theta = 2\cos^2\theta - 1$.]

Figure for 82

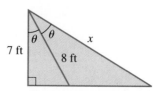

83. **Geometry** In part (a) of the figure, M and N are the midpoints of the sides of a square. Find the exact value of $\cos \theta$. [*Hint:* The solution uses the Pythagorean theorem, the definition of sine and cosine, a half-angle identity, and some auxiliary lines as drawn in part (b) of the figure.]

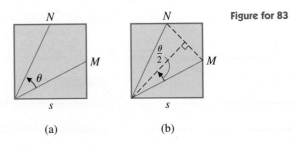

Figure for 83

(a)　　　　(b)

☆4.5 PRODUCT-SUM AND SUM-PRODUCT IDENTITIES

- ◆ **Product-Sum Identities**
- ◆ **Sum-Product Identities**
- ◆ **Application: Music**

In this section, we will develop identities for converting the product of two trigonometric functions into the sum of two trigonometric functions, and vice versa. These identities have many uses, both theoretically and practically. For example, in Exercise 5.4 these identities will be of use in solving a particular type of conditional trigonometric equation. In calculus, knowing how to convert a product into a sum will allow easy solutions to some problems that would otherwise be difficult to solve. A discussion of beat frequencies in music, which appears at the end of this section, demonstrates how converting a sum into a product aids in the analysis of the beat phenomenon in sound. A similar type of analysis is used in modeling certain types of long-range underwater sound propagation for submarine detection.

◆ Product-Sum Identities

The product-sum identities are easily derived from the sum and difference identities developed in Section 4.3. To obtain a product-sum identity, we add, left side to left side and right side to right side, the sum and difference identities for sine:

$$\sin(x + y) = \sin x \cos y + \cos x \sin y$$
$$\sin(x - y) = \sin x \cos y - \cos x \sin y$$
$$\sin(x + y) + \sin(x - y) = 2 \sin x \cos y$$

or

$$\sin x \cos y = \frac{1}{2}[\sin(x + y) + \sin(x - y)]$$

---

* Sections marked with a star may be omitted without loss of continuity.

Similarly, by adding or subtracting appropriate sum and difference identities, we can obtain three other product identities for sines and cosines. These identities are listed below for convenient reference.

PRODUCT-SUM IDENTITIES

For x and y any real numbers or angles in degree or radian measure for which both sides are defined,

$$\sin x \cos y = \frac{1}{2}[\sin(x + y) + \sin(x - y)]$$

$$\cos x \sin y = \frac{1}{2}[\sin(x + y) - \sin(x - y)]$$

$$\sin x \sin y = \frac{1}{2}[\cos(x - y) - \cos(x + y)]$$

$$\cos x \cos y = \frac{1}{2}[\cos(x + y) + \cos(x - y)]$$

◆ **EXAMPLE 1** Using a Product-Sum Identity

Write the product $\cos 3t \sin t$ as a sum or difference.

SOLUTION
$$\cos x \sin y = \frac{1}{2}[\sin(x + y) - \sin(x - y)] \qquad \text{Let } x = 3t \text{ and } y = t.$$

$$\cos 3t \sin t = \frac{1}{2}[\sin(3t + t) - \sin(3t - t)]$$

$$= \frac{1}{2}\sin 4t - \frac{1}{2}\sin 2t$$

◆

MATCHED PROBLEM 1   Write the product $\cos 5\theta \cos 2\theta$ as a sum or difference.

◆ **EXAMPLE 2** Using a Product-Sum Identity

Evaluate $\sin 105° \sin 15°$ exactly using a product-sum identity.

SOLUTION
$$\sin x \sin y = \frac{1}{2}[\cos(x - y) - \cos(x + y)]$$

$$\sin 105° \sin 15° = \frac{1}{2}[\cos(105° - 15°) - \cos(105° + 15°)]$$

$$= \frac{1}{2}[\cos 90° - \cos 120°]$$

$$= \frac{1}{2}\left[0 - \left(-\frac{1}{2}\right)\right] = \frac{1}{4} \qquad \text{or} \qquad 0.25$$

◆

MATCHED PROBLEM 2   Evaluate cos 165° sin 75° exactly using a product-sum identity.

◆ Sum-Product Identities

The product-sum identities can be transformed into equivalent forms called sum-product identities. These identities are used to express sums and differences involving sines and cosines as products involving sines and cosines. We illustrate the transformation for one identity. The other three identities can be obtained by following the same procedure.

Let us start with the product-sum identity

$$\sin \alpha \cos \beta = \frac{1}{2}[\sin(\alpha + \beta) + \sin(\alpha - \beta)] \qquad (1)$$

We would like

$$\alpha + \beta = x \qquad \alpha - \beta = y$$

Solving this system, we have

$$\alpha = \frac{x + y}{2} \qquad \beta = \frac{x - y}{2} \qquad (2)$$

By substituting (2) into identity (1) and simplifying, we obtain

$$\textbf{sin } x + \sin y = 2 \sin \frac{x + y}{2} \cos \frac{x - y}{2}$$

All four sum-product identities are listed below for convenient reference.

SUM-PRODUCT IDENTITIES

For x and y any real numbers or angles in degree or radian measure for which both sides are defined,

$$\sin x + \sin y = 2 \sin \frac{x + y}{2} \cos \frac{x - y}{2}$$

$$\sin x - \sin y = 2 \cos \frac{x + y}{2} \sin \frac{x - y}{2}$$

$$\cos x + \cos y = 2 \cos \frac{x + y}{2} \cos \frac{x - y}{2}$$

$$\cos x - \cos y = -2 \sin \frac{x + y}{2} \sin \frac{x - y}{2}$$

◆ EXAMPLE 3   Using a Sum-Product Identity

Write the difference $\sin 7\theta - \sin 3\theta$ as a product.

SOLUTION

$$\sin x - \sin y = 2 \cos \frac{x + y}{2} \sin \frac{x - y}{2}$$

$$\sin 7\theta - \sin 3\theta = 2 \cos \frac{7\theta + 3\theta}{2} \sin \frac{7\theta - 3\theta}{2}$$

$$= 2 \cos 5\theta \sin 2\theta \qquad \blacklozenge$$

MATCHED PROBLEM 3   Write the sum $\cos 3t + \cos t$ as a product.

◆ EXAMPLE 4   Using a Sum-Product Identity

Find the exact value of $\sin 105° - \sin 15°$ using an appropriate sum-product identity.

SOLUTION

$$\sin x - \sin y = 2 \cos \frac{x + y}{2} \sin \frac{x - y}{2}$$

$$\sin 105° - \sin 15° = 2 \cos \frac{105° + 15°}{2} \sin \frac{105° - 15°}{2}$$

$$= 2 \cos 60° \sin 45°$$

$$= 2 \left(\frac{1}{2}\right)\left(\frac{\sqrt{2}}{2}\right) = \frac{\sqrt{2}}{2} \qquad \blacklozenge$$

MATCHED PROBLEM 4   Find the exact value of $\cos 165° - \cos 75°$ by using an appropriate sum-product identity.

◆ Application: Music

If two tones that have the same loudness and that are close in pitch (frequency) are sounded, one following the other, most people have difficulty in recognizing that the tones are different. However, if the tones are sounded simultaneously, they will react with each other, producing a low warbling sound called a **beat**. The beat or warble will be slow or rapid, depending on how far apart the initial frequencies are. Musicians, when tuning an instrument with other instruments or a tuning fork, listen for these lower beat frequencies and try to eliminate them by adjusting their instruments. The more rapid the beat frequency (warbling) when two instruments play together, the greater the difference in their frequencies and the more out of tune they are. If, after adjustments, no beats are heard, the two instruments are in tune.

What is behind this beat phenomenon? Figure 1(a) shows a tone of 64 Hz (cycles per second), and Figure 1(b) shows a tone of the same loudness but with a frequency of 72 Hz. If both tones are sounded simultaneously and time is started when the two tones are completely out of phase, waves (a) and (b) will interact to form wave (c). Sum-product identities are useful in the mathematical

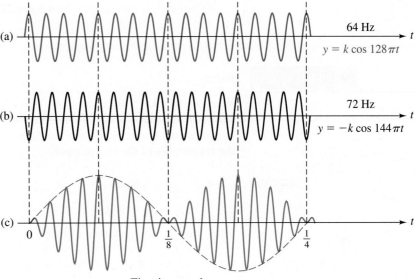

FIGURE 1
Beats

analysis of the beat phenomenon. In particular, we will use the sum-product identity

$$\cos x - \cos y = -2 \sin \frac{x + y}{2} \sin \frac{x - y}{2}$$

in a brief mathematical discussion of the three waves in Figure 1.

We start with the fact that wave (c) is the sum of waves (a) and (b):

$$
\begin{aligned}
y &= k \cos 128\pi t - k \cos 144\pi t \\
&= k(\cos 128\pi t - \cos 144\pi t) \\
&= k\left(-2 \sin \frac{128\pi t + 144\pi t}{2} \sin \frac{128\pi t - 144\pi t}{2}\right) \\
&= -2k \sin 136\pi t \sin(-8\pi t) \\
&= 2k \sin 136\pi t \sin 8\pi t \\
&= 2k \sin 8\pi t \sin 136\pi t
\end{aligned}
$$

The factor $\sin 136\pi t$ represents a sound of frequency 68 Hz, the average of the frequencies of the individual tones. The first factor, $2k \sin 8\pi t$, can be thought of as a time-varying amplitude of the second factor, $\sin 136\pi t$. It is the first factor that describes the slow warble of volume, or the beat, of the combined sounds. This pulsation of sound is described in terms of frequency—that is, it tells how many times per second the maximum loudness or number of beats occurs. Referring to Figure 1(c), we see that there are 2 beats in $\frac{1}{4}$ sec, or 8 beats/sec. Hence, the beat frequency is 8 beats/sec (which is the difference in frequencies of the two original waves).

In general, if two equally loud tones of frequencies f_1 and f_2 are produced simultaneously, and if $f_1 > f_2$, then the beat frequency, f_b, is given by

$$f_b = f_1 - f_2 \qquad \textit{Beat frequency}$$

◆ EXAMPLE 5   Music

When certain keys on a piano are struck, a felt-covered hammer strikes two strings. If the piano is out of tune, the tones from the two strings create a beat, and the sound is sour. If a piano tuner counts 15 beats in 5 sec, how far apart are the frequencies of the two strings?

SOLUTION $$f_b = \frac{15}{5} = 3 \text{ beats/sec}$$

$$f_1 - f_2 = f_b = 3 \text{ Hz}$$

Thus, the two strings are out of tune by 3 cycles/sec.   ◆

MATCHED PROBLEM 5   What is the beat frequency for the two tones in Figure 1?

Answers to   **1.** $\cos 5\theta \cos 2\theta = \frac{1}{2}\cos 7\theta + \frac{1}{2}\cos 3\theta$ **2.** $(-\sqrt{3} - 2)/4$
Matched Problems   **3.** $\cos 3t + \cos t = 2 \cos 2t \cos t$ **4.** $-\sqrt{6}/2$ **5.** $f_b = 8$ Hz

EXERCISE 4.5

A *Write each product as a sum or difference involving sines and cosines.*

1. $\cos 7A \cos 5A$

2. $\sin 3m \cos m$

3. $\cos 2\theta \sin 3\theta$

4. $\sin u \sin 3u$

Write each difference or sum as a product involving sines and cosines.

5. $\cos 7\theta + \cos 5\theta$

6. $\sin 3t + \sin t$

7. $\sin u - \sin 5u$

8. $\cos 5w - \cos 9w$

B *Evaluate each of the following exactly using an appropriate identity.*

9. $\cos 75° \sin 15°$

10. $\sin 195° \cos 75°$

11. $\sin 105° \sin 165°$

12. $\cos 15° \cos 75°$

Evaluate each of the following exactly using an appropriate identity.

13. $\sin 195° + \sin 105°$

14. $\cos 285° + \cos 195°$

15. $\sin 75° - \sin 165°$

16. $\cos 15° - \cos 105°$

Use sum and difference identities from Section 4.3 to establish each of the following.

17. $\sin x \sin y = \frac{1}{2}[\cos(x - y) - \cos(x + y)]$

18. $\cos x \cos y = \frac{1}{2}[\cos(x + y + \cos(x - y)]$

Use appropriate substitutions in the product-sum identities to obtain the following.

19. $\sin x - \sin y = 2 \cos \dfrac{x + y}{2} \sin \dfrac{x - y}{2}$

20. $\cos x - \cos y = -2 \sin \dfrac{x + y}{2} \sin \dfrac{x - y}{2}$

Verify each of the following identities.

21. $\dfrac{\cos t - \cos 3t}{\sin t + \sin 3t} = \tan t$

22. $\dfrac{\sin 2t + \sin 4t}{\cos 2t - \cos 4t} = \cot t$

23. $\dfrac{\sin x + \sin y}{\cos x + \cos y} = \tan \dfrac{x + y}{2}$

24. $\dfrac{\sin x - \sin y}{\cos x - \cos y} = -\cot \dfrac{x + y}{2}$

25. $\dfrac{\cos x - \cos y}{\sin x + \sin y} = -\tan \dfrac{x - y}{2}$

26. $\dfrac{\cos x + \cos y}{\sin x - \sin y} = \cot \dfrac{x - y}{2}$

27. $\dfrac{\sin x + \sin y}{\sin x - \sin y} = \dfrac{\tan \frac{1}{2}(x + y)}{\tan \frac{1}{2}(x - y)}$

28. $\dfrac{\cos x + \cos y}{\cos x - \cos y} = -\cot \dfrac{x + y}{2} \cot \dfrac{x - y}{2}$

C   **29.** $\sin x \sin y \sin z = \frac{1}{4}[\sin(x + y - z)$
$+ \sin(y + z - x)$
$+ \sin(z + x - y)$
$- \sin(x + y + z)]$

30. $\cos x \cos y \cos z = \frac{1}{4}[\cos(x + y - z)$
$+ \cos(y + z - x)$
$+ \cos(z + x - y)$
$+ \cos(x + y + z)]$

Problems 31–34 require the use of a graphing calculator.
(A) Graph y1 and y2 for $0 \le x \le 1$ and $-2 \le y \le 2$.
(B) Convert y1 to a sum or difference and repeat part A.

31. $y1 = 2 \cos(16\pi x) \sin(2\pi x)$, $y2 = 2 \sin(2\pi x)$

32. $y1 = 2 \sin(20\pi x) \cos(2\pi x)$, $y2 = 2 \cos(2\pi x)$

33. $y1 = 2 \sin(24\pi x) \sin(2\pi x)$, $y2 = 2 \sin(2\pi x)$

34. $y1 = 2 \cos(28\pi x) \cos(2\pi x)$, $y2 = 2 \cos(2\pi x)$

Applications

35. **Music**   If one tone is described by $y = k \sin 522\pi t$ and another by $y = k \sin 512\pi t$, write their sum as a product. What is the beat frequency if both notes are sounded together?

36. **Music**   If one tone is described by $y = k \cos 524\pi t$ and another by $y = k \cos 508\pi t$, write their sum as a product. What is the beat frequency if both notes are sounded together?

C   **37.** **Music**   Equations

$$y = 0.3 \cos 72\pi t \quad \text{and} \quad y = -0.3 \cos 88\pi t$$

are equations of sound waves with frequencies 36 and 44 hertz, respectively. If both sounds are emitted simultaneously, a beat frequency results. Use the viewing rectangle $0 \le t \le 0.25$ and $-0.8 \le y \le 0.8$ for each part, (A) through (D).
(A) Graph $y = 0.3 \cos 72\pi t$.
(B) Graph $y = -0.3 \cos 88\pi t$.
(C) Graph $y1 = 0.3 \cos 72\pi t - 0.3 \cos 88\pi t$ and $y2 = 0.6 \sin 8\pi t$.
(D) Convert y1 in (C) to a product and graph the new y1 along with y2 from (C).

C   **38.** **Music**   Equations

$$y = 0.4 \cos 132\pi t \quad \text{and} \quad y = -0.4 \cos 152\pi t$$

are equations of sound waves with frequencies 66 and 76 hertz, respectively. If both sounds are emitted simultaneously, a beat frequency results. Use the viewing rectangle $0 \le t \le 0.2$ and $-0.8 \le y \le 0.8$ for each part, (A) through (D).
(A) Graph $y = 0.4 \cos 132\pi t$.
(B) Graph $y = -0.4 \cos 152\pi t$.
(C) Graph $y1 = 0.4 \cos 132\pi t - 0.4 \cos 152\pi t$ and $y2 = 0.8 \sin 10\pi t$.
(D) Convert y1 in (C) to a product and graph the new y1 along with y2 from (C).

⋆4.6   FROM $M \sin Bt + N \cos Bt$ TO $A \sin(Bt + C)$

In the process of solving certain kinds of problems that require the use of more advanced mathematics—problems dealing with heat flow, electrical circuits, spring-mass systems, and so on—we are led naturally to the form

$$y = M \sin Bt + N \cos Bt \tag{1}$$

⋆ Sections marked with a star may be omitted without loss of continuity.

With a little ingenuity and the use of the sum identity for sine,

$$\sin(x + y) = \sin x \cos y + \cos x \sin y \tag{2}$$

we can convert equation (1) into the form

$$y = A \sin(Bt + C) \tag{3}$$

This form of a solution is often more helpful and convenient than (1), since from it we can easily determine amplitude, period, frequency, and phase shift (and a graph if necessary).

How do we proceed? We start by trying to get the right side of equation (1) to look like the right side of identity (2). Then we use (2), from right to left, to obtain (3). Getting (1) to look like (2) requires constants M and N to be replaced by $\cos C$ and $\sin C$, respectively, where C is an appropriate number. It turns out that C can be any angle (in radians if t is real or in radians) having $P(M, N)$ on its terminal side (see Figure 1).

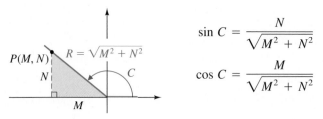

$$\sin C = \frac{N}{\sqrt{M^2 + N^2}}$$

$$\cos C = \frac{M}{\sqrt{M^2 + N^2}}$$

FIGURE 1

In (1), M and N must each be divided by $\sqrt{M^2 + N^2}$ to replace M with $\cos C$ and N with $\sin C$. We accomplish this as follows.

$$y = M \sin Bt + N \cos Bt$$

$$= \frac{\sqrt{M^2 + N^2}}{\sqrt{M^2 + N^2}}(M \sin Bt + N \cos Bt)$$

$$= \sqrt{M^2 + N^2}\left(\frac{M}{\sqrt{M^2 + N^2}} \sin Bt + \frac{N}{\sqrt{M^2 + N^2}} \cos Bt\right)$$

$$= \sqrt{M^2 + N^2}\,(\cos C \sin Bt + \sin C \cos Bt)$$

$$= \sqrt{M^2 + N^2}\,(\sin Bt \cos C + \cos Bt \sin C)$$

$$= \sqrt{M^2 + N^2}\,\sin(Bt + C) \qquad \text{Using identity (2)}$$

We thus have form (3) with $A = \sqrt{M^2 + N^2}$. We summarize the results in the following box for convenient reference.

$$y = M \sin Bt + N \cos Bt = \sqrt{M^2 + N^2} \sin(Bt + C)$$

where C is any angle (in radians if t is real) having $P(M, N)$ on its terminal side.

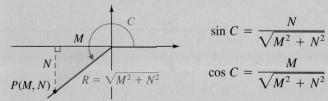

$$\sin C = \frac{N}{\sqrt{M^2 + N^2}}$$

$$\cos C = \frac{M}{\sqrt{M^2 + N^2}}$$

$$\text{Amplitude} = \sqrt{M^2 + N^2} \qquad \text{Frequency} = \frac{B}{2\pi}$$

$$\text{Period} = \frac{2\pi}{B} \qquad \text{Phase shift} = -\frac{C}{B}$$

(Or, find the period and phase shift by solving the two equations, $Bt + C = 0$ and $Bt + C = 2\pi$.)

In converting equation (1) to form (3) you can either go through the steps that led up to the boxed material or use the boxed material directly, whichever is easier. Let us consider several examples.

◆ EXAMPLE 1

From $M \sin Bt + N \cos Bt$ to $A \sin(Bt + C)$

Write

$$y = -\sqrt{3} \sin 2t + \cos 2t$$

in the form $y = A \sin(Bt + C)$, and indicate its amplitude, period, frequency, and phase shift. (Choose C as small as possible, but positive.)

SOLUTION

$$M = -\sqrt{3} \qquad \text{and} \qquad N = 1$$

Locate $P(M, N) = P(-\sqrt{3}, 1)$ to determine C (see Figure 2).

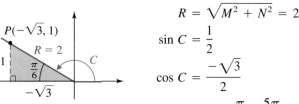

$$R = \sqrt{M^2 + N^2} = 2$$

$$\sin C = \frac{1}{2}$$

$$\cos C = \frac{-\sqrt{3}}{2}$$

$$C = \pi - \frac{\pi}{6} = \frac{5\pi}{6}$$

FIGURE 2

Thus,

$$y = -\sqrt{3} \sin 2t + \cos 2t = 2 \sin\left(2t + \frac{5\pi}{6}\right)$$

Amplitude

$$\text{Amplitude} = |A| = |2| = 2$$

Period and phase shift

$$2t + \frac{5\pi}{6} = 0 \qquad\qquad 2t + \frac{5\pi}{6} = 2\pi$$

$$2t = -\frac{5\pi}{6} \qquad\qquad 2t = -\frac{5\pi}{6} + 2\pi$$

$$t = -\frac{5\pi}{12} \qquad\qquad t = -\frac{5\pi}{12} + \pi$$

$$\text{Period} = \pi \qquad \text{Phase shift} = -\frac{5\pi}{12}$$

Frequency

$$\text{Frequency} = \frac{1}{\text{Period}} = \frac{1}{\pi}$$

MATCHED PROBLEM 1   Repeat Example 1 for $y = -\sin \pi t + \cos \pi t$.

◆ EXAMPLE 2   **From $M \sin Bt + N \cos Bt$ to $A \sin(Bt + C)$**

Write

$$y = \sin \pi t - \sqrt{3} \cos \pi t$$

in the form $y = A \sin(Bt + C)$, where C is chosen (positive or negative) so that $|C|$ is minimum. Indicate amplitude, period, phase shift, and frequency. Graph the equation over the interval $0 \le t \le 3$.

SOLUTION $M = 1 \quad$ and $\quad N = -\sqrt{3}$

Locate $P(M, N) = P(1, -\sqrt{3})$ to determine C (see Figure 3).

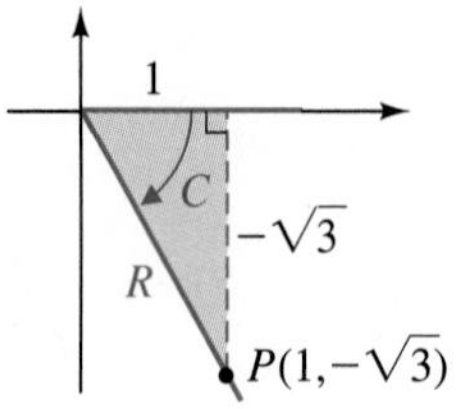

$$R = \sqrt{M^2 + N^2} = 2$$

FIGURE 3

$$\sin C = \frac{-\sqrt{3}}{2}$$

$$\cos C = \frac{1}{2}$$

$$C = -\frac{\pi}{3} \qquad |C| \text{ is minimum for this choice.}$$

Thus,

$$y = 2 \sin\left(\pi t - \frac{\pi}{3}\right)$$

Amplitude

$$\text{Amplitude} = |A| = |2| = 2$$

Period and phase shift

$$\pi t - \frac{\pi}{3} = 0 \qquad\qquad \pi t - \frac{\pi}{3} = 2\pi$$

$$\pi t = \frac{\pi}{3} \qquad\qquad \pi t = \frac{\pi}{3} + 2\pi$$

$$t = \frac{1}{3} \qquad\qquad t = \frac{1}{3} + 2$$

$$\text{Period} = 2 \qquad \text{Phase shift} = \frac{1}{3}$$

Frequency   *Graph*

$$\text{Frequency} = \frac{1}{\text{Period}} = \frac{1}{2}$$

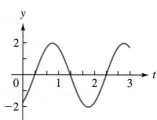

FIGURE 4

MATCHED PROBLEM 2   Repeat Example 2 for $y = \sin \pi t + \sqrt{3} \cos \pi t, \; -1 \le t \le 2$

◆ EXAMPLE 3   From $M \sin Bt + N \cos Bt$ to $A \sin(Bt + C)$

Write

$$y = -3 \sin 2\pi t - 4 \cos 2\pi t$$

in the form $y = A \sin(Bt + C)$, where C is chosen so that $|C|$ is minimum. Compute C to two decimal places.

SOLUTION   Let $M = -3$ and $N = -4$; then locate $P(M, N) = P(-3, -4)$ to determine C (see Figure 5).

FIGURE 5

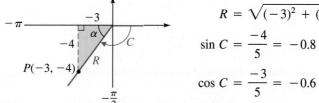

$$R = \sqrt{(-3)^2 + (-4)^2} = 5$$

$$\sin C = \frac{-4}{5} = -0.8$$

$$\cos C = \frac{-3}{5} = -0.6$$

Find the reference angle α. Then

$$C = \alpha - \pi$$

$$\sin \alpha = \frac{4}{5} = 0.8$$

$$\alpha \approx 0.93 \qquad \text{Use a calculator: } \alpha = \sin^{-1} 0.8.$$

Thus,

$$C \approx 0.93 - \pi \approx -2.21 \qquad \text{Use } \pi \approx 3.14.$$

and $|C|$ is minimum for this choice of C. We can now write

$$y = 5 \sin(2\pi t - 2.21)$$

Amplitude

$$\text{Amplitude} = |A| = |5| = 5$$

Period and phase shift

$$2\pi t - 2.21 = 0 \qquad\qquad 2\pi t - 2.21 = 2\pi$$

$$2\pi t = 2.21 \qquad\qquad 2\pi t = 2.21 + 2\pi$$

$$t = \frac{2.21}{2\pi} \qquad\qquad t = \frac{2.21}{2\pi} + 1$$

$$t = 0.35 \qquad\qquad t = 0.35 + 1$$

$$\text{Period} = 1 \qquad \text{Phase shift} \approx 0.35$$

Frequency

$$\text{Frequency} = \frac{1}{\text{Period}} = 1$$

MATCHED PROBLEM 3   Repeat Example 3 for $y = -4 \sin(t/2) + 3 \cos(t/2)$.

Answers to   **1.** $y = \sqrt{2} \sin(\pi t + 3\pi/4)$; Amplitude $= \sqrt{2}$, Period $= 2$,
Matched Problems   Frequency $= \frac{1}{2}$, Phase shift $= -\frac{3}{4}$

2. $y = 2 \sin(\pi t + \pi/3)$; Amplitude $= 2$, Period $= 2$, Frequency $= \frac{1}{2}$,
Phase shift $= -\frac{1}{3}$

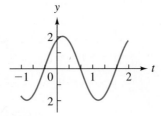

3. $y = 5 \sin(t/2 + 2.50)$; Amplitude $= 5$, Period $= 4\pi$,
Frequency $= \frac{1}{4}\pi$, Phase shift $= -5$

EXERCISE 4.6

In all problems in this exercise, t represents time in seconds.

A *Write each equation in the form $y = A \sin(Bt + C)$. Indicate amplitude, period, frequency, and phase shift. Choose C as small as possible, but positive.*

1. $y = \sin t + \cos t$ **2.** $y = -\sin t + \cos t$

3. $y = -\sin t - \cos t$ **4.** $y = \sin t - \cos t$

5. $y = -\sqrt{3} \sin t + \cos t$ **6.** $y = \sin t - \sqrt{3} \cos t$

7. $y = -\sin t - \sqrt{3} \cos t$ **8.** $y = \sqrt{3} \sin t - \cos t$

B *Write each equation in the form $y = A \sin(Bt + C)$, where C is chosen so that $|C|$ is minimum. Indicate amplitude, period, frequency, and phase shift; then graph the equation for the indicated interval.*

9. $y = \sin t - \cos t, \quad 0 \leq t \leq 3\pi$

10. $y = \sin t + \cos t, \quad -\pi \leq t \leq 2\pi$

11. $y = \sin \pi t + \cos \pi t, \quad -1 \leq t \leq 2$

12. $y = \sin \pi t - \cos \pi t, \quad 0 \leq t \leq 3$

13. $y = \sqrt{3} \sin \pi t - \cos \pi t, \quad 0 \leq t \leq 3$

14. $y = \sqrt{3} \sin \pi t + \cos \pi t, \quad -1 \leq t \leq 2$

15. $y = \sin 2\pi t + \cos 2\pi t, \quad -1 \leq t \leq 1$

16. $y = \sin 2\pi t - \cos 2\pi t, \quad 0 \leq t \leq 2$

17. $y = -\sin 2\pi t - \sqrt{3} \cos 2\pi t, \quad 0 \leq t \leq 2$

18. $y = -\sin 2\pi t + \sqrt{3} \cos 2\pi t, \quad -1 \leq t \leq 1$

C *Write each equation in the form $y = A \sin(Bt + C)$, where C is chosen so that $|C|$ is minimum and C is computed to two decimal places using a calculator. Indicate amplitude, period, frequency, and phase shift.*

19. $y = 4 \sin \pi t - 3 \cos \pi t$

20. $y = -3 \sin \dfrac{t}{4} + 4 \cos \dfrac{t}{4}$

21. $y = -5 \sin 3t + 3 \cos 3t$

22. $y = 2 \sin 8t - 5 \cos 8t$

 Problems 23–26 require the use of a graphing calculator and they are related to Problems 19–22. Graph the given equation over the indicated interval, approximate the x intercepts in this interval to two decimal places, and identify the intercept that corresponds to the phase shift for the form $y = A \sin(Bt + C)$ we determined earlier.

23. From Problem 19,

$$y = 4 \sin \pi x - 3 \cos \pi x, \quad -1 \leq x \leq 1$$

24. From Problem 20,

$$y = -3 \sin \frac{x}{4} + 4 \cos \frac{x}{4}, \quad -4\pi \leq x \leq 4\pi$$

25. From Problem 21,

$$y = -5 \sin 3x + 3 \cos 3x, \quad -\frac{\pi}{3} \leq x \leq \frac{x}{3}$$

26. From Problem 22,

$$y = 2 \sin 8x - 5 \cos 8x, \quad -\frac{\pi}{8} \leq x \leq \frac{\pi}{8}$$

Applications

*27. Physics A weight suspended from a spring, with spring constant 64, is pulled 4 cm below its equilibrium position and is then given a downward thrust to produce an initial downward velocity of 24 cm/sec. In more advanced mathematics (differential equations) the equation of motion (neglecting friction and air resistance) is found to be given approximately by

$$y = -3 \sin 8t - 4 \cos 8t$$

where y is the position on the scale in the figure at time t (y is in centimeters and t is in seconds). Write this equation in the form

$$y = A \sin(Bt + C)$$

and indicate the amplitude, period, frequency, and phase shift of the motion. (Choose the least positive C and keep A positive.)

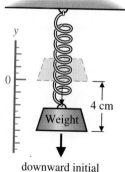

downward initial
velocity = 24 cm/sec

Figure for 27

*28. Music A musical tone is described by

$$y = 0.04 \sin 200\pi t - 0.03 \cos 200\pi t$$

(A) Write the equation in the form $y = A \sin(Bt + C)$. Compute C (to two decimal places) so that $|C|$ is minimum.

(B) Indicate amplitude, period, frequency, and phase shift.

CHAPTER 4 SUMMARY

4.1
FUNDAMENTAL IDENTITIES
AND THEIR USE

An equation in one or more variables is said to be an **identity** if the left side is equal to the right side for all replacements of the variables for which both sides are defined. If the left side is equal to the right side only for certain values of the variables and not for all values for which both sides are defined, then the equation is called a **conditional equation**.

Fundamental Trigonometric Identities

For x any real number or angle in degree or radian measure for which both sides are defined:

Reciprocal identities

$$\csc x = \frac{1}{\sin x} \qquad \sec x = \frac{1}{\cos x} \qquad \cot x = \frac{1}{\tan x}$$

Quotient identities

$$\tan x = \frac{\sin x}{\cos x} \qquad \cot x = \frac{\cos x}{\sin x}$$

Identities for negatives

$$\sin(-x) = -\sin x \qquad \cos(-x) = \cos x \qquad \tan(-x) = -\tan x$$

Pythagorean identities

$$\sin^2 x + \cos^2 x = 1 \qquad \tan^2 x + 1 = \sec^2 x \qquad 1 + \cot^2 x = \csc^2 x$$

4.2
VERIFYING TRIGONOMETRIC
IDENTITIES

When **verifying an identity**, start with the expression on one side and through a sequence of valid steps involving the use of known identities or algebraic manipulation, convert that expression into the expression on the other side. *Do not* add the same quantity to each side, multiply each side by the same nonzero quantity, or square or take the square root of both sides.

Some Suggestions for Verifying Identities

1. Start with the more complicated side of the identity and transform it into the simpler side. In the process, keep the simpler side of the identity in mind; knowing where you are going often suggests steps that should be taken to get there.

2. Specifically, after you have taken any obvious steps, try to
   - Express each function in terms of sine and cosine; or
   - Use algebraic manipulation such as simplifying, multiplying, factoring, combining fractions, splitting single fractions, and so on.

4.3
SUM, DIFFERENCE, AND
COFUNCTION IDENTITIES

For x and y any real numbers or angles in degree or radian measure for which both sides are defined:

Sum identities

$$\sin(x + y) = \sin x \cos y + \cos x \sin y$$
$$\cos(x + y) = \cos x \cos y - \sin x \sin y$$
$$\tan(x + y) = \frac{\tan x + \tan y}{1 - \tan x \tan y}$$

Difference identities

$$\sin(x - y) = \sin x \cos y - \cos x \sin y$$
$$\cos(x - y) = \cos x \cos y + \sin x \sin y$$
$$\tan(x - y) = \frac{\tan x - \tan y}{1 + \tan x \tan y}$$

Cofunction identities

(Replace $\pi/2$ with $90°$ if x is in degree measure.)

$$\sin\left(\frac{\pi}{2} - x\right) = \cos x \qquad \cos\left(\frac{\pi}{2} - x\right) = \sin x \qquad \tan\left(\frac{\pi}{2} - x\right) = \cot x$$

$$\cot\left(\frac{\pi}{2} - x\right) = \tan x \qquad \sec\left(\frac{\pi}{2} - x\right) = \csc x \qquad \csc\left(\frac{\pi}{2} - x\right) = \sec x$$

4.4
DOUBLE-ANGLE AND
HALF-ANGLE IDENTITIES

Double-angle identities

For x any real number or angle in degree or radian measure for which both sides are defined,

$$\sin 2x = 2 \sin x \cos x \qquad \cos 2x = \cos^2 x - \sin^2 x$$
$$\tan 2x = \frac{2 \tan x}{1 - \tan^2 x} \qquad\qquad = 1 - 2 \sin^2 x$$
$$= 2 \cos^2 x - 1$$

Half-angle identities

For x any real number or angle in degree or radian measure for which both sides are defined,

$$\sin\frac{x}{2} = \pm\sqrt{\frac{1 - \cos x}{2}} \qquad \cos\frac{x}{2} = \pm\sqrt{\frac{1 + \cos x}{2}}$$

$$\tan\frac{x}{2} = \pm\sqrt{\frac{1 - \cos x}{1 + \cos x}} = \frac{\sin x}{1 + \cos x} = \frac{1 - \cos x}{\sin x}$$

where the sign is determined by the quadrant in which $x/2$ lies.

Product-sum identities

For x and y any real numbers or angles in degree or radian measure for which both sides are defined,

$$\sin x \cos y = \frac{1}{2}[\sin(x + y) + \sin(x - y)]$$

$$\cos x \sin y = \frac{1}{2}[\sin(x + y) - \sin(x - y)]$$

$$\sin x \sin y = \frac{1}{2}[\cos(x - y) - \cos(x + y)]$$

$$\cos x \cos y = \frac{1}{2}[\cos(x + y) + \cos(x - y)]$$

Sum-product identities

For x and y any real numbers or angles in degree or radian measure for which both sides are defined,

$$\sin x + \sin y = 2 \sin \frac{x + y}{2} \cos \frac{x - y}{2}$$

$$\sin x - \sin y = 2 \cos \frac{x + y}{2} \sin \frac{x - y}{2}$$

$$\cos x + \cos y = 2 \cos \frac{x + y}{2} \cos \frac{x - y}{2}$$

$$\cos x - \cos y = -2 \sin \frac{x + y}{2} \sin \frac{x - y}{2}$$

$$y = M \sin Bt + N \cos Bt = \sqrt{M^2 + N^2} \sin(Bt + C)$$

where C is any angle (in radians if t is real) having $P(M, N)$ on its terminal side.

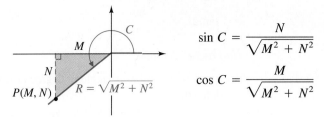

$$\sin C = \frac{N}{\sqrt{M^2 + N^2}}$$

$$\cos C = \frac{M}{\sqrt{M^2 + N^2}}$$

FIGURE 1

$$\text{Amplitude} = \sqrt{M^2 + N^2} \qquad \text{Frequency} = \frac{B}{2\pi}$$

$$\text{Period} = \frac{2\pi}{B} \qquad \text{Phase shift} = -\frac{C}{B}$$

(Or, find the period and phase shift by solving the two equations $Bt + C = 0$ and $Bt + C = 2\pi$.)

CHAPTER 4 REVIEW EXERCISE

Work through all the problems in this chapter review and check the answers. Answers to all review problems appear in the back of the book; following each answer is an italic number that indicates the section in which that type of problem is discussed. Where weaknesses show up, review the appropriate sections in the text.

Choose the function on the right that will make the equation an identity.

1. $\sin x = \begin{cases} \dfrac{1}{\sec x} \\ \dfrac{1}{\csc x} \end{cases}$

2. $\sec x = \begin{cases} \dfrac{1}{\cos x} \\ \dfrac{1}{\sin x} \end{cases}$

3. $\cot x = \begin{cases} \dfrac{\sin x}{\cos x} \\ \dfrac{\cos x}{\sin x} \end{cases}$

4. $\tan x = \begin{cases} \dfrac{\sin x}{\cos x} \\ \dfrac{\cos x}{\sin x} \end{cases}$

5. $\sin^2 x = \begin{cases} 1 - \cos^2 x \\ 1 + \cos^2 x \end{cases}$

6. $\cos(-x) = \begin{cases} -\cos x \\ \cos x \end{cases}$

Verify each identity in Problems 7–15 without looking at a table of identities.

7. $\csc x \sin x = \sec x \cos x$

8. $\cot x \sin x = \cos x$

9. $\tan x = -\tan(-x)$

10. $\dfrac{\sin^2 x}{\cos x} = \sec x - \cos x$

11. $\dfrac{\csc x}{\cos x} = \tan x + \cot x$

12. $(\cos^2 x)(\cot^2 x + 1) = \cot^2 x$

13. $\dfrac{\sin \alpha \csc \alpha}{\cot \alpha} = \tan \alpha$

14. $\dfrac{\sin^2 u - \cos^2 u}{\sin u \cos u} = \tan u - \cot u$

15. $\dfrac{\sec \theta - \csc \theta}{\sec \theta \csc \theta} = \sin \theta - \cos \theta$

16. Using $\cos(x + y) = \cos x \cos y - \sin x \sin y$, show that $\cos(x + 2\pi) = \cos x$

17. Using $\sin(x + y) = \sin x \cos y + \cos x \sin y$, show that $\sin(x + \pi) = -\sin x$

Verify each identity for the indicated value.

18. $\cos 2x = 1 - 2\sin^2 x, \quad x = 30°$

19. $\sin \dfrac{x}{2} = \pm\sqrt{\dfrac{1 - \cos x}{2}}, \quad x = \dfrac{\pi}{2}$

☆ 20. Write $\sin 8t \sin 5t$ as a sum or difference.

☆ 21. Write $\sin w + \sin 5w$ as a product.

☆ 22. Write $-\sqrt{3} \sin t - \cos t$ in the form $A \sin(Bt + C)$. Indicate amplitude, period, frequency, and phase shift. Choose C as small as possible, but positive.

Verify each identity in Problems 23–26.

23. $\dfrac{1 - \cos^2 t}{\sin^3 t} = \csc t$

24. $\dfrac{(\cos \alpha - 1)^2}{\sin^2 \alpha} = \dfrac{1 - \cos \alpha}{1 + \cos \alpha}$

25. $\dfrac{1 - \tan^2 x}{1 - \tan^4 x} = \cos^2 x$

26. $\cot^2 x \cos^2 x = \cot^2 x - \cos^2 x$

B *Verify the identities in Problems 27–41. Use the list of identities inside the front cover if necessary.*

27. $\dfrac{\sin x}{1 - \cos x} = (\csc x)(1 + \cos x)$

28. $\dfrac{1 - \tan^2 x}{1 - \cot^2 x} = 1 - \sec^2 x$

29. $\tan(x + \pi) = \tan x$

30. $1 - (\cos \beta - \sin \beta)^2 = \sin 2\beta$

31. $\dfrac{\sin 2x}{\cot x} = 1 - \cos 2x$

32. $\dfrac{2 \tan x}{1 + \tan^2 x} = \sin 2x$

33. $2 \csc 2x = \tan x + \cot x$

34. $\csc x = \dfrac{\cot(x/2)}{1 + \cos x}$

35. $\dfrac{\sin(x - y)}{\sin(x + y)} = \dfrac{\tan x - \tan y}{\tan x + \tan y}$

36. $\csc 2x = \dfrac{\tan x + \cot x}{2}$

37. $\dfrac{2 - \sec^2 x}{\sec^2 x} = \cos 2x$

38. $\tan \dfrac{x}{2} = \dfrac{\sec x - 1}{\tan x}$

☆ **39.** $\dfrac{\sin t + \sin 5t}{\cos t + \cos 5t} = \tan 3t$

☆ **40.** $\dfrac{\sin x + \sin y}{\cos x - \cos y} = -\cot \dfrac{x - y}{2}$

☆ **41.** $\dfrac{\cos x - \cos y}{\cos x + \cos y} = -\tan \dfrac{x + y}{2} \tan \dfrac{x - y}{2}$

Evaluate each of the following exactly using an appropriate identity.

☆ **42.** $\sin 165° \sin 15°$ ☆ **43.** $\cos 165° - \cos 75°$

44. Use fundamental identities to find the exact values of the remaining trigonometric functions of x, given

$$\cos x = -\frac{2}{3} \quad \text{and} \quad \tan x < 0$$

45. Find the exact values of $\sin 2x$, $\cos 2x$, and $\tan 2x$, given $\tan x = \frac{4}{3}$ and $0 < x < \pi/2$. Do not use a calculator.

46. Find the exact values of $\sin(x/2)$, $\cos(x/2)$, and $\tan(x/2)$, given $\cos x = -\frac{5}{13}$ and $-\pi < x < -\pi/2$. Do not use a calculator.

In Problems 47 and 48, write each equation in the form $y = A \sin(Bt + C)$, where C is chosen so that $|C|$ is as small as possible. Indicate amplitude, period, frequency, and phase shift; then graph the equation for the indicated interval.

☆ **47.** $y = -\sin \pi t + \cos \pi t$, $0 \le t \le 3$

☆ **48.** $y = \sin 2\pi t + \sqrt{3} \cos 2\pi t$, $0 \le t \le 2$

C **49.** Find the exact values of $\sin x$, $\cos x$, and $\tan x$, given $\sec 2x = -\frac{13}{12}$ and $-\pi/2 < x < 0$. Do not use a calculator.

Verify the identities in Problems 50 and 51.

50. $\dfrac{\cot x}{\csc x + 1} = \dfrac{\csc x - 1}{\cot x}$

51. $\cot 3x = \dfrac{3 \tan^2 x - 1}{\tan^3 x - 3 \tan x}$

52. Use the definition of sine, cosine, and tangent on a unit circle to prove that

$$\tan x = \dfrac{\sin x}{\cos x}$$

53. Prove that the cosine function has a period of 2π.

54. Prove that the cotangent function has a period of π.

55. By letting

$$x + y = u \quad \text{and} \quad x - y = v$$

in $\sin x \sin y = \frac{1}{2}[\cos(x - y) - \cos(x + y)]$, show that

$$\cos v - \cos u = 2 \sin \dfrac{u + v}{2} \sin \dfrac{u - v}{2}$$

56. Show that $\sin 3x = 3 \sin x - 4 \sin^3 x$ is an identity.

☆ **57.** Write $y = 1.6 \sin 4t - 1.2 \cos 4t$ in the form $y = A \sin(Bt + C)$ where C is chosen so that $|C|$ is as small as possible and C is computed to two decimal places using a calculator. Indicate amplitude, period, frequency, and phase shift.

Problems 58–67 require the use of a graphing calculator. In Problems 58–62, graph $f(x)$, find a simpler function $g(x)$ that has the same graph as $f(x)$, and verify the identity $f(x) = g(x)$. [Assume $g(x) = k + A\,t(Bx)$ where $t(x)$ is one of the six trigonometric functions.]

58. $f(x) = \dfrac{3 \sin^2 x}{1 - \cos x} + \dfrac{\tan^2 x \cos^2 x}{1 + \cos x}$

59. $f(x) = \dfrac{\sin x}{\cos x - \sin x} + \dfrac{\sin x}{\cos x + \sin x}$

60. $f(x) = 3 \sin^2 x + \cos^2 x$

61. $f(x) = \dfrac{3 - 4 \cos^2 x}{1 - 2 \sin^2 x}$

62. $f(x) = \dfrac{2 + \sin x - 2 \cos x}{1 - \cos x}$

In Problems 63 and 64, graph y1, y2, and y3 = y1 − y2 + 1 in the same viewing rectangle for $-2\pi \le x \le 2\pi$ and state the interval(s) where the graphs of y1 and y2 coincide.

63. $y1 = \tan \dfrac{x}{2}$ $y2 = \sqrt{\dfrac{1 - \cos x}{1 + \cos x}}$

64. $y1 = \tan \dfrac{x}{2}$ $y2 = -\sqrt{\dfrac{1 - \cos x}{1 + \cos x}}$

☆ **65.** Graph $y1 = 2 \cos 30\pi x \sin 2\pi x$ and $y2 = 2 \sin 2\pi x$ for $0 \le x \le 1$ and $-2 \le y \le 2$.

☆ **66.** Repeat Problem 65 after converting y1 to a sum or difference.

☆ **67.** Graph $y = 1.6 \sin 4t - 1.2 \cos 4t$ for $-\pi/4 \le x \le \pi/4$, approximate the x intercepts in this interval to two decimal places, and identify the intercept that corresponds to the phase shift determined in Problem 57.

Reciprocal Identities

$$\csc x = \frac{1}{\sin x} \qquad \sec x = \frac{1}{\cos x} \qquad \cot x = \frac{1}{\tan x}$$

Quotient Identities

$$\tan x = \frac{\sin x}{\cos x} \qquad \cot x = \frac{\cos x}{\sin x}$$

Identities for Negatives

$$\sin(-x) = -\sin x \qquad \cos(-x) = \cos x$$
$$\tan(-x) = -\tan x$$

Pythagorean Identities

$$\sin^2 x + \cos^2 x = 1 \qquad \tan^2 x + 1 = \sec^2 x$$
$$1 + \cot^2 x = \csc^2 x$$

Sum Identities

$$\sin(x + y) = \sin x \cos y + \cos x \sin y$$
$$\cos(x + y) = \cos x \cos y - \sin x \sin y$$
$$\tan(x + y) = \frac{\tan x + \tan y}{1 - \tan x \tan y}$$

Difference Identities

$$\sin(x - y) = \sin x \cos y - \cos x \sin y$$
$$\cos(x - y) = \cos x \cos y + \sin x \sin y$$
$$\tan(x - y) = \frac{\tan x - \tan y}{1 + \tan x \tan y}$$

Cofunction Identities

(Replace $\pi/2$ with $90°$ if x is in degree measure.)

$$\sin\left(\frac{\pi}{2} - x\right) = \cos x \qquad \cos\left(\frac{\pi}{2} - x\right) = \sin x$$
$$\tan\left(\frac{\pi}{2} - x\right) = \cot x \qquad \cot\left(\frac{\pi}{2} - x\right) = \tan x$$
$$\sec\left(\frac{\pi}{2} - x\right) = \csc x \qquad \csc\left(\frac{\pi}{2} - x\right) = \sec x$$

Product-Sum Identities

$$\sin x \cos y = \tfrac{1}{2}[\sin(x + y) + \sin(x - y)]$$
$$\cos x \sin y = \tfrac{1}{2}[\sin(x + y) - \sin(x - y)]$$
$$\sin x \sin y = \tfrac{1}{2}[\cos(x - y) - \cos(x + y)]$$
$$\cos x \cos y = \tfrac{1}{2}[\cos(x + y) + \cos(x - y)]$$

LAWS OF SINES AND COSINES

Sum-Product Identities

$$\sin x + \sin y = 2 \sin \frac{x + y}{2} \cos \frac{x - y}{2}$$
$$\sin x - \sin y = 2 \cos \frac{x + y}{2} \sin \frac{x - y}{2}$$
$$\cos x + \cos y = 2 \cos \frac{x + y}{2} \cos \frac{x - y}{2}$$
$$\cos x - \cos y = -2 \sin \frac{x + y}{2} \sin \frac{x - y}{2}$$

Double-Angle Identities

$$\sin 2x = 2 \sin x \cos x \qquad \cos 2x = \begin{cases} \cos^2 x - \sin^2 x \\ 1 - 2\sin^2 x \\ 2\cos^2 x - 1 \end{cases}$$
$$\tan 2x = \frac{2 \tan x}{1 - \tan^2 x} = \frac{2 \cot x}{\cot^2 x - 1} = \frac{2}{\cot x - \tan x}$$

Half-Angle Identities

$$\sin \frac{x}{2} = \pm\sqrt{\frac{1 - \cos x}{2}}$$
$$\cos \frac{x}{2} = \pm\sqrt{\frac{1 + \cos x}{2}}$$

Sign is determined by quadrant in which $x/2$ lies.

$$\tan \frac{x}{2} = \frac{1 - \cos x}{\sin x} = \frac{\sin x}{1 + \cos x} = \pm\sqrt{\frac{1 - \cos 2x}{1 + \cos 2x}}$$
$$\sin^2 x = \frac{1 - \cos 2x}{2} \qquad \cos^2 x = \frac{1 + \cos 2x}{2}$$
$$\tan^2 x = \frac{1 - \cos 2x}{1 + \cos 2x}$$

Law of Sines

$$\frac{\sin \alpha}{a} = \frac{\sin \beta}{b} = \frac{\sin \gamma}{c}$$

Law of Cosines

$$a^2 = b^2 + c^2 - 2bc \cos \alpha$$
$$b^2 = a^2 + c^2 - 2ac \cos \beta$$
$$c^2 = a^2 + b^2 - 2ab \cos \gamma$$

Do you use graphing calculators in your trigonometry class?

Then you should know about

TRIGONOMETRY ACTIVITIES FOR THE TI-82 AND TI-85 GRAPHING CALCULATORS

Cynthia Dennis
San Jacinto College North

Linda Neal
San Jacinto College South

This paperback volume provides trigonometry-specific exploratory learning activities, using Texas Instruments TI-82 and TI-85 graphing calculators.

An appendix relates the activities in this volume to topics covered in ANALYTIC TRIGONOMETRY WITH APPLICATIONS, Sixth Edition by Raymond A. Barnett & Michael R. Ziegler.

Important Features
• Projects allow students to learn by exploring and discovering concepts.
• Specificity for trigonometry makes for greater depth of topic coverage than in similar but nontrigonometry-specific manuals.

PWS PUBLISHING COMPANY
A Division of International Thomson Publishing Inc. I(T)P

Degrees and Radians

$\frac{1}{360}$ circumference

$1°$

1 radian

$\dfrac{\theta°}{180°} = \dfrac{\theta}{\pi \text{ rad}}$

$\text{Degrees} \times \dfrac{\pi}{180} = \text{Radians}$

$\text{Radians} \times \dfrac{180}{\pi} = \text{Degrees}$

Angles and Arcs

$\dfrac{\theta°}{360°} = \dfrac{s}{C}$

C (circumference)

θ in Degrees

$\dfrac{\theta}{360°} = \dfrac{s}{C}$

θ in Radians

$\theta = \dfrac{s}{R}$

$s = R\theta$

Pythagorean Theorem

$a^2 + b^2 = c^2$

Similar Triangles

$\dfrac{a}{a'} = \dfrac{b}{b'} = \dfrac{c}{c'}$

$\dfrac{a}{a'} = \dfrac{b}{b'} = \dfrac{c}{c'}$

Trigonometric Functions

$\sin x = \dfrac{b}{R}$ $\csc x = \dfrac{R}{b}$

$\cos x = \dfrac{a}{R}$ $\sec x = \dfrac{R}{a}$

$\tan x = \dfrac{b}{a}$ $\cot x = \dfrac{a}{b}$

$R = \sqrt{a^2 + b^2} > 0$

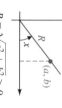

(x in degrees or radians)
For x any real number and T any trigonometric function,
$T(x) = T(x \text{ rad})$
For a unit circle:

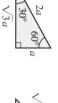

$P(\cos x, \sin x)$

x rad

x units

$(1, 0)$

$(0, 1)$

Special Triangles

30°–60° triangle

45° triangle

Special Values

θ	$\sin\theta$	$\csc\theta$	$\cos\theta$	$\sec\theta$	$\tan\theta$	$\cot\theta$
0° or 0	0	N.D.	1	1	0	N.D.
30° or $\pi/6$	1/2	2	$\sqrt{3}/2$	$2/\sqrt{3}$	$1/\sqrt{3}$	$\sqrt{3}$
45° or $\pi/4$	$1/\sqrt{2}$	$\sqrt{2}$	$1/\sqrt{2}$	$\sqrt{2}$	1	1
60° or $\pi/3$	$\sqrt{3}/2$	$2/\sqrt{3}$	1/2	2	$\sqrt{3}$	$1/\sqrt{3}$
90° or $\pi/2$	1	1	0	N.D.	N.D.	0

N.D. = Not defined

Accuracy for Triangles

Angle to Nearest	Significant Digits for Side Measure
1°	2
10' or 0.1°	3
1' or 0.01°	4
10" or 0.001°	5

Graphing Trigonometric Functions

$y = A\sin(Bx + C)$ $y = A\cos(Bx + C)$

$\text{Amplitude} = |A|$ $\text{Period} = \dfrac{2\pi}{B}$ $\text{Frequency} = \dfrac{B}{2\pi}$

$\text{Phase shift} = -\dfrac{C}{B} \begin{cases} \text{left if } -C/B < 0 \\ \text{right if } -C/B > 0 \end{cases}$

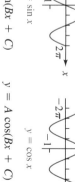

$y = \sin x$

$y = \cos x$

$y = A\tan(Bx + C)$ $y = A\cot(Bx + C)$

$\text{Period } \dfrac{\pi}{B}$ $\text{Phase shift} = -\dfrac{C}{B} \begin{cases} \text{left if } -C/B < 0 \\ \text{right if } -C/B > 0 \end{cases}$

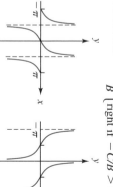

$y = \tan x$

$y = \cot x$

Inverse Trigonometric Functions

$y = \sin^{-1} x$ means $x = \sin y$

where $-\dfrac{\pi}{2} \le y \le \dfrac{\pi}{2}$ and $-1 \le x \le 1$

$y = \cos^{-1} x$ means $x = \cos y$

where $0 \le y \le \pi$ and $-1 \le x \le 1$

$y = \tan^{-1} x$ means $x = \tan y$

where $-\dfrac{\pi}{2} < y < \dfrac{\pi}{2}$ and x is any real number

$y = \cot^{-1} x$ means $x = \cot y$

where $0 < y < \pi$ and x is any real number

$y = \sec^{-1} x$ means $x = \sec y$

where $0 \le y \le \pi, y \ne \dfrac{\pi}{2}$, and $x \le -1$ or $x \ge 1$

$y = \csc^{-1} x$ means $x = \csc y$

where $-\dfrac{\pi}{2} \le y \le \dfrac{\pi}{2}, y \ne 0$,

and $x \le -1$ or $x \ge 1$

Applications

68. Precalculus: Trigonometric Substitution In the expression $\sqrt{u^2 - a^2}$, $a > 0$, let $u = a \sec x$, $-\pi/2 < x < \pi/2$, simplify, and write in a form that is free of radicals.

69. Precalculus: Angle of Intersection of Two Lines Use the results of Problem 57 in Exercise 4.3 to find the acute angle of intersection (to the nearest 0.1°) between the two lines $y = 4x + 5$ and $y = \frac{1}{3}x - 2$.

70. Engineering Find the exact value of x in the figure; then find x and θ to three decimal places.

Figure for 70

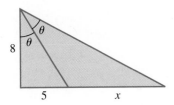

71. Music One tone is given by $y = 0.3 \cos 120\pi t$ and another by $y = -0.3 \cos 140\pi t$. Write their sum as a product. What is the beat frequency if both notes are sounded together?

C ☆ 72. Music Use a graphing calculator with the viewing rectangle set to $0 \le t \le 0.2$ and $-0.8 \le y \le 0.8$ to graph the indicated equations.
(A) $y1 = 0.3 \cos 120\pi t$
(B) $y2 = -0.3 \cos 140 \pi t$
(C) $y3 = y1 + y2$ and $y4 = 0.6 \sin 10\pi t$
(D) Repeat part (C) using the product form of $y3$ from Problem 71.

☆ 73. Physics The equation of motion for a weight suspended from a spring is given by

$$y = -8 \sin 3t - 6 \cos 3t$$

where y is displacement of the weight from its equilibrium position in centimeters and t is time in seconds. Write this equation in the form $y = A \sin(Bt + C)$; keep A positive, choose C positive and as small as possible, and compute C to two decimal places. Indicate the amplitude, period, frequency, and phase shift.

C ☆ 74. Physics Use a graphing calculator to graph

$$y = -8 \sin 3t - 6 \cos 3t$$

for $-2\pi/3 \le t \le 2\pi/3$, approximate the t intercepts in this interval to two decimal places, and identify the intercept that corresponds to the phase shift determined in Problem 73.

INVERSE TRIGONOMETRIC FUNCTIONS; TRIGONOMETRIC EQUATIONS

---

* Sections marked with a star may be omitted without loss of continuity.

Before you begin this chapter, briefly review Appendix B.2 on the general concept of the inverse of a function.

 n Chapter 1 recall that in solving a right triangle as in Figure 1 for α, we wrote

$$\sin \alpha = \frac{3}{5} = 0.6$$

$$\alpha = \sin^{-1} 0.6 \qquad \text{or} \qquad \arcsin 0.6$$

$$\alpha = 0.64 \text{ rad} \qquad \text{or} \qquad 36.87° \qquad \textit{To two decimal places}$$

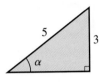

FIGURE 1

In this context, both $\sin^{-1} 0.6$ and $\arcsin 0.6$ represent the acute angle (in either radian or degree measure) whose sine is 0.6. At that time, we said that the concepts behind the inverse function symbols $\sin^{-1}$ (or arcsin), $\cos^{-1}$ (or arccos), and $\tan^{-1}$ (or arctan) would be discussed in greater detail in this chapter. Now we extend the meaning of these symbols so they apply not only to triangle problems, but to a wide variety of problems that have nothing to do with triangles or angles. That is, we will create another set of tools—a new set of functions—so that you can put them in your mathematical tool box for general use on a wide variety of new problems.

After we complete the discussion of inverse trigonometric functions, we will be in a position to solve many types of equations involving trigonometric functions. Trigonometric equations form the subject matter of the last two sections of this chapter.

5.1 INVERSE SINE, COSINE, AND TANGENT FUNCTIONS

- ◆ **Inverse Sine Function**
- ◆ **Inverse Cosine Function**
- ◆ **Inverse Tangent Function**
- ◆ **Inverse Trigonometric Functions with Angle Ranges**
- ◆ **Summary**

In this section we will define the inverse sine, cosine, and tangent functions; look at their graphs; present some basic and useful identities; and consider some applications in Exercise 5.1.

◆ Inverse Sine Function

For a function to have an inverse that is a function, it is necessary that the original function be **one-to-one**. That is, each domain value must correspond to exactly one range value, and each range value must correspond to exactly one domain value. The first condition is satisfied by all functions, but the second condition is not satisfied by some functions. For example, Figure 1(a) illustrates a function that is one-to-one; Figure 1(b) illustrates a function that is not one-to-one.

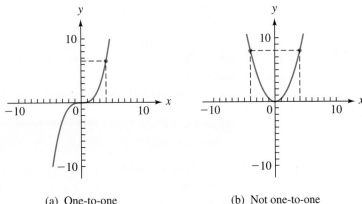

(a) One-to-one   (b) Not one-to-one

FIGURE 1

To form the inverse sine function, we start with the sine function whose graph, domain, and range are indicated in Figure 2. Note that the sine function is not one-to-one. Figure 3 shows that for the range value $y = 0.5$, for example, there are an unlimited number of domain values x such that $\sin x = 0.5$. Each point where the dashed horizontal line passing through $y = 0.5$ crosses the graph corresponds to a domain value whose sine is 0.5.

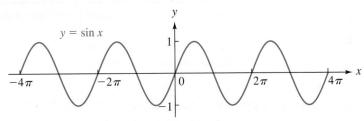

Domain: All real numbers
Range: $-1 \leq y \leq 1$

FIGURE 2
Sine function

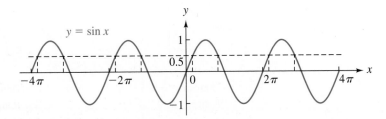

FIGURE 3
$\sin x = 0.5$

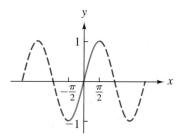

FIGURE 4

$y = \sin x$ is one-to-one for $-\pi/2 \le x \le \pi/2$

How can we restrict the domain of the sine function so that the sine function on this restricted domain becomes one-to-one—that is, so that every horizontal line will pass through at most one point on the graph? Actually, we can do this in an unlimited number of ways. The generally accepted way, however, is illustrated in Figure 4.

We use this restricted sine function to define the inverse sine function.

INVERSE SINE FUNCTION

The **inverse sine function** is defined as the inverse of the restricted sine function $y = \sin x$, $-\pi/2 \le x \le \pi/2$. Thus,

$$\left.\begin{array}{l} y = \arcsin x \\ y = \sin^{-1} x \end{array}\right\} \quad \text{are equivalent to} \quad \sin y = x,$$

$$\text{where} \begin{cases} -\pi/2 \le y \le \pi/2 \\ -1 \le x \le 1 \end{cases}$$

The inverse sine of x is the number or angle y, $-\pi/2 \le y \le \pi/2$, whose sine is x.

To graph $y = \sin^{-1} x$, we take the coordinates of each point on the graph of the restricted sine function and reverse the order. For example, since $(-\pi/2, -1)$, $(0, 0)$, and $(\pi/2, 1)$ are on the graph of the restricted sine function, then $(-1, -\pi/2)$, $(0, 0)$, and $(1, \pi/2)$ are on the graph of the inverse sine function, as shown in Figure 5. Using these three points provides us with a quick way of sketching the graph of the inverse sine function. A more accurate graph can be obtained by using a calculator set in radian mode and a set of domain values

FIGURE 5

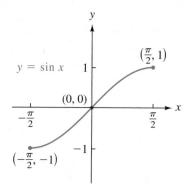

Domain: $-\dfrac{\pi}{2} \le x \le \dfrac{\pi}{2}$
Range: $-1 \le y \le 1$

(a) Restricted sine function

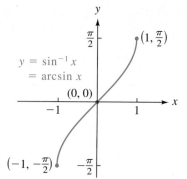

Domain: $-1 \le x \le 1$
Range: $-\dfrac{\pi}{2} \le y \le \dfrac{\pi}{2}$

(b) Inverse sine function

from -1 to 1 (see Problem 45 in Exercise 5.1). An instant graph can be obtained on a graphing calculator.

We state the important sine–inverse sine identities, which follow from the general properties of inverse functions (see Appendix B.2).

SINE–INVERSE SINE IDENTITIES

$$\sin(\sin^{-1} x) = x, \qquad -1 \le x \le 1$$
$$\sin^{-1}(\sin x) = x, \qquad -\pi/2 \le x \le \pi/2$$

$$\sin(\sin^{-1} 0.5) = 0.5 \qquad\qquad \sin(\sin^{-1} 1.5) \ne 1.5$$
$$\sin^{-1}[\sin(-1.3)] = -1.3 \qquad \sin^{-1}[\sin(-3)] \ne -3$$

[*Note:* 1.5 is not in the domain of the inverse sine function, and -3 is not in the restricted domain of the sine function. Try calculating all these examples with your calculator and see what happens!]

◆ EXAMPLE 1   Exact Values

Find exact values without using a calculator.

(A) $\sin^{-1}(\sqrt{3}/2)$ (B) $\arcsin(-\tfrac{1}{2})$

(C) $\sin^{-1}(\sin 1.2)$ (D) $\cos(\sin^{-1} \tfrac{2}{3})$

SOLUTIONS   (A) $y = \sin^{-1}(\sqrt{3}/2)$ is equivalent to $\sin y = \sqrt{3}/2$, $-\pi/2 \le y \le \pi/2$. What y between $-\pi/2$ and $\pi/2$ has sine $\sqrt{3}/2$? This y must be associated with a first quadrant reference triangle (see Figure 6).

$$\sin y = \frac{\sqrt{3}}{2}$$

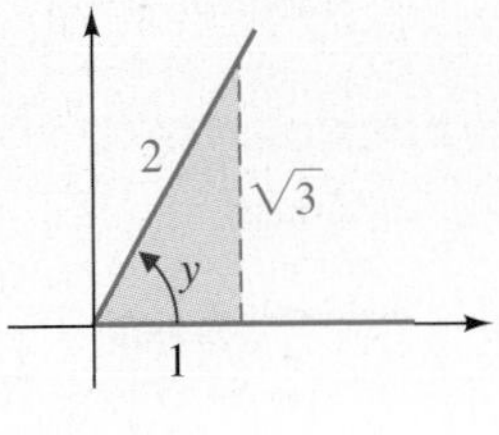

Reference triangle is a special 30°–60° triangle,

$$y = \frac{\pi}{3}$$

FIGURE 6

Thus,

$$\sin^{-1} \frac{\sqrt{3}}{2} = \frac{\pi}{3}$$

since $\pi/3$ is the only number between $-\pi/2$ and $\pi/2$ with sine equal to $\sqrt{3}/2$.

(B) $y = \arcsin(-\tfrac{1}{2})$ is equivalent to $\sin y = -\tfrac{1}{2}$, $-\pi/2 \le y \le \pi/2$. What y between $-\pi/2$ and $\pi/2$ has sine $-\tfrac{1}{2}$? This y must be negative and associated with a fourth quadrant reference triangle (see Figure 7).

$$\sin y = -\frac{1}{2}$$

Reference triangle is a
special 30°–60° triangle,

$$y = -\frac{\pi}{6}$$

FIGURE 7

Thus,

$$\arcsin\left(-\frac{1}{2}\right) = -\frac{\pi}{6}$$

[*Note:* *y* cannot be $11\pi/6$ even though $\sin(11\pi/6) = -\frac{1}{2}$. Why?]

(C) $\sin^{-1}(\sin 1.2) = 1.2$ *Sine–inverse sine identity*

(D) Let $y = \sin^{-1}\frac{2}{3}$; then $\sin y = \frac{2}{3}$, $-\pi/2 \leq y \leq \pi/2$. Draw the reference triangle associated with *y*; then $\cos y = \cos(\sin^{-1}\frac{2}{3})$ can be determined directly from the triangle (after finding the third side) without actually finding *y* (see Figure 8).

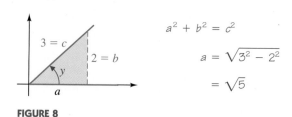

$$a^2 + b^2 = c^2$$

$$a = \sqrt{3^2 - 2^2}$$

$$= \sqrt{5}$$

FIGURE 8

Thus,

$$\cos\left(\sin^{-1}\frac{2}{3}\right) = \cos y = \frac{\sqrt{5}}{3}$$

◆

MATCHED PROBLEM 1   Find exact values without using a calculator.

(A) $\arcsin(\sqrt{2}/2)$ (B) $\sin^{-1}(-1)$
(C) $\sin[\sin^{-1}(-0.4)]$ (D) $\tan[\sin^{-1}(-1/\sqrt{5})]$

◆   EXAMPLE 2   Calculator Values

Find to four significant digits using a calculator.

(A) $\sin^{-1}(0.8432)$ (B) $\arcsin(-0.3042)$
(C) $\sin^{-1} 1.357$ (D) $\cot[\sin^{-1}(-0.1087)]$

SOLUTIONS

[*Note:* Recall that the keys used to obtain $\sin^{-1}$ vary among different brands of calculators. (Read the user's manual for your calculator.) Two common designations are $\boxed{\sin^{-1}}$ and the combination $\boxed{\text{inv}}\ \boxed{\sin}$. For all these problems set your calculator in the radian mode.]

(A) $\sin^{-1}(0.8432) = 1.003$
(B) $\arcsin(-0.3042) = -0.3091$
(C) $\sin^{-1} 1.357 = \text{Error*}$ *1.357 is not in the domain of $\sin^{-1}$.*
(D) $\cot[\sin^{-1}(-0.1087)] = -9.145$ ◆

MATCHED PROBLEM 2

Find to four significant digits using a calculator.

(A) $\arcsin 0.2903$ (B) $\sin^{-1}(-0.7633)$
(C) $\arcsin(-2.305)$ (D) $\sec[\sin^{-1}(-0.3446)]$

◆ **Inverse Cosine Function**

The generally accepted restriction on the cosine function, which ensures that an inverse will exist, is to have domain values so that $0 \le x \le \pi$ (see Figure 9).

FIGURE 9
$y = \cos x$ is one-to-one for
$0 \le x \le \pi$

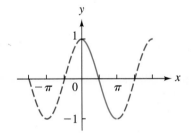

INVERSE COSINE FUNCTION

The **inverse cosine function** is defined as the inverse of the restricted cosine function $y = \cos x$, $0 \le x \le \pi$. Thus,

$$\left. \begin{array}{l} y = \arccos x \\ y = \cos^{-1} x \end{array} \right\} \text{ are equivalent to } \cos y = x, \text{ where } \begin{cases} 0 \le y \le \pi \\ -1 \le x \le 1 \end{cases}$$

The inverse cosine of x is the number or angle y, $0 \le y \le \pi$, whose cosine is x.

---

* Some calculators use a more advanced definition of the inverse sine function involving complex numbers and will display an ordered pair of real numbers as the value of $\sin^{-1} 1.357$. You should interpret such a result as an indication that the number entered is not in the domain of the inverse sine function as we have defined it.

Figure 10 compares the graphs of the restricted cosine function and its inverse. Notice that $(0, 1)$, $(\pi/2, 0)$, and $(\pi, -1)$ are on the restricted cosine graph. Reversing the coordinates gives us three points on the graph of the inverse cosine function. We complete the discussion by giving the cosine–inverse cosine identities.

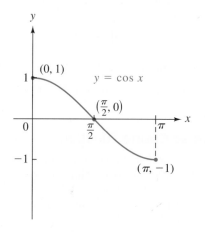

Domain: $0 \le x \le \pi$ Domain: $-1 \le x \le 1$
Range: $-1 \le y \le 1$ Range: $0 \le y \le \pi$

(a) Restricted cosine function   (b) Inverse cosine function

FIGURE 10

COSINE–INVERSE COSINE IDENTITIES

$$\cos(\cos^{-1} x) = x, \qquad -1 \le x \le 1$$
$$\cos^{-1}(\cos x) = x, \qquad 0 \le x \le \pi$$

$$\cos(\cos^{-1} 0.5) = 0.5 \qquad \cos(\cos^{-1} 3) \ne 3$$
$$\cos^{-1}(\cos 3) = 3 \qquad \cos^{-1}[\cos(-1)] \ne -1$$

[*Note:* 3 is not in the domain of the inverse cosine function, and -1 is not in the restricted domain of the cosine function. Try calculating all these examples with your calculator and see what happens!]

◆ EXAMPLE 3   Exact Values

Find exact values without using a calculator.

(A) $\cos^{-1} \frac{1}{2}$ (B) $\arccos(-\sqrt{3}/2)$
(C) $\cos(\cos^{-1} 0.7)$ (D) $\sin[\cos^{-1}(-\frac{1}{3})]$

SOLUTIONS (A) $y = \cos^{-1} \frac{1}{2}$ is equivalent to $\cos y = \frac{1}{2}$, $0 \le y \le \pi$. What y between 0 and π has cosine $\frac{1}{2}$? This y must be associated with a first quadrant reference triangle (see Figure 11).

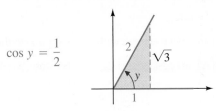

$$\cos y = \frac{1}{2}$$

Reference triangle is a special 30°–60° triangle,

$$y = \frac{\pi}{3}$$

FIGURE 11

Thus,

$$\cos^{-1} \frac{1}{2} = \frac{\pi}{3}$$

(B) $y = \arccos(-\sqrt{3}/2)$ is equivalent to $\cos y = -\sqrt{3}/2$, $0 \le y \le \pi$. What y between 0 and π has cosine $-\sqrt{3}/2$? This y must be associated with a second quadrant reference triangle (see Figure 12).

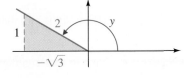

$$\cos y = -\frac{\sqrt{3}}{2}$$

Reference triangle is a special 30°–60° triangle,

$$y = \frac{5\pi}{6}$$

FIGURE 12

Thus,

$$\arccos\left(-\frac{\sqrt{3}}{2}\right) = \frac{5\pi}{6}$$

[*Note:* y cannot be $-5\pi/6$, even though $\cos(-5\pi/6) = -\sqrt{3}/2$. Why?]

(C) $\cos(\cos^{-1} 0.7) = 0.7$ *Cosine–inverse cosine identity*

(D) Let $y = \cos^{-1}(-\frac{1}{3})$; then $\cos y = -\frac{1}{3}$, $0 \le y \le \pi$. Draw a reference triangle associated with y; then $\sin y = \sin[\cos^{-1}(-\frac{1}{3})]$ can be determined directly from the triangle (after finding the third side) without actually finding y (see Figure 13).

FIGURE 13

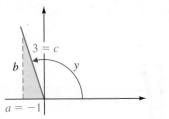

$$a^2 + b^2 = c^2$$

$$b = \sqrt{3^2 - (-1)^2}$$

$$= \sqrt{8} = 2\sqrt{2}$$

Thus,

$$\sin\left[\cos^{-1}\left(-\frac{1}{3}\right)\right] = \sin y = \frac{2\sqrt{2}}{3}$$

◆

MATCHED PROBLEM 3   Find exact values without using a calculator.

(A) $\arccos(\sqrt{2}/2)$ (B) $\cos^{-1}(-1)$
(C) $\cos^{-1}(\cos 3.05)$ (D) $\cot[\cos^{-1}(-1/\sqrt{5})]$

◆ **EXAMPLE 4**   Calculator Values

Find to four significant digits using a calculator.

(A) $\cos^{-1} 0.4325$ (B) $\arccos(-0.8976)$
(C) $\cos^{-1} 2.137$ (D) $\csc[\cos^{-1}(-0.0349)]$

SOLUTIONS   Set your calculator in radian mode.

(A) $\cos^{-1} 0.4325 = 1.124$
(B) $\arccos(-0.8976) = 2.685$
(C) $\cos^{-1} 2.137 = $ Error   *2.137 is not in the domain of $\cos^{-1}$.*
(D) $\csc[\cos^{-1}(-0.0349)] = 1.001$

◆

MATCHED PROBLEM 4   Find to four significant digits using a calculator.

(A) $\arccos 0.6773$ (B) $\cos^{-1}(-0.8114)$
(C) $\arccos(-1.003)$ (D) $\cot[\cos^{-1}(-0.5036)]$

◆ **EXAMPLE 5**   Exact Values

Find the exact value of $\cos(\sin^{-1}\frac{3}{5} - \cos^{-1}\frac{4}{5})$ without using a calculator.

SOLUTION   We use the difference identity for cosine and the procedure outlined in Examples 1(D) and 3(D) to obtain

$$\cos(x - y) = \cos x \cos y + \sin x \sin y$$

$$\cos\left(\sin^{-1}\frac{3}{5} - \cos^{-1}\frac{4}{5}\right) = \cos\left(\sin^{-1}\frac{3}{5}\right)\cos\left(\cos^{-1}\frac{4}{5}\right) + \sin\left(\sin^{-1}\frac{3}{5}\right)\sin\left(\cos^{-1}\frac{4}{5}\right)$$

$$= \left(\frac{4}{5}\right) \cdot \left(\frac{4}{5}\right) + \left(\frac{3}{5}\right) \cdot \left(\frac{3}{5}\right)$$

$$= 1$$

◆

MATCHED PROBLEM 5   Find the exact value of $\sin(2 \cos^{-1}\frac{3}{5})$ without using a calculator.

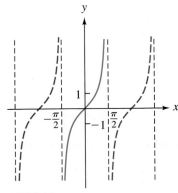

FIGURE 14
$y = \tan x$ is one-to-one for
$-\pi/2 < x < \pi/2$

♦ Inverse Tangent Function

To restrict the tangent function so that every horizontal line will pass through at most one point on its graph, we choose to restrict the domain to the interval $-\pi/2 < x < \pi/2$ (see Figure 14).

We use this restricted tangent function to define the inverse tangent function.

INVERSE TANGENT FUNCTION

The **inverse tangent function** is defined as the inverse of the restricted tangent function $y = \tan x$, $-\pi/2 < x < \pi/2$. Thus,

$$\left. \begin{array}{l} y = \arctan x \\ y = \tan^{-1} x \end{array} \right\} \text{ are equivalent to } \tan y = x,$$

$$\text{where } \begin{cases} -\pi/2 < y < \pi/2 \\ x \text{ is any real number} \end{cases}$$

The inverse tangent of x is the number or angle y, $-\pi/2 < y < \pi/2$, whose tangent is x.

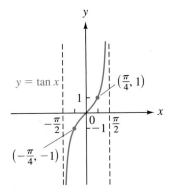

Domain: $-\dfrac{\pi}{2} < x < \dfrac{\pi}{2}$
Range: All real numbers

(a) Restricted tangent function

Figure 15 compares the graphs of the restricted tangent function and its inverse. Notice that $(-\pi/4, -1)$, $(0, 0)$, and $(\pi/4, 1)$ are on the restricted tangent graph. Reversing the coordinates gives us three points on the graph of the inverse tangent function. Also note that the vertical asymptotes become horizontal asymptotes. We now state the tangent–inverse tangent identities.

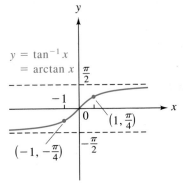

Domain: All real numbers
Range: $-\dfrac{\pi}{2} < y < \dfrac{\pi}{2}$

(b) Inverse tangent function

FIGURE 15

TANGENT–INVERSE TANGENT IDENTITIES

$\tan(\tan^{-1} x) = x$, for all x
$\tan^{-1}(\tan x) = x$, $-\pi/2 < x < \pi/2$

$\tan(\tan^{-1} 25) = 25$ $\qquad \tan[\tan^{-1}(-325)] = -325$
$\tan^{-1}(\tan 1.2) = 1.2$ $\qquad \tan^{-1}[\tan(-\pi)] \neq -\pi$

[**Note:** $-\pi$ is not in the restricted domain of the tangent function. (Try calculating all these examples with your calculator and see what happens!)]

◆ **EXAMPLE 6**   Exact Values

Find exact values without using a calculator.

(A) $\tan^{-1}(-1/\sqrt{3})$ (B) $\tan^{-1}[\tan(-1.2)]$

SOLUTIONS   (A) $y = \tan^{-1}(-1/\sqrt{3})$ is equivalent to $\tan y = -1/\sqrt{3}$, where y satisfies $-\pi/2 < y < \pi/2$. What y between $-\pi/2$ and $\pi/2$ has tangent $-1/\sqrt{3}$? This y must be negative and associated with a fourth quadrant reference triangle (see Figure 16):

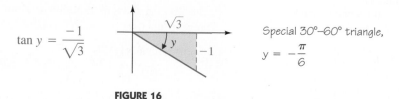

$$\tan y = \frac{-1}{\sqrt{3}}$$

Special 30°–60° triangle,

$$y = -\frac{\pi}{6}$$

FIGURE 16

Thus,

$$\tan^{-1}\left(-\frac{1}{\sqrt{3}}\right) = -\frac{\pi}{6}$$

[*Note:* y cannot be $11\pi/6$. Why?]

(B) $\tan^{-1}[\tan(-1.2)] = -1.2$ *Tangent–inverse tangent identity*   ◆

MATCHED PROBLEM 6   Find exact values without using a calculator.

(A) $\arctan \sqrt{3}$ (B) $\tan(\tan^{-1} 35)$

◆ **EXAMPLE 7**   Calculator Values

Find to four significant digits using a calculator.

(A) $\tan^{-1} 3$ (B) $\arctan(-25.45)$
(C) $\tan^{-1} 1{,}435$ (D) $\sec[\tan^{-1}(-0.1308)]$

SOLUTIONS   Set calculator in radian mode.

(A) $\tan^{-1}3 = 1.249$
(B) $\arctan(-25.45) = -1.532$
(C) $\tan^{-1}1{,}435 = 1.570$
(D) $\sec[\tan^{-1}(-0.1308)] = 1.009$ ◆

MATCHED PROBLEM 7   Find to four significant digits using a calculator.

(A) $\tan^{-1}7$ (B) $\arctan(-13.08)$
(C) $\tan^{-1}735$ (D) $\csc[\tan^{-1}(-1.033)]$

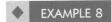
 EXAMPLE 8   Finding an Equivalent Algebraic Expression

Express $\sin(\tan^{-1}x)$ as an algebraic expression in x.

SOLUTION   Let

$$y = \tan^{-1}x, \qquad -\frac{\pi}{2} < y < \frac{\pi}{2}$$

or, equivalently,

$$\tan y = x, \qquad -\frac{\pi}{2} < y < \frac{\pi}{2}$$

The two possible reference triangles for y are shown in Figure 17.

FIGURE 17
Reference triangles for $y = \tan^{-1}x$

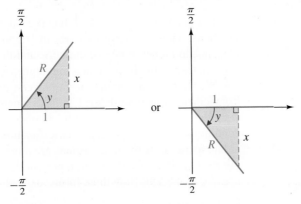

In either case,
$$R = \sqrt{x^2 + 1}$$

Thus,

$$\sin(\tan^{-1}x) = \sin y = \frac{x}{R} = \frac{x}{\sqrt{x^2 + 1}}$$

◆

MATCHED PROBLEM 8   Express $\tan(\arccos x)$ as an algebraic expression in x.

◆ Inverse Trigonometric Functions with Angle Ranges

We first defined trigonometric functions with angle domains, in degree or radian measure, and with real number ranges. Then, we defined the circular functions with real number domains and real number ranges. Technically, the trigonometric functions and the circular functions are not the same: the first are defined in terms of angles and the second in terms of real numbers. The functions, however, are closely related in that every real number in the domain of a circular function can be associated with an angle in either degree or radian measure, and vice versa (Figure 18).

FIGURE 18
Real numbers and angles

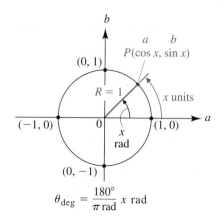

$$\theta_{\text{deg}} = \frac{180°}{\pi \, \text{rad}} \, x \, \text{rad}$$

In common usage, circular functions are also referred to as trigonometric functions. Thus, we have two sets of trigonometric functions, one with angle domains, in radian or degree measure, and the other with real number domains. (We are free to use the particular trigonometric function that best suits our needs.)

We have a similar situation with inverse trigonometric functions. The inverse trigonometric functions defined in the first part of this section are actually inverse circular functions, with real number domains and ranges. Corresponding to these definitions are inverse trigonometric functions with angle ranges, in degree or radian measure. If the range values are angles in degree measure we will often use the Greek letter θ (theta) to represent this value. Thus, for example, we may use any of the following three forms, depending on our interest:

Real number range

$$y = \tan^{-1} x, \qquad -\frac{\pi}{2} < y < \frac{\pi}{2} \qquad \text{\textit{y is a real number}}$$

Angle range in radian measure

$$y = \tan^{-1} x, \qquad -\frac{\pi}{2} < y < \frac{\pi}{2} \qquad \text{\textit{y is an angle in radian measure}}$$

Angle range in degree measure

$$\theta = \tan^{-1} x, \qquad -90° < \theta < 90° \qquad \text{\textit{θ is an angle in degree measure}}$$

Thus, depending on the context, we can write

$$\tan^{-1} 1 = \frac{\pi}{4} \qquad \text{or} \qquad \tan^{-1} 1 = \frac{\pi}{4} \, \text{radians} \qquad \text{or} \qquad \tan^{-1} 1 = 45°$$

⚠ Caution   This discussion does not mean that inverse trigonometric functions are multivalued. If we wish to use the inverse tangent function with a real number range, then $\tan^{-1} 1$ is equal to $\pi/4$ and no other real number. If we wish to use the inverse tangent function with an angle range, then $\tan^{-1} 1$ is the angle with radian measure $\pi/4$ or degree measure 45°, and no other angle.   ◇

◆ EXAMPLE 9   Inverse Trigonometric Functions and Degree Measure

Find the degree measure of θ.
(A) $\theta = \sin^{-1}(1/2)$ (Exact value without a calculator.)
(B) $\theta = \tan^{-1}(-1.3025)$ (To 2 decimal places with a calculator.)

SOLUTONS   (A) $\theta = \sin^{-1}(1/2)$ is equivalent to

$$\sin \theta = \frac{1}{2}, \qquad -90° \leq \theta \leq 90°$$

Thus, $\theta = 30°$
(B) Set calculator in degree mode.

$$\theta = \tan^{-1}(-1.3025) = -52.48°$$ ◆

MATCHED PROBLEM 9   Find the degree measure of θ.
(A) $\theta = \cos^{-1}(1/2)$ (Exact value without a calculator.)
(B) $\theta = \tan^{-1}(25.08)$ (To 2 decimal places with a calculator.)

◆ **Summary**

We summarize the definitions of the inverse trigonometric functions in the box for convenient reference.

INVERSE SINE, COSINE, AND TANGENT FUNCTIONS

$y = \sin^{-1} x$ is equivalent to $x = \sin y$,   where $-1 \leq x \leq 1$ and $-\pi/2 \leq y \leq \pi/2$
$y = \cos^{-1} x$ is equivalent to $x = \cos y$,   where $-1 \leq x \leq 1$ and $0 \leq y \leq \pi$
$y = \tan^{-1} x$ is equivalent to $x = \tan y$,   where x is any real number and $-\pi/2 < y < \pi/2$

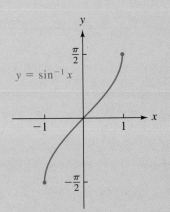

Domain: $-1 \leq x \leq 1$
Range: $-\frac{\pi}{2} \leq y \leq \frac{\pi}{2}$

(a)

Domain: $-1 \leq x \leq 1$
Range: $0 \leq y \leq \pi$

(b)

Domain: All real numbers
Range: $-\frac{\pi}{2} < y < \frac{\pi}{2}$

(c)

Answers to
Matched Problems

1. (A) $\pi/4$ (B) $-\pi/2$ (C) -0.4 (D) $-\frac{1}{2}$
2. (A) 0.2945   (B) -0.8684 (C) Not defined   (D) 1.065
3. (A) $\pi/4$ (B) π (C) 3.05   (D) $-\frac{1}{2}$
4. (A) 0.8267   (B) 2.517   (C) Not defined   (D) -0.5829
5. $\frac{24}{25}$ **6.** (A) $\pi/3$ (B) 35
7. (A) 1.429   (B) -1.494 (C) 1.569   (D) -1.392
8. $\dfrac{\sqrt{1-x^2}}{x}$ **9.** (A) $60°$ (B) $87.72°$

EXERCISE 5.1

A Find exact real number values without using a calculator.

1. $\sin^{-1} 0$
2. $\cos^{-1} 0$
3. $\arccos(\sqrt{3}/2)$
4. $\arcsin(\sqrt{3}/2)$
5. $\tan^{-1} 1$
6. $\arctan \sqrt{3}$
7. $\cos^{-1} \frac{1}{2}$
8. $\sin^{-1}(\sqrt{2}/2)$
9. $\arctan(1/\sqrt{3})$
10. $\arccos 1$
11. $\tan^{-1} 0$
12. $\sin^{-1} \frac{1}{2}$

Evaluate to four significant digits using a calculator.

13. $\cos^{-1} 0.4038$
14. $\sin^{-1} 0.9103$
15. $\tan^{-1} 43.09$
16. $\arctan 103.7$
17. $\arcsin 1.131$
18. $\arccos 3.051$

B Find exact real number values without using a calculator.

19. $\arccos(-\frac{1}{2})$
20. $\arcsin(-\sqrt{2}/2)$
21. $\tan^{-1}(-1)$
22. $\arctan(-\sqrt{3})$
23. $\sin^{-1}(-\sqrt{3}/2)$
24. $\cos^{-1}(-1)$
25. $\cos^{-1}(-\sqrt{3}/2)$
26. $\sin^{-1}(-1)$
27. $\sin[\sin^{-1}(-0.6)]$
28. $\tan(\tan^{-1} 25)$
29. $\tan^{-1}[\tan(-1.5)]$
30. $\cos^{-1}(\cos 2.3)$
31. $\tan(\cos^{-1} \frac{1}{2})$
32. $\sin[\cos^{-1}(\sqrt{3}/2)]$
33. $\cos[\sin^{-1}(-\sqrt{2}/2)]$
34. $\sec[\sin^{-1}(-\sqrt{3}/2)]$
35. $\cot[\cos^{-1}(-\frac{1}{2})]$
36. $\csc[\tan^{-1}(-1)]$

Evaluate to four significant digits using a calculator.

37. $\tan^{-1}(-4.038)$
38. $\arctan(-10.04)$
39. $\sec[\sin^{-1}(-0.0399)]$
40. $\cot[\cos^{-1}(-0.7003)]$
41. $\csc[\tan^{-1}(-4.118)]$
42. $\tan[\sin^{-1}(-0.4618)]$
43. $\sqrt{2} + \tan^{-1} \sqrt[3]{5}$
44. $\sqrt{5} + \cos^{-1}(1 - \sqrt{2})$

Graph Problems 45 and 46 with the aid of a calculator. Plot points using x values -1.0, -0.8, -0.6, -0.4, -0.2, 0.0, 0.2, 0.4, 0.6, 0.8, and 1.0; then join the points with a smooth curve.

45. $y = \sin^{-1} x$
46. $y = \cos^{-1} x$

Find the exact degree measure of θ without a calculator.

47. $\theta = \arccos(-1/2)$
48. $\theta = \arcsin(-\sqrt{2}/2)$
49. $\theta = \tan^{-1}(-1)$
50. $\theta = \arctan(-\sqrt{3})$
51. $\theta = \sin^{-1}(-\sqrt{3}/2)$
52. $\theta = \cos^{-1}(-1)$

Find the degree measure of θ to two decimal places using a calculator.

53. $\theta = \tan^{-1} 3.0413$
54. $\theta = \cos^{-1} 0.7149$
55. $\theta = \arcsin(-0.8107)$
56. $\theta = \arccos(-0.7728)$
57. $\theta = \arctan(-17.305)$
58. $\theta = \tan^{-1}(-0.3031)$

C Find exact real number values without using a calculator.

59. $\sin[\cos^{-1}(-\frac{4}{5}) + \sin^{-1}(-\frac{3}{5})]$
60. $\cos[\sin^{-1}(-\frac{3}{5}) + \cos^{-1} \frac{4}{5}]$
61. $\sin[\arccos \frac{1}{2} + \arcsin(-1)]$
62. $\cos[\cos^{-1}(-\sqrt{3}/2) - \sin^{-1}(-\frac{1}{2})]$
63. $\sin[2 \sin^{-1}(-\frac{4}{5})]$
64. $\sin[2 \cos^{-1}(-\frac{3}{5})]$
65. $\sin\left(\dfrac{\cos^{-1} \frac{1}{5}}{2}\right)$
66. $\cos\left(\dfrac{\cos^{-1} \frac{1}{3}}{2}\right)$

Write each of the following as an algebraic expression in x free of trigonometric or inverse trigonometric functions.

67. $\sin(\cos^{-1} x)$, $-1 \le x \le 1$

68. $\cos(\sin^{-1} x)$, $\quad -1 \leq x \leq 1$

69. $\tan(\arcsin x)$, $\quad -1 \leq x \leq 1$

70. $\cos(\arctan x)$

71. $\sin(2 \tan^{-1} x)$

72. $\sin(2 \cos^{-1} x)$, $\quad -1 \leq x \leq 1$

73. $\sin(\cos^{-1} x - \sin^{-1} x)$, $\quad -1 \leq x \leq 1$

74. $\cos(\sin^{-1} x + \cos^{-1} x)$, $\quad -1 \leq x \leq 1$

In Problems 75 and 76 solve for x in terms of y. State restrictions on y using a double inequality.

75. $y = a \sin(bx + c)$, $\quad a > 0, b > 0$,
$\quad -\pi/2 \leq bx + c \leq \pi/2$

76. $y = a \cos(bx + c)$, $\quad a > 0, b > 0, 0 \leq bx + c \leq \pi$

Problems 77–86 require the use of a graphing calculator.

In Problems 77–82, graph each function over the indicated interval.

77. $y = \sin^{-1}(x - 1)$, $\quad 0 \leq x \leq 2$

78. $y = \cos^{-1}(x + 2)$, $\quad -3 \leq x \leq -1$

79. $y = 3 \cos^{-1} \dfrac{x}{2}$, $\quad -2 \leq x \leq 2$

80. $y = 2 \sin^{-1} \dfrac{x}{3}$, $\quad -3 \leq x \leq 3$

81. $y = \tan^{-1}(2x - 4)$, $\quad -1 \leq x \leq 5$

82. $y = \tan^{-1}(3x + 2)$, $\quad -5 \leq x \leq 4$

83. The identity $\sin(\sin^{-1} x) = x$ is valid for $-1 \leq x \leq 1$.
(A) Graph $y = \sin(\sin^{-1} x)$ for $-1 \leq x \leq 1$.
(B) What happens if you graph y over a larger interval, say $-3 \leq x \leq 3$? Explain.

84. The identity $\cos(\cos^{-1} x) = x$ is valid for $-1 \leq x \leq 1$.
(A) Graph $y = \cos(\cos^{-1} x)$ for $-1 \leq x \leq 1$.
(B) What happens if you graph y over a larger interval, say $-3 \leq x \leq 3$? Explain.

85. The identity $\sin^{-1}(\sin x) = x$ is valid for $-\pi/2 \leq x \leq \pi/2$.
(A) Graph $y = \sin^{-1}(\sin x)$ for $-\dfrac{\pi}{2} \leq x \leq \dfrac{\pi}{2}$.
(B) What happens if you graph y over a larger interval, say $-2\pi \leq x \leq 2\pi$? Explain.

86. The identity $\cos^{-1}(\cos x) = x$ is valid for $0 \leq x \leq \pi$.
(A) Graph $y = \cos^{-1}(\cos x)$ for $0 \leq x \leq \pi$.
(B) What happens if you graph y over a larger interval, say $-2\pi \leq x \leq 2\pi$? Explain.

Applications

87. Sonic Boom An aircraft flying faster than the speed of sound produces sound waves that pile up behind the aircraft in the form of a cone. The cone of a level flying aircraft intersects the ground in the form of a hyperbola and along this curve we experience a *sonic boom* (see the figure). In Section 2.4 it was shown that

$$\sin \frac{\theta}{2} = \frac{\text{Speed of sound}}{\text{Speed of aircraft}} \qquad (1)$$

where θ is the cone angle. The ratio

$$M = \frac{\text{Speed of aircraft}}{\text{Speed of sound}} \qquad (2)$$

is called the **mach number**. A mach number of 2.3 would indicate an aircraft moving at 2.3 times the speed of sound. From (1) and (2) we obtain

$$\sin \frac{\theta}{2} = \frac{1}{M}$$

(A) Write θ in terms of M.
(B) Find θ to the nearest degree for $M = 1.7$ and $M = 2.3$.

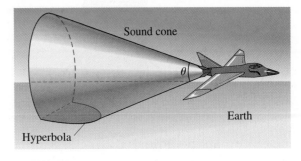

Figure for 87

88. Space Science A spacecraft traveling in a circular orbit h mi above the earth observes horizons in each direction (see figure on p. 258), where R is the radius of the earth.
(A) Express θ in terms of h and R.
(B) Find θ to the nearest degree for $h = 425$ mi and $R = 3,960$ mi.

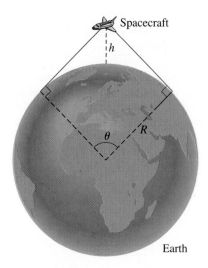

Figure for 88

89. Engineering
(A) Show that the volume of fuel d ft deep in a horizontal circular tank L ft long with radius R, $d < R$, is given by (see the figure)

$$V = \left[R^2 \cos^{-1} \frac{R - d}{R} - (R - d) \sqrt{R^2 - (R - d)^2} \right] L$$

Figure for 89

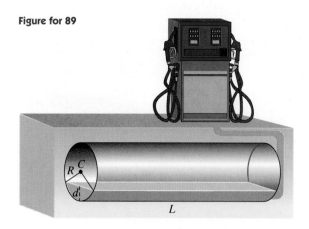

(B) If the fuel in a tank 30 ft long with radius 3 ft is found to be 2 ft deep, how many cubic feet (to the nearest cubic foot) of fuel are in the tank?

90. Engineering Find the volume (to the nearest cubic foot) of fuel in the tank in Problem 89 for a tank with radius 4 ft, length 40 ft, and a fuel depth of 2 ft.

c 91. Engineering   The function

$$y1 = 30 \left[9 \cos^{-1} \frac{3 - x}{3} - (3 - x) \sqrt{9 - (3 - x)^2} \right]$$

represents the volume of fuel x ft deep in the tank in Problem 89. Graph $y1$ and $y2 = 350$ in the same viewing rectangle and use approximation techniques to find the depth to one decimal place when the tank contains 350 cu ft of fuel.

c 92. Engineering   Repeat Problem 91 for the tank in Problem 90.

93. Engineering   The length of the belt around the two pulleys in the figure is given by

$$L = \pi D + (d - D)\theta + 2C \sin \theta$$

where θ (in radians) is given by

$$\theta = \cos^{-1} \frac{D - d}{2C}$$

Verify these formulas and find the length of the belt to one decimal place if $D = 6$ in., $d = 4$ in., and $C = 10$ in.

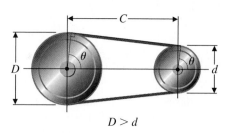

Figure for 93

94. Engineering   For Problem 93, find the length of the belt to one decimal place if $D = 4$ in., $d = 2$ in., and $C = 6$ in.

c 95. Engineering   The function

$$y1 = 6\pi - 2 \cos^{-1} \frac{1}{x} + 2x \sin\left(\cos^{-1} \frac{1}{x} \right)$$

represents the length of the belt around the two pulleys in Problem 93 when the centers of the pulleys are x ft apart. Graph $y1$ and $y2 = 40$ in the same viewing rectangle and use approximation techniques to find the distance between the centers of the pulleys to one decimal place when the belt is 40 ft long.

96. Engineering Repeat Problem 95 for the pulleys in Problem 94.

97. Precalculus

(A) The figure represents a circular courtyard surrounded by a high stone wall. A flood light, located at E, shines into the courtyard. If a person walks x ft away from the center along DC, show that his or her shadow will move a distance given by

$$d = 2R\theta = 2R \tan^{-1} \frac{x}{R}$$

where θ is in radians. [*Hint:* Draw a line from A to C.]

(B) Find d to one decimal place if $R = 50$ ft and $x = 25$ ft.

98. Precalculus

(A) A painting 4 ft high and 8 ft from the floor is being observed in an art gallery (see the figure). If the observer is standing x ft from the wall and his eye is 5 ft from the floor, show that

$$\theta = \arctan \frac{4x}{x^2 + 21}$$

$$\left[\textit{Hint:} \text{ Use the sum identity } \tan(\alpha + \theta) = \frac{\tan \alpha + \tan \theta}{1 - \tan \alpha \tan \theta}. \right]$$

(B) Find θ in decimal degrees to one decimal place for $x = 20$ ft.

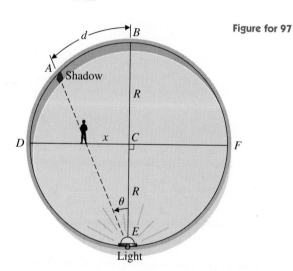

Figure for 97

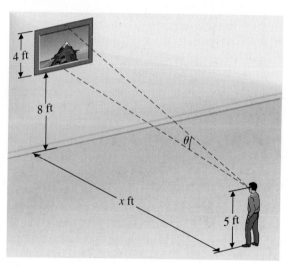

Figure for 98

☆5.2 INVERSE COTANGENT, SECANT, AND COSECANT FUNCTIONS

♦ **Definition of Inverse Cotangent, Secant, and Cosecant Functions**
♦ **Calculator Evaluation**

♦ Definition of Inverse Cotangent, Secant, and Cosecant Functions

Paralleling the development for the inverse sine, cosine, and tangent functions, we define the inverse cotangent, secant, and cosecant functions as follows.

---

* Sections marked with a star may be omitted without loss of continuity.

INVERSE COTANGENT, SECANT, AND COSECANT FUNCTIONS

$y = \cot^{-1} x$ is equivalent to $x = \cot y$,   where $0 < y < \pi$ and x is any real number

$y = \sec^{-1} x$ is equivalent to $x = \sec y$,   where $0 \le y \le \pi$, $y \ne \pi/2$, and $x \le -1$ or $x \ge 1$

$y = \csc^{-1} x$ is equivalent to $x = \csc y$,   where $-\pi/2 \le y \le \pi/2$, $y \ne 0$, and $x \le -1$ or $x \ge 1$

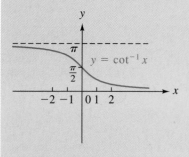

Domain: All real numbers
Range: $0 < y < \pi$

(a)

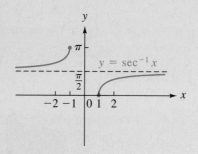

Domain: $x \le -1$ or $x \ge 1$
Range: $0 \le y \le \pi$, $y \ne \dfrac{\pi}{2}$

(b)

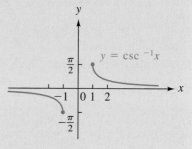

Domain: $x \le -1$ or $x \ge 1$
Range: $-\dfrac{\pi}{2} \le y \le \dfrac{\pi}{2}$, $y \ne 0$

(c)

[**Note:**   The ranges for $\sec^{-1}$ and $\csc^{-1}$ are sometimes selected differently.]

◆ **EXAMPLE 1**   Exact Values

Find exact values without using a calculator.

(A) $\text{arccot}(-1)$ (B) $\sec^{-1}(2/\sqrt{3})$

SOLUTIONS   (A) $y = \text{arccot}(-1)$ is equivalent to $\cot y = -1$, $0 < y < \pi$. What number between 0 and π has cotangent -1? This y must be positive and in the second quadrant (see Figure 1).

$$\cot y = -1 = -\frac{1}{1}$$

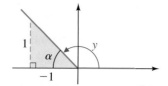

$$\alpha = \frac{\pi}{4}$$

$$y = \frac{3\pi}{4}$$

FIGURE 1

Thus,

$$\text{arccot}(-1) = \frac{3\pi}{4}$$

(B) $y = \sec^{-1}(2/\sqrt{3})$ is equivalent to $\sec y = 2/\sqrt{3}$, $0 \leq y \leq \pi$, $y \neq \pi/2$. What number between 0 and π has secant $2/\sqrt{3}$? This y is positive and in the first quadrant. We draw a reference triangle (see Figure 2).

$$\sec y = \frac{2}{\sqrt{3}}$$ 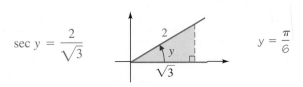 $$y = \frac{\pi}{6}$$

FIGURE 2

Thus,

$$\sec^{-1} \frac{2}{\sqrt{3}} = \frac{\pi}{6}$$

◆

MATCHED PROBLEM 1 Find exact values without using a calculator.

(A) $\operatorname{arccot}(-\sqrt{3})$ (B) $\csc^{-1}(-2)$

◆ EXAMPLE 2 **Exact Values**

Find the exact value of $\tan[\sec^{-1}(-3)]$ without using a calculator.

SOLUTION Let $y = \sec^{-1}(-3)$; then

$$\sec y = -3, \qquad 0 \leq y \leq \pi, \qquad y \neq \pi/2$$

This y is positive and in the second quadrant. Draw a reference triangle, find the third side, and then determine $\tan y$ from the triangle (see Figure 3).

$$\sec y = -3 = -\frac{3}{1}$$

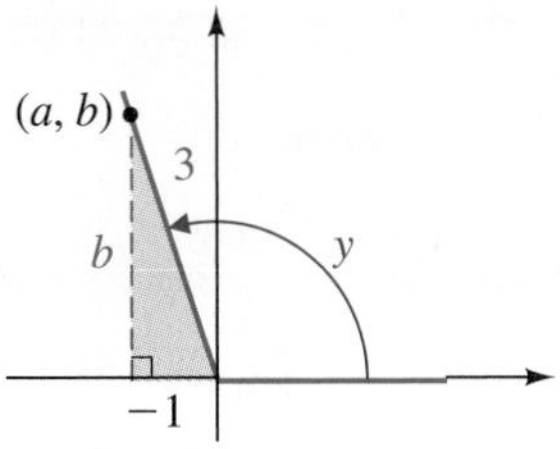

(a, b)

3

b

y

-1

$$b = \sqrt{3^2 - (-1)^2}$$
$$= \sqrt{8} = 2\sqrt{2}$$

FIGURE 3

Thus,

$$\tan[\sec^{-1}(-3)] = \tan y = \frac{b}{a} = \frac{2\sqrt{2}}{-1} = -2\sqrt{2}$$

MATCHED PROBLEM 2 Find the exact value of $\cot[\csc^{-1}(-\frac{5}{3})]$ without using a calculator.

◆ Calculator Evaluation

Many calculators have keys for sin, cos, tan, $\sin^{-1}$, $\cos^{-1}$, $\tan^{-1}$, or their equivalents. To find sec x, csc x, and cot x using a calculator, we use the reciprocal identities

$$\sec x = \frac{1}{\cos x} \qquad \csc x = \frac{1}{\sin x} \qquad \cot x = \frac{1}{\tan x}$$

How can we evaluate $\sec^{-1} x$, $\csc^{-1} x$, and $\cot^{-1} x$ using a calculator with only $\sin^{-1}$, $\cos^{-1}$, and $\tan^{-1}$ keys? The following inverse identities are made to order for this purpose.

INVERSE TRIGONOMETRIC IDENTITIES

$$\cot^{-1} x = \begin{cases} \tan^{-1} \dfrac{1}{x}, & x > 0 \\[2em] \pi + \tan^{-1} \dfrac{1}{x}, & x < 0 \end{cases}$$

$$\sec^{-1} x = \cos^{-1} \frac{1}{x}, \qquad x \geq 1, \qquad \text{or} \qquad x \leq -1$$

$$\csc^{-1} x = \sin^{-1} \frac{1}{x}, \qquad x \geq 1, \qquad \text{or} \qquad x \leq -1$$

We will establish here the first part of the inverse cotangent identity. The inverse secant and cosecant identities are left to you to do (see Problems 47 and 48, Exercise 5.2). Let

$$y = \cot^{-1} x, \qquad x > 0$$

Then

$$\cot y = x, \qquad 0 < y < \pi/2 \qquad\qquad \text{Definition of } \cot^{-1}$$

$$\frac{1}{\tan y} = x, \qquad 0 < y < \pi/2 \qquad\qquad \text{Reciprocal identity}$$

$$\tan y = \frac{1}{x}, \qquad 0 < y < \pi/2 \qquad\qquad \text{Algebra}$$

$$y = \tan^{-1} \frac{1}{x}, \qquad 0 < y < \pi/2 \qquad\qquad \text{Definition of } \tan^{-1}$$

Thus,

$$\cot^{-1} x = \tan^{-1} \frac{1}{x}, \qquad \text{for } x > 0$$

◆ EXAMPLE 3   Calculator Evaluation

Use a calculator to evaluate the following as real numbers to three decimal places.

(A) $\cot^{-1} 4.05$ (B) $\csc^{-1}(-12)$

SOLUTIONS   (A) Put the calculator in radian mode, enter 4.05, take its reciprocal, and then
press the $\tan^{-1}$ key (or its equivalent) to obtain

$$\cot^{-1} 4.05 = \tan^{-1} \frac{1}{4.05} = 0.242 \qquad \boxed{4.05}\ \boxed{1/x}\ \boxed{\tan^{-1}}$$

(B) Put the calculator in radian mode, enter -12, take its reciprocal, and then
press the $\sin^{-1}$ key (or its equivalent) to obtain

$$\csc^{-1}(-12) = \sin^{-1}\left(-\frac{1}{12}\right) \qquad \boxed{12}\ \boxed{+/-}\ \boxed{1/x}\ \boxed{\sin^{-1}}$$
$$= -0.083 \qquad\qquad\qquad ◆$$

MATCHED PROBLEM 3   Use a calculator to evaluate the following as real numbers to three decimal places.

(A) $\cot^{-1} 2.314$ (B) $\sec^{-1}(-1.549)$

Answers to   **1.** (A) $5\pi/6$ (B) $-\pi/6$ **2.** $-\frac{4}{3}$
Matched Problems   **3.** (A) 0.408   (B) 2.273

EXERCISE 5.2

A *Find the exact real number value of each without using a calculator.*

1. $\cot^{-1} \sqrt{3}$ **2.** $\cot^{-1} 0$

3. arccsc 1   **4.** arcsec 2

5. $\sec^{-1} \sqrt{2}$ **6.** $\csc^{-1} 2$

7. $\sin(\cot^{-1} 0)$ **8.** $\cos(\cot^{-1} 1)$

9. $\tan(\csc^{-1} \frac{5}{4})$ **10.** $\cot(\sec^{-1} \frac{5}{3})$

B **11.** $\cot^{-1}(-1)$ **12.** $\sec^{-1}(-1)$

13. $\operatorname{arcsec}(-2)$ **14.** $\operatorname{arccsc}(-\sqrt{2})$

15. $\operatorname{arccsc}(-2)$ **16.** $\operatorname{arccot}(-\sqrt{3})$

17. $\csc^{-1} \frac{1}{2}$ **18.** $\sec^{-1}(-\frac{1}{2})$

19. $\cos[\csc^{-1}(-\frac{5}{3})]$ **20.** $\tan[\cot^{-1}(-1/\sqrt{3})]$

21. $\cot[\sec^{-1}(-\frac{5}{4})]$ **22.** $\sin[\cot^{-1}(-\frac{3}{4})]$

23. $\cos[\sec^{-1}(-2)]$ **24.** $\sin[\csc^{-1}(-2)]$

25. $\cot(\cot^{-1} 33.4)$ **26.** $\sec[\sec^{-1}(-44)]$

27. $\csc[\csc^{-1}(-4)]$ **28.** $\cot[\cot^{-1}(-7.3)]$

Use a calculator to evaluate the following as real numbers to three decimal places.

29. $\cot^{-1} 3.065$ **30.** $\cot^{-1} 7.306$

31. $\sec^{-1}(-1.963)$ **32.** $\sec^{-1} 2.041$

33. $\csc^{-1} 1.172$ **34.** $\csc^{-1}(-1.938)$

35. $\cot^{-1}(-5.104)$ **36.** $\cot^{-1}(-12.236)$

Find the exact degree measure of θ without using a calculator.

37. $\theta = \operatorname{arcsec}(-2)$ **38.** $\theta = \operatorname{arccsc}(-\sqrt{2})$

39. $\theta = \cot^{-1}(-1)$ **40.** $\theta = \operatorname{arccot}(-1/\sqrt{3})$

41. $\theta = \csc^{-1}(-2/\sqrt{3})$ **42.** $\theta = \sec^{-1}(-1)$

Find the degree measure to two decimal places using a calculator.

43. $\theta = \cot^{-1} 0.3288$ **44.** $\theta = \sec^{-1} 1.3989$

45. $\theta = \operatorname{arccsc}(-1.2336)$ **46.** $\theta = \operatorname{arcsec}(-1.2939)$

47. $\theta = \operatorname{arccot}(-0.0578)$ **48.** $\theta = \cot^{-1}(-3.2994)$

C *Find exact values for each problem without using a calculator.*

49. $\tan[\csc^{-1}(-\frac{5}{3}) + \tan^{-1}\frac{1}{4}]$

50. $\tan[\tan^{-1} 4 - \sec^{-1}(-\sqrt{5})]$

51. $\tan[2 \cot^{-1}(-\frac{3}{4})]$

52. $\tan[2 \sec^{-1}(-\sqrt{5})]$

In Problems 53–58 write as an algebraic expression in x free of trigonometric or inverse trigonometric functions.

53. $\sin(\cot^{-1} x)$ **54.** $\cos(\cot^{-1} x)$

55. $\csc(\sec^{-1} x)$ **56.** $\tan(\csc^{-1} x)$

57. $\sin(2 \cot^{-1} x)$ **58.** $\sin(2 \sec^{-1} x)$

59. Show that $\sec^{-1} x = \cos^{-1}(1/x)$ for $x \geq 1$ and $x \leq -1$.

60. Show that $\csc^{-1} x = \sin^{-1}(1/x)$ for $x \geq 1$ and $x \leq -1$.

Problems 61–64 require the use of a graphing calculator. Use an appropriate inverse trigonometric identity to graph each function in the viewing rectangle $-5 \leq x \leq 5$, $-\pi \leq y \leq \pi$.

61. $y = \sec^{-1} x$ **62.** $y = \csc^{-1} x$

63. $y = \cot^{-1} x$ (Use two viewing rectangles, one for $-5 \leq x \leq 0$, and the other for $0 \leq x \leq 5$.)

64. $y = \cot^{-1} x$ [Using one viewing rectangle, $-5 \leq x \leq 5$, graph $y1 = \pi(x < 0) + \tan^{-1}(1/x)$, where $<$ is selected from the TEST menu. The expression $(x < 0)$ assumes a value of 1 for $x < 0$ and 0 for $x \geq 0$.]

5.3 TRIGONOMETRIC EQUATIONS I

♦ **Solving sin x = a, cos x = b, and tan x = c for x over One Period**
♦ **Solving sin x = a, cos x = b, and tan x = c for All x**

In Chapter 4 we considered trigonometric equations called identities. These equations were true for all replacements of the variable(s) for which both sides were defined. Now we consider another class of trigonometric equations, called **conditional equations**, which are true for some replacements of the variable but false for other replacements. For example,

$$\sin x = \cos x$$

is a conditional equation, since it is true for $x = \pi/4$ and false for $x = 0$. In this section and the next we consider some standard techniques for solving conditional trigonometric equations.

Suppose you are asked to find all solutions for the equation

$$\sin x = 0.5$$

Looking at the graph of $y = \sin x$ in Figure 1, we see that because of the periodic nature of the sine function, there are an unlimited number of solutions (that is, values of x such that $y = 0.5$). How can we write an expression that represents all solutions and at the same time enables us to find any particular solution in which we might have an interest? We will start by solving equations of the form $\sin x = a$, $\cos x = b$, and $\tan x = c$ for all solutions over an interval with length one period of the trigonometric function involved. Then, using these solutions, we will be able to write an expression that will represent all possible solutions. If we can represent all possible solutions, then it should be an easy matter to select any particular solutions in which we might have an interest.

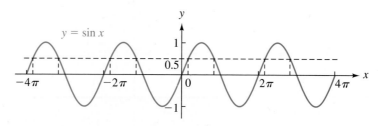

FIGURE 1
sin x = 0.5

◆ Solving sin x = a, cos x = b,
and tan x = c for x over One Period

The best way to proceed is through several examples.

◆ EXAMPLE 1   Solving Trigonometric Equations over a Specified Interval

Find all solutions for $0 \leq x \leq 2\pi$.

(A) $\sin x = \frac{1}{2}$ (Find exact solutions.)
(B) $\sin x = -0.4$ (Compute solutions to four decimal places.)

SOLUTIONS   (A) $\sin x = \dfrac{1}{2}, \qquad 0 \leq x \leq 2\pi$

Sine is positive in the first and second quadrants. Sketch reference triangles and angles corresponding to x as in Figure 2.

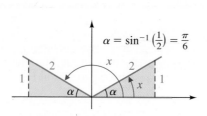
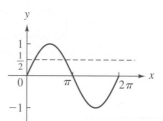

FIGURE 2

$$x = \begin{cases} \alpha \\ \pi - \alpha \end{cases} = \begin{cases} \sin^{-1} \frac{1}{2} \\ \pi - \sin^{-1} \frac{1}{2} \end{cases} = \begin{cases} \pi/6 \\ \pi - (\pi/6) \end{cases} = \begin{cases} \pi/6 \\ 5\pi/6 \end{cases}$$

(B) $\sin x = -0.4 = -\dfrac{0.4}{1} \qquad 0 \leq x \leq 2\pi$

Sine is negative in the third and fourth quadrants. Sketch reference triangles and angles as in Figure 3.

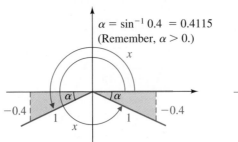

FIGURE 3

$$x = \begin{cases} \pi + \alpha \\ 2\pi - \alpha \end{cases} = \begin{cases} \pi + \sin^{-1} 0.4 \\ 2\pi - \sin^{-1} 0.4 \end{cases} = \begin{cases} \pi + 0.4115 \\ 2\pi - 0.4115 \end{cases} = \begin{cases} 3.5531 \\ 5.8717 \end{cases}$$

◆

MATCHED PROBLEM 1   Find all solutions for $0 \le x \le 2\pi$.

(A) $\cos x = \frac{1}{2}$ (Find exact solutions.)

(B) $\cos x = -0.4$ (Find solutions to four decimal places.)

◆ EXAMPLE 2   **Solving Trigonometric Equations over a Specified Interval**

Find all solutions for $-\pi/2 < x < \pi/2$.

(A) $\tan x = \sqrt{3}$ (Find exact solutions.)

(B) $\tan x = -4$ (Find solutions to four decimal places.)

SOLUTIONS   (A) $\tan x = \sqrt{3} = \dfrac{\sqrt{3}}{1}$, $-\dfrac{\pi}{2} < x < \dfrac{\pi}{2}$

Tangent is positive in the first quadrant. Sketch the reference triangle and angle corresponding to x as in Figure 4.

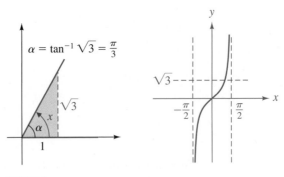

FIGURE 4

Thus,

$$x = \alpha = \tan^{-1}\sqrt{3} = \frac{\pi}{3}$$

(B) $\tan x = -4 = -\dfrac{4}{1}, \qquad -\dfrac{\pi}{2} < x < \dfrac{\pi}{2}$

Tangent is negative in the fourth quadrant. Sketch the reference triangle and angle corresponding to x as in Figure 5.

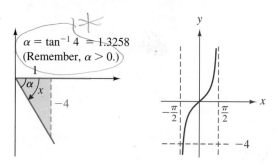

FIGURE 5

Thus,

$$x = -\alpha = -\tan^{-1}4 = -1.3258$$

◆

MATCHED PROBLEM 2   Find all solutions for $-\pi/2 < x < \pi/2$.

(A) $\tan x = -1$ (Find exact solutions.)
(B) $\tan x = 3.5$ (Find solutions to four decimal places.)

◆ EXAMPLE 3   Solving Trigonometric Equations over a Specified Interval

Find all solutions for $\sec\theta = -4.2$, $-180° \leq \theta \leq 180°$. Compute solutions to three decimal places.

SOLUTION   First replace $\sec\theta$ with $1/(\cos\theta)$, and solve for $\cos\theta$.

$$\sec\theta = -4.2$$

$$\frac{1}{\cos\theta} = -4.2$$

$$\cos\theta = \frac{-1}{4.2}$$

Now proceed as before. Cosine is negative in the second and third quadrants. Sketch reference triangles and angles corresponding to θ, $-180° \leq \theta \leq 180°$ as in Figure 6.

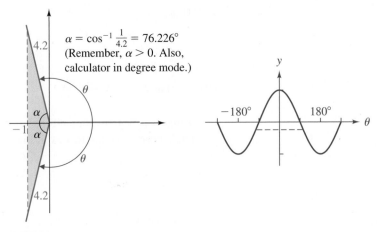

FIGURE 6

Thus,

$$\theta = \begin{cases} 180° - \alpha \\ -180° + \alpha \end{cases} = \begin{cases} 180° - \cos^{-1} \frac{1}{4.2} \\ -180° + \cos^{-1} \frac{1}{4.2} \end{cases}$$

$$= \begin{cases} 180° - 76.226° \\ -180° + 76.226° \end{cases} = \begin{cases} 103.774° \\ -103.774° \end{cases}$$

MATCHED PROBLEM 3   Find all solutions for cot $\theta = -5$, $0° < \theta < 180°$. Compute solutions to two decimal places.

◆ Solving sin $x = a$, cos $x = b$, and tan $x = c$ for All x

To find all solutions for an equation of the form sin $x = a$, cos $x = b$, or tan $x = c$, we find all solutions for one period (any period), and then add to these solutions all possible integer* multiples of the period of the trigonometric function involved. Examples should make the process clear.

◆ EXAMPLE 4   Finding All Solutions for a Trigonometric Equation

Find all solutions:

(A) cos $x = 0$ (Find exactly; x is real.)

(B) sin $\theta = -0.43$ (Evaluate inverse functions to two decimal places; θ in degrees.)

(C) tan $x = -1.60$ (Evaluate inverse functions to two decimal places; x is real.)

---

* Recall that the set of integers is the set of positive and negative whole numbers and zero: $\{\ldots, -4, -3, -2, -1, 0, 1, 2, 3, 4, \ldots\}$.

SOLUTIONS   (A) $\cos x = 0$, x real

We first find all solutions over one period, say $0 \le x \le 2\pi$. (We could just as well choose $-\pi \le x \le \pi$, $-2\pi \le x \le 0$, and so on. Since any period will do, we make the easy choice of $0 \le x \le 2\pi$.) There are no reference triangles (Why?), but from earlier work we know that (as in Figure 7)

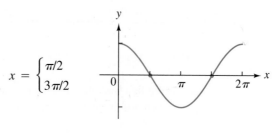

$$x = \begin{cases} \pi/2 \\ 3\pi/2 \end{cases}$$

FIGURE 7

Because cosine has a period of 2π, any integer multiple of 2π added to either of these two solutions is also a solution. Thus, the general solution is given by

$$x = \begin{cases} \pi/2 + 2k\pi \\ 3\pi/2 + 2k\pi \end{cases} \quad k \text{ any integer} \qquad \textit{General solution}$$

or equivalently (and more simply) by

$$x = \frac{\pi}{2} + k\pi, \quad k \text{ any integer}$$

By a **general solution** we mean that any particular solution of the equation $\cos x = 0$ can be obtained from the general solution by a suitable choice of an integer k.

(B) $\sin \theta = -0.43 = -\dfrac{0.43}{1}$, θ degrees

We first find all solutions over one period, say $0° \le \theta \le 360°$; then add integer multiples of $360°$ to these solutions. We proceed as before with reference triangles as in Figure 8.

FIGURE 8

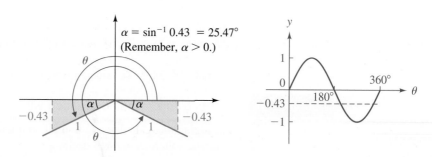

Solutions for $0° \leq \theta \leq 360°$:

$$\theta = \begin{cases} 180° + \alpha \\ 360° - \alpha \end{cases} = \begin{cases} 180° + 25.47° \\ 360° - 25.47° \end{cases} = \begin{cases} 205.47° \\ 334.53° \end{cases}$$

General solution:

$$\theta = \begin{cases} 205.47° + k360° \\ 334.53° + k360° \end{cases} \quad k \text{ any integer}$$

(C) $\tan x = -1.60 = -\dfrac{1.60}{1}, \quad x$ real

We first find all solutions over one period, say $-\pi/2 < x < \pi/2$ (Figure 9); then add integer multiples of π (the period of the tangent function) to these solutions.

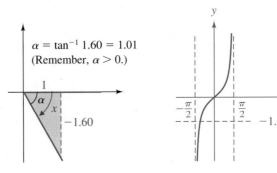

FIGURE 9

Solution for $-\pi/2 < x < \pi/2$:

$$x = -\alpha = -\tan^{-1} 1.60 = -1.01$$

General solution:

$$x = -1.01 + k\pi, \quad k \text{ any integer}$$ ◆

MATCHED PROBLEM 4   Find all solutions:

(A) $\sin x = 1$ (Find exactly; x is real.)
(B) $\cos \theta = -0.32$ (Evaluate inverse functions to two decimal places; θ in degrees.)
(C) $\tan x = 3.00$ (Evaluate inverse functions to two decimal places; x is real.)

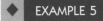

◆ EXAMPLE 5   Solving a Trigonometric Equation with a Graphing Calculator

Use a graphing calculator to find all solutions for $\sin x = 0.36$, $-\pi \leq x \leq \pi$. Compute solutions to three decimal places.

SOLUTION The solutions of the equation $\sin x = 0.36$ are the x coordinates of the points of intersection of the graphs of

$$y1 = \sin x \quad \text{and} \quad y2 = 0.36$$

in Figure 10. Examining the graph, we see that there are two points of intersection. Using the zoom and trace procedures of the calculator, we obtain the following approximations to the x coordinates of these intersection points:

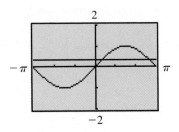

FIGURE 10
$y1 = \sin x, y2 = 0.36$

$$x = \begin{cases} 0.368 \\ 2.773 \end{cases}$$ ◆

MATCHED PROBLEM 5 Use a graphing calculator to find all solutions for $\cos x = -0.74$, $0 \le x \le 2\pi$. Compute solutions to three decimal places.

Answers to
Matched Problems

1. (A) $x = \begin{cases} \cos^{-1}\frac{1}{2} \\ 2\pi - \cos^{-1}\frac{1}{2} \end{cases} = \begin{cases} \pi/3 \\ 5\pi/3 \end{cases}$

(B) $x = \begin{cases} \pi - \cos^{-1} 0.4 \\ \pi + \cos^{-1} 0.4 \end{cases} = \begin{cases} 1.9823 \\ 4.3009 \end{cases}$

2. (A) $x = -\pi/4$ (B) $x = \tan^{-1} 3.5 = 1.2925$

3. $\theta = 180° - \alpha = 180° - \tan^{-1}\frac{1}{5} = 168.69°$

4. (A) $x = \pi/2 + 2k\pi$, k any integer

(B) $\theta = \begin{cases} 180° - \cos^{-1} 0.32 + k360° \\ 180° + \cos^{-1} 0.32 + k360° \end{cases}$

$= \begin{cases} 108.66° + k360° \\ 251.34° + k360° \end{cases}$ k any integer

(C) $x = \tan^{-1} 3.00 + k\pi = 1.25 + k\pi$, k any integer

5. $x = \begin{cases} 2.404 \\ 3.879 \end{cases}$

EXERCISE 5.3

A *Find exact solutions over the indicated intervals (x is a real number and θ is in degrees).*

1. $\sin x = 1/\sqrt{2}, \quad 0 \le x \le 2\pi$

2. $\cos x = 1/\sqrt{2}, \quad 0 \le x \le 2\pi$

3. $\cos x = -\frac{1}{2}, \quad 0 \le x \le 2\pi$

4. $\sin x = -\frac{1}{2}, \quad 0 \le x \le 2\pi$

5. $\tan x = -\sqrt{3}, \quad -\pi/2 < x < \pi/2$

6. $\tan x = 1/\sqrt{3}, \quad -\pi/2 < x < \pi/2$

7. $\cos \theta = \frac{1}{2}, \quad 0° \le \theta \le 360°$

8. $\sin \theta = \frac{1}{2}, \quad 0° \le \theta \le 360°$

9. $\sin \theta = -\sqrt{3}/2, \quad 0° \le \theta \le 360°$

10. $\cos \theta = -\sqrt{3}/2, \quad 0° \le \theta \le 360°$

11. $\tan \theta = -1/\sqrt{3}, \quad -90° < \theta < 90°$

12. $\tan \theta = 1, \quad -90° < \theta < 90°$

Find all solutions for x a real number and θ in degrees. Represent the solutions in exact form. [Hint: Use the results of Problems 1–12.]

13. $\sin x = 1/\sqrt{2}$ 14. $\cos x = 1/\sqrt{2}$

15. $\cos x = -\frac{1}{2}$ 16. $\sin x = -\frac{1}{2}$

17. $\tan x = -\sqrt{3}$ 18. $\tan x = 1/\sqrt{3}$

19. $\cos \theta = \frac{1}{2}$ 20. $\sin \theta = \frac{1}{2}$

21. $\sin \theta = -\sqrt{3}/2$ 22. $\cos \theta = -\sqrt{3}/2$

23. $\tan \theta = -1/\sqrt{3}$ 24. $\tan \theta = 1$

B *Solve to two decimal places over the indicated intervals for x a real number and θ in degrees.*

25. $\sin x = 0.60, \quad 0 \le x \le 2\pi$

26. $\cos x = 0.60, \quad 0 \le x \le 2\pi$

27. $\cos x = -0.46, \quad 0 \le x \le 2\pi$

28. $\sin x = -0.46, \quad 0 \le x \le 2\pi$

29. $\tan x = -22.00, \quad -\pi/2 < x < \pi/2$

30. $\tan x = -0.84, \quad -\pi/2 < x < \pi/2$

31. $\cos \theta = 0.26, \quad 0° \le \theta \le 360°$

32. $\sin \theta = 0.77, \quad 0° \le \theta \le 360°$

33. $\sin \theta = -0.84, \quad 0° \le \theta \le 360°$

34. $\cos \theta = -0.21, \quad 0° \le \theta \le 360°$

35. $\tan \theta = -9.45, \quad -90° < \theta < 90°$

36. $\tan \theta = -12.08, \quad -90° < \theta < 90°$

Find all solutions for x a real number and θ in degrees. [Hint: Use the results of Problems 25–36.] Evaluate inverse functions to two decimal places.

37. $\sin x = 0.60$ 38. $\cos x = 0.60$

39. $\cos x = -0.46$ 40. $\sin x = -0.46$

41. $\tan x = -22.00$ 42. $\tan x = -0.84$

43. $\cos \theta = 0.26$ 44. $\sin \theta = 0.77$

45. $\sin \theta = -0.84$ 46. $\cos \theta = -0.21$

47. $\tan \theta = -9.45$ 48. $\tan \theta = -12.08$

Find all solutions over the indicated intervals for x a real number and θ in degrees. Find exact solutions, if possible; if not, compute inverse functions to three decimal places.

49. $\sin x = 0, \quad$ all real x 50. $\cos x = 1, \quad$ all real x

51. $\cos x = -1, \quad$ all real x 52. $\sin x = -1, \quad$ all real x

53. $\csc x = 2, \quad 0 \le x \le 2\pi$

54. $\sec x = 2, \quad 0 \le x \le 2\pi$

55. $\sec \theta = -5, \quad 0° \le \theta \le 360°$

56. $\csc \theta = -8, \quad 0° \le \theta \le 360°$

C 57. $\sin x = 0.4315, \quad 0 \le x \le 4\pi$

58. $\cos x = 0.2714, \quad 0 \le x \le 4\pi$

59. $\sin x - 0.2243, \quad -2\pi \le x \le 2\pi$

60. $\cos x = -0.5328, \quad -2\pi \le x \le 2\pi$

61. $\tan \theta = 4.083, \quad -360° \le \theta \le 360°$

62. $\tan \theta = -1.119 \quad -360° \le \theta \le 360°$

63. Given $\sin x = a, 0 \le x \le 2\pi$, for what values of a does the equation have the following?
   (A) No solution   (B) Exactly one solution
   (C) Exactly two solutions   (D) More than two solutions

64. Given $\cos x = b, 0 \le x \le 2\pi$, for what values of b does the equation have the following?
   (A) No solution   (B) Exactly one solution
   (C) Exactly two solutions   (D) More than two solutions

 Use a graphing calculator for Problems 65–70. Solve each equation for $-\pi \le x \le \pi$, and compute solutions to three decimal places (see Example 5).

65. $\sin x = -0.21$ 66. $\cos x = 0.42$

67. $\sec x = 4$ 68. $\csc x = -3$

69. $\tan x = 2$ 70. $\cot x = 3$

5.4 TRIGONOMETRIC EQUATIONS II

We are now ready to solve trigonometric equations of a more general nature. No one rule will lead to all the solutions of every trigonometric equation you are likely to encounter. Solving trigonometric equations often requires the use of identities, algebraic manipulation, ingenuity, and perseverance. The following suggestions may help you get started.

SUGGESTIONS FOR SOLVING TRIGONOMETRIC EQUATIONS

1. Regard one particular trigonometric function as an unknown and solve for it. After solving for a trigonometric function, solve for the variable following the procedures we discussed in Section 5.3.
2. If more than one trigonometric function is present, use identities or factoring to try to write the equation in terms of one trigonometric function.
3. If different arguments are present in two or more trigonometric functions (e.g., $\sin 2x$ and $\sin x$), use appropriate identities to write all trigonometric functions in terms of the same argument.
4. Some trigonometric equations, after suitable manipulation, may be recognized as quadratic in form. If so, solve by factoring or by use of the quadratic formula.*
5. In general, if a trigonometric equation can be solved (not all can be explicitly solved), then the solution usually can be found by suitable use of identities and algebraic manipulation.

◆ **EXAMPLE 1**   Exact Solution

Find exact solutions:

$$2 \sin 4x - \sqrt{3} = 0, \qquad 0 \le x \le \pi/2$$

SOLUTION   Note that the only trigonometric function present is $\sin 4x$, and it is present only once. We first solve for $\sin 4x$, then for $4x$, and finally for x.

Step 1   Solve for $\sin 4x$.

$$2 \sin 4x - \sqrt{3} = 0$$
$$2 \sin 4x = \sqrt{3}$$
$$\sin 4x = \frac{\sqrt{3}}{2}$$

---

* Recall that a quadratic equation of the form $ax^2 + bx + c = 0$, with $b^2 - 4ac \ge 0$, has solutions $x = (-b \pm \sqrt{b^2 - 4ac})/2a$.

Step 2   Solve for $4x$.

First note that if $0 \leq x \leq \pi/2$, then

$$4(0) \leq 4x \leq 4\left(\frac{\pi}{2}\right)$$ *Multiply each member of the inequality by 4.**

$$0 \leq 4x \leq 2\pi$$

Referring to Figure 1, we find

$$4x = \begin{cases} \alpha \\ \pi - \alpha \end{cases} = \begin{cases} \sin^{-1}(\sqrt{3}/2) \\ \pi - \sin^{-1}(\sqrt{3}/2) \end{cases} = \begin{cases} \pi/3 \\ 2\pi/3 \end{cases}$$

FIGURE 1

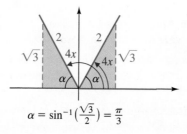

$$\alpha = \sin^{-1}\left(\frac{\sqrt{3}}{2}\right) = \frac{\pi}{3}$$

Step 3   Solve for x.

$$4x = \frac{\pi}{3} \qquad\qquad 4x = \frac{2\pi}{3}$$

$$x = \frac{\pi}{12} \qquad\qquad x = \frac{\pi}{6}$$

Thus, $x = \pi/12, \pi/6$.   ◆

MATCHED PROBLEM 1   Find exact solutions:

$$2\sin(x/3) + 1 = 0, \qquad 0 \leq x \leq 6\pi$$

◆   **EXAMPLE 2**   **All Real Solutions**

Solve $\cos 3x = -0.3105$ for all real x. Compute inverse functions to four decimal places.

SOLUTION   First, let $u = 3x$ and solve

$$\cos u = -0.3105$$

for all real u as in the last section. We start by finding all solutions over one period, say $0 \leq u \leq 2\pi$ (see Figure 2).

* Recall that if $a \leq b \leq c$, then $\begin{cases} da \leq db \leq dc & \text{if } d > 0 \\ da \geq db \geq dc & \text{if } d < 0 \end{cases}$

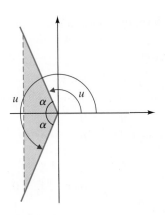 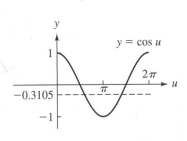

FIGURE 2

$$u = \begin{cases} (\pi - \alpha) + 2k\pi \\ (\pi + \alpha) + 2k\pi \end{cases}$$

$$= \begin{cases} (\pi - \cos^{-1} 0.3105) + 2k\pi \\ (\pi + \cos^{-1} 0.3105) + 2k\pi \end{cases} \quad k \text{ any integer}$$

Now replace u with $3x$ and solve for x:

$$3x = \begin{cases} (\pi - \cos^{-1} 0.3105) + 2k\pi \\ (\pi + \cos^{-1} 0.3105) + 2k\pi \end{cases}$$

$$x = \begin{cases} \dfrac{\pi - \cos^{-1} 0.3105}{3} + \dfrac{2k\pi}{3} \\[2mm] \dfrac{\pi + \cos^{-1} 0.3105}{3} + \dfrac{2k\pi}{3} \end{cases}$$

$$x \approx \begin{cases} 0.6288 + \dfrac{2k\pi}{3} \\[2mm] 1.4656 + \dfrac{2k\pi}{3} \end{cases} \quad k \text{ any integer}$$

◆

MATCHED PROBLEM 2   Solve $\sin 5x = -0.6213$ for all real x. Compute inverse functions to four decimal places.

   **EXAMPLE 3**   Exact Solutions

Find exact solutions:

$$\sin 2x = \sin x, \qquad 0 \le x \le 2\pi$$

SOLUTION   Notice how this problem differs from Example 1. Here we have the sine function occurring twice, once involving a multiple of x, $2x$, and once involving only x.

We try to transform the equation into a form so that all trigonometric functions are in terms of x alone, and not multiples of x. A double-angle identity will accomplish the task.

$$\sin 2x = \sin x$$
$$2 \sin x \cos x = \sin x \qquad \textit{Double-angle identity}$$
$$2 \sin x \cos x - \sin x = 0$$
$$\sin x(2 \cos x - 1) = 0 \qquad \textit{Factor}$$

Therefore, $\sin x = 0$ or $2 \cos x - 1 = 0$:

$$\sin x = 0 \qquad\qquad\qquad 2 \cos x - 1 = 0$$
$$x = 0, \pi, 2\pi$$
$$\cos x = \frac{1}{2} \qquad \textit{Solve as in Section 5.3}$$
$$x = \frac{\pi}{3}, \frac{5\pi}{3}$$

Combining the solutions from both equations, we have $x = 0, \pi/3, \pi,$ $5\pi/3, 2\pi$. ◆

MATCHED PROBLEM 3   Find exact solutions:

$$\sin^2 x = \frac{1}{2} \sin 2x, \qquad 0 \le x \le 2\pi$$

◆ **EXAMPLE 4**   Exact Solutions

Find exact solutions:

$$\cos^2 \theta = \cos \theta + \sin^2 \theta, \qquad 0° \le \theta < 360°$$

SOLUTION   In this problem we have both sine and cosine functions, and we cannot solve by factoring as in Example 3. So we try to transform the equation into one that has only the sine function or the cosine function. We choose the latter, using a Pythagorean identity for $\sin^2 \theta$.

$$\cos^2 \theta = \cos \theta + \sin^2 \theta$$
$$\cos^2 \theta = \cos \theta + 1 - \cos^2 \theta \qquad \textit{Pythagorean identity}$$
$$2 \cos^2 \theta - \cos \theta - 1 = 0 \qquad\qquad \textit{Quadratic in } \cos \theta$$
$$(\cos \theta - 1)(2 \cos \theta + 1) = 0 \qquad\qquad \textit{Factor}$$

Therefore,

$$\cos \theta - 1 = 0 \qquad\qquad 2 \cos \theta + 1 = 0$$
$$\cos \theta = 1$$
$$\theta = 0° \qquad\qquad \cos \theta = -\frac{1}{2} \qquad \textit{Solve as in Section 5.3.}$$
$$\theta = 120°, 240°$$

Thus, $\theta = 0°, 120°, 240°$.

Key Algebraic Steps in Example 4

$$2u^2 - u - 1 = 0$$

$$(u - 1)(2u + 1) = 0$$

$$u - 1 = 0 \quad \text{or} \quad 2u + 1 = 0$$

$$u = 1 \qquad\qquad 2u = -1$$

$$u = -\tfrac{1}{2}$$

◆

MATCHED PROBLEM 4   Find exact solutions:

$$\sin^2 \theta = \cos^2 \theta - \sin \theta, \quad 0° \le \theta < 360°$$

◆ **EXAMPLE 5**   **All Real Solutions**

Solve $\cos 2x = 2(\sin x - 1)$ for all real x. Compute inverse functions to four decimal places.

SOLUTION   Since we have the cosine of a multiple of x and the sine of x alone, we try to transform the equation into a form that is free of multiples of x and is in terms of sine alone or cosine alone. A double-angle identity for $\cos 2x$ accomplishes both objectives.

$$\cos 2x = 2(\sin x - 1) \qquad\qquad \text{\textit{Use double-angle}}$$
$$\text{\textit{identity.}}$$
$$1 - 2\sin^2 x = 2\sin x - 2$$

$$2\sin^2 x + 2\sin x - 3 = 0 \qquad\qquad \text{\textit{Quadratic in sin x}}$$

$$\sin x = \frac{-2 \pm \sqrt{4 - 4(2)(-3)}}{2(2)} \qquad \text{\textit{Use the quadratic}}$$
$$\text{\textit{formula.}}$$

$$\sin x = -1.8228 \quad \text{or} \quad 0.8228$$

Thus,

$$\sin x = -1.8228 \qquad \text{\textit{No solution; why?}}$$

or

$$\sin x = 0.8228 \qquad \text{\textit{Solve as in Section 5.3}}$$

$$x = \begin{cases} \sin^{-1} 0.8228 + 2k\pi \\ \pi - \sin^{-1} 0.8228 + 2k\pi \end{cases} \approx \begin{cases} 0.9665 + 2k\pi \\ 2.1751 + 2k\pi \end{cases} \quad k \text{ any integer}$$

Key Algebraic Steps in Example 5

$$au^2 + bu + c = 0 \qquad\qquad \text{\textit{Quadratic equation}}$$

$$u = \frac{-b \pm \sqrt{b^2 - 4ac}}{2a} \qquad \text{\textit{Quadratic formula}}$$

◆

MATCHED PROBLEM 5   Solve $\cos 2x = 4 \cos x - 2$ for all real x. Compute inverse functions to four decimal places.

Caution

1. In solving any equation, remember to be careful not to divide both sides of the equation by an expression that involves the variable for which you are solving. Solutions of the original equation may be lost by this procedure. For example, consider the equation

$$\sin^2 x = \sin x, \qquad 0 \le x \le 2\pi$$

To solve, *do not* divide both sides by sin x. Instead, collect all nonzero terms on the left side and factor as follows:

$$\sin^2 x - \sin x = 0$$
$$(\sin x)(\sin x - 1) = 0$$

$$\sin x = 0 \qquad \text{or} \qquad \sin x - 1 = 0$$
$$x = 0, \pi, 2\pi \qquad\qquad\qquad \sin x = 1$$
$$x = \frac{\pi}{2}$$

If we had divided both sides of the original equation by sin x, we would have lost three of the four solutions!

2. What happens if we multiply both sides of an equation by an expression that involves the variable for which we are solving, or if we square both sides of an equation? (The latter is actually a useful technique for solving some types of equations.) Contrary to losing solutions, either process may introduce **extraneous solutions** (solutions of the final equation that may not be solutions to the original equation). This is not as serious as losing solutions, since we have only to check the final solutions in the original equation, and discard any that do not satisfy the equation. Example 6 illustrates the procedure.   ◇

◆ **EXAMPLE 6**   Exact Solutions

Find exact solutions:

$$\csc x + \cot x = 1, \qquad 0 \le x < 2\pi$$

SOLUTION   To solve this equation, we square both sides after subtracting cot x from each side. We can then use Pythagorean identities to transform the equation into one involving only one type of trigonometric function. The squaring process may create extraneous solutions, so all solutions must be checked in the original equation to determine whether any must be discarded.

$$\csc x = 1 - \cot x$$
$$\csc^2 x = 1 - 2\cot x + \cot^2 x \qquad \text{Both sides squared}$$
$$1 + \cot^2 x = 1 - 2\cot x + \cot^2 x \qquad \text{Pythagorean identity}$$
$$0 = -2\cot x$$
$$\cot x = 0$$
$$x = \frac{\pi}{2}, \frac{3\pi}{2}$$

Checking these in the original equation, we find that $3\pi/2$ must be discarded. Hence, $x = \pi/2$ is the only solution. ◆

MATCHED PROBLEM 6   Find exact solutions:

$$1 + \cos x = \sin x, \qquad 0 \le x < 2\pi$$

◆ EXAMPLE 7   Solving a Trigonometric Equation with a Graphing Calculator

Use a graphing calculator to solve $\cos x \cos 2x = 0.05$ for $-\pi \le x \le \pi$. Compute solutions to three decimal places.

SOLUTION   Use the double angle identity for $\cos 2x$ to produce an equation that is free of multiples of x and involves only the cosine function.

$$\cos x \cos 2x = 0.05$$
$$\cos x(2 \cos^2 x - 1) = 0.05$$
$$2 \cos^3 x - \cos x - 0.05 = 0 \qquad \text{Cubic in } \cos x$$

Factoring cubic polynomials can be difficult, and there is no simple analog of the quadratic formula for cubic equations. However, we can easily approximate the solutions with a graphing calculator. From Figure 3, we see that the graph of

$$y1 = 2 \cos^3 x - \cos x - 0.05$$

has six x intercepts for $-\pi \le x \le \pi$. Using zoom and trace procedures (or a built-in root approximation procedure if your calculator has one) and the symmetry property of the graph, the solutions to three decimal places are

$$\pm 0.751, \qquad \pm 1.621, \qquad \pm 2.319 \qquad ◆$$

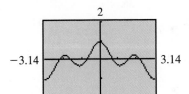

FIGURE 3
$y1 = 2 \cos^3 x - \cos x - 0.05$

MATCHED PROBLEM 7   Use a graphing calculator to solve $\sin x \cos 2x = -0.5$ for $-\pi \le x \le \pi$. Compute solutions to three decimal places.

In Example 7, we transformed the given equation into one involving $\cos x$ alone to illustrate that the algebraic techniques used in earlier examples will not work here. In general, when using a graphing calculator to approximate solutions, we do not need to perform any such transformations. Thus, we could have solved Example 7 by approximating the x intercepts of

$$y1 = \cos x \cos 2x - 0.05$$

or by approximating the x coordinates of the points of intersection of the graphs of

$$y2 = \cos x \cos 2x \qquad \text{and} \qquad y3 = 0.05$$

Answers to   **1.** $7\pi/2, 11\pi/2$ **2.** $x = \begin{cases} 0.7624 + \dfrac{2k\pi}{5} \\[2mm] 1.1226 + \dfrac{2k\pi}{5} \end{cases}$ k any integer
Matched Problems

3. $0, \pi/4, \pi, 5\pi/4, 2\pi$ **4.** $30°, 150°, 270°$

5. $\begin{cases} 1.2735 + 2k\pi \\ -1.2735 + 2k\pi \end{cases}$ k an integer   **6.** $\pi/2, \pi$ **7.** $1.086, 2.056$

EXERCISE 5.4

A *Find exact solutions over the indicated interval.*

1. $1 + \cos 4x = 0, \quad 0 \le x < \pi/2$

2. $1 - \sin 4x = 0, \quad 0 \le x < \pi/2$

3. $1 + \sqrt{2} \sin(\theta/2) = 0, \quad 0° \le \theta < 720°$

4. $1 - \sqrt{2} \cos(\theta/2) = 0, \quad 0° \le \theta < 720°$

5. $\sqrt{3} - \tan 2x = 0, \quad -\pi/4 < x < \pi/4$

6. $1 + \tan 2x = 0, \quad -\pi/4 < x < \pi/4$

7. $4 \cos^2 x - 3 = 0, \quad 0 \le x < 2\pi$

8. $2 \sin^2 x - 1 = 0, \quad 0 \le x < 2\pi$

9. $\sin^2 \theta = \sin \theta, \quad 0° \le \theta \le 180°$

10. $\cos^2 \theta = \cos \theta, \quad 0° \le \theta \le 180°$

B *Find exact solutions over the indicated interval.*

11. $2 \sin 2x = \sqrt{3}, \quad 0 \le x \le \pi$

12. $2 \cos 2x = 1, \quad 0 \le x \le \pi$

13. $\sin x \cos x = \frac{1}{2}, \quad 0 \le x \le \pi$

14. $1 - 2\sqrt{2} \sin x \cos x = 0, \quad 0 \le x \le \pi$

15. $\tan \dfrac{x}{2} - 1 = 0, \quad 0 \le x \le 2\pi$

16. $\sec \dfrac{x}{2} + 2 = 0, \quad 0 \le x \le 2\pi$

17. $\sin 2x = \sin x, \quad$ all real x

18. $\sin 2x + \cos x = 0, \quad$ all real x

19. $\cos^2 \theta = \frac{1}{2} \sin 2\theta, \quad$ all θ

20. $2 \sin^2 \theta + \sin 2\theta = 0, \quad$ all θ

21. $2 \sin^2 x = 3 \sin x - 1, \quad 0 \le x < 2\pi$

22. $2 \cos^2 x + \cos x = 1, \quad 0 \le x < 2\pi$

23. $\sin^2 \theta + 2 \cos \theta = -2, \quad 0° \le \theta < 360°$

24. $2 \cos^2 \theta + 3 \sin \theta = 0, \quad 0° \le \theta < 360°$

25. $\cos 2\theta + \sin^2 \theta = 0, \quad 0° \le \theta < 360°$

26. $\cos 2\theta + \cos \theta = 0, \quad 0° \le \theta < 360°$

27. $4 \cos^2 2x - 4 \cos 2x + 1 = 0, \quad 0 \le x \le \pi$

28. $2 \sin^2(x/2) - 3 \sin(x/2) + 1 = 0, \quad 0 \le x \le 2\pi$

29. $\cos x = \cot x, \quad 0 \le x < 2\pi$

30. $\tan x = -2 \sin x, \quad 0 \le x < 2\pi$

Solve to four significant digits over the indicated interval.

31. $4 \cos^2 \theta = 7 \cos \theta + 2, \quad 0° \le \theta \le 180°$

32. $6 \sin^2 \theta + 5 \sin \theta = 6, \quad 0° \le \theta \le 90°$

33. $\cos 2x + 10 \cos x = 5, \quad 0 \le x < 2\pi$

34. $2 \sin x = \cos 2x, \quad 0 \le x < 2\pi$

Solve for all real number solutions. Compute inverse functions to four significant digits.

35. $\sin 2x = 0.4836$ **36.** $\cos 2x = 0.5824$

37. $\cos 4x = -0.6670$ **38.** $\sin 4x = -0.7119$

39. $\cos^2 x = 3 - 5 \cos x$ **40.** $2 \sin^2 x = 1 - 2 \sin x$

41. $2 \cos 2x = 7 \cos x$ **42.** $7 \sin x = 2 \cos 2x$

C *Find exact solutions over the indicated interval. (Use sum-product identities for Problems 51–54.)*

43. $\sin x = \cos x, \quad$ all real x

44. $\sqrt{3} \sin x - \cos x = 0, \quad$ all real x

45. $2 \sin^2 x + 3 \cos x = 3, \quad$ all real x

46. $2 \cos^2 x = 3 \sin x + 3, \quad$ all real x

47. $\sin x + \cos x = 1, \quad 0 \le x < 2\pi$

48. $\cos x - \sin x = 1, \quad 0 \le x < 2\pi$

49. $\sec x + \tan x = 1, \quad 0 \le x < 2\pi$

50. $\tan x - \sec x = 1, \quad 0 \le x < 2\pi$

51. $\sin 3x + \sin x = 0, \quad 0 \le x \le \pi$

52. $\cos 5x + \cos 3x = 0, \quad 0 \le x \le \pi$

53. $\sin 5x - \sin 3x = \sin x, \quad 0 \le x \le \pi/2$

54. $\cos 3x - \cos x = \sin x, \quad 0 \le x \le \pi$

☆*For those who covered Section 4.6 where*

$$M \sin Bt + N \cos Bt = \sqrt{M^2 + N^2} \sin(Bt + C)$$

was discussed, use this identity to find all solutions to Problems 55–58.

55. $\sin t + \cos t = 1$ **56.** $\sin t - \cos t = 1$

57. $\sin \pi t - \cos \pi t = 1$ **58.** $\sin \pi t + \cos \pi t = 1$

Use a graphing calculator in Problems 59–64 to approximate all solutions over the indicated interval. Compute solutions to three decimal places.

59. $\cos x = x$, all real x

60. $3 \sin x = x$, all real x

61. $\sin 2x + 2 \sin x = 1$, $0 \le x \le 2\pi$

62. $\sin 2x + 2 \cos x = 1$, $0 \le x \le 2\pi$

63. $\sin x \sin 2x = -0.5$, $0 \le x \le 2\pi$

64. $\cos x \sin 2x = 0.5$, $0 \le x \le 2\pi$

Applications

65. Electric Current An alternating current generator produces a current given by the equation

$$I = 30 \sin 120\pi t$$

where t is time in seconds and I is current in amperes. Find the least positive t to four significant digits such that $I = 25$ amp.

66. Electric Current Find the least positive t in Problem 65 to four significant digits, such that $I = -10$ amp.

67. Photography A polarizing filter for a camera contains two parallel plates of polarizing glass, one fixed and the other able to rotate. If θ is the angle of rotation from the position of maximum light transmission, then the intensity of light leaving the filter is $\cos^2 \theta$ times the intensity entering the filter (see the figure). Find the least positive θ (in decimal degrees to two decimal places) so that the intensity of light leaving the filter is 70% of that entering. [*Hint:* Solve $I \cos^2 \theta = 0.70I$.]

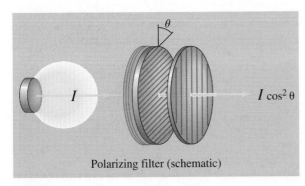

Figure for 67

68. Photography Find θ in Problem 67 so that the light leaving the filter is 40% of that entering.

69. Geometry The area of a segment of a circle in the figure is given by

$$A = \frac{1}{2} r^2 (\theta - \sin \theta)$$

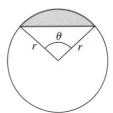

Figure for 69

Use a graphing calculator to find the angle θ subtended by the segment if the radius is 10 m and the area is 40 m². Compute the solution in radians to two decimal places.

70. Geometry Repeat Problem 69 if the radius is 8 m and the area is 48 m².

71. Architecture An arched doorway is formed by placing a circular arc on top of a rectangle (see the figure). If the rectangle is 4 ft wide and 8 ft high and the circular arc is 5 ft long, what is the area of the doorway? Compute answer to two decimal places.

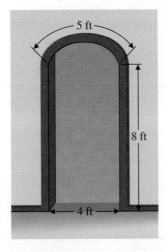

Figure for 71

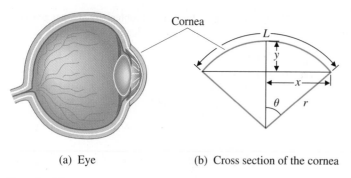

(a) Eye   (b) Cross section of the cornea

Figure for 73

c **72. Architecture**   Repeat Problem 71 if the rectangle is 3 ft wide and 7 ft high and the circular arc is 4 ft long.

c **73. Eye Surgery**   A surgical technique for correcting an astigmatism involves removing small pieces of tissue in order to change the curvature of the cornea.* In the cross section of a cornea shown in part (b) of the figure, the circular arc with radius r and central angle 2θ represents the surface of the cornea.

  (A) If $x = 5.5$ mm and $y = 2.5$ mm, find L, the length of the corneal cross section, correct to four decimal places.

  (B) Reducing the width of the cross section without changing its length has the effect of pushing the cornea outward and giving it a rounder, yet still circular, shape. Use a graphing calculator to approximate y to four decimal places if x is reduced to 5.4 mm and L remains the same as it was in part (A).

c **74. Eye Surgery**   Refer to Problem 73. Increasing the width of a cross section of the cornea without changing its length has the effect of pushing the cornea inward and giving it a flatter, yet still circular, shape. Use a graphing calculator to approximate y to four decimal places if x is increased to 5.6 mm and L remains the same as it was in part (A) of Problem 73.

∗75. Astronomy   The planet Mercury travels around the sun in an elliptical orbit given approximately by

$$r = \frac{3.44 \times 10^7}{1 - 0.206 \cos \theta}$$

---

* Based on the article "The Surgical Correction of Astigmatism" by Sheldon Rothman and Helen Strassberg in *The UMAP Journal,* Vol. V, No. 2 (1984).

(see the figure). Find the least positive θ (in decimal degrees to three significant digits) such that Mercury is 3.78×10^7 mi from the sun.

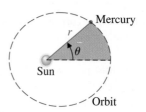

Figure for 75

∗76. Astronomy   Find the least positive θ (in decimal degrees to three significant digits) in Problem 75 such that Mercury is 3.09×10^7 mi from the sun.

Precalculus   *Find simultaneous solutions for each of the following systems of equations for $0° \le \theta \le 360°$. These are polar equations, which will be discussed in Chapter 7.*

∗77. $r = 2 \sin \theta$ **∗78.** $r = 2 \sin \theta$
      $r = 2(1 - \sin \theta)$ $r = \sin 2\theta$

∗79. Precalculus   Given the equation $xy = -2$, replace x and y with

$$x = u \cos \theta - v \sin \theta$$
$$y = u \sin \theta + v \cos \theta$$

and simplify the left side of the resulting equation. Find the least positive θ in degree measure so that the coefficient of the uv term will be 0.

∗80. Precalculus   Repeat Problem 79 for the equation $2xy = 1$.

CHAPTER 5 SUMMARY

5.1
INVERSE SINE, COSINE, AND
TANGENT FUNCTIONS

$y = \sin^{-1} x$ is equivalent to $x = \sin y$,

$\qquad$ where $-1 \le x \le 1$ and $-\pi/2 \le y \le \pi/2$

$y = \cos^{-1} x$ is equivalent to $x = \cos y$,

$\qquad$ where $-1 \le x \le 1$ and $0 \le y \le \pi$

$y = \tan^{-1} x$ is equivalent to $x = \tan y$,

$\qquad$ where x is any real number and $-\pi/2 < y < \pi/2$

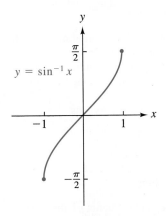

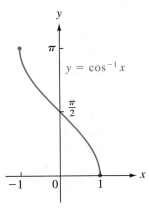

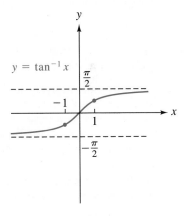

Domain: $-1 \le x \le 1$
Range: $-\dfrac{\pi}{2} \le y \le \dfrac{\pi}{2}$

Domain: $-1 \le x \le 1$
Range: $0 \le y \le \pi$

Domain: All real numbers
Range: $-\dfrac{\pi}{2} < y < \dfrac{\pi}{2}$

The inverse sine, cosine, and tangent functions are also denoted by arcsin x, arccos x, and arctan x, respectively.

Inverse Sine, Cosine, and Tangent Identities

$\sin(\sin^{-1} x) = x, \qquad -1 \le x \le 1$

$\sin^{-1}(\sin x) = x, \qquad -\pi/2 \le x \le \pi/2$

$\cos(\cos^{-1} x) = x, \qquad -1 \le x \le 1$

$\cos^{-1}(\cos x) = x, \qquad 0 \le x \le \pi$

$\tan(\tan^{-1} x) = x, \qquad \text{for all } x$

$\tan^{-1}(\tan x) = x, \qquad -\pi/2 < x < \pi/2$

★5.2
INVERSE COTANGENT,
SECANT, AND COSECANT
FUNCTIONS

$y = \cot^{-1} x$ is equivalent to $x = \cot y$,

$\qquad$ where $0 < y < \pi$ and x is any real number

$y = \sec^{-1} x$ is equivalent to $x = \sec y$,

$\qquad$ where $0 \le y \le \pi$, $y \ne \pi/2$, and $x \le -1$ or $x \ge 1$

$y = \csc^{-1} x$ is equivalent to $x = \csc y$,

$\qquad$ where $-\pi/2 \le y \le \pi/2$, $y \ne 0$, and $x \le -1$ or $x \ge 1$

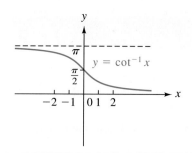

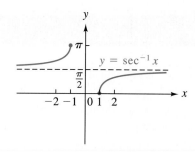

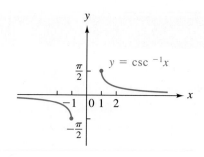

Domain: All real numbers
Range: $0 < y < \pi$

Domain: $x \leq -1$ or $x \geq 1$
Range: $0 \leq y \leq \pi, y \neq \dfrac{\pi}{2}$

Domain: $x \leq -1$ or $x \geq 1$
Range: $-\dfrac{\pi}{2} \leq y \leq \dfrac{\pi}{2}, y \neq 0$

[*Note:* The ranges for $\sec^{-1}$ and $\csc^{-1}$ are sometimes selected differently.]

Inverse Trigonometric Identities

$$\cot^{-1} x = \begin{cases} \tan^{-1} \dfrac{1}{x}, & x > 0 \\[3mm] \pi + \tan^{-1} \dfrac{1}{x}, & x < 0 \end{cases}$$

$$\sec^{-1} x = \cos^{-1} \dfrac{1}{x}, \quad x \geq 1, \text{ or } x \leq -1$$

$$\csc^{-1} x = \sin^{-1} \dfrac{1}{x}, \quad x \geq 1, \text{ or } x \leq -1$$

5.3
TRIGONOMETRIC
EQUATIONS I

An equation that is true for some replacements of the variable, but not for all replacements of the variable, is called a **conditional equation**. To solve conditional equations of the form $\sin x = a$, $\cos x = b$, or $\tan x = c$, first find all solutions as x varies over one period of the function and then extend these particular solutions to the **general solution** by adding all integer multiples of the period of the function.

5.4
TRIGONOMETRIC
EQUATIONS II

Suggestions for Solving Trigonometric Equations

1. Regard one particular trigonometric function as an unknown and solve for it. After solving for a trigonometric function, solve for the variable following the procedures discussed in Section 5.3.

2. If more than one trigonometric function is present, use identities or factoring to try to write the equation in terms of one trigonometric function.

3. If different arguments are present in two or more trigonometric functions (e.g., $\sin 2x$ and $\sin x$), use appropriate identities to write all trigonometric functions in terms of the same argument.

4. Some trigonometric equations, after suitable manipulation, may be recognized as quadratic in form. If so, solve by factoring or by use of the quadratic formula.

5. In general, if a trigonometric equation can be solved (not all can be explicitly solved), then the solution usually can be found by suitable use of identities and algebraic manipulation.

CHAPTER 5 REVIEW EXERCISE

Work through all the problems in this chapter review and check the answers. Answers to all review problems appear in the back of the book; following each answer is an italic number that indicates the section in which that type of problem is discussed. Where weaknesses show up, review the appropriate sections in the text.

A *Evaluate exactly as real numbers.*

1. $\cos^{-1}(\sqrt{3}/2)$ **2.** $\arcsin\frac{1}{2}$

3. $\sin^{-1}(-1)$ **4.** $\cos^{-1}(-1)$

5. $\sin^{-1}(-\sqrt{3}/2)$ **6.** $\arccos(-\frac{1}{2})$

7. $\arctan 1$ **8.** $\tan^{-1}(-\sqrt{3})$

9. $\arccos 0$ **10.** $\sin^{-1} 2$

☆**11.** $\cot^{-1}(-1/\sqrt{3})$ ☆**12.** $\operatorname{arcsec}(-2)$

☆**13.** $\operatorname{arccsc}(-1)$ ☆**14.** $\sec^{-1}(-\sqrt{2})$

Evaluate as a real number to four significant digits using a calculator.

15. $\sin^{-1}(-0.8277)$ **16.** $\cos^{-1} 0.6068$

17. $\arccos(-1.328)$ **18.** $\tan^{-1} 75.14$

☆**19.** $\cot^{-1} 5.632$ ☆**20.** $\operatorname{arccsc}(-3.548)$

Find exact solutions over the indicated interval.

21. $2 \cos x - \sqrt{3} = 0, \quad 0 \le x < 2\pi$

22. $2 \sin^2 \theta = \sin \theta, \quad 0° \le \theta < 360°$

23. $4 \cos^2 x - 3 = 0, \quad 0 \le x < 2\pi$

24. $2 \cos^2 \theta + 3 \cos \theta + 1 = 0, \quad 0° \le \theta < 360°$

25. $\sqrt{2} \sin 4x - 1 = 0, \quad 0 \le x < \pi/2$

26. $\tan(\theta/2) + \sqrt{3} = 0, \quad -180° < \theta < 180°$

B *Find exact values for each.*

27. $\cos(\cos^{-1} 0.315)$ **28.** $\tan^{-1}[\tan(-1.5)]$

29. $\sin[\tan^{-1}(-\frac{3}{4})]$ **30.** $\cot[\arccos(-\frac{2}{3})]$

31. $\sec[\sin^{-1}(-1/\sqrt{5})]$ ☆**32.** $\sec(\sec^{-1} 4)$

☆**33.** $\csc[\cot^{-1}(-\frac{1}{3})]$ ☆**34.** $\cos(\operatorname{arccsc} 5)$

Evaluate to four significant digits using a calculator.

35. $\sin^{-1}(\cos 22.37)$ **36.** $\sin^{-1}(\tan 1.345)$

37. $\sin[\tan^{-1}(-14.00)]$ **38.** $\csc[\cos^{-1}(-0.4081)]$

39. $\sin^{-1}(\sin 2.000)$ ☆**40.** $\sin[\sec^{-1}(-2.987)]$

☆**41.** $\cos(\cot^{-1} 6.823)$ ☆**42.** $\sec[\operatorname{arccsc}(-25.52)]$

Find the exact degree measure of θ without using a calculator.

43. $\theta = \cos^{-1}(-1/\sqrt{2})$ **44.** $\theta = \tan^{-1}(1/\sqrt{3})$

45. $\theta = \arcsin(-1)$

Find the degree measure of θ to two decimal places using a calculator.

46. $\theta = \cos^{-1} 0.3456$ **47.** $\theta = \arctan(-12.45)$

48. $\theta = \sin^{-1} 0.0025$

Find exact solutions over the indicated interval.

49. $\sin^2 \theta = -\cos 2\theta, \quad 0° \le \theta \le 360°$

50. $\sin 2x = \frac{1}{2}, \quad 0 \le x < \pi$

51. $2 \cos x + 2 = -\sin^2 x, \quad -\pi \le x < \pi$

52. $2 \sin^2 \theta - \sin \theta = 0, \quad \text{all } \theta$

53. $\sin 2x = \sqrt{3} \sin x, \quad \text{all real } x$

54. $2 \sin^2 \theta + 5 \cos \theta + 1 = 0, \quad 0° \le \theta < 360°$

55. $3 \sin 2x = -2 \cos^2 2x, \quad 0 \le x \le \pi$

Solve for all real x. Compute inverse functions to four significant digits.

56. $\sin x = 0.7088$ **57.** $\cos x = -0.1187$

58. $\tan x = -4.318$ **59.** $\csc x = 4.138$

Solve for all real number solutions. Compute inverse functions to four significant digits.

60. $\sin^2 x + 2 = 4 \sin x$ **61.** $\tan^2 x = 2 \tan x + 1$

C *Find exact solutions over the indicated interval.*

62. $\cos x = 1 - \sin x, \quad 0 \le x < 2\pi$

63. $\cos^2 2x = \cos 2x + \sin^2 2x, \quad 0 \le x < \pi$

64. $\sin 7x - \sin x = \sin 3x, \quad 0 \le x \le \pi/2$

65. Solve to three significant digits:

$$2 + 2 \sin x = 1 + 2 \cos^2 x, \quad 0 \le x \le 2\pi$$

Find the exact value.

66. $\sin[2 \tan^{-1}(-\frac{3}{4})]$ **67.** $\sin(\sin^{-1}\frac{3}{5} + \cos^{-1}\frac{4}{5})$

☆**68.** $\cot[\frac{1}{2} \sec^{-1}(\frac{13}{5})]$

Write Problems 69–72 as an algebraic expression in x free of trigonometric or inverse trigonometric functions.

69. $\tan(\sin^{-1} x)$ **70.** $\cos(\tan^{-1} x)$

71. $\cos(\tan^{-1} x + \sin^{-1} x)$ ☆**72.** $\cos(2 \sec^{-1} x)$

c *Problems 73–83 require the use of a graphing calculator. In Problems 73–75, graph each function over the indicated interval.*

73. $y = \sin^{-1}(x + 3)$, $-4 \le x \le -2$

74. $y = -2 \cos^{-1} \dfrac{x}{3}$, $-3 \le x \le 3$

75. $y = \tan^{-1}(1 - 2x)$, $-2 \le x \le 3$

In Problems 76–83, approximate all solutions over the indicated interval using a graphing calculator. Compute solutions to three decimal places.

76. $\sin x = 0.25$, $-\pi \le x \le \pi$

77. $\cot x = -4$, $-\pi \le x \le \pi$

78. $\sec x = 2$, $-\pi \le x \le \pi$

79. $\cos x = x^2$,   all real x

80. $\sin x = \sqrt{x}$, $x \ge 0$

81. $2 \sin x \cos 2x = 1$, $0 \le x \le 2\pi$

82. $\sin \dfrac{x}{2} + 3 \sin x = 2$, $0 \le x \le 4\pi$

83. $\sin x + 2 \sin 2x + 3 \sin 3x = 3$, $0 \le x \le 2\pi$

Applications

84. Music   The note A above middle C has a frequency of 440 Hz. If the intensity I of the sound at a certain point t seconds after the sound is made can be described by the equation

$$I = 0.08 \sin 880 \, \pi t$$

find the smallest positive t such that $I = 0.05$. Compute the answer to two significant digits.

85. Electric Current   An alternating current generator produces a current given by the equation

$$I = 30 \sin 120 \pi t$$

where t is time in seconds and I is current in amperes. Find the least positive t to four significant digits such that $I = 20$ amp.

86. Navigation   A small craft is approaching a large vessel on the course shown in the figure.
(A) Express the angle θ subtended by the large vessel in terms of the distance x between the two ships.
(B) Find θ in decimal degrees to one decimal place for $x = 1,200$ ft.

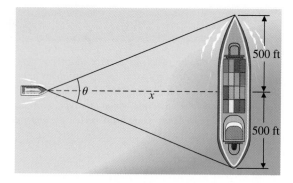

Figure for 86

87. Precalculus   A 14 ft high billboard is located atop a 62 ft high building (see the figure). If an observer is standing x ft from the building and his eye is 6 ft above the ground, express the angle θ in terms of x. Find θ in decimal degrees to two decimal places for $x = 150$ ft.

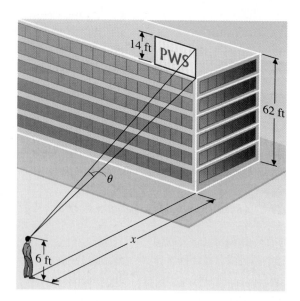

Figure for 87

c **88. Precalculus**   Refer to Problem 87. Set your calculator in degree mode, graph the equation found in Problem 87, and use approximation techniques to find x to one decimal place for $\theta = 6°$.

89. Engineering Specifications for an electromagnet call for a channel formed from a piece of circular steel tubing by cutting off one side (see the figure).

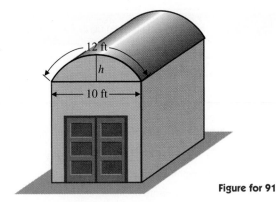

Figure for 91

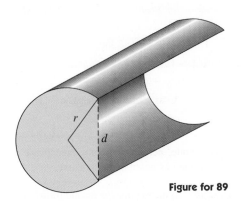

Figure for 89

(A) If r is the radius of the tubing and d is the width of the open side of the channel, show that the cross-sectional area of the channel is given by

$$A = \pi r^2 - r^2 \sin^{-1}\frac{d}{2r} + \frac{d}{4}\sqrt{4r^2 - d^2}$$

(Assume that the larger of the two pieces is used for the channel.)

(B) Find the area to four decimal places if $r = 0.5$ in. and $d = 0.7$ in.

c 90. Engineering Refer to Problem 89. If the channel is cut from steel tubing 1 in. in diameter and the specifications call for a cross-sectional area of 0.75 sq in., approximate the width of the opening to four decimal places.

c 91. Architecture The roof for a 10 ft wide storage shed is formed by bending a 12 ft wide steel panel into a circular arc (see the figure). Approximate to two decimal places the height h of the arc.

c 92. Physics The equation of motion for a weight suspended from a spring is given by

$$y = -1.8 \sin 4t - 2.4 \cos 4t$$

where y is displacement of the weight from its equilibrium position (positive direction upwards) and t is time in seconds.

(A) Graph y for $0 \le t \le \pi/2$.

(B) Approximate to two decimal places the time(s) t, $0 \le t \le \pi/2$, when the weight is 2 in. above the equilibrium position.

(C) Approximate to two decimal places the time(s) t, $0 \le t \le \pi/2$, when the weight is 2 in. below the equilibrium position.

☆93. Physics Refer to Problem 92.

(A) Write y in the form $A \sin(Bt + C)$ where A and C are positive and C is as small as possible. Compute C to two decimal places.

(B) Use the form in part (A) to approximate to two decimal places the time(s) t, $0 \le t \le \pi/2$, when the weight is 2 in. above the equilibrium position.

(C) Use the form in part (A) to approximate to two decimal places the time(s) t, $0 \le t \le \pi/2$, when the weight is 2 in. below the equilibrium position.

Work through all the problems in this cumulative review and check the answers. Answers to all review problems appear in the back of the book; following each answer is an italic number that indicates the section in which that type of problem is discussed. Where weaknesses show up, review the appropriate sections in the text.

A 1. Find the degree measure of 4.21 rad.

2. Find the radian measure of 505°42′.

3. The hypotenuse of a right triangle is 7.6 m and one of the sides is 4.5 m. Find the acute angles and the other side.

4. Find the value of cos θ and cot θ if the terminal side of θ contains $P(-5, -12)$.

5. Evaluate to four significant digits using a calculator:
   (A) cos 67°45′   (B) csc 176.2°   (C) cot 2.05

6. Sketch a graph of each function for $-2\pi \le x \le 2\pi$.
   (A) $y = \cos x$ (B) $y = \csc x$ (C) $y = \cot x$

Verify each identity in Problems 7–10 without looking at a table of identities.

7. $\tan x \csc x = \sec x$

8. $\csc \theta - \sin \theta = \cos \theta \cot \theta$

9. $(\sin^2 u)(\tan^2 u + 1) = \sec^2 u - 1$

10. $\dfrac{\sin^2 \alpha - \cos^2 \alpha}{\sin \alpha \cos \alpha} = \dfrac{\tan \alpha - \cot \alpha}{\tan \alpha \cot \alpha}$

Evaluate Problems 11–20 exactly as real numbers.

11. $\sin \dfrac{5\pi}{6}$

12. $\cos \dfrac{-7\pi}{4}$

13. $\tan \dfrac{7\pi}{3}$

14. $\sec \dfrac{3\pi}{2}$

15. $\sin^{-1}(1/\sqrt{2})$

16. $\arctan 0$

17. $\cos^{-1}(-\sqrt{3}/2)$

18. $\arcsin 3$

☆19. $\operatorname{arccot}(-\sqrt{3})$

☆20. $\sec^{-1} 2$

Evaluate Problems 21–26 as real numbers to four significant digits using a calculator.

21. $\sin^{-1}(0.0505)$

22. $\cos^{-1}(-0.7228)$

23. $\arctan(-9)$

☆24. $\operatorname{arccot}(3)$

☆25. $\sec^{-1} 2.6$

☆26. $\operatorname{arccsc}(-0.5969)$

Find exact solutions for Problems 27–29 over the indicated interval.

27. $2 \sin \theta - 1 = 0$, $\quad 0° \le \theta < 360°$

28. $3 \tan x + \sqrt{3} = 0$, $\quad -\pi/2 < x < \pi/2$

29. $2 \cos x + 2 = 0$, $\quad -\pi \le x < \pi$

☆30. Write $\sin 7u \cos 3u$ as a sum or difference.

☆31. Write $\cos 5w - \cos w$ as a product.

☆32. Write $\sqrt{3} \sin t + \cos t$ in the form $A \sin(Bt + C)$. Indicate amplitude, period, frequency, and phase shift. Choose C as small as possible, but positive.

B 33. Convert 92.462° to degree-minute-second form.

34. Find the degree measure of a central angle subtended by an arc of 12 in. in a circle with circumference 30 in.

35. Find x exactly and θ to the nearest 0.1° in the figure.

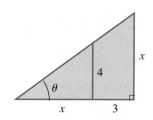

Figure for 35

36. Convert 72° to radian measure in terms of π.

37. Sketch the reference triangle and find the reference angle α for $\theta = 315°$.

38. Find the exact value of each of the other five trigonometric functions if $\tan \theta = \frac{1}{2}$ and $\sin \theta < 0$.

In Problems 39–41, sketch a graph of each function for the indicated interval. State the period and, if applicable, the amplitude and phase shift.

39. $y = 2 - 2 \sin \dfrac{x}{2}$, $\quad -\pi \le x \le 5\pi$

40. $y = 3 \cos(2x - \pi)$, $\quad -\pi \le x \le 2\pi$

41. $y = 2 \tan(\pi x - \pi/4)$, $\quad 0 \le x \le 3$

42. Find an equation of the form $y = k + A \cos Bx$ whose graph is shown in the figure at the top of page 289.

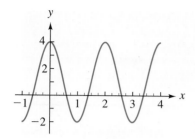

Figure for 42

43. If the sides of a right triangle are 19.4 cm and 41.7 cm, find the hypotenuse and find the acute angles to the nearest 10′.

Verify the identities in Problems 44–49.

44. $\dfrac{\cos x}{1 + \sin x} + \tan x = \sec x$

45. $\dfrac{1 + \cos \theta}{1 + \sin \theta} = (\sec \theta - \tan \theta)(\sec \theta + 1)$

46. $\cot \dfrac{u}{2} = \csc u + \cot u$

47. $\sec^2 \dfrac{x}{2} = 2 + 2 \cot x \,(\cot x - \csc x)$

48. $\dfrac{2}{1 + \sec 2\theta} = 1 - \tan^2 \theta$

49. $\dfrac{\sin x - \sin y}{\cos x + \cos y} = \tan \dfrac{x - y}{2}$

50. Find the exact values of sin (x/2) and cos 2x, given tan x = $\frac{8}{15}$ and $\pi < x < 3\pi/2$.

☆**51.** Evaluate cos 165° + cos 15° exactly using an appropriate identity.

☆**52.** Write $y = \sin \pi t - \sqrt{3} \cos \pi t$ in the form $y = A \sin(Bt + C)$, where C is chosen so that $|C|$ is as small as possible. Indicate amplitude, period, frequency, and phase shift; then graph the equation for $-1 \le t \le 2$.

53. Find exact values for each of the following:
   (A) $\sin(\cos^{-1} 0.4)$ (B) $\sec[\arctan(-\sqrt{5})]$
   (C) $\csc(\sin^{-1} \frac{1}{3})$ ☆(D) $\tan(\sec^{-1} 4)$

54. Evaluate to four significant digits using a calculator:
   (A) $\csc[\arctan(-5.624)]$ (B) $\cos^{-1}(\sec 2.558)$
   (C) $\tan^{-1}(\cos 0.1028)$ ☆(D) $\sin[\text{arcsec}\,(-4.612)]$

55. Find the exact degree value of $\theta = \tan^{-1}\sqrt{3}$ without using a calculator.

56. Find the degree measure of $\theta = \sin^{-1} 0.8989$ to two decimal places using a calculator.

Find exact solutions to Problems 57–59 over the indicated interval.

57. $2 + 3 \sin x = \cos 2x, \quad 0 \le x \le 2\pi$

58. $\sin 2\theta = 2 \cos \theta, \quad$ all θ

59. $4 \tan^2 x - 3 \sec^2 x = 0, \quad$ all real x

Find all real solutions to Problems 60–62. Compute inverse functions to four significant digits.

60. $\sin x = -0.5678$

61. $\sec x = 2.345$

62. $2 \cos 2x = 7 \cos x$

C 63. A point on a circle with radius 5 and center at the origin in a rectangular coordinate system starts at (5, 0) and moves clockwise around the circle until it reaches the point $(-1.4, -4.8)$. Find the length of the arc traversed on the circle by the point. Compute the answer to two decimal places.

64. If θ is a fourth quadrant angle and $\cos \theta = a, 0 < a < 1$, express the other five trigonometric functions of θ in terms of a.

65. Graph $y = 0.5 \csc(4x + \pi), \quad 0 < x < 2\pi$

66. Show that $\tan 3x = \tan x \dfrac{2 \cos 2x + 1}{2 \cos 2x - 1}$ is an identity.

67. Find the exact value of $\cos(2 \sin^{-1} \frac{1}{3})$.

68. Write $\sin(\cos^{-1} x - \tan^{-1} x)$ as an algebraic expression in x free of trigonometric or inverse trigonometric functions.

69. Find exact solutions for all real x,
$$\sin x = 1 + \cos x$$

70. Find solutions to four significant digits,
$$\sin x = \cos^2 x, \qquad 0 \le x \le 2\pi$$

☆**71.** Write $y = -2.4 \sin \dfrac{t}{2} + 3.2 \cos \dfrac{t}{2}$ in the form $y = A \sin(Bt + C)$ where C is chosen so that $|C|$ is as small as possible and C is computed to two decimal places using a calculator. Indicate amplitude, period, frequency, and phase shift.

Problems 72–82 require the use of a graphing calculator. In problems 72–75, graph f(x), find a simpler function g(x) which has the same graph as f(x), and verify the indentity f(x) = g(x). [Assume g(x) = k + A t(Bx) where t(x) is one of the six trigonometric functions.]

72. $f(x) = \dfrac{\sin^2 x}{1 - \cos x} + \dfrac{2 \tan^2 x \cos^2 x}{1 + \cos x}$

73. $f(x) = 2 \sin^2 x + 6 \cos^2 x$

74. $f(x) = \dfrac{2 - 2 \sin^2 x}{2 \cos^2 x - 1}$

75. $f(x) = \dfrac{3 \cos x + \sin x - 3}{\cos x - 1}$

Graph each function over the indicated interval.

76. $y = \cos^{-1}(x + 1), \quad -2 \le x \le 0$

77. $y = 3 \sin^{-1} \dfrac{x}{2}, \quad -2 \le x \le 2$

78. $y = -2 \tan^{-1}(3x - 2), \quad -1 \le x \le 2$

Approximate all solutions over the indicated interval. Compute solutions to three decimal places.

79. $\tan x = 3, \quad -\pi \le x \le \pi$

80. $\cos x = \sqrt{x}, \quad x > 0$

81. $\cos \dfrac{x}{2} - 2 \sin x = 1, \quad 0 \le x \le 4\pi$

☆ **82.** Graph $y = -2.4 \sin \dfrac{t}{2} + 3.2 \cos \dfrac{t}{2}$ for $-2\pi \le t \le 2\pi$, approximate the *t* intercepts in this interval to two decimal places, and identify the intercept that corresponds to the phase shift determined in Problem 71.

Applications

83. Geography/Navigation   Find the distance (to the nearest mile) between Pittsburgh, PA with latitude 40°26′N and Roanoke, VA with latitude 37°32′N. (Both cities have approximately the same longitude.) Use $\pi \approx 3.14$ and $r \approx 3960$ mi for the earth's radius.

84. Navigation   An airplane flying on a level course directly toward a beacon on the ground records two angles of depression as indicated in the figure. Find the altitude *h* of the plane to the nearest meter.

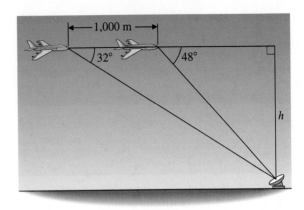

Figure for 84

85. Precalculus: Trigonometric Substitution   In the expression $\sqrt{u^2 - a^2}$, $a > 0$, let $u = a \csc x$, $0 < x < \pi/2$, simplify, and write in a form that is free of radicals.

86. Engineering   Find the exact values of *x* and θ in the figure.

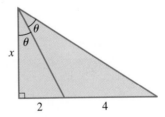

Figure for 86

87. Physics   The equation of motion for a weight suspended from a spring is given by

$$y = -7.2 \sin 5t - 9.6 \cos 5t$$

where *y* is the displacement of the weight from its equilibrium position in centimeters (positive direction upwards) and *t* is time in seconds. Find the smallest positive *t* to three decimal places such that the weight is in the equilibrium position.

C 88. Physics   Refer to Problem 87. Graph *y* for $0 \le t \le \pi/2$ and approximate to three decimal places the time(s) *t* in this interval such that the weight is 5 cm above the equilibrium position.

☆**89. Physics**   Refer to Problem 87.
   (A) Write y in the form $A \sin(Bt + C)$ where A and C are positive and C is as small as possible. Compute C to three decimal places.
   (B) Use the form in part (A) to approximate to three decimal places the time(s) t, $0 \le t \le \pi/2$, when the weight is 2 cm below the equilibrium position.

90. Electrical Circuits   The current I (in amperes) in an electrical circuit is given by

$$I = 60 \sin\left(90\pi t - \frac{\pi}{2}\right)$$

Find the amplitude, period, frequency, and phase shift. Graph the equation for $0 \le t \le \frac{4}{45}$.

91. Electrical Circuits   Refer to Problem 90. Find the smallest positive t to four significant digits such that $I = 35$ amps.

92. Boat Safety   A boat is approaching a 200 ft high vertical cliff (see the figure).
   (A) Write an equation for the distance d from the boat to the base of the cliff in terms of the angle of elevation θ from the boat to the top of the cliff.
   (B) Graph this equation for $0 < \theta \le \pi/2$.

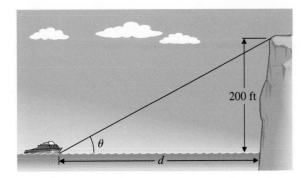

Figure for 92

93. Precalculus   Two guy wires are attached to a radio tower as shown in the figure.
   (A) Show that $\theta = \arctan \dfrac{100x}{x^2 + 20,000}$
   (B) Find θ in decimal degrees to one decimal place for $x = 50$ ft.

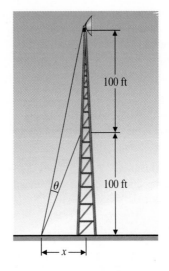

Figure for 93

C **94. Precalculus**   Refer to Problem 93. Set your calculator in degree mode, graph

$$y1 = \tan^{-1} \frac{100x}{x^2 + 20,000} \quad \text{and} \quad y2 = 15$$

in the same viewing rectangle, and use approximation techniques to find x to one decimal place when $\theta = 15°$.

95. Engineering   The chain on a bicycle goes around the pedal sprocket and the rear wheel sprocket (see the figure). The radius of the pedal sprocket is 11.0 cm, the radius of the rear sprocket is 4.00 cm, and the diameter of the rear wheel is 70.0 cm. If the pedal sprocket rotates through an angle of 18π radians, through how many radians does the rear wheel rotate?

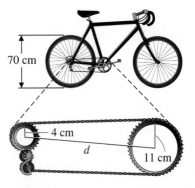

Figure for 95

☆**96. Engineering**   Refer to Problem 95. If the pedal sprocket rotates at 60.0 rpm, how fast is the bicycle traveling in centimeters per minute?

97. Engineering   Refer to Problem 95. If the distance between the centers of the sprockets is $d = 50.0$ cm, how long is the chain?

c **98. Engineering**   Refer to Problem 95. If the length of the chain is 130 cm, what is the distance between the centers of the sprockets?

c **99. Modeling Daylight Duration**   Table 1 gives the duration of daylight on the fifteenth day of each month for one year at Anchorage, Alaska.

(A) Convert the data in Table 1 from hours and minutes to two-place decimal hours. Enter this data for a two year period in your graphing calculator and produce a scatter plot in the following viewing rectangle: $1 \leq x \leq 24, 6 \leq y \leq 20$.

(B) A function of the form $y = k + A \sin(Bx + C)$ can be used to model this data. Use the converted data from Table 1 to determine k and A (to two decimal places), and B (exact value). Use the graph in part (A) to visually estimate C (to one decimal place).

(C) Plot the data from part (A) and the equation from part (B) in the same viewing rectangle. If necessary, adjust your value of C to produce a better fit.

TABLE 1

x (months)	1	2	3	4	5	6	7	8	9	10	11	12
y (daylight duration)	6:31	9:10	11:50	14:48	17:33	19:16	18:27	15:51	12:57	10:09	7:19	5:36

ADDITIONAL TRIANGLE TOPICS; VECTORS

ou have now had quite a bit of experience in solving right triangles. This chapter considers the more general types of triangles that do not contain a right angle. The laws of sines and cosines are the principal tools used to solve these triangles.

The concept of *vector,* a very important and useful form in both pure and applied mathematics, is also developed in this chapter both geometrically and algebraically.

6.1 LAW OF SINES

♦ **Deriving the Law of Sines**

♦ **Solving ASA and AAS Cases**

♦ **Solving the Ambiguous SSA Case**

Until now, we have considered only triangle problems that involved right triangles. We now turn to **oblique triangles**, that is, triangles that contain no right angle. Every oblique triangle is either **acute** (all angles are between 0° and 90°) or **obtuse** (one angle is between 90° and 180°). Figure 1 illustrates an acute triangle and an obtuse triangle.

   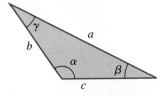

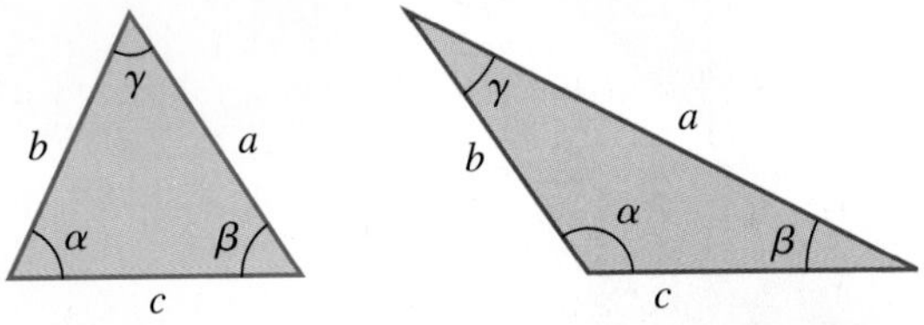

(a) Acute triangle   (b) Obtuse triangle

FIGURE 1

Notice how we labeled the sides and angles of the oblique triangles shown in Figure 1: Side a is opposite angle α, side b is opposite angle β, and side c is opposite angle γ. Note that the largest side of a triangle is opposite the largest angle. Given any three of the six quantities indicated in Figure 1, we are interested in finding the remaining three, if they exist. This process is called **solving the triangle**.

In this section we will develop the *law of sines* and in the next section, the *law of cosines.* These are two basic tools important in the solution of oblique triangles. If the given quantities include an angle and the opposite side, we will use the law of sines; otherwise, we will use the law of cosines.

Before we proceed with specifics, recall the rules governing angle measure and significant digits for side measure (listed inside the front cover for easy reference).

TABLE 1

Angle to nearest	Significant digits for side measure
1°	2
10′ or 0.1°	3
1′ or 0.01°	4
10″ or 0.001°	5

REMARK ON CALCULATIONS

When you solve for a particular side or angle, carry out all operations within the calculator and then round to the appropriate number of significant digits (following the rules in Table 1) at the end of the calculation. Note that your answers still may differ slightly from those in the book, depending on the order in which you solve for the sides and angles.

◆ Deriving the Law of Sines

The law of sines is relatively easy to derive using the right triangle properties we studied earlier. We also use the fact that

$$\sin(180° - x) = \sin x$$

which is obtained from the difference identity for sine. Referring to the triangles in Figure 2, we proceed as follows: For each triangle,

$$\sin \alpha = \frac{h}{b} \quad \text{and} \quad \sin \beta = \frac{h}{a}$$

Therefore,

$$h = b \sin \alpha \quad \text{and} \quad h = a \sin \beta$$

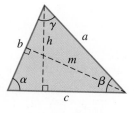

(a) Acute triangle

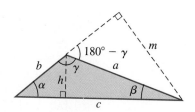

(b) Obtuse triangle

FIGURE 2

Thus,

$$b \sin \alpha = a \sin \beta$$

and

$$\frac{\sin \alpha}{a} = \frac{\sin \beta}{b} \tag{1}$$

Similarly, for each triangle in Figure 2,

$$\sin \alpha = \frac{m}{c}$$

and

$$\sin \gamma = \sin(180° - \gamma)$$

$$= \frac{m}{a}$$

Therefore,

$$m = c \sin \alpha \qquad \text{and} \qquad m = a \sin \gamma$$

Thus,

$$c \sin \alpha = a \sin \gamma$$

and

$$\frac{\sin \alpha}{a} = \frac{\sin \gamma}{c} \tag{2}$$

If we combine equations (1) and (2), we obtain the **law of sines**

LAW OF SINES

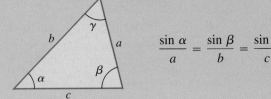

$$\frac{\sin \alpha}{a} = \frac{\sin \beta}{b} = \frac{\sin \gamma}{c}$$

In words, the ratio of the sine of an angle to its opposite side is the same as the ratio of the sine of either of the other angles to its opposite side.

If the given quantities include an angle and the opposite side, we will use the law of sines. Otherwise, we will use the law of cosines.

Thus, the law of sines is used to solve triangles, given

1. Two angles and any side (ASA or AAS), or
2. Two sides and an angle opposite one of them (SSA)

We will start with the law of cosines (Section 6.2) to solve a triangle if we are given

3. Two sides and the included angle (SAS) or
4. Three sides (SSS)

We will apply the law of sines to the easier ASA and AAS cases first, then we will turn to the more troublesome SSA case.

◆ Solving ASA and AAS Cases

If we are given two angles and the included side (ASA case) or two angles and the side opposite one of the angles (AAS case), we first find the measure of the third angle (remember that the sum of the measures of all three angles in any triangle is 180°). We then use the law of sines to find the other two sides.

⚠ **Caution**   Note that for the ASA or AAS case to determine a unique triangle, the sum of the two given angles must be between 0° and 180° (see Figure 3).

FIGURE 3

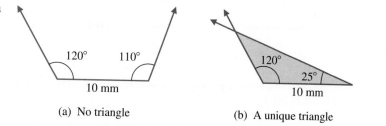

(a) No triangle   (b) A unique triangle   ◇

Examples 1 and 2 illustrate the use of the law of sines in solving the ASA and AAS cases.

◆ **EXAMPLE 1**   Using the Law of Sines (ASA)

Solve the triangle (Figure 4).

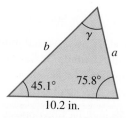

FIGURE 4

SOLUTION   This is an example of the ASA case.

Solve for γ.

$$\alpha + \beta + \gamma = 180°$$
$$\gamma = 180° - (45.1° + 75.8°)$$
$$= 59.1°$$

Solve for a.

$$\frac{\sin \alpha}{a} = \frac{\sin \gamma}{c}$$

$$a = (\sin \alpha)\frac{c}{\sin \gamma}$$

$$= (\sin 45.1°)\frac{10.2}{\sin 59.1°}$$

$$= 8.42 \text{ in.}$$

Solve for b.

$$\frac{\sin \beta}{b} = \frac{\sin \gamma}{c}$$

$$b = (\sin \beta)\frac{c}{\sin \gamma}$$

$$= (\sin 75.8°)\frac{10.2}{\sin 59.1°}$$

$$= 11.5 \text{ in.}$$ ◆

MATCHED PROBLEM 1   Solve the triangle with $\alpha = 28.0°$, $\beta = 45.3°$, and $c = 122$ m.

REMARK   In Example 1, note that in solving for a and b, the quantity $c/(\sin \gamma)$ appears in each calculation. Compute $c/(\sin \gamma)$ and store the result (without rounding) in a memory cell. Recall it and multiply by $\sin \alpha$ to obtain a; then recall it and multiply by $\sin \beta$ to obtain b. The use of memory cells in a calculator often speeds up the process of solving an oblique triangle.   ◇

◆ EXAMPLE 2   Using the Law of Sines (AAS): Sundials

The angle of the gnomon, β, on a sundial must be the same as the latitude where the sundial is used. If the latitude of San Francisco is 38°N, and the angle of elevation of the sun, α, is 63° at noon, how long will the shadow of a 12 in. gnomon be on the face of the sundial (see Figure 5)?

SOLUTION   We are given two angles and the side opposite one of the angles (AAS). First we find the third angle, then we find the length of the shadow using the law of sines.

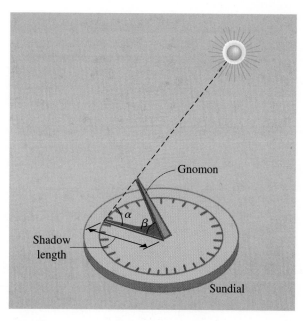

FIGURE 5
Sundial

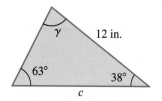

FIGURE 6

It is helpful to make a simple drawing (Figure 6) showing the given and unknown parts.

Solve for γ.

$$\gamma = 180° - (63° + 38°) = 79°$$

Solve for c.

$$\frac{\sin \alpha}{a} = \frac{\sin \gamma}{c}$$

$$c = \frac{a \sin \gamma}{\sin \alpha}$$

$$= \frac{12 \sin 79°}{\sin 63°} = 13 \text{ in.} \qquad \textit{Shadow length}$$

◆

MATCHED PROBLEM 2   Repeat Example 2 for Columbus, Ohio, with a latitude of 40°N, a gnomon of length 15 in., and a 52° angle of elevation for the sun.

◆   Solving the Ambiguous SSA Case

If we are given two sides and an angle opposite one of the sides—the SSA case—then it is possible to have 0, 1, or 2 triangles, depending on the measures of the two sides and the angle. Therefore, we refer to the SSA case as the **ambiguous case**. Table 2 illustrates the various possibilities.

TABLE 2
SSA Variations

	a ($h = b \sin \alpha$)	Number of triangles	Figure	
α Acute	$0 < a < h$	0		(a)
	$a = h$	1		(b)
	$h < a < b$	2		(c)
	$a \geq b$	1		(d)
α Obtuse	$0 < a \leq b$	0		(e)
	$a > b$	1		(f)

Triangles involving the SSA case can be solved without memorizing Table 2. The particular case in the table generally becomes apparent in the solution process. The cases where α is obtuse, (e) and (f) in Table 2, cause little trouble. If α is acute, and in the solution process we end up with $\sin \beta > 1$, we know that β does not exist. Hence, no triangle can exist, and we must have (a) in Table 2. If $\sin \beta = 1$, then $\beta = 90°$ and we must have (b) in Table 2. If $0 < \sin \beta < 1$, then there are two choices for β, one the supplement of the other [recall, $\sin(180° - \beta) = \sin \beta$]. The inverse key on a calculator will give you only the acute angle. To obtain the obtuse supplement, the measure of the acute angle must be subtracted from 180°. If the obtuse choice is added to α and the sum is greater than or equal to 180°, then the obtuse choice must be rejected, leaving only the acute choice, which results in one triangle as illustrated in (d) in Table 2. If the obtuse choice is added to α and the sum is less than 180°, then we have two possible triangles as indicated in (c) in Table 2. Several examples should help to clarify the process of solving the various SSA cases.

◆ **EXAMPLE 3**   Using the Law of Sines (SSA): No Triangle

Find β in the triangle with $\alpha = 34°$, $b = 2.3$ mm, and $a = 1.2$ mm.

SOLUTION   To find β, we go directly to the law of sines.

$$\frac{\sin \beta}{b} = \frac{\sin \alpha}{a} \qquad \text{Law of sines}$$

$$\sin \beta = \frac{b \sin \alpha}{a} \qquad \text{Solve for } \sin \beta$$

$$= \frac{2.3 \sin 34°}{1.2} \qquad \text{Substitute } a, b, \text{ and } \alpha$$

$$\approx 1.0718 \qquad \text{Evaluate}$$

Since $\sin \beta = 1.0718$ has no solution, then no triangle exists with the given measurements. [We could have come to the same conclusion by noting that α is acute and $a < h$ ($a = 1.2$ and $h = b \sin \alpha = 2.3 \sin 3.4 \approx 1.2861$)—case (a) in Table 2.]   ◆

MATCHED PROBLEM 3   Use the law of sines to find β in the triangle with $\alpha = 130°$, $b = 1.3$ m, and $a = 1.2$ m.

◆   EXAMPLE 4   Using the Law of Sines (SSA): One Triangle

Solve a triangle with $\alpha = 47°$, $a = 3.7$ ft, and $b = 3.5$ ft.

SOLUTION   *Solve for* β.

$$\frac{\sin \beta}{b} = \frac{\sin \alpha}{a} \qquad \text{Law of sines}$$

$$\sin \beta = \frac{b \sin \alpha}{a} \qquad \text{Solve for } \sin \beta$$

$$= \frac{3.5 \sin 47°}{3.7} \qquad \text{Substitute } a, b, \text{ and } \alpha$$

$$\approx 0.6918 \qquad \text{Evaluate}$$

$$\beta = \sin^{-1} 0.6918$$

$$= 44° \qquad \text{Use the inverse key(s) on a calculator}$$

The calculator gives the measure of the acute angle whose sine is 0.6918. The other possible choice for β is the supplement of 44°, that is, $180° - 44° = 136°$. But if we add 136° to $\alpha = 47°$, we get 183°, which is greater than 180° (the sum of all measures of the three angles in a triangle). Because it is not possible to form a triangle having two angles with measures 47° and 136°, the supplement must be rejected, and we are left with only one triangle using $\beta = 44°$. (We could have come to the same conclusion noting that α is acute and $a > b$—case (d) in Table 2.) We complete the problem by solving for γ and c.

Solve for γ.

$$\gamma = 180° - (\alpha + \beta)$$

$$= 180° - (47° + 44°) = 89°$$

Solve for c.

$$\frac{\sin \alpha}{a} = \frac{\sin \gamma}{c}$$

$$c = \frac{a \sin \gamma}{\sin \alpha}$$

$$= \frac{3.7 \sin 89°}{\sin 47°} = 5.1 \text{ ft}$$

◆

MATCHED PROBLEM 4 Solve a triangle with $\alpha = 139°$, $a = 42$ yd, and $b = 27$ yd.

◆ **EXAMPLE 5** Using the Law of Sines (SSA): Two Triangles

Solve a triangle with $\alpha = 26°$, $a = 11$ cm, and $b = 18$ cm.

SOLUTION *Solve for β.*

$$\frac{\sin \beta}{b} = \frac{\sin \alpha}{a}$$

$$\sin \beta = \frac{b \sin \alpha}{a}$$

$$= \frac{18 \sin 26°}{11}$$

$$\approx 0.7173$$

$$\beta = \sin^{-1} 0.7173$$

$$= 46°$$

◆

But β could also be the supplement of 46°; that is, $180° - 46° = 134°$. To see if the supplement, 134°, can be an angle in a second triangle, we add 134° to $\alpha = 26°$ to obtain 160°. Since 160° is less than 180°, the sum of the measure of all angles of a triangle, there are two possible triangles meeting the original conditions, one with $\beta = 46°$ and the other with $\beta = 134°$. [We could have arrived at the same conclusion by noting that $h < a < b$—case (c) in Table 2.] Figure 7 illustrates the results: $\beta = 134°$ and $\beta' = 46°$.

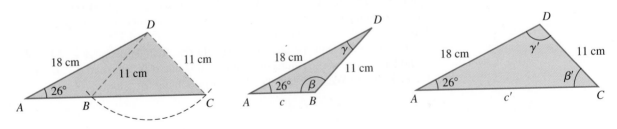

FIGURE 7

Solve for γ and γ'.

$$\gamma = 180° - (26° + 134°) = 20°$$
$$\gamma' = 180° - (26° + 46°) = 108°$$

Solve for c and c'.

$$\frac{\sin \alpha}{a} = \frac{\sin \gamma}{c} \qquad\qquad \frac{\sin \alpha}{a} = \frac{\sin \gamma'}{c'}$$

$$c = \frac{a \sin \gamma}{\sin \alpha} \qquad\qquad c' = \frac{a \sin \gamma'}{\sin \alpha}$$

$$= \frac{11 \sin 20°}{\sin 26°} \qquad\qquad = \frac{11 \sin 108°}{\sin 26°}$$

$$= 8.6 \text{ cm} \qquad\qquad = 24 \text{ cm}$$

◆

MATCHED PROBLEM 5   Solve the triangle(s) with $a = 8.0$ mm, $b = 11$ mm, and $\alpha = 35°$.

We conclude by outlining a strategy for the SSA case.

STRATEGY FOR SOLVING THE SSA CASE

Step	Find	Method
1.	Acute angle opposite given side (If the sine is greater than 1, stop: No triangle is determined.)	Law of sines
2.	Obtuse supplement of the acute angle found in step 1 and number of triangles determined (If the sum of the obtuse angle and the original given angle is less than 180°, two triangles are determined.)	Subtract acute angle from 180°
3.	Third angle (or angles, if two triangles are determined)	Subtract the sum of the given angle and the angle(s) found in step 2 from 180°.
4.	Third side (or sides, if two triangles are determined)	Law of sines

Answers to
Matched Problems

1. $\gamma = 106.7°$, $a = 59.8$ m, $b = 90.5$ m
2. Shadow length = 19″
3. No triangle: α is obtuse and $a \leq b$.
4. $\beta = 25°$, $\gamma = 16°$, $c = 18$ yd
5. $\beta = 128°$, $\beta' = 52°$; $\gamma = 17°$, $\gamma' = 93°$; $c = 4.1$ mm, $c' = 14$ mm

EXERCISE 6.1

Assume all triangles are labeled as in the figure unless stated to the contrary. Your answers may differ slightly from those in the book, depending on the order in which you solve for the sides and angles.

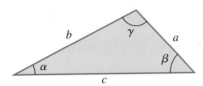

A *Solve each triangle given the indicated measures of angles and sides.*

1. $\alpha = 41°$, $\beta = 33°$, $c = 21$ mi
2. $\alpha = 73°$, $\beta = 28°$, $c = 42$ m
3. $\beta = 43°$, $\gamma = 36°$, $a = 92$ cm
4. $\alpha = 122°$, $\gamma = 18°$, $b = 12$ mm
5. $\beta = 27.5°$, $\gamma = 54.5°$, $a = 9.27$ mm
6. $\alpha = 118.3°$, $\gamma = 12.2°$, $b = 17.3$ km
7. $\alpha = 122.7°$, $\beta = 34.4°$, $b = 18.3$ cm
8. $\alpha = 67.7°$, $\beta = 54.2°$, $b = 123$ ft
9. $\beta = 12°40'$, $\gamma = 100°0'$, $b = 13.1$ km
10. $\alpha = 73°50'$, $\beta = 51°40'$, $a = 36.6$ mm

B *In Problems 11–24 solve each triangle. If a problem has no solution, say so. If a problem involves two triangles, solve both unless stated to the contrary.*

11. $\beta = 52.0°$, $a = 8.00$ cm, $b = 12.0$ cm
12. $\alpha = 54.3°$, $a = 44.5$ in., $b = 30.0$ in.
13. $\alpha = 27.5°$, $a = 15.0$ mm, $b = 36.4$ mm
14. $\alpha = 47.7°$, $a = 8.5$ ft, $b = 12.5$ ft
15. $\alpha = 18.92°$, $a = 48.35$ yd, $b = 105.0$ yd, β acute
16. $\alpha = 12.2°$, $a = 0.668$ m, $b = 2.43$ m, β acute
17. $\alpha = 18.92°$, $a = 48.35$ yd, $b = 105.0$ yd, β obtuse
18. $\alpha = 12.2°$, $a = 0.668$ m, $b = 2.43$ m, β obtuse
19. $\alpha = 135°20'$, $a = 14.6$ m, $b = 18.3$ m
20. $\alpha = 122°40'$, $a = 105$ mi, $b = 152$ mi
21. $\beta = 33°50'$, $a = 673$ ft, $b = 1{,}240$ ft
22. $\beta = 29°30'$, $a = 43.2$ in., $b = 56.5$ in.

23. $\beta = 27.3°$, $a = 244$ ft, $b = 135$ ft
24. $\beta = 38.9°$, $a = 42.7$ cm, $b = 30.0$ cm

C 25. Mollweide's equation,

$$(a - b) \cos \frac{\gamma}{2} = c \sin \frac{\alpha - \beta}{2}$$

is often used to check the final solution of a triangle since all six parts of a triangle are involved in the equation. If, after substitution, the left side does not equal the right side, then an error has been made in solving a triangle. Use this equation to check Problem 1 to two decimal places. (Remember that rounding may not produce exact equality, but the left and right sides of the equation should be close.)

26. Use Mollweide's equation (see Problem 25) to check Problem 3 to two decimal places.

27. Use the law of sines and suitable identities to show that for any triangle,

$$\frac{a - b}{a + b} = \frac{\tan \dfrac{\alpha - \beta}{2}}{\tan \dfrac{\alpha + \beta}{2}}$$

28. Verify (to three decimal places) the formula in Problem 27 with values from Problem 1 and its solution.

29. Let $\beta = 46.8°$ and $a = 66.8$ yd. Determine a value k so that if $0 < b < k$, there is no solution; if $b = k$, there is one solution; if $k < b < a$, there are two solutions.

30. Let $\beta = 36.6°$ and $b = 12.2$ m. Determine a value k so that if $0 < b < k$, there is no solution; if $b = k$, there is one solution; if $k < b < a$, there are two solutions.

Applications

31. **Surveying**   To determine the distance across the Grand Canyon in Arizona, a 1.00 mi baseline, *AB*, is established along the southern rim of the canyon. Sightings are then made from each end of the baseline *A* and *B* to a point *C* across the canyon (see the figure). Find the distance from *A* to *C* if $\angle BAC = 118.1°$ and $\angle ABC = 58.1°$.

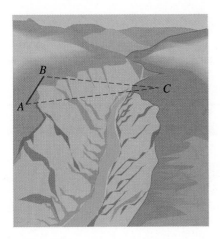

Figure for 31

32. Surveying   Refer to Problem 31. The Grand Canyon was surveyed at a narrower part of the canyon using a similar 1.00 mi baseline along the southern rim. Find the length of AC if $\angle BAC = 28.5°$ and $\angle ABC = 144.6°$.

33. Fire Spotting   A fire at F is spotted from two fire lookout stations, A and B, which are located 10.3 mi apart. If station B reports the fire at angle $ABF = 52.6°$, and station A reports the fire at angle $BAF = 25.3°$, how far is the fire from station A? From station B?

Figure for 33

34. Coast Patrol   Two lookout posts, A and B, which are located 12.4 mi apart, are established along a coast to watch for illegal foreign fishing boats coming within the

3 mi limit. If post A reports a ship S at angle $BAS = 37.5°$, and post B reports the same ship at angle $ABS = 19.7°$, how far is the ship from post A? How far is the ship from the shore (assuming the shore is along the line joining the two observation posts)?

35. Surveying   An underwater telephone cable is to cross a shallow lake from point A to point B (see the figure). Stakes are located at A, B, and C. Distance AC is measured to be 112 m, $\angle CAB$ to be 118.4°, and $\angle ABC$ to be 19.2°. Find the distance AB.

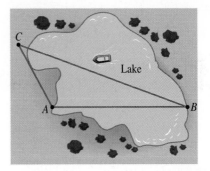

Figure for 35

36. Surveying   A suspension bridge is to cross a river from point B to point C (see the figure). Distance AB is measured to be 0.652 mi, $\angle ABC$ to be 81.3°, and $\angle BCA$ to be 41.4°. Compute the distance BC.

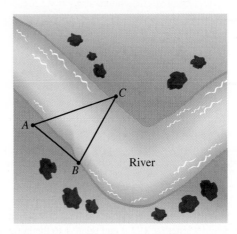

Figure for 36

*37. Coastal Piloting A boat is traveling along a coast at night. A flashing buoy marks a reef. While proceeding on the same course, the navigator of the boat sights the buoy twice, 4.6 nautical mi apart, and forms triangles as in the figure. If the boat continues on course, by how far will it miss the reef?

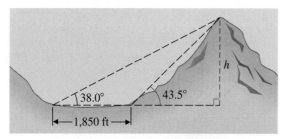

Figure for 37

*38. Surveying Find the height of the mountain above the valley in the figure.

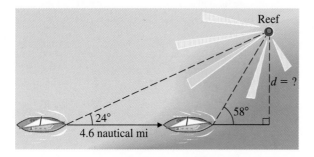

How high is the mountain?

Figure for 38

*39. Tree Height A tree growing on a hillside casts a 157 ft shadow straight down the hill (see the figure). Find the vertical height of the tree if relative to the horizontal, the hill slopes 11.0° and the angle of elevation of the sun is 42.0°.

*40. Tree Height Find the height of the tree in Problem 39 if the shadow length is 102 ft and relative to the horizontal, the hill slopes 15.0° and the angle of elevation of the sun is 62.0°.

Figure for 39

*41. Aircraft Design Find the measures of β and c for the sweptback wing of a supersonic jet, given the information in the figure.

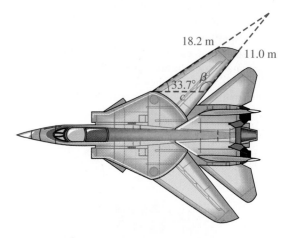

Figure for 41

*42. Aircraft Design Find the measures of β and c in Problem 41 if the extended leading and trailing edges of the wing are 20.0 m and 15.5 m, respectively, and the measure of the given angle is 35.3° instead of 33.7°. (Refer to the figure for Problem 41.)

*43. Eye A cross section of the cornea of an eye, a circular arc, is shown in the figure. With the given information in the figure, find the radius of the arc r and the length of the arc s.

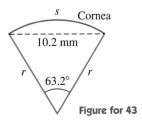

S   Cornea

10.2 mm

r

63.2°

r

Figure for 43

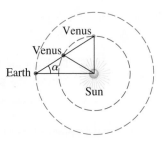

Venus

Venus

Earth

α

Sun

Distance from Earth to Venus

Figure for 47

*44. **Eye**   Repeat Problem 43 using a central angle of 98.9° and a chord of length 11.8 mm.

*45. **Space Science**   When a satellite is directly over tracking station B (see the figure), tracking station A measures the angle of elevation θ of the satellite (from the horizon line) to be 24.9°. If the tracking stations, both on the equator, are 504 mi apart (that is, the arc AB is 504 mi long) and the radius of the earth is 3,964 mi, how high is the satellite above B? [*Hint:*   Find all angles for the triangle ACS first.]

48. **Astronomy**   In Problem 47 find the maximum value of α. [*Hint:*   The value of α is maximum when a straight line joining Earth and Venus in the figure is tangent to Venus's orbit.]

49. **Engineering**   A 12 cm piston rod joins a piston to a 4.2 cm crankshaft (see the figure). What is the longest distance d of the piston from the center of the crankshaft when the rod makes an angle of 8.0°?

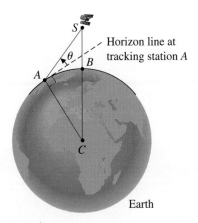

S

θ

B

Horizon line at tracking station A

A

C

Earth

Figure for 45

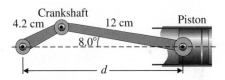

Crankshaft

4.2 cm

12 cm

Piston

8.0°

d

Figure for 49

50. **Engineering**   See Problem 49. What is the shortest distance d of the piston from the center of the crankshaft when the rod makes an angle of 8.0°?

*51. **Surveying**   The scheme illustrated in the figure is used to determine inaccessible heights when d, α, β, and γ can be measured. Show that

$$h = d \sin \alpha \csc(\alpha + \beta) \tan \gamma$$

*46. **Space Science**   For the satellite in Problem 45, compute its height above tracking station B if the stations are 632 mi apart and the angle of elevation θ of the satellite (above the horizon line) is 26.2° at tracking station A.

47. **Astronomy**   The orbits of the earth and Venus are approximately circular, with the sun at the center (see the figure). A sighting of Venus is made from Earth and the angle α is found to be 18°40′. If the diameter of the orbit of the earth is 2.99×10^8 km and the diameter of the orbit of Venus is 2.17×10^8 km, what are the possible distances from Earth to Venus?

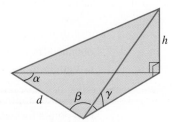

h

α

d

β

γ

Figure for 51

6.2 LAW OF COSINES

◆   **Deriving the Law of Cosines**
◆   **Solving the SAS Case**
◆   **Solving the SSS Case**

In Figure 1(a) we are given two sides and the included angle (SAS), and in Figure 1(b) we are given three sides (SSS). In neither case do we have an angle and a side opposite the angle; hence, the law of sines cannot be used. In this section we develop the *law of cosines*, which can be used for these two cases.

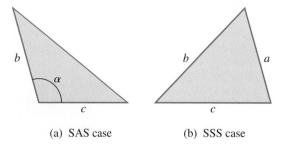

(a) SAS case   (b) SSS case

FIGURE 1

◆   Deriving the Law of Cosines

The following is the law of cosines.

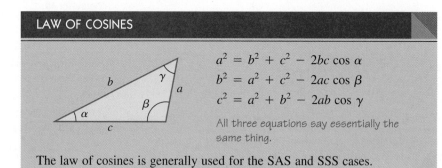

LAW OF COSINES

$$a^2 = b^2 + c^2 - 2bc \cos \alpha$$
$$b^2 = a^2 + c^2 - 2ac \cos \beta$$
$$c^2 = a^2 + b^2 - 2ab \cos \gamma$$

All three equations say essentially the same thing.

The law of cosines is generally used for the SAS and SSS cases.

We will derive this law for the first case only. (The other cases can then be obtained from the first case simply by relabeling the figure.) We start by locating a triangle in a rectangular coordinate system. Figure 2 shows three typical triangles.

For an arbitrary triangle located as in Figure 2, we use the distance-between-two-points formula to obtain

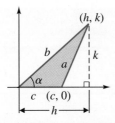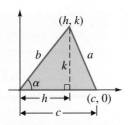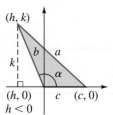

FIGURE 2

$$a = \sqrt{(h - c)^2 + (k - 0)^2}$$

or, squaring both sides,

$$a^2 = (h - c)^2 + k^2$$
$$= h^2 - 2hc + c^2 + k^2 \qquad (1)$$

From Figure 2 we note that

$$b^2 = h^2 + k^2$$

Substituting b^2 for $h^2 + k^2$ in equation (1), we obtain

$$a^2 = b^2 + c^2 - 2hc \qquad (2)$$

But

$$\cos \alpha = \frac{h}{b}$$
$$h = b \cos \alpha$$

Thus, by replacing h in (2) with $b \cos \alpha$, we reach our objective,

$$a^2 = b^2 + c^2 - 2bc \cos \alpha$$

[*Note:*   If α is acute, then $\cos \alpha > 0$; if α is obtuse, then $\cos \alpha < 0$.]

◆ Solving the SAS Case

In this case we start by using the law of cosines to find the side opposite the given angle. We can then use either the law of cosines or the law of sines to find a second angle. Because of the simpler computation, we will generally use the law of sines.

For the second angle we choose the angle opposite the shorter of the two given sides, which guarantees an acute angle. An obtuse angle, if present, must always be opposite the longest side in the triangle; of course, if the given angle is obtuse, then both remaining angles must be acute. The reason for choosing an acute angle when using the law of sines is that the inverse sine function in a calculator will give us the measure of the angle directly—which is not the case for an obtuse angle. This discussion leads to the following strategy for solving the SAS case.

STRATEGY FOR SOLVING THE SSA CASE

Step	Find	Method
1.	Side opposite given angle	Law of cosines
2.	Second angle (Find the angle opposite the shorter of the two given sides— this angle will always be acute.)	Law of sines or law of cosines
3.	Third angle	Subtract the sum of the measures of the given angle and the angle found in step 2 from 180°.

◆ EXAMPLE 1   Using the Law of Cosines (SAS)

Solve the triangle (Figure 3).

FIGURE 3

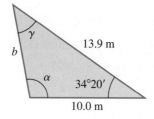

SOLUTION   *Solve for b.* Use the law of cosines:

$$b^2 = a^2 + c^2 - 2ac \cos \beta \qquad \text{Solve for } b$$

$$b = \sqrt{a^2 + c^2 - 2ac \cos \beta}$$
$$= \sqrt{(13.9)^2 + (10.0)^2 - 2(13.9)(10.0) \cos (34°20')}$$
$$= 7.98 \text{ m}$$

Solve for γ. Since side c is shorter than side a, γ must be acute, so we use the law of sines to solve for γ.

$$\frac{\sin \gamma}{c} = \frac{\sin \beta}{b} \qquad \text{Solve for } \sin \gamma$$

$$\sin \gamma = \frac{c \sin \beta}{b} \qquad \text{Solve for } \gamma$$

$$= \frac{10.0 \sin 34°20'}{7.98}$$

$$\gamma = \sin^{-1}\left(\frac{10.0 \sin 34°20'}{7.98}\right) \qquad \begin{array}{l}\text{Since } \gamma \text{ is acute, we can use the} \\ \text{inverse sine key on a calculator} \\ \text{to find the measure of } \gamma.\end{array}$$

$$= 45.0° \text{ or } 45°0'$$

Solve for α.

$$\alpha = 180° - (\beta + \gamma)$$
$$= 180° - (34°20' + 45°0') = 100°40'$$ ◆

MATCHED PROBLEM 1

Solve a triangle with $\alpha = 86.0°$, $b = 15.0$ m, and $c = 24.0$ m (angles in decimal degrees).

◆   Solving the SSS Case

When we start with three sides of a triangle, our problem is to find the three angles. Although the law of cosines can be used to find any of the three angles, we will always start with the angle opposite the longest side to get an obtuse angle, if present, out of the way at the start. Using the law of cosines to find the only angle that might be obtuse simplifies the subsequent calculations. Note that if $0 < \theta < 180°$ and $0 < t < 1$, then the equation $\cos \theta = t$ always has a unique solution, which may be acute or obtuse and which is easily obtained from a calculator. A second angle, which must be acute, can be found using either law, although computations are simpler with the law of sines. This discussion leads to the following strategy for solving the SSS case.

STRATEGY FOR SOLVING THE SSS CASE		
Step	**Find**	**Method**
1.	Angle opposite longest side (This will take care of an obtuse angle, if present.)	Law of cosines
2.	Either of the remaining angles (Always acute, since a triangle cannot have more than one obtuse angle.)	Law of sines or law of cosines
3.	Third angle	Subtract the sum of the measures of the angles found in steps 1 and 2 from 180°.

◆ EXAMPLE 2

Solving the SSS Case: Surveying

A triangular plot of land has sides $a = 21.9$ m, $b = 24.6$ m, and $c = 12.0$ m. Find the measures of all three angles in decimal degrees.

SOLUTION

Find the measure of the angle opposite the longest side first using the law of cosines, which in this problem is angle β. We make a rough sketch to keep the various parts of the triangle straight (Figure 4).

Solve for β. It is not clear whether β is acute, obtuse, or 90°; we do not need to know beforehand, because the law of cosines will automatically tell us in the solution process.

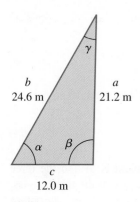

b
24.6 m

a
21.2 m

γ

α β

c
12.0 m

FIGURE 4

$$b^2 = a^2 + c^2 - 2ac \cos \beta \qquad \text{Solve for } \cos \beta$$

$$\cos \beta = \frac{a^2 + c^2 - b^2}{2ac} \qquad \text{Solve for } \beta$$

$$\beta = \cos^{-1}\left(\frac{a^2 + c^2 - b^2}{2ac}\right)$$

$$= \cos^{-1}\left(\frac{(21.2)^2 + (12.0)^2 - (24.6)^2}{2(21.2)(12.0)}\right)$$

$$= 91.3° \qquad\qquad\qquad \text{Obtuse}$$

Solve for α. Both α and γ must be acute, since β is obtuse. We arbitrarily choose to find α first using the law of sines. Since α is acute the inverse sine function in a calculator will give us the measure of this angle directly.

$$\frac{\sin \alpha}{a} = \frac{\sin \beta}{b} \qquad \text{Solve for } \sin \alpha$$

$$\sin \alpha = \frac{a \sin \beta}{b} \qquad \text{Solve for } \alpha$$

$$\alpha = \sin^{-1}\left(\frac{a \sin \beta}{b}\right)$$

$$= \sin^{-1}\left(\frac{21.2 \sin 91.3}{24.6}\right)$$

$$= 59.5°$$

Solve for γ.

$$\gamma = 180° - (\alpha + \beta)$$

$$= 180° - (59.5° + 91.3°) = 29.2° \qquad\qquad\qquad \blacklozenge$$

MATCHED PROBLEM 2   A triangular plot of land has sides $a = 217$ ft, $b = 362$ ft, and $c = 345$ ft. Find the measures of all three angles to the nearest ten feet.

Answers to   **1.** $a = 27.4$ m, $\beta = 33.1°$, $\gamma = 60.9°$
Matched Problems   **2.** $\alpha = 35°40'$, $\beta = 76°30'$, $\gamma = 67°50'$

EXERCISE 6.2

All triangles are assumed labeled as in Figure 1, Section 6.1, unless stated to the contrary. Your answers may differ slightly from those in the book, depending on the order in which you solve for the sides and angles.

A *Solve each triangle.*

1. $\alpha = 50°40'$, $b = 7.03$ mm, $c = 7.00$ mm

2. $\alpha = 71°0'$, $b = 5.32$ cm, $c = 5.00$ cm

3. $\gamma = 134.0°$, $a = 20.0$ m, $b = 8.00$ m

4. $\alpha = 120.0°$, $b = 5.00$ km, $c = 10.0$ km

B **5.** $a = 9.00$ yd, $b = 6.00$ yd, $c = 10.0$ yd   (Decimal degrees)

6. $a = 5.00$ km, $b = 5.50$ km, $c = 6.00$ km   (Decimal degrees)

7. $a = 420.0$ km, $b = 770.0$ km, $c = 860.0$ km (Degrees and minutes)

8. $a = 15.0$ cm, $b = 12.0$ cm, $c = 10.0$ cm   (Degrees and minutes)

Problems 9–30 represents a variety of problems involving the first two sections of this chapter. Solve each triangle using the law of sines or the law of cosines (or both). If a problem does not have a solution, say so.

9. $\beta = 132.4°$, $\gamma = 17.3°$, $b = 67.6$ ft

10. $\alpha = 57.2°$, $\gamma = 112.0°$, $c = 24.8$ ft

11. $\beta = 66.5°$, $a = 13.7$ m, $c = 20.1$ m

12. $\gamma = 54.2°$, $a = 112$ ft, $b = 87.2$ ft

13. $\beta = 84.4°$, $\gamma = 97.8°$, $a = 12.3$ cm

14. $\alpha = 95.6°$, $\gamma = 86.3°$, $b = 43.5$ cm

15. $\gamma = 66.4°$, $b = 25.5$ yd, $c = 25.5$ yd

16. $\beta = 38.4°$, $a = 11.5$ m, $b = 14.0$ m

17. $a = 10.5$ in., $b = 5.23$ in., $c = 9.66$ in. (Decimal degrees)

18. $a = 15.0$ ft, $b = 18.0$ ft, $c = 22.0$ ft (Decimal degrees)

19. $\alpha = 112.4°$, $b = 10.2$ cm, $c = 18.7$ cm

20. $\beta = 104.5°$, $a = 17.2$ mm, $c = 11.7$ mm

21. $\gamma = 80.3°$, $a = 14.5$ mm, $c = 10.0$ mm

22. $\beta = 63.4°$, $b = 50.5$ in., $c = 64.4$ in.

23. $\alpha = 46.3°$, $\gamma = 105.5°$, $b = 643$ m

24. $\beta = 123.6°$, $\gamma = 21.9°$, $a = 108$ cm

25. $a = 12.2$ m, $b = 16.7$ m, $c = 30.0$ m

26. $a = 28.2$ yd, $b = 52.3$ yd, $c = 22.0$ yd

27. $\alpha = 46.7°$, $a = 18.1$ yd, $b = 22.6$ yd

28. $\gamma = 58.4°$, $b = 7.23$ cm, $c = 6.54$ cm

29. $\alpha = 36.5°$, $\beta = 72.4°$, $\gamma = 71.1°$

30. $\alpha = 29°20'$, $\beta = 32°50'$, $\gamma = 117°50'$

C **31.** Using the law of cosines, show that if $\beta = 90°$, then $b^2 = c^2 + a^2$ (the Pythagorean theorem).

32. Using the law of cosines, show that if $b^2 = c^2 + a^2$, then $\beta = 90°$.

33. Check Problem 1 using Mollweide's equation (see Problem 25, Exercise 6.1),

$$(a - b) \cos \frac{\gamma}{2} = c \sin \frac{\alpha - \beta}{2}$$

34. Check Problem 3 using Mollweide's equation (see Problem 33).

35. Show that for any triangle with standard labeling (Figure 1, Section 6.1),

$$c = b \cos \alpha + a \cos \beta$$

36. Show that for any triangle with standard labeling (Figure 1, Section 6.1),

$$\frac{a^2 + b^2 + c^2}{2abc} = \frac{\cos \alpha}{a} + \frac{\cos \beta}{b} + \frac{\cos \gamma}{c}$$

Applications

37. **Surveying**   A geologist wishes to determine the distance *CB* across the base of a volcanic cinder cone (see the figure). Distances *AC* and *AB* are measured to be 425 m and 384 m, respectively, and $\angle CAB$ to be 98.3°. Find the approximate distance across the base of the cinder cone.

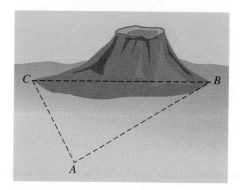

Figure for 37

38. **Surveying**   To estimate the length *CB* of the lake in the figure, a surveyor measures *AB* and *AC* to be 89 m and 74 m, respectively, and $\angle CAB$ to be 95°. Find the approximate length of the lake.

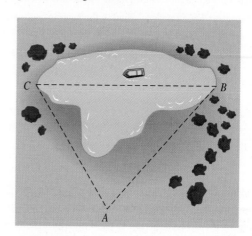

Figure for 38

39. Geometry: Engineering   Find the measure in decimal degrees of a central angle subtended by a chord of length 13.8 cm in a circle of radius 8.26 cm.

40. Geometry: Engineering   Find the measure in decimal degrees of a central angle subtended by a chord of length 112 ft in a circle of radius 72.8 ft.

41. Search and Rescue   At midnight two search planes set out from New York to find a boat in distress. Plane *A* travels due east at 400 km/hr, and plane *B* travels northeast at 500 km/hr. At 2 AM plane *A* spots a flare from the boat and radios plane *B* to come and assist in the rescue. How far is plane *B* from plane *A* at this time? Compute the answer to two significant digits.

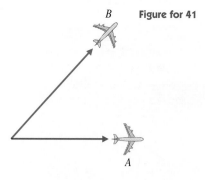

Figure for 41

42. Navigation   Los Angeles and San Francisco are approximately 600 km apart. A pilot flying from Los Angeles to San Francisco finds that after she is 200 km from Los Angeles the plane is 20° off course. How far is the plane from San Francisco at this time (to two significant digits)?

∗43. Geometry: Engineering   A 58.3 cm chord of a circle subtends a central angle of 27.8°. Find the radius of the circle to three significant digits using the law of cosines.

44. Geometry: Engineering   Find to the nearest centimeter the perimeter of a regular pentagon inscribed in a circle with radius 5 cm (see the figure).

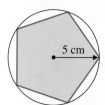

Figure for 44

∗45. Geometry: Engineering   Three circles of radius 2 cm, 3 cm, and 8 cm are tangent to each other (see the figure). Find to the nearest 10′ the three angles formed by the lines joining their centers.

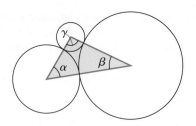

Figure for 45

∗46. Engineering   A tunnel for hydroelectric power is to be constructed through a mountain from one reservoir to another at a lower level (see the figure). The distance from the top of the mountain to the lower end of the tunnel is 5.32 mi, and from the top of the mountain to the upper end of the tunnel is 2.63 mi. The angles of depression of the two slopes of the mountain are 42.7° and 48.8°, respectively.
(A) What is the length of the tunnel?
(B) What angle does the tunnel make with the horizontal?

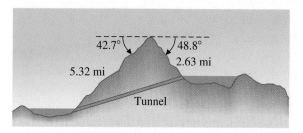

Figure for 46

∗47. Engineering: Construction   A fire lookout is to be constructed as indicated in the figure. Support poles *AD* and

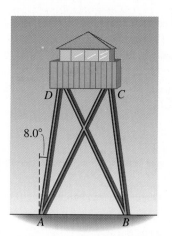

Figure for 47

BC are each 18.0 ft long and tilt 8.0° inward from the vertical. The distance between the tops of the poles *DC* is 12.0 ft. Find the length of a brace *AC* and the distance *AB* between the supporting poles at ground level.

*48. **Engineering: Construction** Repeat Problem 47 with support poles *AD* and *BC* making an angle of 11.0° with the vertical instead of 8.0°.

*49. **Space Science** A satellite *S*, in circular orbit around the earth, is sighted by a tracking station *T* (see the figure). The radar-determined distance *TS* is 1,034 mi, and the angle of elevation above the horizon is 32.4°. How high is the satellite above the earth at the time of the sighting? The radius of the earth is $R = 3,964$ mi.

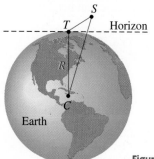

Figure for 49

*50. **Space Science** For communications between a space shuttle and the White Sands Missile Range in southern New Mexico, two satellites are placed in geostationary orbit, 130° apart, each 22,300 mi above the surface of the earth (see the figure). (When a satellite is in **geostationary orbit** it remains stationary above a fixed point on the earth's surface.) Radio signals are sent from an orbiting shuttle by way of the satellites to the White Sands facility, and vice versa. This arrangement enables the White Sands facility to maintain radio contact with a shuttle over most of the earth's surface. How far (to the nearest 100 mi) is

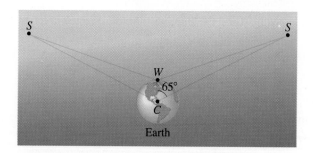

Figure for 50

one of these geostationary satellites from the White Sands facility (W)? The radius of the earth is 3,964 mi.

*51. **Geometry: Engineering** A rectangular solid has sides of 6.0 cm, 3.0 cm, and 4.0 cm (see the figure). Find $\angle ABC$ in the figure in decimal degrees.

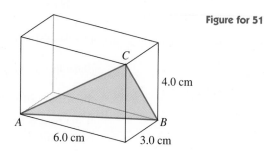

Figure for 51

*52. **Geometry: Engineering** Refer to Problem 51. Find $\angle ACB$ in decimal degrees.

*53. **Surveying** A plot of land was surveyed with the resulting information shown in the figure. Find the length of *DC*.

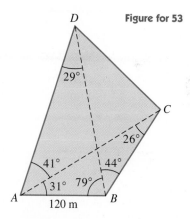

Figure for 53

*54. **Surveying** A plot of land was surveyed with the resulting information shown in the figure. Find the length of *BC*.

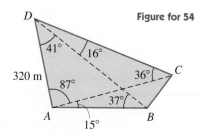

Figure for 54

☆ **6.3**   AREAS OF TRIANGLES

 ◆   **Base and Height Given**
 ◆   **Two Sides and Included Angle Given**
 ◆   **Three Sides Given (Heron's Formula)**

In this section, we discuss three frequently used methods of finding areas of triangles given the indicated information. The derivation of Heron's formula also illustrates a significant use of identities.

◆   Base and Height Given

If the base b and height h of a triangle are given (see Figure 1), then the area A is one-half the area of a parallelogram with the same base and height:

$$A = \frac{1}{2} bh$$

FIGURE 1

$A = \dfrac{1}{2} bh$

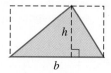
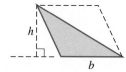

◆   Two Sides and Included Angle Given

Given two sides, a and b, and an included angle θ (see Figure 2), the above formula can be converted into the following form.

$$A = \frac{ab}{2} \sin \theta$$

FIGURE 2

$A = \dfrac{ab}{2} \sin \theta$

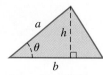
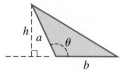

This conversion is accomplished by using the sine function to express h in terms of θ and a.

$$h = a \sin \theta$$

which holds whether θ is acute or obtuse, since $\sin(180° - \theta) = \sin \theta$. Thus,

$$A = \frac{1}{2} bh = \frac{1}{2} ba \sin \theta = \frac{ab}{2} \sin \theta$$

☆ Sections marked with a star may be omitted without loss of continuity.

◆ EXAMPLE 1　　Finding the Area of a Triangle Given Two Sides and the Included Angle

Find the area of the triangle (Figure 3).

FIGURE 3

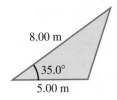

8.00 m

35.0°

5.00 m

SOLUTION
$$A = \frac{ab}{2} \sin \theta = \frac{1}{2}(8.00)(5.00) \sin 35.0°$$
$$= 11.5 \text{ m}^2$$
◆

MATCHED PROBLEM 1　Find the area of a triangle with $a = 12.0$ cm, $b = 7.00$ cm, and included angle $\theta = 125.0°$.

◆ **Three Sides Given (Heron's Formula)**

A famous formula from the Greek philosopher–mathematician Heron of Alexandria (75 AD) enables us to compute the area of a triangle directly when given only the lengths of the three sides of the triangle.

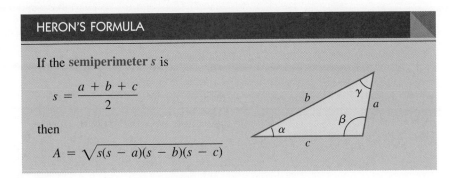

HERON'S FORMULA

If the **semiperimeter** s is

$$s = \frac{a + b + c}{2}$$

then

$$A = \sqrt{s(s - a)(s - b)(s - c)}$$

Heron's formula is obtained from

$$A = \frac{bc}{2} \sin \alpha \tag{1}$$

by expressing $\sin \alpha$ in terms of the sides a, b, and c. Several identities and the law of cosines play a central role in the derivation of the formula. We will first get $\sin(\alpha/2)$ and $\cos(\alpha/2)$ in terms of a, b, and c, then we will use a double-angle identity in the form

$$\sin \alpha = 2 \sin \frac{\alpha}{2} \cos \frac{\alpha}{2} \tag{2}$$

to write $\sin \alpha$ in terms of a, b, and c. We start with a half-angle identity for sine in the form

$$\sin^2 \frac{\alpha}{2} = \frac{1 - \cos \alpha}{2} \qquad (3)$$

The following version of the law of cosines involves $\cos \alpha$ and all three sides of the triangle a, b, and c.

$$a^2 = b^2 + c^2 - 2bc \cos \alpha$$

Solving for $\cos \alpha$, we obtain

$$\cos \alpha = \frac{b^2 + c^2 - a^2}{2bc} \qquad (4)$$

Substituting (4) into (3), we can write $\sin^2(\alpha/2)$ in terms of a, b, and c.

$$\sin^2 \frac{\alpha}{2} = \frac{1 - \dfrac{b^2 + c^2 - a^2}{2bc}}{2}$$

$$= \frac{2bc\left(1 - \dfrac{b^2 + c^2 - a^2}{2bc}\right)}{2bc(2)} \qquad \textit{Convert to simple fraction.}$$

$$= \frac{2bc - b^2 - c^2 + a^2}{4bc} \qquad \textit{Numerator factors (not obvious)}$$

$$= \frac{(a + b - c)(a - b + c)}{4bc} \qquad (5)$$

To bring the semiperimeter

$$s = \frac{a + b + c}{2} \qquad (6)$$

into the picture, we write (5) in the form

$$\sin^2 \frac{\alpha}{2} = \frac{(a + b + c - 2b)(a + b + c - 2c)}{4bc}$$

$$= \frac{2\left(\dfrac{a + b + c}{2} - b\right)2\left(\dfrac{a + b + c}{2} - c\right)}{4bc}$$

$$= \frac{\left(\dfrac{a + b + c}{2} - b\right)\left(\dfrac{a + b + c}{2} - c\right)}{bc} \qquad (7)$$

Substituting (6) into (7) and solving for $\sin(\alpha/2)$, we obtain

$$\sin \frac{\alpha}{2} = \sqrt{\frac{(s - b)(s - c)}{bc}} \qquad (8)$$

If we repeat this reasoning starting with a half-angle identity for cosine in the form

$$\cos^2 \frac{\alpha}{2} = \frac{1 + \cos \alpha}{2} \tag{9}$$

we obtain

$$\cos \frac{\alpha}{2} = \sqrt{\frac{s(s-a)}{bc}} \tag{10}$$

Substituting (8) and (10) into identity (2) produces

$$\sin \alpha = 2\sqrt{\frac{(s-b)(s-c)}{bc}}\sqrt{\frac{s(s-a)}{bc}} = \frac{2}{bc}\sqrt{s(s-a)(s-b)(s-c)} \tag{11}$$

And we are almost there. We now substitute (11) into formula (1) to obtain Heron's formula.

$$A = \frac{bc}{2}\left[\frac{2}{bc}\sqrt{s(s-a)(s-b)(s-c)}\right]$$

$$= \sqrt{s(s-a)(s-b)(s-c)}$$

The derivation of Heron's formula provides a good illustration of the importance of identities. Try to imagine a derivation of this formula without identities.

◆   EXAMPLE 2   Finding the Area of a Triangle Given Three Sides

Find the area of a triangle with sides $a = 12.0$ cm, $b = 8.0$ cm, and $c = 6.0$ cm.

SOLUTION   First find the semiperimeter s.

$$s = \frac{a + b + c}{2} = \frac{12.0 + 8.0 + 6.0}{2} = 13.0 \text{ cm}$$

Then

$$s - a = 13.0 - 12.0 = 1.0 \text{ cm}$$
$$s - b = 13.0 - 8.0 = 5.0 \text{ cm}$$
$$s - c = 13.0 - 6.0 = 7.0 \text{ cm}$$

Thus,

$$A = \sqrt{s(s-a)(s-b)(s-c)}$$

$$= \sqrt{13.0(1.0)(5.0)(7.0)}$$

$$= 21 \text{ cm}^2 \qquad \text{To two significant digits}$$

[*Note:*   The computed area has the same number of significant digits as the side with the least number of significant digits.]   ◆

MATCHED PROBLEM 2   Find the area of a triangle with $a = 6.0$ m, $b = 10.0$ m, and $c = 8.0$ m.

Answers to   **1.** 34.4 cm^2
Matched Problems   **2.** 24 m^2

EXERCISE 6.3

A Find the area of the triangle matching the information given in Problems 1–12.

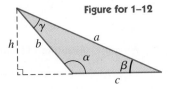 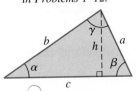

Figure for 1–12

1. $h = 12.0$ m, $c = 17.0$ m
2. $h = 7.0$ ft, $c = 10.0$ ft
3. $\alpha = 30.0°$, $b = 6.0$ cm, $c = 8.0$ cm
4. $\alpha = 45°$, $b = 5.0$ m, $c = 6.0$ m
5. $a = 4.00$ in., $b = 6.00$ in., $c = 8.00$ in.
6. $a = 4.00$ ft, $b = 10.00$ ft, $c = 12.00$ ft

B **7.** $\alpha = 23°20'$, $b = 403$ ft, $c = 512$ ft
8. $\alpha = 58°40'$, $b = 28.2$ in., $c = 6.40$ in.
9. $\alpha = 132.67°$, $b = 12.1$ cm, $c = 10.2$ cm
10. $\alpha = 147.5°$, $b = 125$ mm, $c = 67.0$ mm
11. $a = 12.7$ m, $b = 20.3$ m, $c = 24.4$ m
12. $a = 5.24$ cm, $b = 3.48$ cm, $c = 6.04$ cm

C **13.** Obtain equation (10) in the text following the same type of reasoning that was used to obtain equation (8) in the derivation of Heron's formula.

14. If $s = (a + b + c)/2$ is the semiperimeter of a triangle with sides, a, b, and c (see the figure), show that the radius R of an inscribed circle is given by

$$R = \sqrt{\frac{(s - a)(s - b)(s - c)}{s}}$$

Figure for 14

15. Show that the diagonals of a parallelogram divide the figure into four triangles all having the same area.

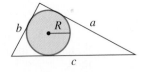

Figure for 15

6.4 VECTORS: GEOMETRICALLY DEFINED

- ◆   **Geometric Vectors; Vector Addition**
- ◆   **Velocity Vectors**
- ◆   **Force Vectors**
- ◆   **Resolution of a Vector into Components**
- ◆   **Centrifugal Force**

Scalar quantities are physical quantities (length, area, and volume, etc.) that can be completely specified by a single real number. Directed distances, velocities, and forces are quantities that require both a magnitude and direction for their complete specification. We call these **vector quantities**.

Vector forms are widely used in both pure and applied mathematics. They are used extensively in the physical sciences and engineering, and are seeing increased use in the social and life sciences.

In this section we limit our development to intuitive notions of *geometric vectors in a plane.* In Section 6.5 we present an algebraic treatment of vectors, which is just the first step of a development and generalization that can fill whole books.

◆ Geometric Vectors; Vector Addition

If A and B are two points in the plane, then the **directed line segment** from A to B, denoted by $\overrightarrow{AB}$, is the line segment joining A to B with an arrowhead placed at B to indicate the direction is from A to B (see Figure 1). Point A is called the **initial point**, and point B is called the **terminal point**.

A **geometric vector in the plane** is a quantity that possesses both a length and a direction and can be represented by a directed line segment. The directed line segment in Figure 1 represents a vector that is denoted by $\overrightarrow{AB}$. A vector also may be denoted by a boldface letter, such as **v** or **F**, or by a letter with an arrow over it, such as $\vec{v}$. Letters with arrows over them are easier to write by hand, but boldface letters are easier to recognize in print. We will generally use boldface letters for vector quantities in this book, but when you write a vector by hand you should include the arrow above it as a reminder that the quantity is a vector.

The **magnitude** of the vector $\overrightarrow{AB}$, denoted by $|\overrightarrow{AB}|$, $|\vec{v}|$, or $|\mathbf{v}|$, is the length of the directed line segment. Two vectors have the **same direction** if they are parallel and point in the same direction. Two vectors have **opposite direction** if they are parallel and point in opposite directions. The **zero vector**, denoted by $\vec{0}$ or **0**, has a magnitude of zero and an arbitrary direction. Two vectors are **equal** if they have the same magnitude and direction. Thus, a vector may be **translated** from one location to another as long as the magnitude and direction do not change.

The **sum of two vectors u** and **v** can be defined using the **tail-to-tip rule**: Translate **v** so that its tail end (initial point) is at the tip end (terminal point) of **u**. Then, the vector from the tail end of **u** to the tip end of **v** is the sum, denoted by **u** + **v**, of the vectors **u** and **v** (see Figure 2). The sum of two nonparallel vectors also can be defined using the **parallelogram rule**: The **sum of two nonparallel vectors u** and **v** is the diagonal of the parallelogram formed using **u** and **v** as adjacent sides (see Figure 3). (If **u** and **v** are parallel, use the tail-to-tip rule.) Of course, both rules give the same sum, as they should. The choice of which rule to use depends on the situation.

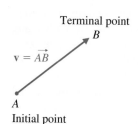

Terminal point
$\mathbf{v} = \overrightarrow{AB}$

A
Initial point

FIGURE 1
Vector $\overrightarrow{AB}$ = **v**

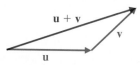

FIGURE 2
Tail-to-tip rule for addition

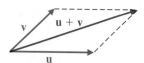

FIGURE 3
Parallelogram rule for addition

The vector **u** + **v** is also called the **resultant** of the two vectors **u** and **v**, and **u** and **v** are called **components** of **u** + **v**. It is useful to observe that vector addition is **commutative**; that is, **u** + **v** = **v** + **u**.

◆   Velocity Vectors

A **velocity vector** is a vector that represents the speed and direction of an object in motion. Vector methods often can be applied to problems involving objects in motion.

 **EXAMPLE 1**   Resultant Velocity

A power boat traveling at 24 km/hr relative to the water has a compass heading (the direction the boat is pointing) of 95°. A strong tidal current, with a heading of 35°, is flowing at 12 km/hr. The velocity of the boat relative to the water is called **apparent velocity**, and the velocity relative to the ground is called the **resultant** or **actual velocity**. The resultant velocity is the vector sum of the apparent velocity and the current velocity. Find the resultant velocity; that is, find the actual speed and direction of the boat relative to the ground. [*Note:* A navigational compass is marked clockwise in degrees starting at north as indicated in Figure 4.]

FIGURE 4

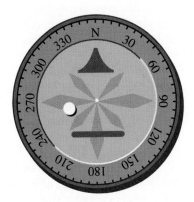

SOLUTION   We can use geometric vectors (see Figure 5(a)) to represent the apparent velocity vector and the current velocity vector. We add the two vectors using the tail-to-

FIGURE 5

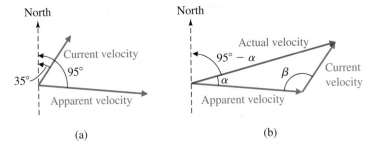

(a)   (b)

tip method of addition of vectors to obtain the resultant (actual) velocity vector as indicated in Figure 5(b). From this vector diagram we obtain the triangle in Figure 6 and solve for β, b, and α.

FIGURE 6

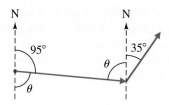

Boat's actual heading: $95° - \alpha$
Boat's actual speed: b

Solve for β. Using the apparent velocity heading (95°) and the current velocity heading (35°), we can find β using Figure 7.

FIGURE 7

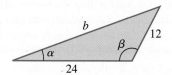

$$\theta = 180° - 95° = 85°$$
$$\beta = \theta + 35° = 85° + 35° = 120°$$

Solve for b. Use the law of cosines:

$$b^2 = a^2 + c^2 - 2ac \cos \beta$$
$$= 12^2 + 24^2 - 2(12)(24) \cos 120°$$
$$b = \sqrt{12^2 + 24^2 - 2(12)(24) \cos 120°}$$
$$= 32 \text{ km/hr} \qquad \text{\textit{Actual speed relative to the ground}}$$

Solve for α. Use the law of sines:

$$\frac{\sin \alpha}{a} = \frac{\sin \beta}{b}$$
$$\sin \alpha = a\left(\frac{\sin \beta}{b}\right)$$
$$= 12\left(\frac{\sin 120°}{32}\right)$$
$$\alpha = \sin^{-1}\left[12\left(\frac{\sin 120°}{32}\right)\right]$$
$$= 19°$$

Actual heading $= 95° - \alpha = 95° - 19° = 76°$ ◆

MATCHED PROBLEM 1   Repeat Example 1 with a current of 8.0 km/hr at 22° and the speedometer on the boat reading 35 km/hr with a compass heading of 85°.

◆ Force Vectors

A **force vector** is a vector that represents the direction and magnitude of an applied force. If an object is subjected to two forces, then the sum of these two forces (the **resultant force**) is a single force. If the resultant force replaced the original two forces, it would act on the object in the same way as the two original forces taken together. In physics it is shown that the resultant force vector can be obtained using vector addition to add the two individual force vectors. It seems natural to use the parallelogram rule for adding force vectors, as illustrated in the next example.

◆ EXAMPLE 2   Resultant Force

Figure 8 shows a man and a horse pulling on a large piece of granite. The man is pulling with a force of 200 lb, and the horse is pulling with a force of 1,000 lb 50° away from the direction of pull of the man. The length of the main diagonal of the parallelogram will be the actual magnitude of the resultant force on the stone, and its direction will be the direction of motion of the stone (if it moves). Find the magnitude and direction α of the resultant force. (The magnitudes of the forces are measured to two significant digits and the angle to the nearest degree.)

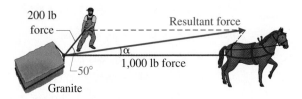

FIGURE 8

SOLUTION   From Figure 8 we obtain the triangle in Figure 9 and then solve for α and b.

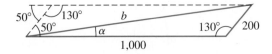

FIGURE 9

Solve for b.

$$b^2 = 1,000^2 + 200^2 - 2(1,000)(200) \cos 130°$$
$$= 1,297,115.04\ldots$$
$$b = \sqrt{1,297,115.04\ldots} = 1,100 \text{ lb}$$

Solve for α.

$$\frac{\sin \alpha}{200} = \frac{\sin 130°}{1,100}$$

$$\sin \alpha = \frac{200}{1,100} \sin 130°$$

$$\alpha = \sin^{-1}\left(\frac{200}{1,100} \sin 130°\right) = 8°$$

◆

MATCHED PROBLEM 2   Repeat Example 2 but change the angle between the force vector from the horse and the force vector from the man to 45° instead of 50°, and change the magnitude of the force vector from the horse to 800 lb instead of 1,000 lb.

◆ **Resolution of a Vector into Components**

Instead of adding vectors, many problems require the breaking down of vectors into components. Whenever a vector is expressed as a resultant of two vectors, these two vectors are called **components** of the given vector. For example, to find the horizontal and vertical components of the vector **v** in Figure 10, we find the magnitudes (the direction of the horizontal and vertical lines are already known) of these components using the sine and cosine functions (see Figure 11).

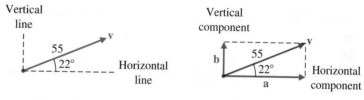

FIGURE 10

FIGURE 11

Magnitude of horizontal component:

$$\cos 22° = \frac{|\mathbf{a}|}{55}$$

$$|\mathbf{a}| = 55 \cos 22° \qquad \text{Horizontal projection of } \mathbf{v}$$

$$= 51$$

Magnitude of vertical component:

$$\sin 22° = \frac{|\mathbf{b}|}{55}$$

$$|\mathbf{b}| = 55 \sin 22° \qquad \text{Vertical projection of } \mathbf{v}$$

$$= 21$$

◆ Centrifugal Force

Ideally, a race car going around a curve will not slide sideways if the track is banked appropriately. How much should a track be banked for a given speed v and a given curve of radius R to eliminate sideways forces? Figure 12 shows the relevant forces.

FIGURE 12

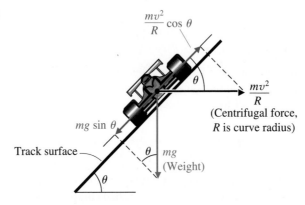

To avoid sideways forces, the components of the forces parallel to the track surface due to the mass of the car and its movement around the curve (centrifugal force) must be zero. That is,

$$mg \sin \theta = \frac{mv^2}{R} \cos \theta$$

$$\frac{\sin \theta}{\cos \theta} = \frac{mv^2}{mgR}$$

$$\tan \theta = \frac{v^2}{gR}$$

$$\theta = \tan^{-1} \frac{v^2}{gR} \qquad \begin{array}{l} v \text{ in meters per second,} \\ R \text{ in meters} \\ g \approx 9.81 \text{ m/sec}^2 \end{array}$$

◆ **EXAMPLE 3**   Race Track Design

At what angle θ must a track be banked at a curve to eliminate sideways forces parallel to the track, given that the curve has a radius of 525 m and the racing car is moving at 85.0 m/sec (about 190 mph)?

SOLUTION $\theta = \tan^{-1} \dfrac{v^2}{gR}$

$= \tan^{-1} \dfrac{(85.0)^2}{(9.81)(525)}$

$= 54.5°$ ◆

MATCHED PROBLEM 3   Repeat Example 3 with a car moving at 25.0 m/sec (about 56 mph) around a curve with a radius of 125 m.

Answers to
Matched Problems
1. Actual speed: 39 km/hr; Actual heading: 74°
2. Resultant force: $b = 950$ lb, $\alpha = 9°$
3. $\theta = 27.0°$

EXERCISE 6.4

Express all angles in decimal degrees. Remember that in navigational problems a compass is divided clockwise into 360° starting at north (see the figure).

A *In Problems 1–6 find $|u + v|$ and θ, given $|u|$ and $|v|$ in parts (a) and (b) of the figure.*

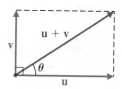 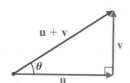

(a) Parallelogram rule   (b) Tail-to-tip rule

Figure for 1–6

1. $|u| = 62$ km/hr, $|v| = 34$ km/hr
2. $|u| = 37$ km/hr, $|v| = 45$ km/hr
3. $|u| = 48$ lb, $|v| = 31$ lb
4. $|u| = 38$ lb, $|v| = 53$ lb
5. $|u| = 143$ knots, $|v| = 57.4$ knots*
6. $|u| = 434$ knots, $|v| = 105$ knots

_____
* A knot is 1 nautical mile per hour.

In Problems 7–12 find the horizontal and vertical components, H and V, respectively, of the vector v given $|v|$ and θ in the figure.

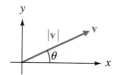

Figure for 7–12

7. $|v| = 42$ lb, $\theta = 34°$
8. $|v| = 250$ lb, $\theta = 67°$
9. $|v| = 23$ knots, $\theta = 62°$
10. $|v| = 48$ knots, $\theta = 27°$
11. $|v| = 244$ km/hr, $\theta = 43.2°$
12. $|v| = 84.0$ km/hr, $\theta = 28.6°$

B *In Problems 13–18, find $|u + v|$ and α, given $|u|$, $|v|$, and θ in parts (a) and (b) of the figure.*

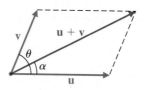 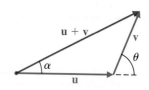

(a) Parallelogram rule   (b) Tail-to-tip rule

Figure for 13–18

13. $|u| = 125$ lb, $|v| = 84$ lb, $\theta = 44°$
14. $|u| = 66$ lb, $|v| = 22$ lb, $\theta = 68°$
15. $|u| = 8.0$ knots, $|v| = 2.0$ knots, $\theta = 64°$
16. $|u| = 21$ knots, $|v| = 3.2$ knots, $\theta = 53°$

C 17. $|u| = 655$ mi/hr, $|v| = 97.3$ mi/hr, $\theta = 66.8°$
18. $|u| = 487$ mi/hr, $|v| = 74.2$ mi/hr, $\theta = 37.4°$

Applications

In Problems 19–22, assume the north, east, south, and west directions are exact.

19. **Navigation**   A river is flowing east (90°) at 3.0 km/hr. A boat crosses the river with a compass heading of 180° (south). If the speedometer on the boat reads 4.0 km/hr, speed relative to the water, what is the boat's actual speed and direction (resultant velocity) relative to the river bottom?

20. **Navigation**   A boat capable of traveling 12 knots on still water maintains a westward compass heading (270°) while crossing a river. If the river is flowing southward (180°) at 4.0 knots, what is the velocity (magnitude and direction) of the boat relative to the river bottom?

*21. **Navigation**   An airplane can cruise at 255 mi/hr in still air. If a steady wind of 46 mi/hr is blowing from the west, what compass heading should the pilot fly in order for the true course of the plane to be north (0°)? Compute the ground speed for this course.

*22. **Navigation**   Two docks are directly opposite each other on a southward flowing river. A boat pilot wishes to go in a straight line from the east dock to the west dock in a ferryboat with a cruising speed of 8.0 knots. If the river's current is 2.5 knots, what compass heading should be maintained while crossing the river? What is the actual speed of the boat relative to land?

23. **Resultant Force**   Two tugs are trying to pull a barge off a shoal as indicated in the figure. Find the magnitude of the resulting force and its direction relative to $\mathbf{F_1}$.

$|F_1| = 1{,}500 \text{ lb}$
$|F_2| = 1{,}100 \text{ lb}$

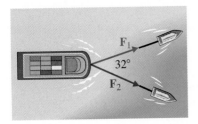

Figure for 23

24. **Resultant Force**   Repeat Problem 23 with $|\mathbf{F_1}| = 1{,}300$ lb and the angle between the force vectors 45°.

25. **Centrifugal Force**   At what angle must a freeway be banked at a curve to eliminate sideways forces parallel to the road, given that the curve has a radius of 138 m and cars are expected to move at 29 m/sec (about 65 mph)?

26. **Centrifugal Force**   Repeat Problem 25 with an expected car speed of 24.6 m/sec (about 55 mph) around a curve with a radius of 105 m.

27. **Resolution of Forces**   A car weighing 2,500 lb is parked on a hill inclined 15° to the horizontal (see the figure). Neglecting friction, what magnitude of force parallel to the hill will keep the car from rolling down the hill? What is the force magnitude perpendicular to the hill?

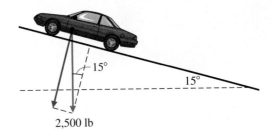

2,500 lb

Figure for 27

28. **Resolution of Forces**   Repeat Problem 27 with the car weighing 4,200 lb and the hill inclined 12°.

*29. **Resolution of Forces**   If two weights are fastened together and placed on inclined planes as indicated in the figure, neglecting friction, which way will they slide? (Weights are accurate to two significant digits, and angles are measured to the nearest degree.)

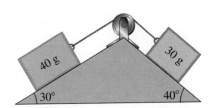

Figure for 29

6.5 VECTORS: ALGEBRAICALLY DEFINED

- ◆ **From Geometric Vectors to Algebraic Vectors**
- ◆ **Vector Addition and Scalar Multiplication**
- ◆ **Unit Vectors**
- ◆ **Algebraic Properties**
- ◆ **Application—Static Equilibrium**

In Section 6.4 we introduced vectors in a geometric setting and considered their velocity and force applications. We discussed geometric vectors in a plane, but they are readily extended to three-dimensional space. However, for the generalization of vector concepts to "higher-dimensional" abstract spaces, it is essential to define the vector concept algebraically. This process is done in such a way that geometric vectors become special cases of the more general algebraic vector. It also turns out that for the solution of certain types of problems in two- and three-dimensional space, algebraic vectors have an advantage over geometric vectors. Static equilibrium problems, which are discussed at the end of this section, provide an example of this type of problem.

We will restrict our development of algebraic vectors to the plane, but the development is readily extended to three- and higher-dimensional spaces. This extension forms the subject matter for whole courses.

◆ From Geometric Vectors to Algebraic Vectors

We start with an arbitrary geometric vector $\overrightarrow{AB}$ in a rectangular coordinate system. A geometric vector $\overrightarrow{AB}$ translated so that its initial point is at the origin is said to be in **standard position**, and the vector $\overrightarrow{OP}$ (as shown in Figure 1) such that $\overrightarrow{OP} = \overrightarrow{AB}$ is said to be the **standard vector** for $\overrightarrow{AB}$. The standard vector is also called the **position vector** or the **radius vector**. Infinitely many geometric vectors have $\overrightarrow{OP}$ in Figure 1 as a standard vector—all vectors with the same magnitude and direction as $\overrightarrow{OP}$.

FIGURE 1
$\overrightarrow{OP}$ is the standard vector for $\overrightarrow{AB}$ ($\overrightarrow{OP} = \overrightarrow{AB}$)

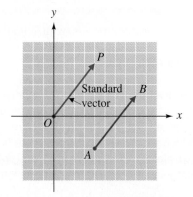

How do we find the standard vector $\overrightarrow{OP}$ for a geometric vector $\overrightarrow{AB}$? The process is not difficult. We need only find the coordinates of P, since the coordinates of O, the origin, are known. If the coordinates of A are (x_a, y_a) and the coordinates of B are (x_b, y_b), then the coordinates of $P(x, y)$ are given by

$$x = x_b - x_a \qquad y = y_b - y_a$$

The process is interpreted geometrically in Example 1.

EXAMPLE 1   Finding a Standard Vector for a Given Geometric Vector

Given the geometric vector $\overrightarrow{AB}$, with initial point $A(8, -3)$ and terminal point $B(4, 5)$, find the standard vector $\overrightarrow{OP}$ for $\overrightarrow{AB}$; that is, find the coordinates of the point P so that $\overrightarrow{OP} = \overrightarrow{AB}$.

SOLUTION   The coordinates (x, y) of P are given by

$$x = x_b - x_a = 4 - 8 = -4$$
$$y = y_b - y_a = 5 - (-3) = 8$$

Figure 2 illustrates these results geometrically.

FIGURE 2

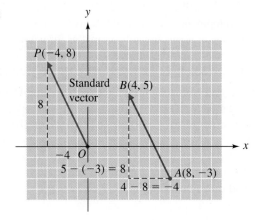

MATCHED PROBLEM 1   Given the geometric vector $\overrightarrow{AB}$, with initial point $A(4, 2)$ and terminal point $B(2, -3)$, find the standard vector $\overrightarrow{OP}$ for $\overrightarrow{AB}$; that is, find the coordinates of the point P so that $\overrightarrow{OP} = \overrightarrow{AB}$.

Let's extend our discussion to get another way of looking at vectors. Since, given any geometric vector $\overrightarrow{AB}$, there always exists a point $P(x, y)$ such that $\overrightarrow{OP} = \overrightarrow{AB}$, the point $P(x, y)$ completely determines the vector $\overrightarrow{AB}$ (except for its position, which we are not concerned about because we are free to translate $\overrightarrow{AB}$ anywhere we please). Conversely, given any point $P(x, y)$ in the plane, the directed line segment joining O to P forms the geometric vector $\overrightarrow{OP}$. Thus,

Every geometric vector in a plane corresponds to an ordered pair of real numbers, and every ordered pair of real numbers corresponds to a geometric vector.

This leads us to a definition of an **algebraic vector** as an ordered pair of real numbers. To avoid confusing a point (a, b) with a vector (a, b), we use special brackets to write the symbol $\langle a, b \rangle$, which represents an algebraic vector. The algebraic vector $\langle a, b \rangle$ corresponds to the standard (geometric) vector $\overrightarrow{OP}$ with terminal point $P(a, b)$ and initial point $O(0, 0)$, as indicated in Figure 3.

The real numbers a and b are **components** of the vector $\langle a, b \rangle$. Two vectors $\mathbf{u} = \langle a, b \rangle$ and $\mathbf{v} = \langle c, d \rangle$ are said to be **equal** if their components are equal—that is, if $a = c$ and $b = d$. The **zero vector** is denoted by $\mathbf{0} = \langle 0, 0 \rangle$ and has arbitrary direction.

Geometric vectors are limited to two- and three-dimensional spaces we can visualize. Algebraic vectors do not have such restrictions. The following are algebraic vectors from two-, three-, four-, and five-dimensional spaces:

$$\langle 4, -2 \rangle \qquad \langle 8, 1, 3 \rangle \qquad \langle 12, 2, -6, 1 \rangle \qquad \langle 5, 11, 3, -8, 4 \rangle$$

Our development in this book is limited to algebraic vectors in two-dimensional space (a plane). Higher-dimensional vector spaces form the subject matter for more advanced treatments of the subject. The *magnitude* of an algebraic vector is defined as follows:

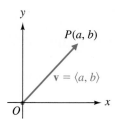

FIGURE 3
Algebraic vector $\langle a, b \rangle$; geometric vector $\overrightarrow{OP}$

MAGNITUDE* OF A VECTOR $\mathbf{v} = \langle a, b \rangle$

The **magnitude** (or **norm**) of a vector $\mathbf{v} = \langle a, b \rangle$, denoted by $|\mathbf{v}|$, is given by

$$|\mathbf{v}| = \sqrt{a^2 + b^2}$$

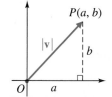

FIGURE 4
$|\mathbf{v}| = \sqrt{a^2 + b^2}$

Geometrically, $\sqrt{a^2 + b^2}$ is the length of the standard (geometric) vector $\overrightarrow{OP}$ associated with the algebraic vector $\langle a, b \rangle$ (see Figure 4).

◆ **EXAMPLE 2** The Magnitude of a Vector

Find the magnitude of the vector $\mathbf{v} = \langle -2, 4 \rangle$.

SOLUTION $|\mathbf{v}| = \sqrt{(-2)^2 + 4^2} = \sqrt{20} \qquad \text{or} \qquad 2\sqrt{5}$ ◆

MATCHED PROBLEM 2 Find the magnitude of the vector $\mathbf{v} = \langle 4, -5 \rangle$.

* The definition of magnitude is readily generalized to higher-dimensional vector spaces. For example, if $\mathbf{v} = \langle a, b, c, d, e \rangle$, then the norm of $\mathbf{v}$ is given by $|\mathbf{v}| = \sqrt{a^2 + b^2 + c^2 + d^2 + e^2}$. But now we are no longer able to interpret the result in terms of geometric vectors.

◆ Vector Addition and Scalar Multiplication

Adding two algebraic vectors is very easy: We simply add the corresponding components of the two vectors.

VECTOR ADDITION

If $\mathbf{u} = \langle a, b \rangle$ and $\mathbf{v} = \langle c, d \rangle$, then

$$\mathbf{u} + \mathbf{v} = \langle a + c, b + d \rangle$$

Since the algebraic vectors in the definition can be associated with geometric vectors in a plane, we can interpret the definition geometrically. The definition is consistent with the parallelogram and tail-to-tip rules for adding geometric vectors given in Section 6.4. For example, if $\mathbf{u} = \langle -3, 2 \rangle$ and $\mathbf{v} = \langle 7, 3 \rangle$, then

$$\mathbf{u} + \mathbf{v} = \langle -3 + 7, 2 + 3 \rangle = \langle 4, 5 \rangle$$

Figure 5 shows that the terminal point of the resultant vector (the diagonal of the parallelogram) has (4, 5) as its coordinates.

FIGURE 5
Vector addition

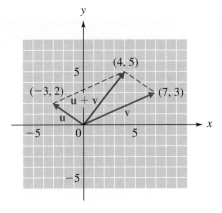

To multiply a vector by a scalar (a real number), multiply each component by the scalar.

SCALAR MULTIPLICATION

If $\mathbf{u} = \langle a, b \rangle$ and k is a scalar, then

$$k\mathbf{u} = k\langle a, b \rangle = \langle ka, kb \rangle$$

Geometrically, if a vector $\mathbf{v}$ is multiplied by a scalar k, the magnitude of the vector $\mathbf{v}$ is multiplied by $|k|$. If k is positive, then $k\mathbf{v}$ has the same direction as $\mathbf{v}$; if k is negative, then $k\mathbf{v}$ has the opposite direction as $\mathbf{v}$. Figure 6 illustrates several cases.

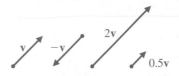

FIGURE 6
Scalar multiplication

◆ **EXAMPLE 3**　　Arithmetic of Algebraic Vectors

Let $\mathbf{u} = \langle -5, 3 \rangle$, $\mathbf{v} = \langle 4, -6 \rangle$, and $\mathbf{w} = \langle -2, 0 \rangle$. Find

(A) $\mathbf{u} + \mathbf{v}$　　(B) $-3\mathbf{u}$　　(C) $3\mathbf{u} - 2\mathbf{v}$　　(D) $2\mathbf{u} - \mathbf{v} + 3\mathbf{w}$

SOLUTIONS　　(A) $\mathbf{u} + \mathbf{v} = \langle -5, 3 \rangle + \langle 4, -6 \rangle = \langle -1, -3 \rangle$
(B) $-3\mathbf{u} = -3\langle -5, 3 \rangle = \langle 15, -9 \rangle$
(C) $3\mathbf{u} - 2\mathbf{v} = 3\langle -5, 3 \rangle - 2\langle 4, -6 \rangle$
　　　　　　　$= \langle -15, 9 \rangle + \langle -8, 12 \rangle = \langle -23, 21 \rangle$
(D) $2\mathbf{u} - \mathbf{v} + 3\mathbf{w} = 2\langle -5, 3 \rangle - \langle 4, -6 \rangle + 3\langle -2, 0 \rangle$
　　　　　　　　　　$= \langle -10, 6 \rangle + \langle -4, 6 \rangle + \langle -6, 0 \rangle$
　　　　　　　　　　$= \langle -20, 12 \rangle$　　◆

MATCHED PROBLEM 3　　Let $\mathbf{u} = \langle 4, -3 \rangle$, $\mathbf{v} = \langle 2, 3 \rangle$, and $\mathbf{w} = \langle 0, -5 \rangle$. Find

(A) $\mathbf{u} + \mathbf{v}$　　(B) $-2\mathbf{u}$　　(C) $2\mathbf{u} - 3\mathbf{v}$　　(D) $3\mathbf{u} + 2\mathbf{v} - \mathbf{w}$

◆ **Unit Vectors**

A vector $\mathbf{v}$ is called a **unit vector** if its magnitude is 1; that is, if $|\mathbf{v}| = 1$. Special unit vectors play an important role in certain applications of vectors. A unit vector can be formed from an arbitrary nonzero vector as follows:

FORMING A UNIT VECTOR

If $\mathbf{v}$ is a nonzero vector, then

$$\mathbf{u} = \frac{1}{|\mathbf{v}|}\,\mathbf{v}$$

is a unit vector with the same direction as $\mathbf{v}$.

◆ EXAMPLE 4   Unit Vector

Find a unit vector **u** with the same direction as the vector $\mathbf{v} = \langle 3, 1 \rangle$.

SOLUTION $|\mathbf{v}| = \sqrt{3^2 + 1^2} = \sqrt{10}$

$$\mathbf{u} = \frac{1}{|\mathbf{v}|} \mathbf{v} = \frac{1}{\sqrt{10}} \langle 3, 1 \rangle$$

$$= \left\langle \frac{3}{\sqrt{10}}, \frac{1}{\sqrt{10}} \right\rangle$$

✔Check $|\mathbf{u}| = \sqrt{\left(\frac{3}{\sqrt{10}}\right)^2 + \left(\frac{1}{\sqrt{10}}\right)^2}$

$$= \sqrt{\frac{9}{10} + \frac{1}{10}} = \sqrt{1} = 1$$

Thus, **u** is a unit vector with the same direction as **v**.   ◆

MATCHED PROBLEM 4   Find a unit vector **u** with the same direction as the vector $\mathbf{v} = \langle 1, -2 \rangle$.

Two very important unit vectors are the **i** and **j** unit vectors, which we now define.

THE i AND j UNIT VECTORS

$\mathbf{i} = \langle 1, 0 \rangle$
$\mathbf{j} = \langle 0, 1 \rangle$

One of the reasons the **i** and **j** unit vectors are important is that any vector $\mathbf{v} = \langle a, b \rangle$ can be expressed in terms of these two vectors.

$$\mathbf{v} = \langle a, b \rangle = \langle a, 0 \rangle + \langle 0, b \rangle$$
$$= a \langle 1, 0 \rangle + b \langle 0, 1 \rangle$$
$$= a\mathbf{i} + b\mathbf{j}$$

Thus,

$$\mathbf{v} = \langle a, b \rangle = a\mathbf{i} + b\mathbf{j}$$

This result is illustrated geometrically in Figure 7.

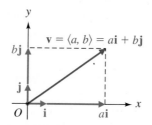

FIGURE 7

◆ EXAMPLE 5   Expressing a Vector in Terms of **i** and **j** Unit Vectors

Express each vector in terms of the **i** and **j** unit vectors.

(A) $\langle 3, -4 \rangle$ (B) $\langle 7, 0 \rangle$ (C) $\langle 0, -2 \rangle$

SOLUTIONS   (A) $\langle 3, -4 \rangle = 3\mathbf{i} - 4\mathbf{j}$ (B) $\langle 7, 0 \rangle = 7\mathbf{i} + 0\mathbf{j} = 7\mathbf{i}$
(C) $\langle 0, -2 \rangle = 0\mathbf{i} - 2\mathbf{j} = -2\mathbf{j}$ ◆

MATCHED PROBLEM 5   Express each vector in terms of the **i** and **j** unit vectors.

(A) $\langle -8, -1 \rangle$ (B) $\langle 0, -3 \rangle$ (C) $\langle 6, 0 \rangle$

◆ Algebraic Properties

Vector addition and scalar multiplication possess algebraic properties similar to the real numbers. These properties enable us to manipulate symbols representing vectors and scalars in much the same way we manipulate symbols that represent real numbers in algebra. These properties are listed below for convenient reference.

ALGEBRAIC PROPERTIES OF VECTORS

(A) The following **addition properties** are satisfied for all vectors **u**, **v**, and **w**:

1. $\mathbf{u} + \mathbf{v} = \mathbf{v} + \mathbf{u}$ Commutative property
2. $\mathbf{u} + (\mathbf{v} + \mathbf{w}) = (\mathbf{u} + \mathbf{v}) + \mathbf{w}$ Associative property
3. $\mathbf{u} + \mathbf{0} = \mathbf{0} + \mathbf{u} = \mathbf{u}$ Additive identity
4. $\mathbf{u} + (-\mathbf{u}) = (-\mathbf{u}) + \mathbf{u} = \mathbf{0}$ Additive inverse

(B) The following **scalar multiplication properties** are satisfied for all vectors **u** and **v** and all scalars m and n:

1. $m(n\mathbf{u}) = (mn)\mathbf{u}$ Associative property
2. $m(\mathbf{u} + \mathbf{v}) = m\mathbf{u} + m\mathbf{v}$ Distributive property
3. $(m + n)\mathbf{u} = m\mathbf{u} + n\mathbf{u}$ Distributive property
4. $1\mathbf{u} = \mathbf{u}$ Multiplication identity

When vectors are represented in terms of the **i** and **j** unit vectors, the algebraic properties listed in the box provide us with efficient procedures for performing algebraic operations with vectors.

◆ **EXAMPLE 6**   Algebraic Operations Involving Unit Vectors

If $\mathbf{u} = 2\mathbf{i} - \mathbf{j}$ and $\mathbf{v} = 4\mathbf{i} + 5\mathbf{j}$, compute each of the following.

(A) $\mathbf{u} + \mathbf{v}$ (B) $\mathbf{u} - \mathbf{v}$ (C) $3\mathbf{u} - 2\mathbf{v}$

SOLUTIONS   (A) $\mathbf{u} + \mathbf{v} = (2\mathbf{i} - \mathbf{j}) + (4\mathbf{i} + 5\mathbf{j})$
$$= 2\mathbf{i} - \mathbf{j} + 4\mathbf{i} + 5\mathbf{j}$$
$$= 2\mathbf{i} + 4\mathbf{i} - \mathbf{j} + 5\mathbf{j}$$
$$= (2 + 4)\mathbf{i} + (-1 + 5)\mathbf{j}$$
$$= 6\mathbf{i} + 4\mathbf{j}$$

(B) $\mathbf{u} - \mathbf{v} = (2\mathbf{i} - \mathbf{j}) - (4\mathbf{i} + 5\mathbf{j})$
$$= 2\mathbf{i} - \mathbf{j} - 4\mathbf{i} - 5\mathbf{j}$$
$$= 2\mathbf{i} - 4\mathbf{i} - \mathbf{j} - 5\mathbf{j}$$
$$= (2 - 4)\mathbf{i} + (-1 - 5)\mathbf{j}$$
$$= -2\mathbf{i} - 6\mathbf{j}$$

(C) $3\mathbf{u} - 2\mathbf{v} = 3(2\mathbf{i} - \mathbf{j}) - 2(4\mathbf{i} + 5\mathbf{j})$
$$= 6\mathbf{i} - 3\mathbf{j} - 8\mathbf{i} - 10\mathbf{j}$$
$$= 6\mathbf{i} - 8\mathbf{i} - 3\mathbf{j} - 10\mathbf{j}$$
$$= (6 - 8)\mathbf{i} + (-3 - 10)\mathbf{j}$$
$$= -2\mathbf{i} - 13\mathbf{j}$$ ◆

MATCHED PROBLEM 6   If $\mathbf{u} = \mathbf{i} - 2\mathbf{j}$ and $\mathbf{v} = 5\mathbf{i} + 2\mathbf{j}$, compute each of the following.

(A) $\mathbf{u} + \mathbf{v}$ (B) $\mathbf{u} - \mathbf{v}$ (C) $2\mathbf{u} + 3\mathbf{v}$

◆ Application—Static Equilibrium

Now we will show how algebraic vectors can be used to solve certain physics and engineering problems. We start with two basic principles regarding forces and objects subject to these forces.

CONDITIONS FOR STATIC EQUILIBRIUM

1. An object at rest is said to be in **static equilibrium.**
2. For an object located at the origin in a rectangular coordinate system to remain in static equilibrium (at rest), it is necessary that the sum of all the force vectors acting on the object be zero (the zero vector).

Examples 7 and 8 illustrate how important engineering/physics problems can be solved using vector methods and the conditions just stated.

◆ EXAMPLE 7   Cable Tension

Two skiers on a stalled chair lift deflect the cable relative to the horizontal as indicated in Figure 8. If the skiers and chair weigh 325 lb (neglect cable weight), what is the tension in each cable running to each tower?

FIGURE 8

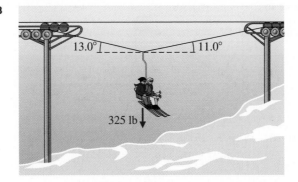

SOLUTION   ***Step 1***   Form a force diagram with all force vectors in standard position at the origin (Figure 9). Our objective is to find $|\mathbf{u}|$ and $|\mathbf{v}|$.

Step 2   Write each force vector in terms of $\mathbf{i}$ and $\mathbf{j}$ unit vectors.

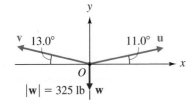

FIGURE 9
Force diagram

$$\mathbf{u} = |\mathbf{u}|(\cos 11.0°)\mathbf{i} + |\mathbf{u}|(\sin 11.0°)\mathbf{j}$$
$$\mathbf{v} = |\mathbf{v}|(-\cos 13.0°)\mathbf{i} + |\mathbf{v}|(\sin 13.0°)\mathbf{j}$$
$$\mathbf{w} = -325\mathbf{j}$$

Step 3   Set the sum of all force vectors equal to the zero vector, and solve for any unknown quantities. For the system to be in static equilibrium, the sum of the force vectors must be zero. That is,

$$\mathbf{u} + \mathbf{v} + \mathbf{w} = \mathbf{0}$$

Writing the vectors as in step 2, we have

$$[|\mathbf{u}|(\cos 11.0°) + |\mathbf{v}|(-\cos 13.0°)]\mathbf{i} + [|\mathbf{u}|(\sin 11.0°) + |\mathbf{v}|(\sin 13.0°) - 325]\mathbf{j} = 0\mathbf{i} + 0\mathbf{j}$$

Now, since two vectors are equal if and only if their corresponding components are equal (that is, $a\mathbf{i} + b\mathbf{j} = 0\mathbf{i} + 0\mathbf{j}$ if and only if $a = 0$ and $b = 0$), we are led to the following system of two equations in the two variables $|\mathbf{u}|$ and $|\mathbf{v}|$:

$$(\cos 11.0°)|\mathbf{u}| + (-\cos 13.0°)|\mathbf{v}| = 0$$
$$(\sin 11.0°)|\mathbf{u}| + (\sin 13.0°)|\mathbf{v}| - 325 = 0$$

Solving this system by standard methods, we find that

$$|\mathbf{u}| = 779 \text{ lb} \qquad \text{and} \qquad |\mathbf{v}| = 784 \text{ lb}$$

You probably did not guess that the tension in each part of the cable is more than the weight hanging from the cable!   ◆

MATCHED PROBLEM 7   Repeat Example 7 with 13.0° replaced by 14.5°, 11.0° replaced by 9.5°, and 325 lb replaced by 435 lb.

◆ EXAMPLE 8   Static Equilibrium

A 1,250 lb weight is hanging from a hoist as indicated in Figure 10. What is the tension in the cable, and what is the compression (force) on the bar?

FIGURE 10

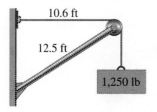

SOLUTION   **Step 1**   Form a force diagram with all force vectors in standard position at the origin (Figure 11).

$$\theta = \cos^{-1} \frac{10.6}{12.5} = 32.0°$$

FIGURE 11

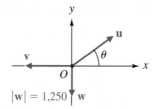

Our objective is to find the compression $|\mathbf{u}|$ and the tension $|\mathbf{v}|$.

Step 2   Write each force vector in terms of $\mathbf{i}$ and $\mathbf{j}$ unit vectors.

$$\mathbf{u} = |\mathbf{u}|(\cos 32.0°)\mathbf{i} + |\mathbf{u}|(\sin 32.0°)\mathbf{j}$$
$$\mathbf{v} = -|\mathbf{v}|\mathbf{i}$$
$$\mathbf{w} = -1{,}250\mathbf{j}$$

Step 3   Set the sum of all force vectors equal to the zero vector and solve for any unknown quantities.

$$\mathbf{u} + \mathbf{v} + \mathbf{w} = \mathbf{0}$$

Writing the vectors as in step 2, we have

$$[|\mathbf{u}|(\cos 32.0°) - |\mathbf{v}|]\mathbf{i} + [|\mathbf{u}|(\sin 32.0°) - 1{,}250]\mathbf{j} = 0\mathbf{i} + 0\mathbf{j}$$

We are thus led to the following system of two equations in the variables $|\mathbf{u}|$ and $|\mathbf{v}|$.

$$(\cos 32.0°)|\mathbf{u}| - |\mathbf{v}| = 0$$
$$(\sin 32.0°)|\mathbf{u}| \qquad - 1{,}250 = 0$$

Solving this system by standard methods, we find that

$$|\mathbf{u}| = 2,360 \text{ lb} \quad \text{and} \quad |\mathbf{v}| = 2,000 \text{ lb} \qquad \blacklozenge$$

MATCHED PROBLEM 8   A 450 lb sign is suspended as shown in Figure 12. Find the compression force on the 2.0 ft rod and the tension on the 4.0 ft rod.

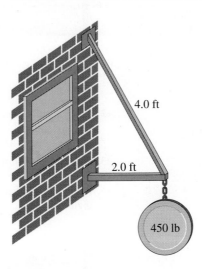

4.0 ft

2.0 ft

450 lb

FIGURE 12

Answers to
Matched Problems

1. $(-2, -5)$ **2.** $\sqrt{41}$
3. (A) $\langle 6, 0 \rangle$ (B) $\langle -8, 6 \rangle$ (C) $\langle 2, -15 \rangle$ (D) $\langle 16, 2 \rangle$
4. $\langle 1/\sqrt{5}, -2/\sqrt{5} \rangle$
5. (A) $-8\mathbf{i} - \mathbf{j}$ (B) $0\mathbf{i} - 3\mathbf{j} = -3\mathbf{j}$ (C) $6\mathbf{i} + 0\mathbf{j} = 6\mathbf{i}$
6. (A) $6\mathbf{i}$ (B) $-4\mathbf{i} - 4\mathbf{j}$ (C) $17\mathbf{i} + 2\mathbf{j}$
7. $|\mathbf{u}| = 1,040 \text{ lb}, |\mathbf{v}| = 1,050 \text{ lb}$
8. $|\mathbf{u}| = 260 \text{ lb (compression)}, |\mathbf{v}| = 520 \text{ lb (tension)}$

EXERCISE 6.5

A *In Problems 1–4 draw the vector $\overrightarrow{AB}$ in a rectangular coordinate system. Draw the standard vector $\overrightarrow{OP}$ so that $\overrightarrow{OP} = \overrightarrow{AB}$. Indicate the coordinates of P.*

1. $A(2, -3);\quad B(5, 1)$
2. $A(1, -2);\quad B(4, 1)$
3. $A(-1, 3);\quad B(-3, -1)$
4. $A(-1, -2);\quad B(-3, 2)$

In Problems 5–8 represent each geometric vector $\overrightarrow{AB}$ as an algebraic vector $\langle a, b \rangle$.

5. $A(-1, -2);\quad B(3, 0)$ **6.** $A(-1, 2);\quad B(2, 3)$
7. $A(0, 2);\quad B(4, -2)$ **8.** $A(-4, 0);\quad B(-2, -4)$

In Problems 9–12 find the magnitude of each vector $\langle a, b \rangle$.

9. $\langle -3, 4 \rangle$ **10.** $\langle 4, -3 \rangle$
11. $\langle -5, -2 \rangle$ **12.** $\langle 3, -5 \rangle$

B *In Problems 13–16 find:*

(A) **u** + **v**   (B) **u** − **v**
(C) 2**u** − 3**v**   (D) 3**u** − **v** + 2**w**

13. $\mathbf{u} = \langle 1, 4 \rangle;\quad \mathbf{v} = \langle -3, 2 \rangle;\quad \mathbf{w} = \langle 0, 4 \rangle$

14. $\mathbf{u} = \langle -1, 3 \rangle;\quad \mathbf{v} = \langle 2, -3 \rangle;\quad \mathbf{w} = \langle -2, 0 \rangle$

15. $\mathbf{u} = \langle 2, -3 \rangle;\quad \mathbf{v} = \langle -1, -3 \rangle;\quad \mathbf{w} = \langle -2, 0 \rangle$

16. $\mathbf{u} = \langle -1, -4 \rangle;\quad \mathbf{v} = \langle 3, 3 \rangle;\quad \mathbf{w} = \langle 0, -3 \rangle$

*In Problems 17–20 form a unit vector **u** with the same direction as **v**.*

17. $\mathbf{v} = \langle 4, -3 \rangle$ **18.** $\mathbf{v} = \langle -3, 4 \rangle$

19. $\mathbf{v} = \langle 2, -3 \rangle$ **20.** $\mathbf{v} = \langle -5, 3 \rangle$

*In Problems 21–26 express **v** in terms of the **i** and **j** unit vectors.*

21. $\mathbf{v} = \langle 3, -2 \rangle$ **22.** $\mathbf{v} = \langle -5, 6 \rangle$

23. $\mathbf{v} = \langle 0, 4 \rangle$ **24.** $\mathbf{v} = \langle -3, 0 \rangle$

25. $\mathbf{v} = \overrightarrow{AB};\quad A(-2, -1);\quad B(0, 2)$

26. $\mathbf{v} = \overrightarrow{AB};\quad A(2, 3);\quad B(-3, 1)$

In Problems 27–32 let $\mathbf{u} = 2\mathbf{i} - 3\mathbf{j}$, $\mathbf{v} = 3\mathbf{i} + 4\mathbf{j}$, and $\mathbf{w} = 5\mathbf{j}$, and perform the indicated operations.

27. **u** − **v**   **28.** **u** + **v**

29. 3**u** − 2**v**   **30.** 2**u** + 3**v**

31. **u** − 2**v** + 2**w**   **32.** 3**u** − **v** − 3**w**

C *In Problems 33–40 let $\mathbf{u} = \langle a, b \rangle$, $\mathbf{v} = \langle c, d \rangle$, and $\mathbf{w} = \langle e, f \rangle$ be vectors, and let m and n be scalars. Prove each of the following vector properties using appropriate properties of real numbers.*

33. **u** + **v** = **v** + **u**

34. **u** + (**v** + **w**) = (**u** + **v**) + **w**

35. **v** + (−**v**) = **0**

36. **v** + **0** = **v**

37. m(**u** + **v**) = m**u** + m**v**

38. (m + n)**v** = m**v** + n**v**

39. 1**v** = **v**

40. (mn)**v** = m(n**v**)

Applications

41. Static Equilibrium A tightrope walker weighing 112 lb deflects a rope as indicated in the figure. How much tension is in each part of the rope?

Figure for 41

42. Static Equilibrium Repeat Problem 41 but change the left angle to 5.5°, the right angle to 6.2°, and the weight of the person to 155 lb.

43. Static Equilibrium A weight of 5,000 lb is supported as indicated in the figure. What are the magnitudes of the forces on the members *AB* and *BC* to two significant digits?

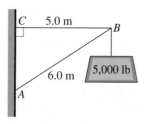

Figure for 43

44. Static Equilibrium A weight of 1,000 lb is supported as indicated in the figure. What are the magnitudes of the forces on the members *AB* and *BC* to two significant digits?

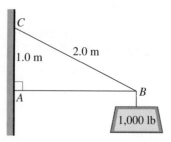

Figure for 44

☆ **6.6** THE DOT PRODUCT

- ◆ **The Dot Product of Two Vectors**
- ◆ **Angle Between Two Vectors**
- ◆ **Scalar Component of One Vector on Another**
- ◆ **Work**

In Section 6.5 we saw that the sum of two vectors is a vector and the scalar multiple of a vector is a vector. In this section we introduce another operation on vectors, called the *dot product*. This product of two vectors, however, is a scalar (a real number), not a vector. The dot product is also called the *scalar product* or the *inner product*. Dot products are used to find angles between vectors, to define concepts and solve problems in physics and engineering, and in the solution of geometric problems.

◆ The Dot Product of Two Vectors

We define the dot product as follows:

THE DOT PRODUCT

The **dot product** of the two vectors

$$\mathbf{u} = \langle a, b \rangle = a\mathbf{i} + b\mathbf{j} \quad \text{and} \quad \mathbf{v} = \langle c, d \rangle = c\mathbf{i} + d\mathbf{j}$$

denoted by $\mathbf{u} \cdot \mathbf{v}$, is the scalar given by

$$\mathbf{u} \cdot \mathbf{v} = ac + bd$$

EXAMPLE 1   Finding Dot Products

Find each dot product.

(A) $\langle 4, 2 \rangle \cdot \langle 1, -3 \rangle$ (B) $(6\mathbf{i} + 2\mathbf{j}) \cdot (\mathbf{i} - 4\mathbf{j})$
(C) $\langle 5, 2 \rangle \cdot \langle 2, -5 \rangle$ (D) $\mathbf{j} \cdot \mathbf{i}$

SOLUTIONS   (A) $\langle 4, 2 \rangle \cdot \langle 1, -3 \rangle = 4 \cdot 1 + 2 \cdot (-3) = -2$
(B) $(6\mathbf{i} + 2\mathbf{j}) \cdot (\mathbf{i} - 4\mathbf{j}) = 6 \cdot 1 + 2 \cdot (-4) = -2$
(C) $\langle 5, 2 \rangle \cdot \langle 2, -5 \rangle = 5 \cdot 2 + 2 \cdot (-5) = 0$
(D) $\mathbf{j} \cdot \mathbf{i} = (0\mathbf{i} + 1\mathbf{j}) \cdot (1\mathbf{i} + 0\mathbf{j}) = 0 \cdot 1 + 1 \cdot 0 = 0$ ◆

MATCHED PROBLEM 1   Find each dot product.

(A) $\langle 3, -2 \rangle \cdot \langle 4, 1 \rangle$ (B) $(5\mathbf{i} - 2\mathbf{j}) \cdot (3\mathbf{i} + 4\mathbf{j})$
(C) $\langle 4, -3 \rangle \cdot \langle 3, 4 \rangle$ (D) $\mathbf{i} \cdot \mathbf{j}$

☆ Sections marked with a star may be omitted without loss of continuity.

Some important properties of the dot product are listed below. All these properties follow directly from the definitions of the vector operations involved and the properties of real numbers.

PROPERTIES OF THE DOT PRODUCT

For all vectors **u**, **v**, and **w**, and k a real number:

1. $\mathbf{u} \cdot \mathbf{u} = |\mathbf{u}|^2$
2. $\mathbf{u} \cdot \mathbf{v} = \mathbf{v} \cdot \mathbf{u}$
3. $\mathbf{u} \cdot (\mathbf{v} + \mathbf{w}) = \mathbf{u} \cdot \mathbf{v} + \mathbf{u} \cdot \mathbf{w}$
4. $(\mathbf{u} + \mathbf{v}) \cdot \mathbf{w} = \mathbf{u} \cdot \mathbf{w} + \mathbf{v} \cdot \mathbf{w}$
5. $k(\mathbf{u} \cdot \mathbf{v}) = (k\mathbf{u}) \cdot \mathbf{v} = \mathbf{u} \cdot (k\mathbf{v})$
6. $\mathbf{u} \cdot \mathbf{0} = 0$

The proofs of these properties are left to Problems 25–30 in Exercise 6.6.

◆ Angle Between Two Vectors

An important application of the dot product is related to the angle between two vectors. If **u** and **v** are two nonzero vectors, we can position **u** and **v** so that their initial points coincide. The **angle between the vectors u and v** is defined to be the angle θ, $0 \le \theta \le \pi$, formed by the vectors (see Figure 1).

FIGURE 1
Angle between two vectors,
$0 \le \theta \le \pi$

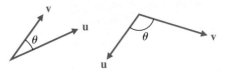

If $\theta = \pi/2$, the vectors **u** and **v** are said to be **orthogonal**, or **perpendicular**. If $\theta = 0$ or $\theta = \pi$, the vectors are said to be **parallel**.

We now show how the dot product can be used to find the angle between two vectors.

ANGLE BETWEEN TWO VECTORS

If **u** and **v** are two nonzero vectors and θ is the angle between **u** and **v**, then

$$\cos \theta = \frac{\mathbf{u} \cdot \mathbf{v}}{|\mathbf{u}||\mathbf{v}|}$$

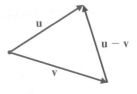

FIGURE 2

We sketch a proof of this result for nonparallel vectors. Given two nonzero, nonparallel vectors **u** and **v**, we can position **u**, **v**, and **u** − **v** so that they form a triangle where θ is the angle between **u** and **v** (see Figure 2). Applying the law of cosines, we obtain

$$|\mathbf{u} - \mathbf{v}|^2 = |\mathbf{u}|^2 + |\mathbf{v}|^2 - 2|\mathbf{u}||\mathbf{v}| \cos \theta \tag{1}$$

Next, we apply some of the properties of the dot product listed earlier to $|\mathbf{u} - \mathbf{v}|^2$.

$$
\begin{aligned}
|\mathbf{u} - \mathbf{v}|^2 &= (\mathbf{u} - \mathbf{v}) \cdot (\mathbf{u} - \mathbf{v}) && \text{Property 1}\\
&= (\mathbf{u} - \mathbf{v}) \cdot \mathbf{u} - (\mathbf{u} - \mathbf{v}) \cdot \mathbf{v} && \text{Property 3}\\
&= \mathbf{u}\,(\mathbf{u} - \mathbf{v}) - \mathbf{v} \cdot (\mathbf{u} - \mathbf{v}) && \text{Property 2}\\
&= \mathbf{u} \cdot \mathbf{u} - \mathbf{u} \cdot \mathbf{v} - \mathbf{v} \cdot \mathbf{u} + \mathbf{v} \cdot \mathbf{v} && \text{Property 3}\\
&= |\mathbf{u}|^2 - 2(\mathbf{u} \cdot \mathbf{v}) + |\mathbf{v}|^2 && \text{Properties 1 and 2}
\end{aligned}
$$

Now substitute this last expression in the left side of (1).

$$|\mathbf{u}|^2 - 2(\mathbf{u} \cdot \mathbf{v}) + |\mathbf{v}|^2 = |\mathbf{u}|^2 + |\mathbf{v}|^2 - 2|\mathbf{u}||\mathbf{v}| \cos \theta$$

This equation simplifies to

$$2|\mathbf{u}||\mathbf{v}| \cos \theta = 2(\mathbf{u} \cdot \mathbf{v})$$

Since **u** and **v** are nonzero, we can solve for $\cos \theta$.

$$\cos \theta = \frac{\mathbf{u} \cdot \mathbf{v}}{|\mathbf{u}||\mathbf{v}|}$$

◆ **EXAMPLE 2**   Angle Between Two Vectors

Find the angle, in decimal degrees to one decimal place, between each pair of vectors.

(A) $\mathbf{u} = \langle 2, 3 \rangle$; $\mathbf{v} = \langle 4, 1 \rangle$ (B) $\mathbf{u} = \langle 3, 1 \rangle$; $\mathbf{v} = \langle -3, -3 \rangle$
(C) $\mathbf{u} = -\mathbf{i} + 3\mathbf{j}$; $\mathbf{v} = 4\mathbf{i} - \mathbf{j}$

SOLUTIONS   (A) $\cos \theta = \dfrac{\mathbf{u} \cdot \mathbf{v}}{|\mathbf{u}||\mathbf{v}|} = \dfrac{11}{\sqrt{13}\sqrt{17}}$

$\theta = \cos^{-1} \dfrac{11}{\sqrt{13}\sqrt{17}} = 42.3°$

(B) $\cos \theta = \dfrac{\mathbf{u} \cdot \mathbf{v}}{|\mathbf{u}||\mathbf{v}|} = \dfrac{-12}{\sqrt{10}\sqrt{18}}$

$\theta = \cos^{-1} \dfrac{-12}{\sqrt{10}\sqrt{18}} = 153.4°$

(C) $\cos \theta = \dfrac{\mathbf{u} \cdot \mathbf{v}}{|\mathbf{u}||\mathbf{v}|} = \dfrac{-7}{\sqrt{10}\sqrt{17}}$

$\theta = \cos^{-1} \dfrac{-7}{\sqrt{10}\sqrt{17}} = 122.5°$

◆

MATCHED PROBLEM 2   Find the angle, in decimal degrees to one decimal place, between each pair of vectors.

(A) $\mathbf{u} = \langle 4, 2 \rangle$; $\mathbf{v} = \langle 3, -2 \rangle$ (B) $\mathbf{u} = \langle -3, -1 \rangle$; $\mathbf{v} = \langle 2, 4 \rangle$
(C) $\mathbf{u} = -4\mathbf{i} - 2\mathbf{j}$; $\mathbf{v} = -2\mathbf{i} + \mathbf{j}$

If $\theta = \pi/2$, then $\cos \theta = 0$ and it follows that $\mathbf{u} \cdot \mathbf{v} = 0$. Conversely, if $\mathbf{u} \cdot \mathbf{v} = 0$, then $\cos \theta = 0$ and $\theta = \pi/2$. Thus, to test for orthogonality, we need only compute $\mathbf{u} \cdot \mathbf{v}$.

TEST FOR ORTHOGONAL VECTORS

Two vectors $\mathbf{u}$ and $\mathbf{v}$ are orthogonal if and only if

$$\mathbf{u} \cdot \mathbf{v} = 0$$

Note: The zero vector is orthogonal to every vector.

EXAMPLE 3   Determining Orthogonality

Determine which of the following pairs of vectors are orthogonal.

(A) $\mathbf{u} = \langle 3, -2 \rangle$; $\mathbf{v} = \langle 4, 6 \rangle$ (B) $\mathbf{u} = 2\mathbf{i} - \mathbf{j}$; $\mathbf{v} = \mathbf{i} - 2\mathbf{j}$

SOLUTIONS   (A) $\mathbf{u} \cdot \mathbf{v} = \langle 3, -2 \rangle \cdot \langle 4, 6 \rangle = 12 + (-12) = 0$
         Thus, $\mathbf{u}$ and $\mathbf{v}$ are orthogonal.
      (B) $\mathbf{u} \cdot \mathbf{v} = (2\mathbf{i} - \mathbf{j}) \cdot (\mathbf{i} - 2\mathbf{j}) = 2 + 2 = 4 \neq 0$
         Thus, $\mathbf{u}$ and $\mathbf{v}$ are not orthogonal.   ◆

MATCHED PROBLEM 3   Determine which of the following pairs of vectors are orthogonal.

(A) $\mathbf{u} = \langle -2, 4 \rangle$; $\mathbf{v} = \langle -2, 1 \rangle$ (B) $\mathbf{u} = 2\mathbf{i} - \mathbf{j}$; $\mathbf{v} = \mathbf{i} + 2\mathbf{j}$

◆   Scalar Component of One Vector on Another

An important use of the dot product is in the determination of the *projection* of a vector $\mathbf{u}$ onto a vector $\mathbf{v}$. To project a vector $\mathbf{u}$ onto an arbitrary nonzero vector $\mathbf{v}$, locate $\mathbf{u}$ and $\mathbf{v}$ so they have the same initial point O; then drop a perpendicular line from the terminal point of $\mathbf{u}$ to the line that contains $\mathbf{v}$. The **vector projection of u onto v** is the vector $\mathbf{p}$ that lies on the line containing $\mathbf{v}$, as shown in Figure 3. The vector $\mathbf{p}$ has the same direction as $\mathbf{v}$ if θ is acute and the opposite direction

FIGURE 3
Projection of **u** onto **v**

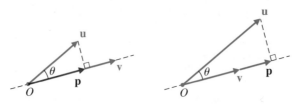

as $\mathbf{v}$ if θ is obtuse. The vector $\mathbf{p}$ is the zero vector if $\mathbf{u}$ and $\mathbf{v}$ are orthogonal. Problems 31–36 in Exercise 6.6 ask you to compute some projections.

The **scalar component of u on v**, denoted by $\text{Comp}_\mathbf{v}\ \mathbf{u}$, is given by $|\mathbf{u}|\cos\theta$. The quantity is positive if θ is acute, 0 if $\theta = \pi/2$, and negative if θ is obtuse. ✳ The scalar component of $\mathbf{u}$ on $\mathbf{v}$ can be expressed in terms of the dot product as follows:

$$\text{Comp}_\mathbf{v}\ \mathbf{u} = |\mathbf{u}|\cos\theta \qquad \text{Multiply numerator and denominator by } |\mathbf{v}|.$$

$$= \frac{|\mathbf{u}||\mathbf{v}|\cos\theta}{|\mathbf{v}|}$$

$$= \frac{\mathbf{u}\cdot\mathbf{v}}{|\mathbf{v}|} \qquad \text{Since } \cos\theta = \frac{\mathbf{u}\cdot\mathbf{v}}{|\mathbf{u}||\mathbf{v}|}$$

THE SCALAR COMPONENT OF **u** ON **v**

Let θ, $0 \leq \theta \leq \pi$, be the angle between the two vectors $\mathbf{u}$ and $\mathbf{v}$. The **scalar component of u on v**, denoted by $\text{Comp}_\mathbf{v}\ \mathbf{u}$, is given by

$$\text{Comp}_\mathbf{v}\ \mathbf{u} = |\mathbf{u}|\cos\theta$$

or, in terms of the dot product, by

$$\text{Comp}_\mathbf{v}\ \mathbf{u} = \frac{\mathbf{u}\cdot\mathbf{v}}{|\mathbf{v}|}$$

◆ **EXAMPLE 4**   Finding the Scalar Component of **u** on **v**

Find $\text{Comp}_\mathbf{v}\ \mathbf{u}$ for each pair of vectors $\mathbf{u}$ and $\mathbf{v}$. Compute answers to three significant digits.

(A) $\mathbf{u} = \langle 2, 3 \rangle$; $\mathbf{v} = \langle 4, 1 \rangle$ (B) $\mathbf{u} = \langle 3, 2 \rangle$; $\mathbf{v} = \langle 4, -6 \rangle$
(C) $\mathbf{u} = -3\mathbf{i} + 3\mathbf{j}$; $\mathbf{v} = 3\mathbf{i} + \mathbf{j}$

SOLUTIONS   (A) $\text{Comp}_\mathbf{v}\ \mathbf{u} = \dfrac{\mathbf{u}\cdot\mathbf{v}}{|\mathbf{v}|} = \dfrac{11}{\sqrt{17}} = 2.67$

(B) $\text{Comp}_\mathbf{v}\ \mathbf{u} = \dfrac{\mathbf{u}\cdot\mathbf{v}}{|\mathbf{v}|} = \dfrac{0}{\sqrt{52}} = 0$

(C) $\text{Comp}_\mathbf{v}\ \mathbf{u} = \dfrac{\mathbf{u}\cdot\mathbf{v}}{|\mathbf{v}|} = \dfrac{-6}{\sqrt{10}} = -1.90$ ◆

MATCHED PROBLEM 4   Find $\text{Comp}_\mathbf{v}\ \mathbf{u}$ for each pair of vectors $\mathbf{u}$ and $\mathbf{v}$. Compute answers to three significant digits.

(A) $\mathbf{u} = \langle 3, 3 \rangle$; $\mathbf{v} = \langle 6, 2 \rangle$ (B) $\mathbf{u} = \langle -4, 2 \rangle$; $\mathbf{v} = \langle 1, 2 \rangle$
(C) $\mathbf{u} = \mathbf{i} + 4\mathbf{j}$; $\mathbf{v} = 4\mathbf{i} - 4\mathbf{j}$

◆ Work

Intuitively, we know work is done when a force causes an object to be moved a certain distance. One example is illustrated in Figure 4, where work is done by the force that the harness exerts on a dog sled to move the sled a certain distance.

FIGURE 4

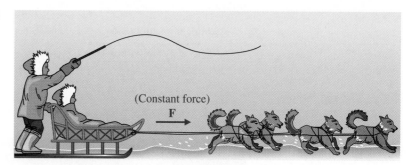

(Constant force)

F

The concept of work is very important in science studies dealing with energy and requires a definition that is quantitative: If the motion of an object takes place in a straight line in the direction of a constant force **F** causing the motion, then we define work W done by the force **F** to be the product of the magnitude of the force $|\mathbf{F}|$ and the distance d through which the object moves. Symbolically,

$$W = |\mathbf{F}|d$$

For example, suppose the dogsled in Figure 4 is moved a distance of 200 ft under a constant force of 100 lb. Then the work done by the force is

$$W = (100 \text{ lb})(200 \text{ ft}) = 20{,}000 \text{ ft-lb*}$$

It is more common that the motion of an object is caused by a constant force acting in a direction different than the line of motion. For example, a person pushing a lawn mower across a level lawn exerts a force downward along the handle (see Figure 5). In this case, only the component of force in the direction of motion (the horizontal component) is responsible for the motion. Suppose the lawn mower is pushed 52 ft across a level lawn with a constant force of 61 lb directed down the handle. If the handle makes a constant angle of 38° relative to the horizontal, then the work done by the force is

$$W = \begin{pmatrix} \text{Component of force in} \\ \text{the direction of motion} \end{pmatrix} (\text{Displacement})$$

$$= (61 \cos 38°)(52)$$

$$= 2{,}500 \text{ ft-lb}$$

A vector **d** is a **displacement vector** for an object that is moved in a straight line if **d** points in the direction of motion and if $|\mathbf{d}|$ is the distance moved. If the

---

* In scientific work, the basic unit of force is the **newton**, the basic unit of linear measure is the **meter**, and the basic unit of work is the **newton-meter** or **joule**. In the British engineering system, which we will use, the basic unit of work is the **foot-pound (ft-lb)**.

FIGURE 5

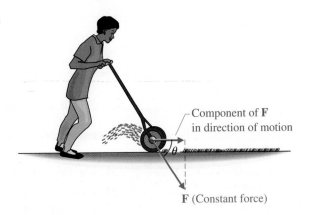

Component of **F**
in direction of motion

θ

F (Constant force)

displacement vector of an object is **0**, then no matter how much force is applied, no work is done. Also, for an object to move in the direction of the displacement vector, the angle between the force vector and the displacement vector must be acute. We now give a more general definition of work.

WORK

If the displacement vector is **d** for an object moved by a force **F**, then the **work** done, W, is the product of the component of force in the direction of motion and the actual displacement. Symbolically,

$$W = (\text{Comp}_{\mathbf{d}}\,\mathbf{F})|\mathbf{d}| = \frac{\mathbf{F} \cdot \mathbf{d}}{|\mathbf{d}|}|\mathbf{d}| = \mathbf{F} \cdot \mathbf{d}$$

◆ **EXAMPLE 5**   Work Done by a Force

How much work is done by a force $\mathbf{F} = \langle 6, 4 \rangle$ that moves an object from the origin to the point $P(8, 2)$? (Force is in pounds and displacement is in feet.)

SOLUTION $W = \mathbf{F} \cdot \mathbf{d}$

$\qquad = \langle 6, 4 \rangle \cdot \langle 8, 2 \rangle$ $d = \langle 8, 2 \rangle$

$\qquad = 56$ ft-lb   ◆

MATCHED PROBLEM 5   How much work is done by a force $\mathbf{F} = -8\mathbf{i} + 4\mathbf{j}$ that moves an object from the origin to the point $P(-12, -4)$? (Force is in pounds and displacement is in feet.)

Answers to   **1.** (A) 10   (B) 7   (C) 0   (D) 0
Matched Problems   **2.** (A) 60.3°   (B) 135.0°   (C) 53.1°
                   3. (A) Not orthogonal   (B) Orthogonal
                   4. (A) $24/\sqrt{40} = 3.79$ (B) $0/\sqrt{5} = 0$ (C) $-12/\sqrt{32} = -2.12$
                   5. 80 ft-lb

EXERCISE 6.6

A In Problems 1–8 find each dot product.

1. $\langle 5, 3 \rangle \cdot \langle -2, 3 \rangle$

2. $\langle 4, -3 \rangle \cdot \langle 2, 2 \rangle$

3. $(5\mathbf{i} - 3\mathbf{j}) \cdot (-2\mathbf{i} + 3\mathbf{j})$

4. $(-3\mathbf{i} + 2\mathbf{j}) \cdot (4\mathbf{i} + \mathbf{j})$

5. $\langle 2, 8 \rangle \cdot \langle 12, -3 \rangle$

6. $\langle -2, 3 \rangle \cdot \langle 9, 6 \rangle$

7. $3\mathbf{i} \cdot 4\mathbf{j}$

8. $-2\mathbf{j} \cdot 6\mathbf{i}$

In Problems 9–14 find the angle, in decimal degrees to one decimal place, between each pair of vectors.

9. $\mathbf{u} = \langle -3, 2 \rangle$; $\mathbf{v} = \langle 0, 4 \rangle$

10. $\mathbf{u} = \langle -2, -4 \rangle$; $\mathbf{v} = \langle -3, 0 \rangle$

11. $\mathbf{u} = \langle 3, 3 \rangle$; $\mathbf{v} = \langle 2, -5 \rangle$

12. $\mathbf{u} = \langle -2, 3 \rangle$; $\mathbf{v} = \langle -1, -4 \rangle$

13. $\mathbf{u} = 2\mathbf{i} - 3\mathbf{j}$; $\mathbf{v} = 6\mathbf{i} + 4\mathbf{j}$

14. $\mathbf{u} = 8\mathbf{i} + 2\mathbf{j}$; $\mathbf{v} = -3\mathbf{i} + 12\mathbf{j}$

B In Problems 15–18 determine which pairs of vectors are orthogonal.

15. $\mathbf{u} = 2\mathbf{i} + \mathbf{j}$; $\mathbf{v} = \mathbf{i} - 2\mathbf{j}$

16. $\mathbf{u} = 5\mathbf{i} - 4\mathbf{j}$; $\mathbf{v} = -4\mathbf{i} - 5\mathbf{j}$

17. $\mathbf{u} = \langle 1, 3 \rangle$; $\mathbf{v} = \langle -3, -1 \rangle$

18. $\mathbf{u} = \langle 4, -3 \rangle$; $\mathbf{v} = \langle 6, -8 \rangle$

In Problems 19–24 find $\text{Comp}_{\mathbf{v}} \mathbf{u}$, the scalar component of $\mathbf{u}$ on $\mathbf{v}$. Compute answers to three significant digits.

19. $\mathbf{u} = \langle -2, 4 \rangle$; $\mathbf{v} = \langle -3, -1 \rangle$

20. $\mathbf{u} = \langle 5, -2 \rangle$; $\mathbf{v} = \langle 6, 1 \rangle$

21. $\mathbf{u} = \langle -2, -4 \rangle$; $\mathbf{v} = \langle 6, -3 \rangle$

22. $\mathbf{u} = \langle 2, -3 \rangle$; $\mathbf{v} = \langle -9, -6 \rangle$

23. $\mathbf{u} = 7\mathbf{i} - 2\mathbf{j}$; $\mathbf{v} = \mathbf{i} + \mathbf{j}$

24. $\mathbf{u} = \mathbf{i} - \mathbf{j}$; $\mathbf{v} = 4\mathbf{i} + 6\mathbf{j}$

C Given $\mathbf{u} = \langle a, b \rangle$, $\mathbf{v} = \langle c, d \rangle$, $\mathbf{w} = \langle e, f \rangle$, and k a scalar, prove Problems 25–30 using the definitions of the given operations and the properties of real numbers.

25. $\mathbf{u} \cdot \mathbf{u} = |\mathbf{u}|^2$

26. $\mathbf{u} \cdot \mathbf{v} = \mathbf{v} \cdot \mathbf{u}$

27. $\mathbf{u} \cdot (\mathbf{v} + \mathbf{w}) = \mathbf{u} \cdot \mathbf{v} + \mathbf{u} \cdot \mathbf{w}$

28. $(\mathbf{u} + \mathbf{v}) \cdot \mathbf{w} = \mathbf{u} \cdot \mathbf{w} + \mathbf{v} \cdot \mathbf{w}$

29. $k(\mathbf{u} \cdot \mathbf{v}) = (k\mathbf{u}) \cdot \mathbf{v} = \mathbf{u} \cdot (k\mathbf{v})$

30. $\mathbf{u} \cdot \mathbf{0} = 0$

The vector projection of $\mathbf{u}$ onto $\mathbf{v}$, denoted by $\text{Proj}_{\mathbf{v}} \mathbf{u}$, is given by

$$\text{Proj}_{\mathbf{v}} \mathbf{u} = (\text{Comp}_{\mathbf{v}} \mathbf{u}) \frac{\mathbf{v}}{|\mathbf{v}|} = \frac{\mathbf{u} \cdot \mathbf{v}}{\mathbf{v} \cdot \mathbf{v}} \mathbf{v}$$

In Problems 31–36 find $\text{Proj}_{\mathbf{v}} \mathbf{u}$.

31. $\mathbf{u} = \langle 3, 4 \rangle$; $\mathbf{v} = \langle 4, 0 \rangle$

32. $\mathbf{u} = \langle 3, -4 \rangle$; $\mathbf{v} = \langle 0, -3 \rangle$

33. $\mathbf{u} = -6\mathbf{i} + 3\mathbf{j}$; $\mathbf{v} = -3\mathbf{i} - 2\mathbf{j}$

34. $\mathbf{u} = -2\mathbf{i} - 3\mathbf{j}$; $\mathbf{v} = \mathbf{i} - 6\mathbf{j}$

35. $\mathbf{u} = -2\mathbf{i} + 2\mathbf{j}$; $\mathbf{v} = -\mathbf{i} - 4\mathbf{j}$

36. $\mathbf{u} = -2\mathbf{i} - 2\mathbf{j}$; $\mathbf{v} = 5\mathbf{i} + \mathbf{j}$

Applications

37. **Work**   A parent pulls a child in a wagon (see the figure) for one block (440 ft). If a constant force of 15 lb is exerted along the handle, how much work is done?

Figure for 37

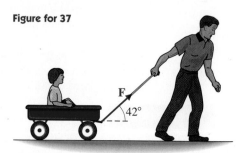

38. **Work**   Repeat Problem 37 using a constant force of 12 lb, a distance of 5,300 ft (approximately a mile), and an angle of 35° relative to the horizontal.

Work   *In Problems 39–44 determine how much work is done by a force $\mathbf{F}$ moving an object from the origin to the point P. (Force is in pounds and displacement is in feet.)*

39. $\mathbf{F} = \langle 10, 5 \rangle$; $P(8, 1)$

40. $\mathbf{F} = \langle 5, 1 \rangle$; $P(8, 5)$

41. $\mathbf{F} = -2\mathbf{i} + 3\mathbf{j}$; $P(-3, 1)$

42. $\mathbf{F} = 3\mathbf{i} - 4\mathbf{j}$; $P(5, 0)$

43. $\mathbf{F} = 10\mathbf{i} + 10\mathbf{j}$; $P(1, 1)$

44. $\mathbf{F} = 8\mathbf{i} - 4\mathbf{j}$; $P(2, -1)$

*45. Geometry Use the accompanying figure with the indicated vector assignments and appropriate properties of the dot product to prove that an angle inscribed in a semicircle is a right angle. [*Note:* $|\mathbf{a}| = |\mathbf{c}|$ = Radius]

Figure for 45

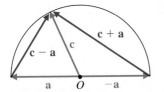

*46. Geometry Use the accompanying figure with the indicated vector assignments and appropriate properties of the dot product to prove that the diagonals of a rhombus are perpendicular. [*Note:* $|\mathbf{a}| = |\mathbf{b}|$]

Figure for 46

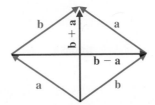

CHAPTER 6 SUMMARY

6.1
LAW OF SINES

An **oblique triangle** is a triangle without a right angle. An oblique triangle is **acute** if all angles are between 0° and 90° and **obtuse** if one angle is between 90° and 180°.

FIGURE 1

(a) Acute triangle

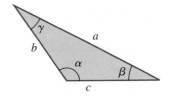

(b) Obtuse triangle

Solving a triangle involves finding three of the six quantities indicated in Figure 1 when given the other three. Table 1 is used to determine the accuracy of these computations.

TABLE 1

Angle to nearest	Significant digits for side measure
1°	2
10′ or 0.1°	3
1′ or 0.01°	4
10″ or 0.001°	5

If the given quantities include an angle and the opposite side (ASA, AAS, or SSA), the **law of sines** is used to solve the triangle in Figure 2.

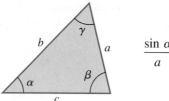

$$\frac{\sin \alpha}{a} = \frac{\sin \beta}{b} = \frac{\sin \gamma}{c}$$

FIGURE 2

The SSA case, often called the **ambiguous case,** has a number of variations (see Table 2).

TABLE 2

SSA variations

	a ($h = b \sin \alpha$)	Number of triangles	Figure	
α Acute	$0 < a < h$	0		(a)
	$a = h$	1		(b)
	$h < a < b$	2		(c)
	$a \geq b$	1		(d)
α Obtuse	$0 < a \leq b$	0		(e)
	$a > b$	1		(f)

6.2 We stated the **law of cosines** for the triangle in Figure 3 as follows.

LAW OF COSINES

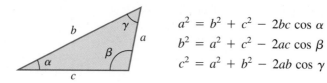

$$a^2 = b^2 + c^2 - 2bc \cos \alpha$$
$$b^2 = a^2 + c^2 - 2ac \cos \beta$$
$$c^2 = a^2 + b^2 - 2ab \cos \gamma$$

FIGURE 3

The law of cosines is generally used as the first step in solving the SAS and SSS cases for oblique triangles. After a side or angle is found using the law of cosines, it is usually easier to use the law of sines to find a second angle.

*6.3
AREAS OF TRIANGLES

The area of a triangle (Figure 4) can be determined from the base and altitude, two sides and the included angle, or all three sides (**Heron's Formula**).

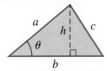 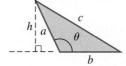

FIGURE 4

$$A = \frac{1}{2} bh = \frac{ab}{2} \sin \theta = \sqrt{s(s - a)(s - b)(s - c)}$$

where $s = \frac{1}{2}(a + b + c)$ is the **semiperimeter**.

6.4
VECTORS: GEOMETRICALLY DEFINED

A **scalar** is a real number. The **directed line segment** from point A to point B in the plane, denoted $\overrightarrow{AB}$, is the line segment from A to B with the arrowhead placed at B to indicate the direction (see Figure 5). Point A is called the **initial point** and point B is called the **terminal point**.

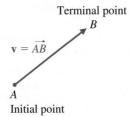

FIGURE 5
Vector $\overrightarrow{AB} = \mathbf{v}$

A **geometric vector** in a plane, denoted $\mathbf{v}$, is a quantity that possesses both a length and a direction and can be represented by directed line segment as indicated in Figure 5. The **magnitude** of the vector $\overrightarrow{AB}$, denoted by $|\overrightarrow{AB}|$, $|\vec{v}|$, or $|\mathbf{v}|$, is the length of the directed line segment. Two vectors have the **same direction** if they are parallel and point in the same direction. Two vectors have **opposite direction** if they are parallel and point in opposite directions. The **zero vector**, denoted by $\vec{0}$ or $\mathbf{0}$, has a magnitude of zero and an arbitrary direction. Two vectors are **equal** if they have the same magnitude and direction. Thus, a vector may be **translated** from one location to another as long as the magnitude and direction do not change.

The **sum of two vectors u and v** can be defined using the **tail-to-tip rule**. The sum of two nonparallel vectors also can be defined using the **parallelogram rule**. Both forms are shown in Figure 6.

FIGURE 6
Vector addition

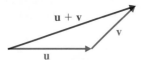
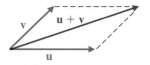

Tail-to-tip rule for addition   Parallelogram rule for addition

The vector **u + v** is also called the **resultant** of the two vectors **u** and **v**, and **u** and **v** are called **vector components** of **u + v**. Vector addition is com-mutative. That is, **u + v = v + u**.

A **velocity vector** represents the direction and speed of an object in motion. The velocity of a boat relative to the water is called **apparent velocity**, and the velocity relative to the ground is called the **resultant** or **actual velocity**. The resultant velocity is the vector sum of the apparent velocity and the current velocity. Similar statements apply to objects in air subject to winds.

A **force vector** represents the direction and magnitude of an applied force. If an object is subjected to two forces, then the sum of these two forces, the **resultant force**, is a single force acting on the object in the same way as the two original forces taken together.

6.5
VECTORS: ALGEBRAICALLY
DEFINED

A geometric vector $\overrightarrow{AB}$ in a rectangular coordinate system translated so that its initial point is at the origin is said to be in **standard position**. The vector $\overrightarrow{OP}$ such that $\overrightarrow{OP} = \overrightarrow{AB}$ is said to be the **standard vector** for $\overrightarrow{AB}$. This is shown in Figure 7.

Note that the vector $\overrightarrow{OP}$ in Figure 7 is the standard vector for infinitely many vectors: all vectors with the same magnitude and direction as $\overrightarrow{OP}$. Refer to Figure 7. If the coordinates of A are (x_a, y_a) and the coordinates of B are (x_b, y_b), then the coordinates of P are given by

$$(x_p, y_p) = (x_b - x_a, y_b - y_a)$$

Every geometric vector in a plane corresponds to an ordered pair of real numbers, and every ordered pair of real numbers corresponds to a geometric vector. This leads to the definition of an **algebraic vector** as an ordered pair of real numbers, denoted by $\langle a, b \rangle$. The real numbers a and b are **components** of the vector $\langle a, b \rangle$.

Two vectors $\mathbf{u} = \langle a, b \rangle$ and $\mathbf{v} = \langle c, d \rangle$ are said to be **equal** if their components are equal. The **zero vector** is denoted by $\mathbf{O} = \langle 0, 0 \rangle$ and has arbitrary direction.

The **magnitude**, or **norm**, of a vector $\mathbf{v} = \langle a, b \rangle$, denoted by $|\mathbf{v}|$, is given by

$$|\mathbf{v}| = \sqrt{a^2 + b^2}$$

FIGURE 7
$\overrightarrow{OP}$ is the standard vector
for $\overrightarrow{AB}$ ($\overrightarrow{OP} = \overrightarrow{AB}$)

Geometrically, $\sqrt{a^2 + b^2}$ is the length of the standard geometric vector $\overrightarrow{OP}$ associated with the algebraic vector $\langle a, b \rangle$.

If $\mathbf{u} = \langle a, b \rangle$, $\mathbf{v} = \langle c, d \rangle$, and k is a scalar, then the **sum** of $\mathbf{u}$ and $\mathbf{v}$ is given by

$$\mathbf{u} + \mathbf{v} = \langle a + c, b + d \rangle$$

and **scalar multiplication** of $\mathbf{u}$ by k is given by

$$k\mathbf{u} = k\langle a, b \rangle = \langle ka, kb \rangle$$

If $\mathbf{v}$ is a nonzero vector, then

$$\mathbf{u} = \frac{1}{|\mathbf{v}|}\,\mathbf{v}$$

is **a unit vector with the same direction as** $\mathbf{v}$. The $\mathbf{i}$ and $\mathbf{j}$ unit vectors are defined as follows (Figure 8).

$$\mathbf{i} = \langle 1, 0 \rangle$$
$$\mathbf{j} = \langle 0, 1 \rangle$$

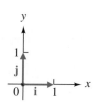

FIGURE 8

Any vector $\mathbf{v}$ can be expressed in terms of $\mathbf{i}$ and $\mathbf{j}$:

$$\mathbf{v} = \langle a, b \rangle = a\mathbf{i} + b\mathbf{j}$$

The following algebraic properties of vector addition and scalar multiplication enable us to manipulate symbols representing vectors and scalars in much the same way we manipulate symbols that represent real numbers in algebra.

(A) The following **addition properties** are satisfied for all vectors $\mathbf{u}$, $\mathbf{v}$, and $\mathbf{w}$:

1. $\mathbf{u} + \mathbf{v} = \mathbf{v} + \mathbf{u}$ Commutative property
2. $\mathbf{u} + (\mathbf{v} + \mathbf{w}) = (\mathbf{u} + \mathbf{v}) + \mathbf{w}$ Associative property
3. $\mathbf{u} + \mathbf{0} = \mathbf{0} + \mathbf{u} = \mathbf{u}$ Additive identity
4. $\mathbf{u} + (-\mathbf{u}) = (-\mathbf{u}) + \mathbf{u} = \mathbf{0}$ Additive inverse

(B) The following **scalar multiplication properties** are satisfied for all vectors $\mathbf{u}$ and $\mathbf{v}$ and all scalars m and n:

1. $m(n\mathbf{u}) = (mn)\mathbf{u}$ Associative property
2. $m(\mathbf{u} + \mathbf{v}) = m\mathbf{u} + m\mathbf{v}$ Distributive property
3. $(m + n)\mathbf{u} = m\mathbf{u} + n\mathbf{u}$ Distributive property
4. $1\mathbf{u} = \mathbf{u}$ Multiplication identity

Algebraic vectors can be used to solve **static equilibrium** problems. The conditions for static equilibrium are:

1. An object at rest is said to be in **static equilibrium.**

2. For an object located at the origin in a rectangular coordinate system to remain in static equilibrium (at rest), it is necessary that the sum of all the force vectors acting on the object be zero (the zero vector).

The **dot product** of the two vectors

$$\mathbf{u} = \langle a, b \rangle = a\mathbf{i} + b\mathbf{j} \qquad \text{and} \qquad \mathbf{v} = \langle c, d \rangle = c\mathbf{i} + d\mathbf{j}$$

denoted by $\mathbf{u} \cdot \mathbf{v}$, is the scalar given by

$$\mathbf{u} \cdot \mathbf{v} = ac + bd$$

For all vectors $\mathbf{u}$, $\mathbf{v}$, and $\mathbf{w}$, and k a real number:

1. $\mathbf{u} \cdot \mathbf{u} = |\mathbf{u}|^2$
2. $\mathbf{u} \cdot \mathbf{v} = \mathbf{v} \cdot \mathbf{u}$
3. $\mathbf{u} \cdot (\mathbf{v} + \mathbf{w}) = \mathbf{u} \cdot \mathbf{v} + \mathbf{u} \cdot \mathbf{w}$
4. $(\mathbf{u} + \mathbf{v}) \cdot \mathbf{w} = \mathbf{u} \cdot \mathbf{w} + \mathbf{v} \cdot \mathbf{w}$
5. $k(\mathbf{u} \cdot \mathbf{v}) = (k\mathbf{u}) \cdot \mathbf{v} = \mathbf{u} \cdot (k\mathbf{v})$
6. $\mathbf{u} \cdot \mathbf{0} = 0$

The **angle between the vectors u and v** is the angle θ, $0 \leq \theta \leq \pi$, formed by positioning the vectors so that their initial points coincide (Figure 9). The vectors $\mathbf{u}$ and $\mathbf{v}$ are **orthogonal** or **perpendicular** if $\theta = \pi/2$ and **parallel** if $\theta = 0$ or $\theta = \pi$. If $\mathbf{u}$ and $\mathbf{v}$ are nonzero then

$$\cos \theta = \frac{\mathbf{u} \cdot \mathbf{v}}{|\mathbf{u}||\mathbf{v}|}$$

FIGURE 9
Angle between two vectors

Two vectors $\mathbf{u}$ and $\mathbf{v}$ are orthogonal if and only if

$$\mathbf{u} \cdot \mathbf{v} = 0$$

[*Note:*   The zero vector is orthogonal to every vector.]

To project a vector $\mathbf{u}$ onto an arbitrary nonzero vector $\mathbf{v}$, locate $\mathbf{u}$ and $\mathbf{v}$ so that they have the same initial point O and drop a perpendicular from the terminal point of $\mathbf{u}$ to the line that contains $\mathbf{v}$. The **vector projection of u onto v** is the vector $\mathbf{p}$ that lies on the line containing $\mathbf{v}$, as shown in Figure 10.

FIGURE 10
Projection of **u** onto **v**

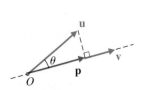
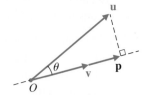
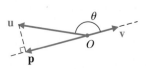

The **scalar component of u on v** is given by

$$\text{Comp}_v \ \mathbf{u} = |\mathbf{u}| \cos \theta$$

or, in terms of the dot product, by

$$\text{Comp}_v \ \mathbf{u} = \frac{\mathbf{u} \cdot \mathbf{v}}{|\mathbf{v}|}$$

A vector **d** is a **displacement vector** for an object that is moved in a straight line if **d** points in the direction of motion and if $|\mathbf{d}|$ is the distance moved. If the displacement vector is **d** for an object moved by a force **F**, then the **work** done, W, is the product of the component of force in the direction of motion and the actual displacement. Symbolically,

$$W = (\text{Comp}_\mathbf{d} \ \mathbf{F})|\mathbf{d}| = \frac{\mathbf{F} \cdot \mathbf{d}}{|\mathbf{d}|} |\mathbf{d}| = \mathbf{F} \cdot \mathbf{d}$$

CHAPTER 6 REVIEW EXERCISE

Work through all the problems in this chapter review and check the answers. Answers to all review problems appear in the back of the book; following each answer is an italic number that indicates the section in which that type of problem is discussed. Where weaknesses show up, review the appropriate sections in the text.

Where applicable, quantities in the problems refer to a triangle labeled as in the figure.

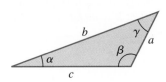

A *In Problems 1–12 determine whether the information in each problem allows you to construct 0, 1, 2, or infinitely many triangles. Use a rough sketch where possible. Do not solve the triangle.*

1. $\alpha = 32°$, $\beta = 74°$, $\gamma = 74°$
2. $\alpha = 47°$, $\beta = 62°$, $c = 130$ m
3. $\alpha = 85°$, $\beta = 99°$, $c = 12$ m
4. $\beta = 105°$, $\gamma = 32°$, $b = 67$ ft
5. $\beta = 43°$, $a = 12$ cm, $b = 4$ cm
6. $\alpha = 43°$, $a = 12$ cm, $b = 9$ cm
7. $\alpha = 43°$, $a = 12$ cm, $b = 14$ cm

8. $\alpha = 110°$, $a = 12$ cm, $b = 11$ cm
9. $\alpha = 110°$, $a = 12$ cm, $b = 16$ cm
10. $\beta = 112°$, $a = 210$ ft, $c = 470$ ft
11. $a = 14$ m, $b = 12$ m, $c = 17$ m
12. $a = 5$ cm, $b = 3$ cm, $c = 10$ cm

In Problems 13–16 solve each triangle given the indicated measures of angles and sides.

13. $\alpha = 53°$, $\gamma = 105°$, $b = 42$ cm
14. $\alpha = 66°$, $\beta = 32°$, $b = 12$ m
15. $\alpha = 49°$, $b = 22$ in., $c = 27$ in.
16. $\alpha = 62°$, $a = 14$ cm, $b = 12$ cm

☆ 17. Find the area of the triangle in Problem 15.

☆ 18. Find the area of the triangle in Problem 16.

19. Two vectors **u** and **v** are located in a coordinate system, as indicated in the figure. Find the direction (relative to the x axis) and magnitude of **u** + **v** if $|\mathbf{u}| = 8.0$ and $|\mathbf{v}| = 5.0$.

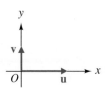

Figure for 19

20. Find the magnitude of the horizontal and vertical components of the vector **v** located in a coordinate system as indicated in the figure.

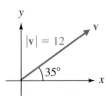

Figure for 20

21. Given $A(-3, 2)$ and $B(-1, -3)$, represent the geometric vector $\overrightarrow{AB}$ as an algebraic vector $\langle a, b \rangle$.

22. Find the magnitude of the vector $\langle -5, 12 \rangle$.

☆ **23.** $\langle 2, -1 \rangle \cdot \langle -3, 2 \rangle = ?$

☆ **24.** $(2\mathbf{i} + \mathbf{j}) \cdot (3\mathbf{i} - 2\mathbf{j}) = ?$

☆ **25.** Find the angle, in decimal degrees to one decimal place, between $\mathbf{u} = \langle 4, 3 \rangle$ and $\mathbf{v} = \langle 3, 0 \rangle$.

☆ **26.** Find the angle, in decimal degrees to one decimal place, between $\mathbf{u} = 5\mathbf{i} + \mathbf{j}$ and $\mathbf{v} = -2\mathbf{i} + 2\mathbf{j}$.

B *In Problems 27–30 solve each triangle given the indicated measures of angles and sides.*

27. $\alpha = 65.0°, \quad b = 103$ m, $\quad c = 72.4$ m

28. $\alpha = 35°20', \quad a = 13.2$ in., $\quad b = 15.7$ in., $\quad \beta$ acute

29. $\alpha = 35°20', \quad a = 13.2$ in., $\quad b = 15.7$ in., $\quad \beta$ obtuse

30. $a = 43$ mm, $\quad b = 48$ mm, $\quad c = 53$ mm

☆ **31.** Find the area of the triangle in Problem 27.

☆ **32.** Find the area of the triangle in Problem 30.

33. Given the vector diagram, find $|\mathbf{u} + \mathbf{v}|$ and θ.

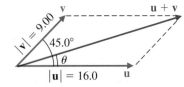

Figure for 33

In Problems 34 and 35 find,

(A) $\mathbf{u} + \mathbf{v}$ (B) $\mathbf{u} - \mathbf{v}$
(C) $3\mathbf{u} - 2\mathbf{v}$ (D) $2\mathbf{u} - 3\mathbf{v} + \mathbf{w}$

34. $\mathbf{u} = \langle 4, 0 \rangle; \quad \mathbf{v} = \langle -2, -3 \rangle; \quad \mathbf{w} = \langle 1, -1 \rangle$

35. $\mathbf{u} = 3\mathbf{i} - \mathbf{j}; \quad \mathbf{v} = 2\mathbf{i} - 3\mathbf{j}; \quad \mathbf{w} = -2\mathbf{j}$

36. Form a unit vector **u** with the same direction as **v** $= \langle -8, 15 \rangle$.

37. Express **v** in terms of **i** and **j** unit vectors.
(A) $\mathbf{v} = \langle -5, 7 \rangle$ (B) $\mathbf{v} = \langle 0, -3 \rangle$
(C) $\mathbf{v} = \overrightarrow{AB}; \quad A(4, -2); \quad B(0, -3)$

☆ **38.** Determine which vector pairs are orthogonal using properties of the dot product.
(A) $\mathbf{u} = \langle -12, 3 \rangle; \quad \mathbf{v} = \langle 2, 8 \rangle$
(B) $\mathbf{u} = -4\mathbf{i} + \mathbf{j}; \quad \mathbf{v} = -\mathbf{i} + 4\mathbf{j}$

☆ **39.** Find $\text{Comp}_\mathbf{v}\, \mathbf{u}$, the scalar component of **u** on **v**. Compute answers to three significant digits.
(A) $\mathbf{u} = \langle 4, 5 \rangle; \quad \mathbf{v} = \langle 3, 1 \rangle$
(B) $\mathbf{u} = -\mathbf{i} + 4\mathbf{j}; \quad \mathbf{v} = 3\mathbf{i} - \mathbf{j}$

C   **40.** An oblique triangle is solved, with the following results:

$$\alpha = 45.1° \quad a = 8.42 \text{ cm}$$
$$\beta = 75.8° \quad b = 11.5 \text{ cm}$$
$$\gamma = 59.1° \quad c = 10.2 \text{ cm}$$

Allowing for rounding, use Mollweide's equation to check these results.

$$(a - b) \cos \frac{\gamma}{2} = c \sin \frac{\alpha - \beta}{2}$$

41. Given an oblique triangle with $\alpha = 52.3°$ and $b = 12.7$ cm, determine a value k so that if $0 < a < k$, there is no solution; if $a = k$, there is one solution; and if $k < a < b$, there are two solutions.

In Problems 42–47 let $\mathbf{u} = \langle a, b \rangle$, $\mathbf{v} = \langle c, d \rangle$, *and* $\mathbf{w} = \langle e, f \rangle$ *be vectors, and let m and n be scalars. Prove each property using the definitions of the operations involved and properties of real numbers. (Although some of these problems appeared in Exercises 6.5 and 6.6, it is useful to consider them again as part of this review.)*

42. $\mathbf{u} + \mathbf{v} = \mathbf{v} + \mathbf{u}$

☆ **43.** $\mathbf{u} \cdot \mathbf{v} = \mathbf{v} \cdot \mathbf{u}$

44. $(mn)\mathbf{v} = m(n\mathbf{v})$

☆ **45.** $m(\mathbf{u} \cdot \mathbf{v}) = (m\mathbf{u}) \cdot \mathbf{v} = \mathbf{u} \cdot (m\mathbf{v})$

☆ **46.** $\mathbf{u} \cdot (\mathbf{v} + \mathbf{w}) = \mathbf{u} \cdot \mathbf{v} + \mathbf{u} \cdot \mathbf{w}$

☆ **47.** $\mathbf{u} \cdot \mathbf{u} = |\mathbf{u}|^2$

Applications

48. Geometry   Find the lengths of the sides of a parallelogram with diagonals 20.0 cm and 16.0 cm long intersecting at 36.4°.

49. Geometry Find *h* to three significant digits in the triangle:

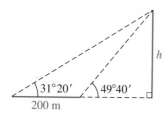

Figure for 49

∗50. Geometry A chord of length 34 cm subtends a central angle of 85° in a circle of radius *r*. Find the radius of the circle.

∗51. Surveying A plot of land has been surveyed with the resulting information shown in the figure. Find the length of *CD*.

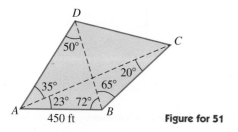

Figure for 51

☆52. Surveying Refer to Problem 51. Find the area of the plot.

53. Engineering A tunnel for a highway is to be constructed through a mountain, as indicated in the figure. How long is the tunnel?

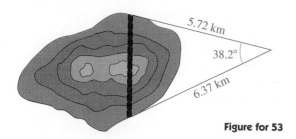

Figure for 53

∗54. Space Science A satellite *S*, in circular orbit around the earth, is sighted by a tracking station *T* (see the figure). The radar-determined distance *TS* is 1,147 mi, and the angle of elevation above the horizon is 28.6°. How high is the satellite above the earth at the time of the sighting? The radius of the earth is *R* = 3,964 mi.

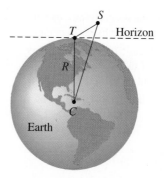

Figure for 54

∗55. Space Science When a satellite is directly over tracking station *B* (see the figure), tracking station *A* measures the angle of elevation θ of the satellite (from the horizon line) to be 21.7°. If the tracking stations, both on the equator, are 632 mi apart (that is, the arc *AB* is 632 mi long) and the radius of the earth is 3,964 mi, how high is the satellite above *B*? [*Hint:* Find all angles for the triangle *ACS* first.]

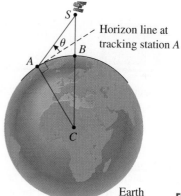

Figure for 55

56. Navigation An airplane flies with a speed of 230 km/hr and a compass heading of 68°. If a 55 km/hr wind is blowing in the direction of 5°, what is the plane's actual direction (relative to north) and ground speed?

57. Resultant Force Two forces act on an object as indicated in the figure.

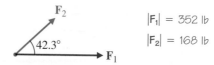

$|F_1| = 352$ lb

$|F_2| = 168$ lb

Figure for 57

Find the magnitude of the resultant force and its direction relative to the horizontal force $\mathbf{F_1}$.

☆ **58. Work**   Refer to Problem 57. How much work is done by the resultant force if the object is moved 22 ft in the direction of the resultant force?

59. Engineering   A 260 lb sign is hung as shown in the figure. Determine the compression force on AB and the tension on CB.

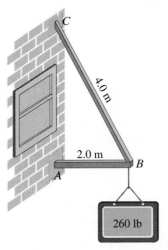

Figure for 59

60. Space Science   To simulate a reduced-gravity environment such as that found on the surface of the moon, NASA constructed an inclined plane and a harness suspended on a cable fastened to a movable track, as indicated in the figure. A person suspended in the harness walks at a right angle to the inclined plane.

   (A) Using the accompanying force diagram, find $|\mathbf{u}|$, the force on the astronaut's feet, in terms of $|\mathbf{w}|$ and θ.

Also find $|\mathbf{v}|$, the tension on the cable, in terms of $|\mathbf{w}|$ and θ.

   (B) What is the force against the feet of a woman astronaut weighing 130 lb ($|\mathbf{w}| = 130$ lb) if $\theta = 72°$? What is the tension on the cable? Compute answers to the nearest pound.

   (C) Find θ, to the nearest degree, so that the force on an astronaut's feet will be the same as the force on the moon (about one-sixth of that on earth).

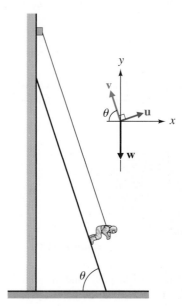

Figure for 60

☆ **61. Engineering: Physics**   Determine how much work is done by the force $\mathbf{F} = \langle -5, 8 \rangle$ moving an object from the origin to the point $P(-8, 2)$. (Force is in pounds and displacement is in feet.)

POLAR COORDINATES; COMPLEX NUMBERS

he rectangular coordinate system is the most widely used coordinate system; in this chapter we will discuss the second most widely used coordinate system: the *polar coordinate system*. After graphing some polar equations, we will use the polar coordinate system to represent complex numbers in *polar or trigonometric form*. The polar form of a complex number is used in the statement of De Moivre's famous theorem, which leads to a process of finding all roots of a given number. You may think that the only root of $x^3 = 1$ is 1. This is true if we restrict solutions to real numbers, but if we allow complex numbers, there are three roots—one real and two imaginary.

7.1 POLAR AND RECTANGULAR COORDINATES

♦ **Polar Coordinate System**
♦ **From Polar Form to Rectangular Form and Vice Versa**

Until now we have had only one way of associating ordered pairs of numbers with points in a plane—namely, the rectangular coordinate system. Another type of coordinate system is widely used: the *polar coordinate system*. Of the many different kinds of coordinate systems that are possible, the polar coordinate system ranks second in importance only to the rectangular coordinate system.

♦ Polar Coordinate System

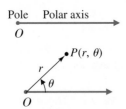

FIGURE 1
Polar coordinate system

To form a **polar coordinate system** in a plane (see Figure 1), we start with a fixed point O and call it the **pole**, or **origin**. From this point we draw a half line (usually horizontal and to the right) and we call this line the **polar axis**.

If P is an arbitrary point in a plane, then we associate **polar coordinates** $(\mathbf{r}, \boldsymbol{\theta})$ with it as follows: Starting with the polar axis as the initial side of an angle, we rotate the terminal side until it, or the extension of it through the pole, passes through the point. The θ coordinate in (r, θ) is this angle, in degree or radian measure. The angle θ is positive if the rotation is counterclockwise and negative if the rotation is clockwise. The r coordinate in (r, θ) is the directed distance from the pole to the point P, which is positive if measured from the pole along the terminal side of θ and negative if measured along the terminal side extended through the pole. Figure 2 illustrates how one point in a polar coordinate system can have many—in fact, an unlimited number of—polar coordinates.

We can now see a major difference between the rectangular coordinate system and the polar coordinate system. In the former, each point has exactly one set of rectangular coordinates; in the latter, a point may have infinitely many polar coordinates.

The pole itself has polar coordinates of the form $(0, \theta)$, where θ is arbitrary. For example, $(0, 37°)$ and $(0, -\pi/4)$ are both polar coordinates of the pole, and there are infinitely many others.

FIGURE 2

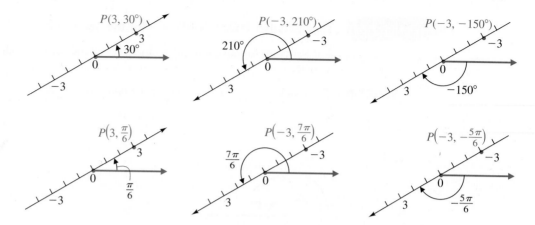

Just as graph paper is readily available for work related to rectangular co-ordinate systems, polar graph paper is available for work related to polar coordinates. The following examples illustrate its use.

◆ **EXAMPLE 1**   Plotting Points in a Polar Coordinate System

Plot the following points in a polar coordinate system.

(A) $A(4, 45°)$; $B(-6, 45°)$; $C(7, 240°)$; $D(3, -75°)$
(B) $A(8, \pi/6)$; $B(5, -3\pi/4)$; $C(-7, 2\pi/3)$; $D(-9, -\pi/6)$

SOLUTIONS

(A)   (B)

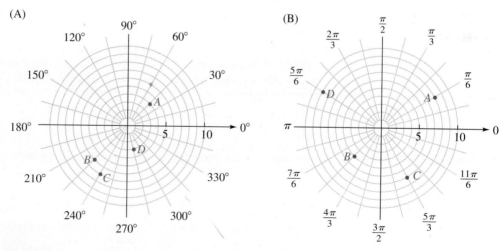

FIGURE 3   **FIGURE 4**

MATCHED PROBLEM 1   Plot the following points in a polar coordinate system.

(A) $A(7, 30°)$; $B(-6, 165°)$; $C(-9, -90°)$
(B) $A(10, \pi/3)$; $B(8, -7\pi/6)$; $C(-5, -5\pi/4)$

◆ **EXAMPLE 2**   Finding Other Polar Coordinates for the Same Point

For the point (6, 60°), find three other sets of coordinates such that $-360° \leq \theta \leq 360°$.

SOLUTION   Three other sets of coordinates for the point (6, 60°) are (see Figure 5)

$$(-6, 240°) \qquad (-6, -120°) \qquad (6, -300°)$$

FIGURE 5

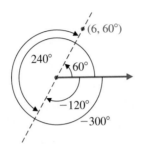

MATCHED PROBLEM 2   For the point (7, $\pi/6$), find three other sets of coordinates such that $-2\pi \leq \theta \leq 2\pi$.

◆ **From Polar Form to Rectangular Form and Vice Versa**

It is frequently convenient to be able to transform coordinates or equations in rectangular form into polar form, or vice versa. The following polar-rectangular relationships are useful in this regard.

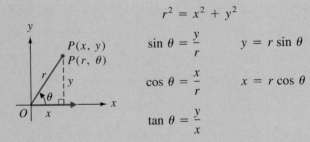

POLAR-RECTANGULAR RELATIONSHIPS

$$r^2 = x^2 + y^2$$

$$\sin \theta = \frac{y}{r} \qquad y = r \sin \theta$$

$$\cos \theta = \frac{x}{r} \qquad x = r \cos \theta$$

$$\tan \theta = \frac{y}{x}$$

[***Note:***   The signs of x and y determine the quadrant for θ. The angle θ is usually chosen so that $|\theta|$ is minimum.]

◆ **EXAMPLE 3**   From Polar to Rectangular Form

Change $A(5, \pi/6)$, $B(-3, 3\pi/4)$, and $C(-2, -5\pi/6)$ to rectangular coordinates.

SOLUTION   Use $x = r \cos \theta$ and $y = r \sin \theta$.

For A:

FIGURE 6

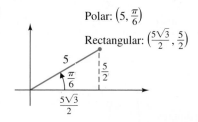

Polar: $\left(5, \frac{\pi}{6}\right)$

Rectangular: $\left(\frac{5\sqrt{3}}{2}, \frac{5}{2}\right)$

$$x = 5 \cos \frac{\pi}{6} = 5\left(\frac{\sqrt{3}}{2}\right) = \frac{5\sqrt{3}}{2}$$

$$y = 5 \sin \frac{\pi}{6} = 5\left(\frac{1}{2}\right) = \frac{5}{2}$$

Rectangular coordinates: $\left(\frac{5\sqrt{3}}{2}, \frac{5}{2}\right)$

For B:

$$x = -3 \cos \frac{3\pi}{4} = (-3)\left(\frac{-\sqrt{2}}{2}\right) = \frac{3\sqrt{2}}{2}$$

$$y = -3 \sin \frac{3\pi}{4} = (-3)\left(\frac{\sqrt{2}}{2}\right) = \frac{-3\sqrt{2}}{2}$$

Rectangular coordinates: $\left(\frac{3\sqrt{2}}{2}, \frac{-3\sqrt{2}}{2}\right)$

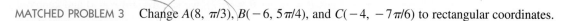

For C:

$$x = -2 \cos\left(\frac{-5\pi}{6}\right) = (-2)\left(\frac{-\sqrt{3}}{2}\right) = \sqrt{3}$$

$$y = -2 \sin\left(\frac{-5\pi}{6}\right) = (-2)\left(\frac{-1}{2}\right) = 1$$

Rectangular coordinates: $(\sqrt{3}, 1)$ ◆

MATCHED PROBLEM 3   Change $A(8, \pi/3)$, $B(-6, 5\pi/4)$, and $C(-4, -7\pi/6)$ to rectangular coordinates.

◆ **EXAMPLE 4**   From Rectangular to Polar Form

Change $A(1, \sqrt{3})$ and $B(-\sqrt{3}, -1)$ into polar form with $r \geq 0$ and $0 \leq \theta < 2\pi$.

SOLUTION   Use $r^2 = x^2 + y^2$ and $\tan \theta = y/x$.

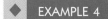

For A:

$$r^2 = 1^2 + (\sqrt{3})^2 = 4$$

$$r = 2$$

$$\tan \theta = \frac{\sqrt{3}}{1}$$

$$\theta = \frac{\pi}{3} \qquad \text{Since A is in the first quadrant}$$

Polar coordinates: $\left(2, \dfrac{\pi}{3}\right)$

For B:

$$r^2 = (-\sqrt{3})^2 + (-1)^2 = 4$$

$$r = 2$$

$$\tan \theta = \frac{-1}{-\sqrt{3}}$$

$$\theta = \frac{7\pi}{6} \qquad \text{Since B is in the third quadrant}$$

Polar coordinates: $\left(2, \dfrac{7\pi}{6}\right)$ ◆

MATCHED PROBLEM 4   Change $A(\sqrt{3}, 1)$ and $B(1, -\sqrt{3})$ to polar form with $r \geq 0$ and $0 \leq \theta < 2\pi$.

 EXAMPLE 5   From Rectangular to Polar Form

Change $x^2 + y^2 - 2x = 0$ to polar form.

SOLUTION   Use $r^2 = x^2 + y^2$ and $x = r \cos \theta$:

$$x^2 + y^2 - 2x = 0$$

$$r^2 - 2r \cos \theta = 0$$

$$r(r - 2 \cos \theta) = 0$$

$$r = 0 \qquad \text{or} \qquad r - 2 \cos \theta = 0$$

The graph of $r = 0$ is the pole, and since the pole is included as a solution of $r - 2 \cos \theta = 0$ (let $\theta = \pi/2$), we can discard $r = 0$ and keep only

$$r - 2 \cos \theta = 0$$

or

$$r = 2 \cos \theta$$ ◆

MATCHED PROBLEM 5   Change $x^2 + y^2 - 2y = 0$ to polar form.

◆ **EXAMPLE 6** From Polar to Rectangular Form

Change $r + 3 \sin \theta = 0$ to rectangular form.

SOLUTION The conversion of this equation, as it stands, to rectangular form gets messy. A simple trick, however, makes the conversion easy: We multiply both sides by r, which simply adds the pole to the graph. In this case, the pole is already included as a solution of $r + 3 \sin \theta = 0$ (let $\theta = 0$), so we have not actually changed anything by doing this. Thus,

$$r + 3 \sin \theta = 0 \qquad \text{Multiply both sides by } r$$
$$r^2 + 3r \sin \theta = 0 \qquad r^2 = x^2 + y^2, \quad y = r \sin \theta$$
$$x^2 + y^2 + 3y = 0$$

◆

MATCHED PROBLEM 6 Change $r = -8 \cos \theta$ to rectangular form.

Answers to
Matched Problems

1. (A)

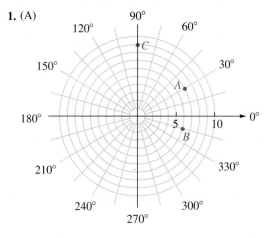

(B)

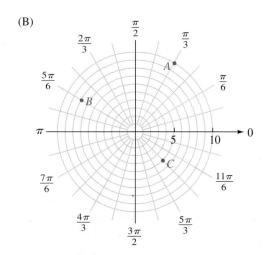

2. $(-7, 7\pi/6)$, $(-7, -5\pi/6)$, $(7, -11\pi/6)$

3. $A(4, 4\sqrt{3})$, $B(3\sqrt{2}, 3\sqrt{2})$, $C(2\sqrt{3}, -2)$

4. $A(2, \pi/6)$, $B(2, 5\pi/3)$

5. $r = 2 \sin \theta$

6. $x^2 + y^2 + 8x = 0$

EXERCISE 7.1

A *Plot in a polar coordinate system.*

1. $A(8, 0°)$; $B(5, 90°)$; $C(6, 30°)$

2. $A(4, 0°)$; $B(7, 180°)$; $C(9, 45°)$

3. $A(-8, 0°)$; $B(-5, 90°)$; $C(-6, 30°)$

4. $A(-4, 0°)$; $B(-7, 180°)$; $C(-9, 45°)$

5. $A(5, -30°)$; $B(4, -45°)$; $C(9, -90°)$

6. $A(8, -45°)$; $B(6, -60°)$; $C(4, -30°)$

7. $A(-5, -30°)$; $B(-4, -45°)$; $C(-9, -90°)$

8. $A(-8, -45°)$; $B(-6, -60°)$; $C(-4, -30°)$

9. $A(6, \pi/6)$; $B(5, \pi/2)$; $C(8, \pi/4)$

10. $A(8, \pi/3)$; $B(4, \pi/4)$; $C(10, 0)$

11. $A(-6, \pi/6)$; $B(-5, \pi/2)$; $C(-8, \pi/4)$

12. $A(-8, \pi/3)$; $B(-4, \pi/4)$; $C(-10, 0)$

13. $A(6, -\pi/6)$; $B(5, -\pi/2)$; $C(8, -\pi/4)$

14. $A(8, -\pi/3)$; $B(4, -\pi/4)$; $C(10, -\pi/6)$

15. $A(-6, -\pi/2)$; $B(-5, -\pi/3)$; $C(-8, -\pi/4)$

16. $A(-6, -\pi/6)$; $B(-5, -\pi/2)$; $C(-8, -\pi/3)$

Change to rectangular coordinates.

17. $(8, \pi/3)$ **18.** $(4, \pi/4)$

19. $(-9, \pi/2)$ **20.** $(-8, \pi)$

21. $(-4, \pi/4)$ **22.** $(-6, \pi/6)$

23. $(10, 5\pi/6)$ **24.** $(8, 7\pi/6)$

25. $(6, -7\pi/6)$ **26.** $(4, -7\pi/4)$

27. $(-4, -\pi/6)$ **28.** $(-5, -\pi/3)$

Change to polar coordinates with $r \geq 0$ and $0 \leq \theta < 2\pi$.

29. $(2\sqrt{3}, 2)$ **30.** $(3, 3\sqrt{3})$

31. $(-4\sqrt{2}, 4\sqrt{2})$ **32.** $(-6\sqrt{3}, 6)$

33. $(-4, -4\sqrt{3})$ **34.** $(5\sqrt{2}, -5\sqrt{2})$

35. $(0, -7)$ **36.** $(-10, 0)$

B *Plot in a polar coordinate system.*

37. $A(5, 210°)$; $B(-5, 210°)$; $C(-5, -210°)$

38. $A(9, 120°)$; $B(-9, 120°)$; $C(-9, -120°)$

39. $A(7, 7\pi/4)$; $B(-7, 7\pi/4)$; $C(-7, -7\pi/4)$

40. $A(6, 4\pi/3)$; $B(-6, 4\pi/3)$; $C(-6, -4\pi/3)$

Change to polar form.

41. $6x - x^2 = y^2$ **42.** $y^2 = 5y - x^2$

43. $2x + 3y = 5$ **44.** $3x - 5y = -2$

45. $x^2 + y^2 = 9$ **46.** $y = x$

47. $2xy = 1$ **48.** $y^2 = 4x$

49. $4x^2 - y^2 = 4$ **50.** $x^2 + 9y^2 = 9$

Change to rectangular form.

51. $r(2 \cos \theta + \sin \theta) = 4$

52. $r(3 \cos \theta - 4 \sin \theta) = -1$

53. $r = 8 \cos \theta$ **54.** $r = -2 \sin \theta$

55. $r^2 \cos 2\theta = 4$ **56.** $r^2 \sin 2\theta = 2$

57. $r = 4$ **58.** $r = -5$

59. $\theta = 30°$ **60.** $\theta = \pi/4$

C   **61.** Change $r = 3/(\sin \theta - 2)$ into rectangular form.

      62. Change $(y - 3)^2 = 4(x^2 + y^2)$ into polar form.

Applications

63. Precalculus: Analytic Geometry   A distance formula for the distance between two points in a polar coordinate system follows directly from the law of cosines (see the figure on page 367):

$$d^2 = r_1^2 + r_2^2 - 2r_1 r_2 \cos(\theta_2 - \theta_1)$$
$$d = \sqrt{r_1^2 + r_2^2 - 2r_1 r_2 \cos(\theta_2 - \theta_1)}$$

Find the distance between the two points $P_1(2, 30°)$ and $P_2(3, 60°)$. Compute your answer to four significant digits.

64. Precalculus: Analytic Geometry   Refer to Problem 63 and find the distance between the two points $P_1(4, \pi/4)$ and $P_2(1, \pi/2)$. Compute your answer to four significant digits.

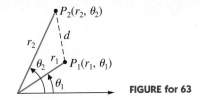

FIGURE for 63

7.2   SKETCHING POLAR EQUATIONS

- ◆   **Point-by-Point Sketching**
- ◆   **Rapid Sketching**
- ◆   **A Table of Standard Polar Curves**
- ◆   **Application**

Just as in rectangular coordinate systems where we sketched graphs of equations involving the variables x and y, in a polar coordinate system we graph equations involving the variables r and θ. We shall see that certain curves have simpler representations in polar coordinates and other curves have simpler representations in rectangular coordinates. Likewise, some applications have simpler solutions in polar coordinates while other applications have simpler solutions in rectangular coordinates. In this section you will gain experience in recognizing and sketching some standard polar graphs.

◆   Point-by-Point Sketching

To graph a polar equation such as $r = 2\theta$ or $r = 4 \sin 2\theta$ in a polar coordinate system, we locate all points with coordinates that satisfy the equation. A sketch of a graph can be obtained (just as in rectangular coordinates) by making a table of values that satisfy the equation, plotting these points, and then joining them with a smooth curve. A calculator can be used to generate the table.

The graph of the polar equation

$$r = a\theta, \qquad a > 0$$

is called **Archimedes' spiral**. Example 1 illustrates how to obtain a particular spiral by point-by-point plotting.

◆   EXAMPLE 1   Spiral

Graph $r = 2\theta$, $0 \le \theta \le 5\pi/3$ in a polar coordinate system using point-by-point plotting (θ in radians).

SOLUTION   We form a table using multiples of $\pi/6$; then plot the points and join them with a smooth curve (see Figure 1).

FIGURE 1

θ	r
0	0.00
$\pi/6$	1.05
$\pi/3$	2.09
$\pi/2$	3.14
$2\pi/3$	4.19
$5\pi/6$	5.24
π	6.28
$7\pi/6$	7.33
$4\pi/3$	8.38
$3\pi/2$	9.42
$5\pi/3$	10.47

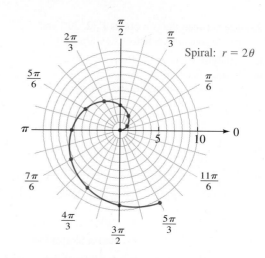

Spiral: $r = 2\theta$

MATCHED PROBLEM 1   Graph $r = 3\theta$, $0 \le \theta \le 7\pi/6$ in a polar coordinate system using point-by-point plotting (θ in radians).

◆   EXAMPLE 2   Circle

Graph $r = 4 \cos \theta$, θ in radians.

SOLUTION   We start with multiples of $\pi/6$ and continue until the graph begins to repeat (see Figure 2). (In intervals of uncertainty, add more points.)

FIGURE 2

θ	r
0	4.00
$\pi/6$	3.46
$\pi/3$	2.00
$\pi/2$	0.00
$2\pi/3$	−2.00
$5\pi/6$	−3.46
π	−4.00
$7\pi/6$	−3.46

Graph is repeating.

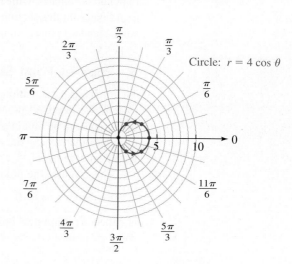

Circle: $r = 4 \cos \theta$

The graph is a circle with radius 2 and center at (2, 0).   ◆

MATCHED PROBLEM 2   Graph $r = 4 \sin \theta$, θ in radians.

If Example 2 is graphed using degrees instead of radians, we get the same graph except that degrees are marked around the polar coordinate system instead of radians (see Figure 3).

FIGURE 3

θ	r
0°	4.00
30°	3.46
60°	2.00
90°	0.00
120°	−2.00
150°	−3.46
180°	−4.00
210°	−3.46

Graph is repeating.

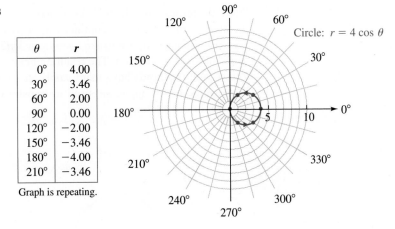

Circle: $r = 4 \cos \theta$

◆ Rapid Sketching

As polar curves become more involved, point-by-point plotting becomes more tedious so we turn to another process, which at first seems complicated. After a little experience, however, you will find that the new approach produces results rather quickly because parts of the process can be done mentally or on scratch paper. Example 3 illustrates this alternative approach.

◆ EXAMPLE 3   Cardioid

Sketch $r = 5 + 5 \cos \theta$, θ in radians.

SOLUTION   Since you have now had a lot of experience in graphing trigonometric functions in a rectangular coordinate system, start by graphing $r = 5 + 5 \cos \theta$ in a rectangular coordinate system (Figure 4). From this graph we can easily observe how r varies as θ varies over particular intervals. We summarize this variation in a short table and sketch the polar curve from the information in this table. [*Note:* After a little experience, many will choose not to actually draw the rectangular graph.]

FIGURE 4

θ	r
0 to $\pi/2$	10 to 5
$\pi/2$ to π	5 to 0
π to $3\pi/2$	0 to 5
$3\pi/2$ to 2π	5 to 10

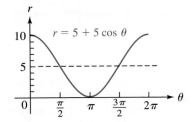

Rectangular coordinate system

Think of r as a "position vector" with a pencil on its terminal end. The initial end is always at the pole and the terminal end changes as θ changes over the indicated intervals. In particular, starting at the polar point (10, 0), we see that r decreases from 10 to 5 as θ increases from 0 to $\pi/2$. As θ continues to increase from $\pi/2$ to π, r continues to decrease from 5 to 0; then the process reverses as θ goes from π to $3\pi/2$ and from $3\pi/2$ to π. The polar graph is repeated as θ continues to increase. The resulting heart-shaped curve in Figure 5 is called a **cardioid**. Cardioids have many applications in engineering and science. Cardioid cams, for example, are used to transform uniform rotational motion into uniform linear motion.

FIGURE 5
Cardioid

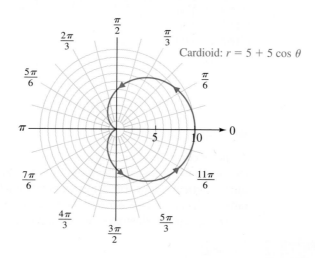

MATCHED PROBLEM 3   Sketch $r = 4 + 4 \sin \theta$, θ in radians.

◆   **EXAMPLE 4**   Four-Leaved Rose

Sketch $r = 6 \sin 2\theta$, θ in radians.

SOLUTION   We proceed as in Example 3 by graphing $r = 6 \sin 2\theta$ in rectangular coordinates first. The variable θ must increase from 0 to 2π before the polar graph repeats (see Figure 6).

FIGURE 6

θ	r
0 to $\pi/4$	0 to 6
$\pi/4$ to $\pi/2$	6 to 0
$\pi/2$ to $3\pi/4$	0 to -6
$3\pi/4$ to π	-6 to 0
π to $5\pi/4$	0 to 6
$5\pi/4$ to $3\pi/2$	6 to 0
$3\pi/2$ to $7\pi/4$	0 to -6
$7\pi/4$ to 2π	-6 to 0

Polar graph repeats

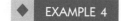

Rectangular coordinate system

Using the information in the table, we complete the polar graph, which is called a **four-leaved rose** (see Figure 7).

FIGURE 7
Four-leaved rose

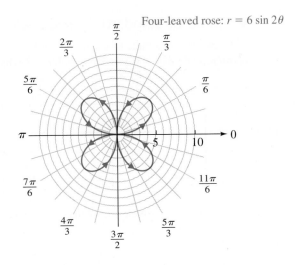

Four-leaved rose: $r = 6 \sin 2\theta$

MATCHED PROBLEM 4   Sketch $r = 4 \cos 3\theta$, θ in radians.

The graph of $r = 6 \sin 2\theta$ in Example 4 is a four-leaved rose, and the graph of $r = 4 \cos 3\theta$ in Matched Problem 4 is a three-leaved rose. In general, it can be shown that if n is even, then the graphs of $r = a \sin n\theta$ and $r = a \cos n\theta$ are $2n$-leaved roses. If n is odd, then the graphs of $r = a \sin n\theta$ and $r = a \cos n\theta$ are n-leaved roses. Other looping curves, such as the **lemniscate** (two leaves), are also common.

In a rectangular coordinate system the simplest type of equations to graph are

$$x = a \qquad \text{and} \qquad y = b$$

The graphs are straight lines: $x = a$ is a vertical line and $y = b$ is a horizontal line. A look ahead to Table 1 will show you that horizontal and vertical lines do not have such simple expressions in polar coordinates.

The easiest type of equations to graph in a polar coordinate system are found by setting the polar variables r and θ equal to constants,

$$r = a \qquad \text{and} \qquad \theta = b$$

Figure 8 illustrates two particular cases. Notice that a circle centered at the origin has a simpler expression in polar coordinates than in rectangular coordinates.

FIGURE 8

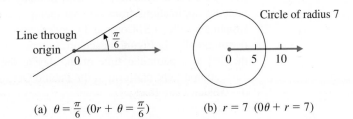

(a) $\theta = \frac{\pi}{6}$ ($0r + \theta = \frac{\pi}{6}$)

(b) $r = 7$ ($0\theta + r = 7$)

◆ Table of Standard Polar Curves

We now present a table of a few standard polar curves. Graphing polar equations is often made easier if you have an idea of what the shape of the curve will be.

TABLE 1
Some standard polar curves

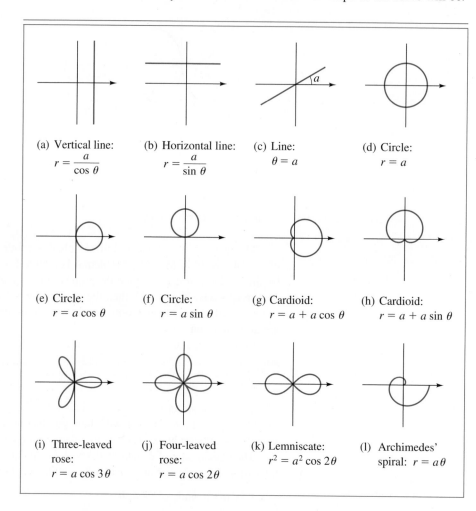

(a) Vertical line:
$$r = \frac{a}{\cos \theta}$$

(b) Horizontal line:
$$r = \frac{a}{\sin \theta}$$

(c) Line:
$$\theta = a$$

(d) Circle:
$$r = a$$

(e) Circle:
$$r = a \cos \theta$$

(f) Circle:
$$r = a \sin \theta$$

(g) Cardioid:
$$r = a + a \cos \theta$$

(h) Cardioid:
$$r = a + a \sin \theta$$

(i) Three-leaved rose:
$$r = a \cos 3\theta$$

(j) Four-leaved rose:
$$r = a \cos 2\theta$$

(k) Lemniscate:
$$r^2 = a^2 \cos 2\theta$$

(l) Archimedes' spiral: $r = a\theta$

◆ Application

Polar coordinate systems are useful in many types of applications: Exercise 7.2 includes applications from sailboat racing and astronomy; Figure 9, which was supplied by the U.S. Coast and Geodetic Survey, illustrates an application from oceanography. In Figure 9 each arrow represents the direction and magnitude of the tide at a particular time of the day at the San Francisco Light Station (a navigational light platform in the Pacific Ocean about 12 nautical miles west of the Golden Gate Bridge).

FIGURE 9
Tidal current curve, San Francisco
Light Station [*Note:* LL + 3ʰ means
3 hr after low low tide, HL − 1ʰ
means 1 hr before high low tide,
and so on.]

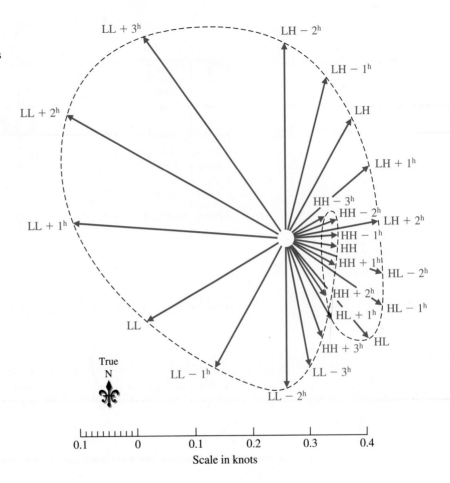

Scale in knots

Answers to
Matched Problems

1.

θ	r
0	0.00
$\pi/6$	1.57
$\pi/3$	3.14
$\pi/2$	4.71
$2\pi/3$	6.28
$5\pi/6$	7.85
π	9.42
$7\pi/6$	11.00

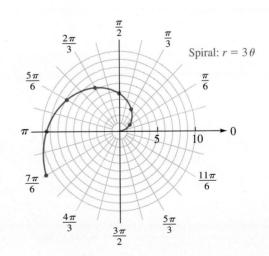

Spiral: $r = 3\theta$

2.

θ	r
0	0.00
$\pi/6$	2.00
$\pi/3$	3.46
$\pi/2$	4.00
$2\pi/3$	3.46
$5\pi/6$	2.00
π	0.00
$7\pi/6$	-2.00

Graph is repeating.

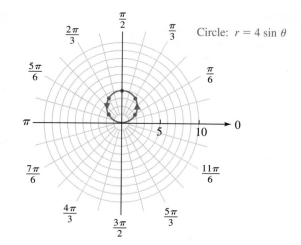

Circle: $r = 4 \sin \theta$

3.

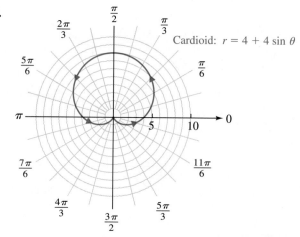

Cardioid: $r = 4 + 4 \sin \theta$

4.

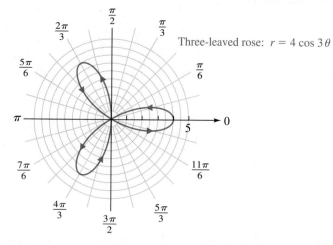

Three-leaved rose: $r = 4 \cos 3\theta$

EXERCISE 7.2

A *In Problems 1–8 use a calculator as an aid to point-by-point plotting.*

1. Graph $r = 10 \cos \theta$ by assigning θ the values 0, $\pi/6$, $\pi/4$, $\pi/3$, $\pi/2$, $2\pi/3$, $3\pi/4$, $5\pi/6$, and π. Then join the resulting points with a smooth curve.

2. Repeat Problem 1 for $r = 8 \sin \theta$, using the same set of values for θ.

3. Graph $r = 3 + 3 \cos \theta, 0° \leq \theta \leq 360°$, using multiples of 30° starting at 0.

4. Graph $r = 4 + 4 \sin \theta, 0° \leq \theta \leq 360°$, using multiples of 30° starting at 0.

5. Graph $r = \theta, 0 \leq \theta \leq 2\pi$, using multiples of $\pi/6$ for θ starting at $\theta = 0$.

6. Graph $r = \theta/2, 0 \leq \theta \leq 2\pi$, using multiples of $\pi/2$ for θ starting at $\theta = 0$.

7. Graph $r = 3/(\cos \theta), -\pi/2 < \theta < \pi/2$, using multiples of $\pi/6$.

8. Graph $r = 2/(\sin \theta), 0 < \theta < \pi$, using multiples of $\pi/6$.

In Problems 9–12 graph each polar equation.

9. $r = 5$
10. $r = 8$
11. $\theta = \pi/4$
12. $\theta = \pi/3$

B *In Problems 13–26 sketch each polar equation using rapid sketching techniques.*

13. $r = 4 \cos \theta$
14. $r = 4 \sin \theta$
15. $r = 8 \cos 2\theta$
16. $r = 10 \sin 2\theta$
17. $r = 6 \sin 3\theta$
18. $r = 5 \cos 3\theta$
19. $r = 3 + 3 \cos \theta$
20. $r = 2 + 2 \sin \theta$
21. $r = 2 + 4 \cos \theta$
22. $r = 2 + 4 \sin \theta$
23. $r = 4 - 2 \sin \theta$
24. $r = 4 - 2 \cos \theta$
25. $r = \theta/\pi, \quad \theta \geq 0$
26. $r\theta = \pi, \quad \theta \geq 0$

C *In Problems 27–30 sketch each polar equation using rapid sketching techniques.*

27. $r = 5 + 5 \cos(\theta/2)$
28. $r = 5 + 5 \sin(\theta/2)$
29. $r^2 = 64 \cos 2\theta$
30. $r^2 = 64 \sin 2\theta$

Precalculus *In Problems 31–34 graph each system of equations in the same polar coordinate system; then solve the system simultaneously. Note that any solution (r_1, θ_1) to the system must satisfy each equation in the system and thus identifies a point of intersection of the two graphs. However,* since each point in a polar coordinate system has infinitely many polar coordinates (see Section 7.1), graphs of two polar equations may intersect with no set of polar coordinates satisfying both equations simultaneously—that is, each equation may have a different set of polar coordinates at the point of intersection. For example, the polar equation $\theta = \pi/6$ goes through the pole at $(0, \pi/6)$, while $r = 4 \sin \theta$ goes through the pole at $(0, 0)$ and $(0, \pi)$, as shown in the figure. The simultaneous solution to this system of equations is $(2, \pi/6)$, which can easily be checked. This is a major difference between systems in polar coordinates and systems in rectangular coordinates.

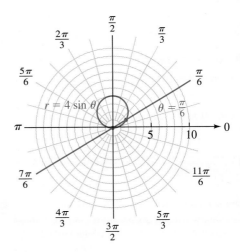

Figure for 31–34

31. $r = 2 \cos \theta$
    $r = 2 \sin \theta$
    $0 \leq \theta \leq \pi$

32. $r = 4 \cos \theta$
    $r = -4 \sin \theta$
    $0 \leq \theta \leq \pi$

33. $r = 8 \sin \theta$
    $r = 8 \cos 2\theta$
    $0° \leq \theta \leq 360°$

34. $r = 6 \cos \theta$
    $r = 6 \sin 2\theta$
    $0° \leq \theta \leq 360°$

In Problems 35–46 use a graphing calculator to graph each polar equation for the given information. Consult the manual to determine the proper procedures for drawing polar curves on your calculator.

35. "The graph of $r = 4 \cos n\theta$ is a $2n$-leaved rose when n is even." Verify this statement by creating polar graphs for $n = 2, 4,$ and 6.

36. "The graph of $r = 4 \sin n\theta$ is a $2n$-leaved rose when n is even." Verify this statement by creating polar graphs for $n = 2, 4,$ and 6.

37. "The graph of $r = 4 \sin^1 n\theta$ is an n-leaved rose for n odd." Verify this statement by creating polar graphs for $n = 3, 5,$ and 7.

38. "The graph of $r = 4 \cos n\theta$ is an n-leaved rose for n odd." Verify this statement by creating polar graphs for $n = 3, 5,$ and 7.

39. $r = 5 \cos \dfrac{\theta}{2}, \quad 0 \le \theta \le 4\pi$

40. $r = 5 \sin \dfrac{\theta}{2}, \quad 0 \le \theta \le 4\pi$

41. $r = 5 \sin \dfrac{\theta}{4}, \quad 0 \le \theta \le 8\pi$

42. $r = 5 \cos \dfrac{\theta}{4}, \quad 0 \le \theta \le 8\pi$

43. $r = \sqrt{\theta}, \quad 0 \le \theta \le 12\pi$

44. $r = 2\sqrt[3]{\theta}, \quad 0 \le \theta \le 10\pi$

45. $r = 4 \tan \theta, \quad 0 \le \theta \le 2\pi$

46. $r = 4 \cot \theta, \quad 0 \le \theta \le 2\pi$

Applications

Polar diagrams are used extensively by serious sailboat racers. The polar diagram in the figure shows the theoretical speeds that boats in the America's Cup competition (1991)

and the older 12-meter boats should have been able to achieve at different points of sail relative to a 16 knot wind. Problems 47 and 48 refer to this polar diagram.

47. Sailboat Racing   How fast, to the nearest half a knot, should the 1991 America's Cup boats have been able to sail going in the following directions relative to the wind?
(A) $30°$ (B) $60°$ (C) $90°$ (D) $120°$

48. Sailboat Racing   How fast, to the nearest half a knot, should the older 12-meter boats have been able to sail going in the following directions relative to the wind?
(A) $30°$ (B) $60°$ (C) $90°$ (D) $120°$

`C` **49. Conic Sections**   Using a graphing caculator, graph the equation

$$r = \frac{8}{1 - e \cos \theta}$$

for the following values of e and identify each curve as a hyperbola, ellipse, or a parabola.
(A) $e = 0.5$ (B) $e = 1$ (C) $e = 2$

`C` **50. Conic Sections**   Using a graphing calculator, graph the equation

$$r = \frac{4}{1 - e \cos \theta}$$

for the following values of e and identify each curve as a hyperbola, ellipse, or a parabola.
(A) $e = 0.7$ (B) $e = 1$ (C) $e = 1.3$

`C` **∗51. Astronomy**
(A) The planet Mercury travels around the sun in an elliptical orbit given approximately by

$$r = \frac{3.44 \times 10^7}{1 - 0.206 \cos \theta}$$

where r is in miles. Graph the orbit with the sun at the pole. Find the distance from Mercury to the sun at **aphelion** (greatest distance from the sun) and at **perihelion** (shortest distance from the sun).

(B) Kepler (1571–1630) showed that a line joining a planet to the sun swept out equal areas in space in equal intervals in time (see the figure). Use this information to determine whether a planet travels faster or slower at aphelion than at perihelion.

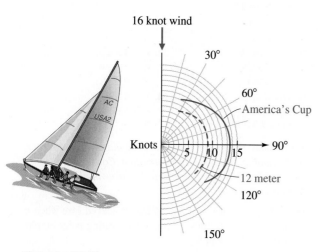

16 knot wind

30°

60°
America's Cup

Knots   90°
5   10   15

12 meter
120°

150°

Figure for 47, 48

Planet

Sun

Figure for 51

7.3 COMPLEX NUMBERS IN RECTANGULAR AND POLAR FORMS

- ◆ **Complex Numbers in Rectangular Form**
- ◆ **Complex Numbers in Polar Form**
- ◆ **Products and Quotients in Polar Form**
- ◆ **Historical Note**

◆ Complex Numbers in Rectangular Form

Since a complex number is any number that can be written in the form

$$a + bi$$

where a and b are real and i is the imaginary unit (see Appendix A.2), each complex number can be associated with a unique ordered pair of real numbers and vice versa. For example,

$$2 - 3i \quad \text{corresponds to} \quad (2, -3)$$

In general,

$$a + bi \quad \text{corresponds to} \quad (a, b)$$

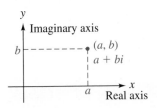

FIGURE 1
Complex plane

Therefore, each complex number can be associated with a unique point in a rectangular coordinate system, and each point in a rectangular coordinate system can be associated with a unique complex number (see Figure 1). The cartesian plane, with points associated with ordered pairs (a, b), is often called the **complex plane** when points (a, b) are associated with complex numbers $a + bi$. The x axis is then called the **real axis** and the y axis is called the **imaginary axis**. The complex numbers use all the points in a plane; the real numbers use all the points on a line.

 EXAMPLE 1   Plotting Complex Numbers

Plot each number in a complex plane.

(A) $3 + 4i$ (B) $-3 - 2i$ (C) -5 (D) $-3i$

SOLUTIONS

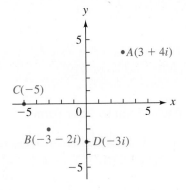

FIGURE 2

◆

MATCHED PROBLEM 1   Plot each number in a complex plane.
(A) $4 + 2i$ (B) $3 - 3i$ (C) -4 (D) $3i$

◆ Complex Numbers in Polar Form

Complex numbers can be changed from rectangular form to **polar (or trigonometric) form** by using the relationships developed in Section 7.1.

Thus, using

$$x = r \cos \theta \qquad \text{and} \qquad y = r \sin \theta$$

we can write (see Figure 3)

$$x + iy = r \cos \theta + ir \sin \theta$$
$$= r(\cos \theta + i \sin \theta)$$

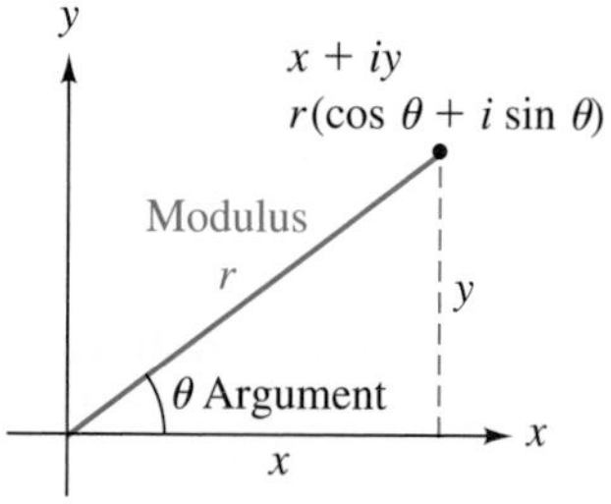

Since sine and cosine each have periods of 2π, we can write

$$\sin(\theta + 2k\pi) = \sin \theta, \qquad k \text{ any integer}$$
$$\cos(\theta + 2k\pi) = \cos \theta, \qquad k \text{ any integer}$$

Hence, we have the following more general polar forms.

FIGURE 3
Polar and rectangular forms

FROM RECTANGULAR TO POLAR FORM

$$z = x + iy = r[\cos(\theta + 2k\pi) + i \sin(\theta + 2k\pi)], \qquad k \text{ any integer}$$

The quadrant for θ is determined by x and y.

The number r is called the **modulus** or **absolute value** of z, denoted by **mod** z or $|z|$. The polar angle that the line joining z to the origin makes with the positive x axis is called the **argument** of z, denoted **arg** z. From Figure 3 we can see the following relationships.

MODULUS AND ARGUMENT FOR $z = x + iy$

$$\text{mod } z = r = |z| = \sqrt{x^2 + y^2}$$
$$\text{arg } z = \theta + 2k\pi, \qquad k \text{ any integer}$$

[**Note:** To write a complex number in polar form, we usually take the smallest positive angle for arg z.]

◆ EXAMPLE 2   From Rectangular Form to Polar Form

Write the following in polar form.

(A) $z_1 = 1 + i\sqrt{3}$ (B) $z_2 = -1 - i$ (C) $z_3 = 2i$

SOLUTIONS   It helps to graph these numbers in a rectangular coordinate system first; then if x and y are associated with special angles we can often determine r and θ by inspection.

(A) $z_1 = 2\left(\cos \dfrac{\pi}{3} + i \sin \dfrac{\pi}{3}\right)$

$r^2 = 1^2 + \sqrt{3}\,^2$
$r = \sqrt{4}$
$r = 2$

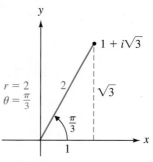

FIGURE 4

(B) $z_2 = \sqrt{2}\left(\cos \dfrac{5\pi}{4} + i \sin \dfrac{5\pi}{4}\right)$

$r^2 = -1^2 + -1^2$
$r = \sqrt{2}$

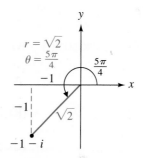

FIGURE 5

(C) $z_3 = 2\left(\cos \dfrac{\pi}{2} + i \sin \dfrac{\pi}{2}\right)$

$r^2 = 0^2 + 2^2$
$r = \sqrt{4}$
$r = 2$

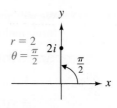

FIGURE 6

MATCHED PROBLEM 2   Write the following in polar form.

(A) $z_1 = \sqrt{3} + i$ (B) $z_2 = -1 + i$ (C) $z_3 = -5i$

◆ EXAMPLE 3   From Polar Form to Rectangular Form

Write the following in rectangular form.

(A) $z_1 = \sqrt{2}\left(\cos \dfrac{\pi}{4} + i \sin \dfrac{\pi}{4}\right)$ (B) $z_2 = 2(\cos 150° + i \sin 150°)$

(C) $z_3 = 5(\cos \pi + i \sin \pi)$

SOLUTIONS   (A) $z_1 = \sqrt{2}\left(\cos\dfrac{\pi}{4} + i\sin\dfrac{\pi}{4}\right)$

$\qquad = \sqrt{2}\left(\dfrac{1}{\sqrt{2}} + i\dfrac{1}{\sqrt{2}}\right)$

$\qquad = 1 + i$

(B) $z_2 = 2(\cos 150° + i\sin 150°)$

$\qquad = 2\left(-\dfrac{\sqrt{3}}{2} + i\dfrac{1}{2}\right)$

$\qquad = -\sqrt{3} + i$

(C) $z_3 = 5(\cos\pi + i\sin\pi)$

$\qquad = 5(-1 + 0i)$

$\qquad = -5$

MATCHED PROBLEM 3   Write the following in rectangular form.

(A) $z_1 = 4\left(\cos\dfrac{\pi}{6} + i\sin\dfrac{\pi}{6}\right)$

(B) $z_2 = 3\sqrt{2}(\cos 315° + i\sin 315°)$

(C) $z_3 = 6\left(\cos\dfrac{3\pi}{2} + i\sin\dfrac{3\pi}{2}\right)$

◆ **Products and Quotients in Polar Form**

We will now see a particular advantage in representing complex numbers in polar form: multiplication and division become easy. We start with multiplication.

$z_1 z_2 = [r_1(\cos\theta_1 + i\sin\theta_1)][r_2(\cos\theta_2 + i\sin\theta_2)]$

$\qquad = r_1 r_2(\cos\theta_1 + i\sin\theta_1)(\cos\theta_2 + i\sin\theta_2)$

$\qquad = r_1 r_2(\cos\theta_1\cos\theta_2 + i\cos\theta_1\sin\theta_2 + i\sin\theta_1\cos\theta_2 - \sin\theta_1\sin\theta_2)$

$\qquad = r_1 r_2[(\cos\theta_1\cos\theta_2 - \sin\theta_1\sin\theta_2) + i(\cos\theta_1\sin\theta_2 + \sin\theta_1\cos\theta_2)]$

$\qquad = r_1 r_2[\cos(\theta_1 + \theta_2) + i\sin(\theta_1 + \theta_2)]$ *Sum identities*

Thus, to multiply two complex numbers in polar form, multiply r_1 and r_2 and add θ_1 and θ_2. Similarly, it can be shown that

$\dfrac{z_1}{z_2} = \dfrac{r_1(\cos\theta_1 + i\sin\theta_1)}{r_2(\cos\theta_2 + i\sin\theta_2)}$

$\qquad = \dfrac{r_1}{r_2}[\cos(\theta_1 - \theta_2) + i\sin(\theta_1 - \theta_2)]$

That is, to divide the complex number z_1 by z_2, divide r_1 by r_2 and subtract θ_2 from θ_1. The proof of this quotient form is left to Problem 56 in Exercise 7.3. The product and quotient forms are summarized below for convenient reference.

PRODUCTS AND QUOTIENTS IN POLAR FORM

Given

$$z_1 = r_1(\cos \theta_1 + i \sin \theta_1)$$

and

$$z_2 = r_2(\cos \theta_2 + i \sin \theta_2),$$

then

$$z_1 z_2 = r_1 r_2 [\cos(\theta_1 + \theta_2) + i \sin(\theta_1 + \theta_2)]$$

$$\frac{z_1}{z_2} = \frac{r_1}{r_2} [\cos(\theta_1 - \theta_2) + i \sin(\theta_1 - \theta_2)]$$

EXAMPLE 4

Products and Quotients

If $z_1 = 8(\cos 50° + i \sin 50°)$ and $z_2 = 4(\cos 30° + i \sin 30°)$, find $z_1 z_2$ and z_1/z_2.

SOLUTIONS

$$z_1 z_2 = (8 \cdot 4)[\cos(50° + 30°) + i \sin(50° + 30°)]$$

$$= 32(\cos 80° + i \sin 80°)$$

$$\frac{z_1}{z_2} = \frac{8}{4} [\cos(50° - 30°) + i \sin(50° - 30°)]$$

$$= 2(\cos 20° + i \sin 20°)$$

MATCHED PROBLEM 4

If $z_1 = 21(\cos 140° + i \sin 140°)$ and $z_2 = 3(\cos 105° + i \sin 105°)$, find $z_1 z_2$ and z_1/z_2.

◆ Historical Note

There is hardly an area in mathematics that does not have some imprint of the famous Swiss mathematician Leonhard Euler (1707–1783) who spent most of his productive life at the New St. Petersburg Academy in Russia and the Prussian Academy in Berlin. Probably the most prolific writer in the history of the subject, he is credited with making the following familiar notations standard.

$f(x)$ function notation

e natural logarithmic base

i imaginary unit, $\sqrt{-1}$

For our immediate interest, he is also responsible for the extraordinary relationship

$$e^{i\theta} = \cos \theta + i \sin \theta$$

If we let $\theta = \pi$, we obtain an equation that relates five of the most important numbers in the history of mathematics:

$$e^{i\pi} + 1 = 0$$

Answers to
Matched Problems

1.

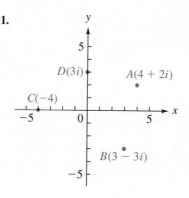

2. (A) $z_1 = 2\left(\cos \dfrac{\pi}{6} + i \sin \dfrac{\pi}{6}\right)$

(B) $z_2 = \sqrt{2}\left(\cos \dfrac{3\pi}{4} + i \sin \dfrac{3\pi}{4}\right)$

(C) $z_3 = 5\left(\cos \dfrac{3\pi}{2} + i \sin \dfrac{3\pi}{2}\right)$

3. (A) $z_1 = 2\sqrt{3} + 2i$ (B) $z_2 = 3 - 3i$ (C) $z_3 = -6i$

4. $z_1 z_2 = 63(\cos 245° + i \sin 245°)$
$z_1/z_2 = 7(\cos 35° + i \sin 35°)$

EXERCISE 7.3

A *Plot each set of complex numbers in a complex plane.*

1. $A = 4 + 5i$; $B = -3 + 4i$; $C = -3i$

2. $A = 3 - 2i$; $B = -4 - 2i$; $C = 5i$

3. $A = 4 + i$; $B = -2 + 3i$; $C = -4$

4. $A = -3 - i$; $B = 5 - 4i$; $C = 4$

5. $A = 8[\cos(\pi/4) + i \sin(\pi/4)]$
$B = 6[\cos(\pi/2) + i \sin(\pi/2)]$
$C = 3[\cos(\pi/6) + i \sin(\pi/6)]$

6. $A = 5[\cos(5\pi/6) + i \sin(5\pi/6)]$
$B = 3[\cos(3\pi/2) + i \sin(3\pi/2)]$
$C = 4[\cos(7\pi/4) + i \sin(7\pi/4)]$

7. $A = 5(\cos 270° + i \sin 270°)$
$B = 4(\cos 60° + i \sin 60°)$
$C = 8(\cos 150° + i \sin 150°)$

8. $A = 3(\cos 310° + i \sin 310°)$
$B = 4(\cos 180° + i \sin 180°)$
$C = 5(\cos 210° + i \sin 210°)$

Find z_1z_2 and z_1/z_2. Write answers in polar form.

9. $z_1 = 10(\cos 45° + i \sin 45°)$
$z_2 = 5(\cos 32° + i \sin 32°)$

10. $z_1 = 4(\cos 62° + i \sin 62°)$
$z_2 = 2(\cos 51° + i \sin 51°)$

11. $z_1 = 7(\cos 163° + i \sin 163°)$
$z_2 = 3(\cos 102° + i \sin 102°)$

12. $z_1 = 5(\cos 204° + i \sin 204°)$
$z_2 = 2(\cos 34° + i \sin 34°)$

B *Change to polar form.*

13. $1 + i$ **14.** $-1 - i$ **15.** $-\sqrt{3} + i$

16. $1 + i\sqrt{3}$ **17.** $4i$ **18.** -4

19. $-1 - i\sqrt{3}$ **20.** $\sqrt{3} - i$ **21.** $2 - i2\sqrt{3}$

22. $-3 + 3i$ **23.** $-8i$ **24.** 10

Change to rectangular form.

25. $\sqrt{2}(\cos 45° + i \sin 45°)$

26. $4[\cos(\pi/3) + i \sin(\pi/3)]$

27. $\sqrt{2}(\cos 135° + i \sin 135°)$

28. $6(\cos 150° + i \sin 150°)$

29. $8(\cos \pi + i \sin \pi)$

30. $7(\cos 0 + i \sin 0)$

31. $12(\cos 90° + i \sin 90°)$

32. $11(\cos 270° + i \sin 270°)$

33. $6[\cos(4\pi/3) + i \sin(4\pi/3)]$

34. $2\sqrt{2}[\cos(5\pi/4) + i \sin(5\pi/4)]$

35. $4(\cos 330° + i \sin 330°)$

36. $8(\cos 300° + i \sin 300°)$

Find each of the following directly and by using polar forms. Write answers in $a + bi$ form and in $r(\cos \theta + i \sin \theta)$ form.

37. $(1 + i)^2$ **38.** $(-1 + i)^2$

39. $(1 + i\sqrt{3})(\sqrt{3} + i)$ **40.** $(-1 + i)(1 + i)$

C *Change to polar form, with $r \geq 0$ and $0° \leq \theta < 360°$. Compute θ to one decimal place and leave r in exact radical form.*

41. $3 + 5i$ **42.** $5 + 6i$ **43.** $-7 + 3i$

44. $-4 - 9i$ **45.** $6 - 5i$ **46.** $3 - 7i$

Change Problems 47–52 to rectangular form $a + bi$. Compute a and b to three significant digits.

47. $9(\cos 37°20' + i \sin 37°20')$

48. $7(\cos 23°40' + i \sin 23°40')$

49. $5(\cos 197.2° + i \sin 197.2°)$

50. $3(\cos 133.8° + i \sin 133.8°)$

51. $11(\cos 321°20' + i \sin 321°20')$

52. $10(\cos 305°30' + i \sin 305°30')$

53. If $z = r(\cos \theta + i \sin \theta)$, show that

$$z^2 = r^2(\cos 2\theta + i \sin 2\theta)$$
$$z^3 = r^3(\cos 3\theta + i \sin 3\theta)$$

What do you think z^n is for n a natural number?

54. Show that $r^{1/2}[\cos(\theta/2) + i \sin(\theta/2)]$ is a square root of $r(\cos \theta + i \sin \theta)$. [*Hint:* Square the first expression to see what happens.]

55. Show that $r^{1/n}[\cos(\theta/n) + i \sin(\theta/n)]$, for n a natural number, is an nth root of $r(\cos \theta + i \sin \theta)$, assuming $z^n = r^n(\cos n\theta + i \sin n\theta)$.

56. Prove that

$$\frac{z_1}{z_2} = \frac{r_1(\cos \theta_1 + i \sin \theta_1)}{r_2(\cos \theta_2 + i \sin \theta_2)}$$

$$= \frac{r_1}{r_2}[\cos(\theta_1 - \theta_2) + i \sin(\theta_1 - \theta_2)]$$

  Applications

57. Resultant Force   An object is located at the pole, and two forces F_1 and F_2 act upon the object. Let the forces be vectors going from the pole to the complex numbers $8(\cos 0° + i \sin 0°)$ and $6(\cos 30° + i \sin 30°)$, respectively. [Force F_1 has a magnitude of 1 lb at a direction of $0°$, and force F_2 has a magnitude of 6 lb at a direction of $30°$.)

(A) Convert the trigonometric forms of these complex numbers to rectangular form and add.

(B) Convert the sum from part (A) back to trigonometric form (to three significant digits).

(C) The vector going from the pole to the complex number in part (B) is the resultant of the two original forces. What is its magnitude and direction?

58. Resultant Force   Repeat Problem 57 with forces F_1 and F_2 associated with the complex numbers

$$20(\cos 0° + i \sin 0°) \quad \text{and} \quad 10(\cos 60° + i \sin 60°)$$

7.4 DE MOIVRE'S THEOREM AND THE nTH ROOT THEOREM

◆   **De Moivre's Theorem**

◆   **The nth Root Theorem**

In this section we come to one of the great theorems in mathematics, *De Moivre's theorem.* De Moivre (1667–1754), of French birth, spent most of his life in London doing private tutoring, writing, and publishing mathematics. He became a close friend of Isaac Newton and belonged to many prestigious professional societies in England, France, and Germany.

De Moivre's theorem enables us to find the power of a complex number in polar (or trigonometric) form very easily. More importantly, the theorem is the basis for the *nth root theorem,* which enables us to find all n roots of any complex number, real or imaginary. How many roots does the equation $x^3 = 1$ have? As was mentioned in the introduction to this chapter, you may think that 1 is the only root. The equation actually has three roots—one real and two imaginary (see Example 3).

◆ De Moivre's Theorem

We now apply the product formula for polar forms to powers z^2, z^3, . . .

$$z^2 = (x + iy)^2$$
$$= [r(\cos \theta + i \sin \theta)][r(\cos \theta + i \sin \theta)]$$
$$= r^2(\cos 2\theta + i \sin 2\theta)$$
$$z^3 = (x + iy)^3$$
$$= [r^2(\cos 2\theta + i \sin 2\theta)][r(\cos \theta + i \sin \theta)]$$
$$= r^3(\cos 3\theta + i \sin 3\theta)$$

You can probably guess what z^4 would be. If you guessed

$$z^4 = r^4(\cos 4\theta + i \sin 4\theta)$$

you are right! Perhaps you are brave enough to jump to the general case for z^n, n any natural number. If so, you have discovered De Moivre's famous theorem:

DE MOIVRE'S THEOREM

For n a natural number,

$$z^n = (x + iy)^n$$
$$= [r(\cos \theta + i \sin \theta)]^n$$
$$= r^n(\cos n\theta + i \sin n\theta)$$

(A general proof of De Moivre's theorem requires a technique called *mathematical induction,* a topic usually considered in an advanced algebra course.)

♦ **EXAMPLE 1**   Finding a Power of a Complex Number

Find $(\sqrt{3} + i)^{13}$ and write the answer in the form $a + bi$.

SOLUTION   Write $\sqrt{3} + i$ in polar form, use De Moivre's theorem, and then convert back to rectangular form.

$$
\begin{aligned}
(\sqrt{3} + i)^{13} &= [2(\cos 30° + i \sin 30°)]^{13} \\
&= 2^{13}(\cos 390° + i \sin 390°) \\
&= 8{,}192(\cos 30° + i \sin 30°) \qquad \text{Why?} \\
&= 8{,}192\left(\frac{\sqrt{3}}{2} + i\frac{1}{2}\right) \\
&= 4{,}096\sqrt{3} + 4{,}096i
\end{aligned}
$$

♦

MATCHED PROBLEM 1   Find $(-1 + i)^6$ and write the answer in the form $a + bi$.

♦ **The *n*th Root Theorem**

Now let us take a look at roots of complex numbers. We say *w* **is an *n*th root of *z*,** *n* a natural number, if

$$w^n = z$$

For example, if $w^2 = z$, then *w* is a square root of *z*; if $w^3 = z$, then *w* is a cube root of *z*; and so on. In particular, if

$$z = r(\cos \theta + i \sin \theta) \tag{1}$$

then

$$r^{1/2}\left(\cos \frac{\theta}{2} + i \sin \frac{\theta}{2}\right) \tag{2}$$

is a square root of *z*. We simply square expression (2) using De Moivre's theorem to obtain (1):

$$
\left[r^{1/2}\left(\cos \frac{\theta}{2} + i \sin \frac{\theta}{2}\right)\right]^2 = (r^{1/2})^2\left(\cos 2 \cdot \frac{\theta}{2} + i \sin 2 \cdot \frac{\theta}{2}\right)
$$
$$
= r(\cos \theta + i \sin \theta)
$$

We can proceed in the same way to show that

$$r^{1/n}\left(\cos \frac{\theta}{n} + i \sin \frac{\theta}{n}\right)$$

is an *n*th root of $r(\cos \theta + i \sin \theta)$. But we can do even better than this: The following theorem shows us how to find all the *n*th roots of a complex number.

nTH ROOT THEOREM

$$r^{1/n}\left(\cos \frac{\theta + k360°}{n} + i \sin \frac{\theta + k360°}{n}\right) \qquad k = 0, 1, \ldots, (n - 1)$$

are n distinct nth roots of $z = r(\cos \theta + i \sin \theta)$, and there are no others.

The proof of this theorem is left to Problems 25 and 26 in Exercise 7.4.

◆ **EXAMPLE 2**   Roots of a Complex Number

Find the six distinct sixth roots of $1 + i\sqrt{3}$ and graph them.

SOLUTION   First, write $1 + i\sqrt{3}$ in polar form.

$$1 + i\sqrt{3} = 2(\cos 60° + i \sin 60°)$$

All six roots are given by

$$2^{1/6}\left(\cos \frac{60° + k360°}{6} + i \sin \frac{60° + k360°}{6}\right), \qquad k = 0, 1, 2, 3, 4, 5$$

Thus,

$$w_1 = 2^{1/6}\left(\cos \frac{60° + \mathbf{0} \cdot 360°}{6} + i \sin \frac{60° + \mathbf{0} \cdot 360°}{6}\right)$$
$$= 2^{1/6}(\cos 10° + i \sin 10°)$$

$$w_2 = 2^{1/6}\left(\cos \frac{60° + \mathbf{1} \cdot 360°}{6} + i \sin \frac{60° + \mathbf{1} \cdot 360°}{6}\right)$$
$$= 2^{1/6}(\cos 70° + i \sin 70°)$$

$$w_3 = 2^{1/6}\left(\cos \frac{60° + \mathbf{2} \cdot 360°}{6} + i \sin \frac{60° + \mathbf{2} \cdot 360°}{6}\right)$$
$$= 2^{1/6}(\cos 130° + i \sin 130°)$$

$$w_4 = 2^{1/6}\left(\cos \frac{60° + \mathbf{3} \cdot 360°}{6} + i \sin \frac{60° + \mathbf{3} \cdot 360°}{6}\right)$$
$$= 2^{1/6}(\cos 190° + i \sin 190°)$$

$$w_5 = 2^{1/6}\left(\cos \frac{60° + \mathbf{4} \cdot 360°}{6} + i \sin \frac{60° + \mathbf{4} \cdot 360°}{6}\right)$$
$$= 2^{1/6}(\cos 250° + i \sin 250°)$$

$$w_6 = 2^{1/6}\left(\cos \frac{60° + \mathbf{5} \cdot 360°}{6} + i \sin \frac{60° + \mathbf{5} \cdot 360°}{6}\right)$$
$$= 2^{1/6}(\cos 310° + i \sin 310°)$$

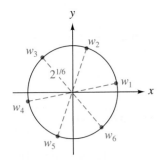

FIGURE 1

The six roots are equally spaced around a circle in the complex plane with radius $2^{1/6}$ (see Figure 1).   ◆

MATCHED PROBLEM 2   Find the three distinct cube roots of $1 + i$.

◆   **EXAMPLE 3**   Solving a Cubic Equation

Solve $x^3 - 1 = 0$. Write final answers in rectangular form.

SOLUTION $x^3 - 1 = 0$

$$x^3 = 1$$

Therefore, x is a cube root of 1, and there are three of them. First, we write 1 in polar form.

$$1 = 1 + 0i = 1(\cos 0° + i \sin 0°)$$

All three cube roots of 1 are given by

$$1^{1/3}\left(\cos \frac{0° + k360°}{3} + i \sin \frac{0° + k360°}{3}\right), \quad k = 0, 1, 2$$

Thus,

$$w_1 = 1^{1/3}\left(\cos \frac{0° + \mathbf{0} \cdot 360°}{3} + i \sin \frac{60° + \mathbf{0} \cdot 360°}{3}\right)$$

$$= 1(\cos 0° + i \sin 0°)$$

$$= 1$$

$$w_2 = 1^{1/3}\left(\cos \frac{0° + \mathbf{1} \cdot 360°}{3} + i \sin \frac{0° + \mathbf{1} \cdot 360°}{3}\right)$$

$$= (\cos 120° + i \sin 120°)$$

$$= -\frac{1}{2} + \frac{\sqrt{3}}{2}i$$

$$w_3 = 1^{1/3}\left(\cos \frac{0° + \mathbf{2} \cdot 360°}{3} + i \sin \frac{0° + \mathbf{2} \cdot 360°}{3}\right)$$

$$= (\cos 240° + i \sin 240°)$$

$$= -\frac{1}{2} - \frac{\sqrt{3}}{2}i$$

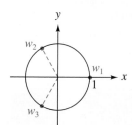

FIGURE 2

The three roots are equally spaced around a unit circle in the complex plane (see Figure 2).   ◆

MATCHED PROBLEM 3   Solve $x^3 + 1 = 0$. Write final answers in rectangular form.

We have only touched on a subject that has far-reaching consequences. The theory of functions of a complex variable provides a powerful tool for engineers, scientists, and mathematicians.

Answers to
Matched Problems

1. $8i$

2. $w_1 = 2^{1/6}(\cos 15° + i \sin 15°)$
$w_2 = 2^{1/6}(\cos 135° + i \sin 135°)$
$w_3 = 2^{1/6}(\cos 255° + i \sin 255°)$

3. $w_1 = \dfrac{1}{2} + \dfrac{\sqrt{3}}{2} i, \quad w_2 = -1, \quad w_3 = \dfrac{1}{2} - \dfrac{\sqrt{3}}{2} i$

EXERCISE 7.4

A *Find the value of each expression using De Moivre's theorem. Leave your answer in polar form.*

1. $[3(\cos 15° + i \sin 15°)]^3$

2. $[2(\cos 30° + i \sin 30°)]^8$

3. $[\sqrt{2}(\cos 45° + i \sin 45°)]^{10}$

4. $[\sqrt{2}(\cos 60° + i \sin 60°)]^8$

5. $(\sqrt{3} + i)^6$ **6.** $(1 + i\sqrt{3})^3$

B *Find the value of each expression using De Moivre's theorem, and write the result in the form $a + bi$.*

7. $(-1 + i)^4$ **8.** $(-\sqrt{3} - i)^4$

9. $(-\sqrt{3} + i)^5$ **10.** $(1 - i)^8$

11. $\left(-\dfrac{1}{2} - \dfrac{\sqrt{3}}{2} i\right)^3$ **12.** $\left(-\dfrac{1}{2} + \dfrac{\sqrt{3}}{2} i\right)^3$

For n and z as indicated, find all nth roots of z. Leave your answer in polar form.

13. $z = 4(\cos 30° + i \sin 30°), \quad n = 2$

14. $z = 16(\cos 60° + i \sin 60°), \quad n = 2$

15. $z = 8(\cos 90° + i \sin 90°), \quad n = 3$

16. $z = 27(\cos 120° + i \sin 120°), \quad n = 3$

17. $z = -1 + i, \quad n = 5$ **18.** $z = 1 - i, \quad n = 5$

19. $z = 1, \quad n = 6$ **20.** $z = i, \quad n = 3$

In Problems 21–24 solve each equation for all roots. Write final answers in rectangular form $a + bi$, where a and b are in exact form.

21. $x^3 - 8 = 0$ **22.** $x^3 + 8 = 0$

23. $x^3 + 27 = 0$ **24.** $x^3 - 27 = 0$

C **25.** Show that

$$\left[r^{1/n}\left(\cos \frac{\theta + k360°}{n} + i \sin \frac{\theta + k360°}{n}\right)\right]^n = r(\cos \theta + i \sin \theta)$$

for n any natural number and k any integer.

26. Show that

$$r^{1/n}\left(\cos \frac{\theta + k360°}{n} + i \sin \frac{\theta + k360°}{n}\right)$$

is the same number for $k = 0$ and $k = n$.

Solve each equation for all roots. Write final answers in rectangular form $a + bi$, where a and b are computed to three decimal places. A calculator will be helpful.

27. $x^5 - 1 = 0$ **28.** $x^4 + 1 = 0$

29. $x^3 + 5 = 0$ **30.** $x^5 - 6 = 0$

CHAPTER 7 SUMMARY

7.1
POLAR AND RECTANGULAR
COORDINATES

Figure 1 illustrates a **polar coordinate** system superimposed on a rectangular coordinate system. The fixed point O is called the **pole** or **origin** and the horizontal half-line is the **polar axis**. The **polar-rectangular relationships** are used to transform coordinates and equations from one system to the other.

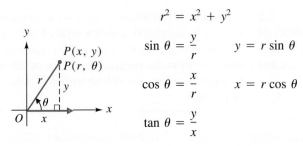

FIGURE 1
Polar-rectangular
relationships

$$r^2 = x^2 + y^2$$

$$\sin \theta = \frac{y}{r} \qquad y = r \sin \theta$$

$$\cos \theta = \frac{x}{r} \qquad x = r \cos \theta$$

$$\tan \theta = \frac{y}{x}$$

[*Note:* The signs of x and y determine the quadrant for θ. The angle θ is usually chosen so that $|\theta|$ is minimum.]

7.2

SKETCHING POLAR EQUATIONS

Polar graphs can be obtained by **point-by-point** plotting of points on the graph, which are usually computed with a calculator. **Rapid graphing techniques** use the uniform variation of $\sin \theta$ and $\cos \theta$ to quickly produce rough sketches of a polar graph. Table 1 illustrates some standard polar curves.

TABLE 1
Some standard polar curves

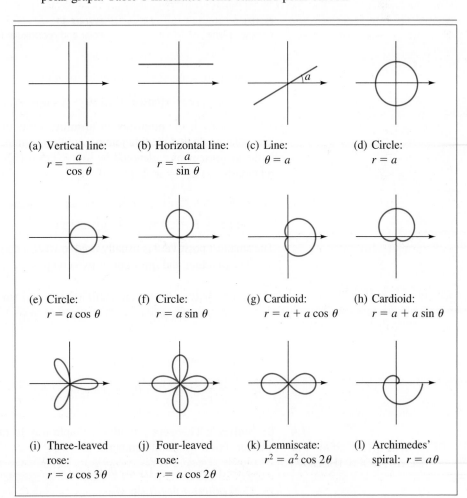

(a) Vertical line:
$$r = \frac{a}{\cos \theta}$$

(b) Horizontal line:
$$r = \frac{a}{\sin \theta}$$

(c) Line:
$$\theta = a$$

(d) Circle:
$$r = a$$

(e) Circle:
$$r = a \cos \theta$$

(f) Circle:
$$r = a \sin \theta$$

(g) Cardioid:
$$r = a + a \cos \theta$$

(h) Cardioid:
$$r = a + a \sin \theta$$

(i) Three-leaved rose:
$$r = a \cos 3\theta$$

(j) Four-leaved rose:
$$r = a \cos 2\theta$$

(k) Lemniscate:
$$r^2 = a^2 \cos 2\theta$$

(l) Archimedes'
spiral: $r = a\theta$

7.3
COMPLEX NUMBERS IN
RECTANGULAR AND POLAR
FORM

If a and b are real numbers and i is the imaginary unit, then the **complex number** $a + bi$ is associated with the ordered pair (a, b). The cartesian plane is called the **complex plane**, the x axis is the **real axis**, the y axis is the **imaginary axis**, and $a + bi$ is the **rectangular form** of the number, as illustrated in Figure 2. Complex numbers can also be written in **polar, or trigonometric, form** as shown in Figure 3.

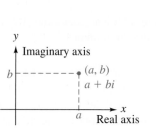

FIGURE 2
Complex plane

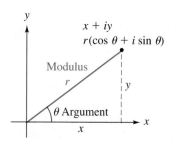

FIGURE 3
Polar and rectangular forms

The more general polar form is given by

$$z = x + iy = r[\cos(\theta + 2k\pi) + i\sin(\theta + 2k\pi)], \qquad k \text{ any integer}$$

The number r is the **modulus**, or **absolute value** of z, denoted by **mod** z or $|z|$. The polar angle that the line joining z to the origin makes with the positive x axis is the **argument** of z, denoted by **arg** z. From Figure 3 we have the following relationships for $z = x + iy$:

$$\text{mod } z = r = |z| = \sqrt{x^2 + y^2}$$

$$\arg z = \theta + 2k\pi, \qquad k \text{ any integer}$$

The smallest positive θ is usually used to write a complex number in polar form.

The **product** and **quotient** of two complex numbers in polar form are given by

$$z_1 z_2 = [r_1(\cos \theta_1 + i\sin \theta_1)][r_2(\cos \theta_2 + i\sin \theta_2)]$$
$$= r_1 r_2[\cos(\theta_1 + \theta_2) + i\sin(\theta_1 + \theta_2)]$$

$$\frac{z_1}{z_2} = \frac{r_1(\cos \theta_1 + i\sin \theta_1)}{r_2(\cos \theta_2 + i\sin \theta_2)}$$

$$= \frac{r_1}{r_2}[\cos(\theta_1 - \theta_2) + i\sin(\theta_1 - \theta_2)]$$

7.4
DE MOIVRE'S THEOREM AND
THE nTH ROOT THEOREM

De Moivre's Theorem provides a simple way to find the power of a complex number. For n a natural number,

$$z^n = (x + iy)^n = [r(\cos \theta + i\sin \theta)]^n$$
$$= r^n(\cos n\theta + i\sin n\theta)$$

The *n*th root theorem states that

$$r^{1/n}\left(\cos\frac{\theta + k360°}{n} + i\sin\frac{\theta + k360°}{n}\right), \qquad k = 0, 1, \ldots, (n-1)$$

are *n* distinct *n*th roots of $z = r(\cos\theta + i\sin\theta)$, and there are no others.

CHAPTER 7 REVIEW EXERCISE

Work through all the problems in this chapter review and check the answers. Answers to all review problems appear in the back of the book; following each answer is an italic number that indicates the section in which that type of problem is discussed. Where weaknesses show up, review the appropriate sections in the text.

A *In Problems 1–4 plot in a polar coordinate system.*

1. $A(5, 210°)$; $B(-7, 180°)$; $C(-5, -45°)$

2. $r = 5\sin\theta$

3. $r = 4 + 4\cos\theta$

4. $r = 8$

5. Change $(2\sqrt{2}, \pi/4)$ to rectangular coordinates.

6. Change $(-\sqrt{3}, 1)$ to polar coordinates $(r \geq 0, 0 \leq \theta < 2\pi)$.

7. Graph $-3 - 2i$ in a rectangular coordinate system.

8. Graph $z = 5(\cos 60° + i\sin 60°)$ in a polar coordinate system.

9. Find $z_1 z_2$ and z_1/z_2 for $z_1 = 9(\cos 42° + i\sin 42°)$ and $z_2 = 3(\cos 37° + i\sin 37°)$. Leave answers in polar form.

10. Find $[2(\cos 10° + i\sin 10°)]^4$ using De Moivre's theorem.

B *In Problems 11–14 plot in a polar coordinate system.*

11. $A(-5, \pi/4)$; $B(5, -\pi/3)$; $C(-8, 4\pi/3)$

12. $r = 8\sin 3\theta$

13. $r = 4\sin 2\theta$

14. $\theta = \pi/6$

15. Change $8x - y^2 = x^2$ to polar form.

16. Change $r(3\cos\theta - 2\sin\theta) = -2$ to rectangular form.

17. Change $r = -3\cos\theta$ to rectangular form.

18. Convert $-\sqrt{3} - i$ to polar form $(r \geq 0, 0° \leq \theta < 360°)$.

19. Convert $3\sqrt{2}[\cos(3\pi/4) + i\sin(3\pi/4)]$ to rectangular form.

20. Convert $(2 + i2\sqrt{3})(-\sqrt{2} + i\sqrt{2})$ to polar form and evaluate.

21. Divide $(-\sqrt{2} + i\sqrt{2})/(2 + i2\sqrt{3})$ by converting to polar form first.

22. Find $(-1 - i)^4$ using De Moivre's theorem, and write the result in the form $a + bi$.

23. Solve $x^3 - 64 = 0$ for all roots, and write answers in the form $a + bi$, where a and b are exact.

24. Find all cube roots of $-4\sqrt{3} - 4i$. Write answers in polar form.

25. Show that $2(\cos 30° + i\sin 30°)$ is a square root of $2 + i2\sqrt{3}$.

C 26. Graph $r = 6/(2 - \cos\theta)$, $0 \leq \theta \leq 2\pi$, using multiples of $\pi/6$ starting at 0.

27. Plot $r = 4 + 4\cos(\theta/2)$ in a polar coordinate system.

28. Change $r(\sin\theta - 2) = 3$ to rectangular form.

29. Find all roots of $x^3 - 12 = 0$. Write answers in rectangular form $a + bi$, where a and b are computed to three decimal places. (Use a calculator.)

30. Show that

$$\left[r^{1/3}\left(\cos\frac{\theta + k360°}{3} + i\sin\frac{\theta + k360°}{3}\right)\right]^3$$
$$= r(\cos\theta + i\sin\theta), \qquad k = 0, 1, 2$$

c *In Problems 31–34, use a graphing calculator to produce the indicated polar graphs.*

31. Graph $r = 5\cos\dfrac{\theta}{5}$ for $0 \leq \theta \leq 5\pi$.

32. Graph $r = 5\cos\dfrac{\theta}{7}$ for $0 \leq \theta \leq 7\pi$.

33. Graph $r = 5(\sin\theta)^{2n}$, for $n = 1, 2$, and 3. How many leaves do you expect the graph will have for arbitrary n?

34. Graph $r = 2/(1 - e\sin\theta)$ for the following values of e and identify each curve as a hyperbola, an ellipse, or a parabola.
    (A) $e = 1.6$ (B) $e = 1$ (C) 0.4

Work through all the problems in this cumulative review and check the answers. Answers to all review problems appear in the back of the book; following each answer is an italic number that indicates the section in which that type of problem is discussed. Where weaknesses show up, review the appropriate sections in the text.

Where applicable, quantities in the problems refer to a triangle labeled as in the figure.

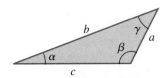

A 1. Which angle has the larger measure: $\alpha = 2\pi/7$ or $\beta = 51°25'40''$?

2. The sides of a right triangle are 1.27 cm and 4.65 cm. Find the hypotenuse and the acute angles in degree measure.

3. Find the values of $\sec\theta$ and $\tan\theta$ if the terminal side of θ contains $P(7, -24)$.

4. Verify the following identities without looking at a table of identities.
  (A) $\cot x \sec x \sin x = 1$
  (B) $\tan\theta + \cot\theta = \sec\theta \csc\theta$

Evaluate Problems 5–8 exactly as real numbers.

5. $\sin\dfrac{11\pi}{6}$

6. $\tan\dfrac{-5\pi}{3}$

7. $\cos^{-1}(-0.5)$

8. $\csc^{-1}(\sqrt{2})$

Evaluate Problems 9–12 as real numbers to four significant digits using a calculator.

9. $\sin 43°22'$

10. $\cot\dfrac{2\pi}{5}$

11. $\sin^{-1}(0.8)$

☆ **12.** $\sec^{-1}(4.5)$

☆ **13.** Write $\sin 3t + \sin t$ as a product.

In Problems 14–17 solve each triangle, given the indicated measures of angles and sides.

14. $\alpha = 52°, \quad \beta = 47°, \quad c = 28$ cm

15. $\beta = 110°, \quad \gamma = 42°, \quad b = 68$ m

16. $\beta = 34°, \quad a = 16$ in., $\quad c = 24$ in.

17. $a = 18$ ft, $\quad b = 23$ ft, $\quad c = 32$ ft

☆ **18.** Find the area of the triangle in Problem 16.

☆ **19.** Find the area of the triangle in Problem 17.

20. Find the magnitude of the horizontal and vertical components of the vector **v** located in a coordinate system as indicated in the figure.

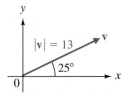

Figure for 20

21. Two vectors **u** and **v** are located in a coordinate system, as indicated in the figure. Find the magnitude and the direction (relative to the x axis) of $\mathbf{u} + \mathbf{v}$ if $|\mathbf{u}| = 6.4$ and $|\mathbf{v}| = 3.9$.

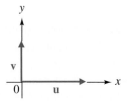

Figure for 21

22. Given $A(4, -2)$ and $B(-3, 7)$, represent the geometric vector $\overrightarrow{AB}$ as an algebraic vector and find its magnitude.

☆ **23.** Find the angle, in decimal degrees to one decimal place, between $\mathbf{u} = 2\mathbf{i} - 7\mathbf{j}$ and $\mathbf{v} = 3\mathbf{i} + 8\mathbf{j}$.

24. Plot in a polar coordinate system:
$A(6, 240°) \qquad B(-4, 225°)$
$C(9, -45°) \qquad D(-7, -60°)$

25. Plot in a polar coordinate system: $r = 5 + 5\sin\theta$.

26. Change $(3\sqrt{2}, 3\pi/4)$ to rectangular coordinates.

27. Change $(-2\sqrt{3}, 2)$ to polar coordinates.

28. Change $z = 2 - 2i$ to polar form.

29. Change $z = 3[\cos(3\pi/2) + i\sin(3\pi/2)]$ to rectangular form.

30. Find z_1z_2 and z_1/z_2 for $z_1 = 3(\cos 50° + i \sin 50°)$ and $z_2 = 5(\cos 15° + i \sin 15°)$. Leave answers in polar form.

31. Find $[3(\cos 25° + i \sin 25°)]^4$ using De Moivre's theorem. Leave answer in polar form.

B 32. Find x exactly and θ to the nearest $10'$ in the figure.

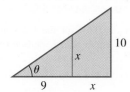

10

x

θ

9   x **Figure for 32**

33. Find the exact values of $\csc \theta$ and $\cos \theta$ if $\cot \theta = 4$ and $\sin \theta < 0$.

34. Graph $y = 1 + 2 \sin(2x + \pi)$ for $-\pi \le x \le 2\pi$. State the amplitude, period, frequency, and phase shift.

Verify the identities in Problems 35–37.

35. $\dfrac{\cos x}{1 - \sin x} + \dfrac{\cos x}{1 + \sin x} = 2 \sec x$

36. $\tan \dfrac{\theta}{2} = \dfrac{1}{\csc \theta + \cot \theta}$

37. $\dfrac{\cos x - \sin x}{\cos x + \sin x} = \sec 2x - \tan 2x$

38. Find the exact values of $\tan (x/2)$ and $\sin 2x$, given $\cos x = \frac{24}{25}$ and $\sin x > 0$.

☆**39.** Write $y = -\sin 2\pi t + \cos 2\pi t$ in the form $y = A \sin (Bt + C)$, where C is chosen so that $|C|$ is as small as possible. Indicate amplitude, period, frequency, and phase shift; then graph the equation for $-1 \le t \le 1$.

40. Find the exact value of $\sec(\sin^{-1} \frac{3}{4})$.

41. Evaluate to four significant digits using a calculator:
(A) $\sec(\sin^{-1} 0.25)$ (B) $\tan^{-1}(\csc 3.75)$
☆(C) $\tan(\operatorname{arccsc} 3.75)$

42. Find exactly all real solutions for
$$\sin 2x + \sin x = 0$$

43. Find all real solutions to four significant digits for
$$2 \cos 2x = 5 \sin x - 4$$

44. Solve the triangle with $b = 17.4$ cm and $\alpha = 49°30'$ for each of the following values of a.
(A) $a = 11.5$ cm   (B) $a = 14.7$ cm
(C) $a = 21.1$ cm

45. Given the vector diagram, find $|\mathbf{u} + \mathbf{v}|$ and θ.

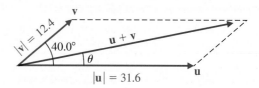

$\mathbf{v}$

$|\mathbf{v}| = 12.4$

$40.0°$

$\mathbf{u} + \mathbf{v}$

θ

$|\mathbf{u}| = 31.6$ $\mathbf{u}$

Figure for 45

46. Find $3\mathbf{u} - 4\mathbf{v}$ for
(A) $\mathbf{u} = \langle 1, -2 \rangle$, $\mathbf{v} = \langle 0, 3 \rangle$
(B) $\mathbf{u} = 2\mathbf{i} + 3\mathbf{j}$, $\mathbf{v} = -\mathbf{i} + 5\mathbf{j}$

47. Find a unit vector $\mathbf{u}$ with the same direction as $\mathbf{v} = \langle 7, -24 \rangle$.

48. Express $\mathbf{v}$ in terms of $\mathbf{i}$ and $\mathbf{j}$ unit vectors if $\mathbf{v} = \overrightarrow{AB}$, $A(-3, 2)$, $B(-1, 5)$.

☆**49.** Determine which vector pairs are orthogonal using properties of the dot product.
(A) $\mathbf{u} = \langle 4, 0 \rangle$; $\mathbf{v} = \langle 0, -5 \rangle$
(B) $\mathbf{u} = \langle 3, 2 \rangle$; $\mathbf{v} = \langle -3, 4 \rangle$
(C) $\mathbf{u} = \mathbf{i} - 2\mathbf{j}$; $\mathbf{v} = 6\mathbf{i} + 3\mathbf{j}$

50. Plot $r = 8 \cos 2\theta$ in a polar coordinate system.

51. Change $x^2 = 6y$ to polar form.

52. Change $r = 4 \sin \theta$ to rectangular form.

53. Convert $(3 + 3i)(-1 + i\sqrt{3})$ to polar form and evaluate. Leave answer in polar form.

54. Convert $(-1 + i\sqrt{3})/(3 + 3i)$ to polar form and evaluate. Leave answer in polar form.

55. Use De Moivre's theorem to evaluate $(1 - i)^6$ and write the result in the form $a + bi$.

56. Find all cube roots of $-i$. Write answers in the form $a + bi$.

C 57. In the circle shown, find the exact radian measure of θ and the coordinates of B to four significant digits if the arc length s is 8 units.

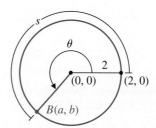

s

θ

2

$(0, 0)$ $(2, 0)$

$B(a, b)$

Figure for 57

58. Refer to the figure for Problem 57. Find θ and s to four significant digits if $(a, b) = (-1.6, -1.2)$.

59. Graph $y = 2 \sec(\pi x + \pi/4)$ for $-1 \le x \le 3$.

60. Express $\sec(2 \tan^{-1} x)$ as an algebraic expression in x free of trigonometric and inverse trigonometric functions.

61. Verify the identity, $\tan 3x = \dfrac{3 - \tan^2 x}{\cot x - 3 \tan x}$.

62. Change $r(\cos \theta + 1) = 1$ to rectangular form.

63. Plot $r = 5 - 3 \sin(\theta/2)$ in a polar coordinate system.

64. Find all roots of $x^3 - 4 = 0$. Write answer in the form $a + bi$ where a and b are computed to three decimal places. (Use a calculator.)

65. De Moivre's theorem can be used to derive multiple angle identities. For example, consider

$$\cos 2\theta + i \sin 2\theta = (\cos \theta + i \sin \theta)^2$$
$$= \cos^2 \theta - \sin^2 \theta + 2i \sin \theta \cos \theta$$

Equating the real and imaginary parts of the left side with the real and imaginary parts of the right side yields two familiar identities:

$$\cos 2\theta = \cos^2 \theta - \sin^2 \theta \qquad \sin 2\theta = 2 \sin \theta \cos \theta$$

(A) Apply De Moivre's theorem to $(\cos \theta + i \sin \theta)^3$ to derive identities for $\cos 3\theta$ and $\sin 3\theta$.

(B) Use identities from Chapter 4 to verify the identities obtained in part (A).

C *Problems 66–76 require the use of a graphing calculator.*

In Problems 66–68, graph f(x), find a simpler function g(x) which has the same graph as f(x), and verify the identity f(x) = g(x). [Assume g(x) = k + A t(Bx) where t(x) is one of the six trigonometric functions.]

66. $f(x) = 2 \cos^2 x - 4 \sin^2 x$

67. $f(x) = \dfrac{6 \sin^2 x - 2}{2 \cos^2 x - 1}$

68. $f(x) = \dfrac{\sin x + \cos x - 1}{1 - \cos x}$

In Problems 69–71 graph each function over the indicated interval.

69. $y = -2 \cos^{-1}(2x - 1), \quad 0 \le x \le 1$

70. $y = 2 \sin^{-1} 2x, \quad -\dfrac{1}{2} \le x \le \dfrac{1}{2}$

71. $y = \tan^{-1}(2x + 5), \quad -5 \le x \le 0$

In Problems 72–74 approximate all solutions over the indicated interval. Compute solutions to three decimal places.

72. $\tan x = 5, \quad -\pi \le x \le \pi$

73. $\cos x = \sqrt[3]{x}, \quad$ all real x

74. $3 \sin 2x \cos 3x = 2, \quad 0 \le x \le 2\pi$

75. Graph $r = 5 \sin(\theta/4)$ in a polar coordinate system for $0 \le \theta \le 8\pi$

76. Graph $r = 2/[1 - e \sin(\theta + 0.6)]$ for the following values of e and identify each curve as a hyperbola, an ellipse, or a parabola.
(A) $e = 0.7$ (B) $e = 1$ (C) $e = 1.5$

Applications

77. Surveying   To determine the length CB of the lake in the figure, a surveyor makes the following measurements: $AC = 520$ ft, $\angle BCA = 52°$, and $\angle CAB = 77°$. Find the approximate length of the lake.

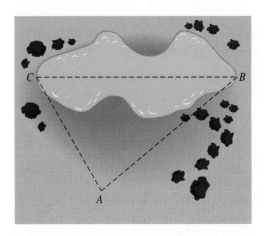

Figure for 77

78. Surveying   Refer to Problem 77. Find the approximate length of a similar lake using the following measurements: $AC = 430$ ft, $AB = 580$ ft, $\angle CAB = 64°$.

79. Tree Height   A tree casts a 35-ft shadow when the angle of elevation of the sun is 54°.
(A) Find the height of the tree if its shadow is cast on level ground.

(B) Find the height of the tree if its shadow is cast straight down a hillside that slopes 11° relative to the horizontal.

80. Navigation Two tracking stations located 4 mi apart on a straight coastline sight a ship (see the figure). How far is the ship from each station?

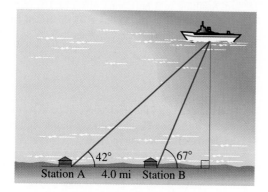

Figure for 80

81. Electrical Circuits The voltage E in an electrical circuit is given by an equation of the form $E = 110 \cos Bt$.
 (A) If the frequency is 70 Hz, what is the period? What is the value of B?
 (B) If the period is 0.0125 sec, what is the frequency? What is the value of B?
 (C) If $B = 100\pi$, what is the period? What is the frequency?

☆ **82. Water Waves** A wave with an amplitude of exactly 2 ft and a period of exactly 4 sec has an equation of the form $y = A \sin Bt$ at a fixed position. How high is the wave from trough to crest? What is its wavelength in feet? How fast is it traveling in feet per second? Compute answers to the nearest foot.

*Problems 83–86 refer to the region OCBA inside the first quadrant portion of a unit circle centered at the origin as shown in the figure.**

83. Precalculus Express the area of $OCBA$ in terms of θ.

84. Precalculus Express the area of $OCBA$ in terms of x.

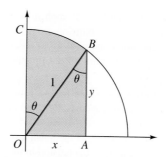

Figure for 83–86

c 85. Precalculus If the area of $OCBA = 0.5$, use the result of Problem 83 and a graphing calculator to find θ to three decimal places.

c 86. Precalculus If the area of $OCBA = 0.4$, use the result of Problem 84 and a graphing calculator to find x to three decimal places.

87. Solar Energy[†] A truncated conical solar collector is aimed directly at the sun as shown in part (a) of the figure. An analysis of the amount of solar energy absorbed by the collecting disk requires certain equations relating the quantities shown in part (b) of the figure. Find an equation that expresses
 (A) R in terms of h, H, r, and α
 (B) β in terms of h, H, r, and R

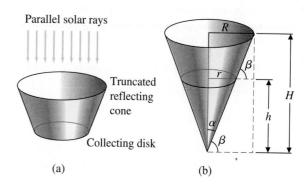

Figure for 87

---

* See "On the Derivatives of Trigonometric Functions" by M. R. Speigle in the *American Mathematical Monthly,* Vol. 63 (1956).

[†] Based on the article "The Solar Concentrating Properties of a Conical Reflector" by Don Leake in *The UMAP Journal,* Vol. 8, No. 4 (1987).

*88. **Navigation**   An airplane can cruise at 265 mph in still air. If a steady wind of 81.5 mph is blowing from the east, what compass heading should the pilot fly in order for the true course of the plane to be north (0°)? Compute the ground speed for this course.

*89. **Static Equilibrium**   A weight suspended from a cable deflects the cable relative to the horizontal as indicated in the figure.

(A) If w is the weight of the object, T_L is the tension in the left side of the cable, and T_R is the tension in the right side, show that

$$T_L = \frac{w \cos \alpha}{\sin(\alpha + \beta)} \quad \text{and} \quad T_R = \frac{w \cos \beta}{\sin(\alpha + \beta)}$$

(B) If $\alpha = \beta$, show that $T_L = T_R = \frac{1}{2}w \csc \alpha$.

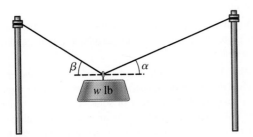

Figure for 89

90. **Railroad Construction**   North American railroads do not express specifications for circular track in terms of the radius. The common practice is to define the degree of a track curve as the degree measure of the central angle θ subtended by a 100-ft chord of the circular arc of track (see the figure).

(A) Find the radius of a 10° track curve to the nearest foot.

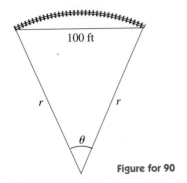

100 ft

r r

θ

Figure for 90

(B) If the radius of a track curve is 2,000 ft, find the degree of the track curve to the nearest 0.1°.

c 91. **Railroad Construction**   Actually using a compass to lay out circular arcs for railroad track is impractical. Instead, track layers work with the distance that curved track is offset from straight track.

(A) Suppose a 50-ft length of rail is bent into a circular arc, resulting in a horizontal offset of 1 ft (see the figure). Show that the radius r of this track curve satisfies the equation

$$r \cos^{-1}\left(\frac{r-1}{r}\right) = 50$$

and approximate r to the nearest foot.

(B) Find the degree of the track curve in part (A) to the nearest 0.1°. (See Problem 90 for the definition of the degree of a track curve.)

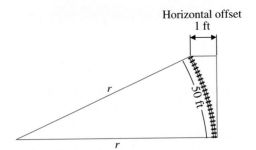

Horizontal offset
1 ft

r

r

50 ft

Figure for 91

92. **Engineering**   A circular log of radius r is cut lengthwise into three pieces by two parallel cuts equidistant from the center of the log (see the figure). Express the cross-sectional area of the center piece in terms of r and θ.

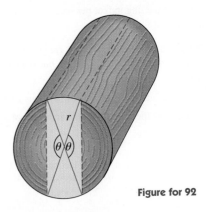

r

θ θ

Figure for 92

C 93. **Engineering** If all three pieces of the log in Problem 92 have the same cross-sectional area, approximate θ to four decimal places.

C 94. **Modeling Twilight Duration** Periods of twilight occur just before sunrise and just after sunset. The length of these periods varies with the time of year and latitudinal position. At the equator, twilight lasts about an hour all year long. Near the poles, there will be times during the year when twilight lasts for 24 hours and other times when it does not occur at all. Table 1 gives the duration of twilight on the eleventh of each month for one year at 20° N latitude.

(A) Convert the data in Table 1 from hours and minutes to two-place decimal hours. Enter this data for a two-year period in your graphing calculator and produce a scatter plot in the following viewing rectangle: $1 \le x \le 24$, $1 \le y \le 5$.

(B) A function of the form $y = k + A \sin(Bx + C)$ can be used to model this data. Use the converted data from Table 1 to determine k and A (to two decimal places), and B (exact value). Use the graph in part (A) to visually estimate C (to one decimal place).

(C) Plot the data from part (A) and the equation from part (B) in the same viewing rectangle. If necessary, adjust your value of C to produce a better fit.

Table 1

x (months)	1	2	3	4	5	6	7	8	9	10	11	12
y (twilight duration)	1:37	1:49	2:21	2:59	3:33	4:07	4:03	3:30	2:48	2:13	1:48	1:34

COMMENTS ON NUMBERS

A.1 Real Numbers

A.2 Complex Numbers

A.3 Significant Digits

A.1 REAL NUMBERS

♦ **The Real Number System**
♦ **The Real Number Line**
♦ **Basic Real Number Properties**

This appendix provides a brief review of the real number system and some of its basic properties. The real number system and its properties are fundamental to the study of a great deal of mathematics.

♦ The Real Number System

The real number system is the number system you have used most of your life. Informally, a **real number** is any number that has a decimal representation. Table 1 describes the set of real numbers and some of its important subsets. Figure 1 illustrates how these various sets of numbers are related to each other.

FIGURE 1
The real number system

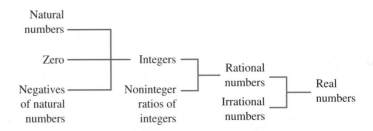

TABLE 1
The set of real numbers

Name	Description	Examples
Natural numbers	Counting numbers (also called positive integers)	1, 2, 3, . . .
Integers	Natural numbers, their negatives, and zero	. . . , −2, −1, 0, 1, 2, . . .
Rational numbers	Numbers that can be represented as *a/b*, where *a* and *b* are integers, *b* ≠ 0; decimal representations are repeating or terminating	−7, 0, 1, 36, $\frac{2}{3}$, 3.8, −0.666 6$\overline{6}$,* 6.383 8$\overline{38}$
Irrational numbers	Numbers that can be represented as nonrepeating and nonterminating decimal numbers	$\sqrt{3}$, $\sqrt[3]{7}$, π, 3.273 93 . . .
Real numbers	Rational numbers and irrational numbers	

* The bar over the 6 means that 6 repeats infinitely, −0.666666666. . . . Similarly, $\overline{38}$ means that 38 is repeated infinitely.

The set of **integers** contains all the **natural numbers** and something else—their negatives and zero. The set of **rational numbers** contains all the integers and something else—noninteger ratios of integers. And the set of *real numbers* contains all the rational numbers and something else—the **irrational numbers**.

◆   The Real Number Line

A one-to-one correspondence exists between the set of real numbers and the set of points on a line. That is, each real number corresponds to exactly one point, and each point corresponds to exactly one real number. A line with a real number associated with each point, and vice versa, as in Figure 2, is called a **real number line**, or simply a **real line**. Each number associated with a point is called the **coordinate** of the point. The point with coordinate 0 is called the **origin**. The arrow on the right end of the real line in Figure 2 indicates a positive direction. The coordinates of all points to the right of the origin are called **positive real numbers**, and those to the left of the origin are called **negative real numbers**.

FIGURE 2
A real number line

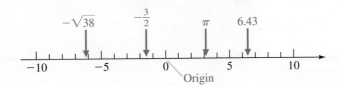

◆ Basic Real Number Properties

We now list some basic properties of the real number system that enable us to convert algebraic expressions into equivalent forms.

BASIC PROPERTIES OF THE SET OF REAL NUMBERS

Let R be the set of real numbers, and let x, y, and z be arbitrary elements of R.

Addition properties

Closure	$x + y$ is a unique element in R.
Associative	$(x + y) + z = x + (y + z)$
Commutative	$x + y = y + x$
Identity	0 is the additive identity; that is, for all x in R, $0 + x = x + 0 = x$, and 0 is the only element in R with this property.
Inverse	For each x in R, $-x$ is its unique additive inverse; that is, $x + (-x) = (-x) + x = 0$, and $-x$ is the only element in R relative to x with this property.

Multiplication properties

Closure	xy is a unique element in R.
Associative	$(xy)z = x(yz)$
Commutative	$xy = yx$
Identity	1 is the multiplicative identity; that is, for all x in R, $(1)x = x(1) = x$, and 1 is the only element in R with this property.
Inverse	For each x in R, $x \neq 0$, $1/x$ is its unique multiplicative inverse; that is, $x(1/x) = (1/x)x = 1$, and $1/x$ is the only element in R relative to x with this property.

Combined property

Distributive	$x(y + z) = xy + xz$
	$(x + y)z = xz + yz$

Do not let the names of these properties intimidate you. Most of the ideas presented are quite simple. In fact, you have been using many of these properties in arithmetic for a long time. We summarized these properties here for convenient reference because they represent some of the basic rules of the "game of algebra," and you cannot play the game very well unless you know the rules.

EXERCISE A.1

1. Give an example of a negative integer, an integer that is neither positive nor negative, and a positive integer.

2. Give an example of a negative rational number, a rational number that is neither positive nor negative, and a positive rational number.

3. Give an example of a rational number that is not an integer.

4. Give an example of an integer that is not a natural number.

5. Indicate which of the following are true:
   (A) All natural numbers are integers.
   (B) All real numbers are irrational.
   (C) All rational numbers are real numbers.

6. Indicate which of the following are true:
   (A) All integers are natural numbers.
   (B) All rational numbers are real numbers.
   (C) All natural numbers are rational numbers.

In Problems 7 and 8, express each number in decimal form to the capacity of your calculator. Observe the repeating decimal representation of the rational numbers and the apparent nonrepeating decimal representation of the irrational numbers. Indicate whether each number is rational or irrational.

7. (A) $\frac{4}{11}$ (B) $\frac{7}{9}$ (C) $\sqrt{7}$ (D) $\frac{13}{8}$

8. (A) $\frac{19}{6}$ (B) $\sqrt{23}$ (C) $\frac{9}{16}$ (D) $\frac{31}{111}$

9. Each of the following real numbers lies between two successive integers on a real number line. Indicate which two.
   (A) $\frac{26}{9}$ (B) $-\frac{19}{5}$ (C) $-\sqrt{23}$

10. Each of the following real numbers lies between two successive integers on a real number line. Indicate which two.
   (A) $\frac{13}{4}$ (B) $-\frac{5}{3}$ (C) $-\sqrt{8}$

In Problems 11–24, replace each question mark with an appropriate expression that will illustrate the use of the indicated real number property.

11. Commutative property $(+)$: $3 + y = ?$

12. Commutative property $(+)$: $u + v = ?$

13. Associative property $(\cdot)$: $3(2x) = ?$

14. Associative property $(\cdot)$: $7(4z) = ?$

15. Commutative property $(\cdot)$: $x7 = ?$

16. Commutative property $(\cdot)$: $yx = ?$

17. Associative property $(+)$: $5 + (7 + x) = ?$

18. Associative property $(+)$: $9 + (2 + m) = ?$

19. Identity property $(+)$: $3m + 0 = ?$

20. Identity property $(+)$: $0 + (2x + 3) = ?$

21. Identity property $(\cdot)$: $1(u + v) = ?$

22. Identity property $(\cdot)$: $1(xy) = ?$

23. Distributive property: $2x + 3x = ?$

24. Distributive property: $7(x + y) = ?$

25. Indicate whether true (T) or false (F), and for each false statement find real number replacements for a and b that will illustrate its falseness. For all real numbers a and b,
   (A) $a + b = b + a$ (B) $a - b = b - a$
   (C) $ab = ba$ (D) $a/b = b/a$

26. Indicate whether true (T) or false (F), and for each false statement find real number replacements for a, b, and c that will illustrate its falseness. For all real numbers a, b, and c,
   (A) $(a + b) + c = a + (b + c)$
   (B) $(a - b) - c = a - (b - c)$
   (C) $a(bc) = (ab)c$
   (D) $(a/b)/c = a/(b/c)$

A.2 COMPLEX NUMBERS

The Pythagoreans (500–275 B.C.) found that the simple equation

$$x^2 = 2 \tag{1}$$

had no rational number solutions. (A **rational number** is any number that can be expressed as P/Q, where P and Q are integers and $Q \neq 0$.) If equation (1) were to have a solution, then a new kind of number had to be invented—the **irrational number**.

The irrational numbers $\sqrt{2}$ and $-\sqrt{2}$ are both solutions to equation (1). The invention of these irrational numbers evolved over a period of about 2,000 years, and it was not until the nineteenth century that they were finally put on a rigorous foundation. The rational numbers and irrational numbers together constitute the real number system.

Is there any need to extend the real number system still further? Yes, if we want the simple equation

$$x^2 = -1$$

to have a solution. Since $x^2 \geq 0$ for any real number x, this equation has no real number solutions. Once again we are forced to invent a new kind of number, a number that has the possibility of being negative when it is squared. These new numbers are called *complex numbers*. The complex numbers, like the irrational numbers, evolved over a long period of time, dating mainly back to Cardano (1545). But it was not until the nineteenth century that they were firmly established as numbers in their own right. A **complex number** is defined to be a number of the form

$$a + bi$$

where a and b are real numbers, and i is called the **imaginary unit**. Thus,

$$5 + 3i \qquad \tfrac{1}{3} - 6i \qquad \sqrt{7} + \tfrac{1}{4}i \qquad 0 + 6i \qquad \tfrac{1}{2} + 0i \qquad 0 + 0i$$

are all complex numbers. Particular kinds of complex numbers are given special names, as follows.

$a + 0i = a$	Real number
$a + bi \quad b \neq 0$	Imaginary number
$0 + bi = bi$	Pure imaginary number
$0 + 0i = 0$	Zero
$1i = i$	Imaginary unit
$a - bi$	Conjugate of $a + bi$

Thus, we see that just as every integer is a rational number, every real number is a complex number.

To use complex numbers we must know how to add, subtract, multiply, and divide them. We start by defining equality, addition, and multiplication.

BASIC DEFINITIONS FOR COMPLEX NUMBERS

Equality: $a + bi = c + di$ if and only if $a = c$ and $b = d$

Addition: $(a + bi) + (c + di) = (a + c) + (b + d)i$

Multiplication: $(a + bi)(c + di) = (ac - bd) + (ad + bc)i$

These definitions, particularly the one for multiplication, may seem a little strange to you. But it turns out that if we want many of the same basic properties that hold for real numbers to hold for complex numbers, and if we also want negative real numbers to have square roots, then we must define addition and multiplication as shown. Let us use the definition of multiplication to see what happens to i when it is squared:

$$i^2 = (0 + 1i)(0 + 1i)$$
$$= (0 \cdot 0 - 1 \cdot 1) + (0 \cdot 1 + 1 \cdot 0)i$$
$$= -1 + 0i$$
$$= -1$$

Thus

$$i^2 = -1$$

This is an important result. We can also write

$$i = \sqrt{-1} \quad \text{and} \quad -i = -\sqrt{-1}$$

Fortunately, you do not have to memorize the definitions of addition and multiplication. We can show that the complex numbers, under these definitions, are closed, associative, and commutative, and multiplication distributes over addition. As a consequence, we can manipulate complex numbers as if they were binomial forms in real number algebra, with the exception that i^2 is to be replaced with -1. Example 1 illustrates the mechanics of carrying out addition, subtraction, multiplication, and division.

◆ EXAMPLE 1   Performing Operations with Complex Numbers

Write each of the following in the form $a + bi$.

(A) $(2 + 3i) + (3 - i)$ (B) $(2 + 3i) - (3 - i)$
(C) $(2 + 3i)(3 - i)$ (D) $(2 + 3i)/(3 - i)$

SOLUTIONS   We treat these as we would ordinary binomials in elementary algebra, with one exception: Whenever i^2 turns up, we replace it with -1.

(A) $(2 + 3i) + (3 - i) = 2 + 3i + 3 - i$
$$= 2 + 3 + 3i - i$$
$$= 5 + 2i$$

(B) $(2 + 3i) - (3 - i) = 2 + 3i - 3 + i$
$$= 2 - 3 + 3i + i$$
$$= -1 + 4i$$

(C) $(2 + 3i)(3 - i) = 6 + 7i - 3i^2$
$$= 6 + 7i - 3(-1)$$
$$= 6 + 7i + 3$$
$$= 9 + 7i$$

(D) To eliminate i from the denominator, we multiply the numerator and denominator by the complex conjugate of $3 - i$, namely, $3 + i$. [Recall from elementary algebra that $(a - b)(a + b) = a^2 - b^2$.]

$$\frac{2 + 3i}{3 - i} \cdot \frac{3 + i}{3 + i} = \frac{6 + 11i + 3i^2}{9 - i^2}$$
$$= \frac{6 + 11i - 3}{9 + 1}$$
$$= \frac{3 + 11i}{10} = \frac{3}{10} + \frac{11}{10}i$$

           ◆

MATCHED PROBLEM 1   Write each of the following in the form $a + bi$.

(A) $(3 - 2i) + (2 + i)$ (B) $(3 - 2i) - (2 - i)$
(C) $(3 - 2i)(2 - i)$ (D) $(3 - 2i)/(2 - i)$

    At this time, your experience with complex numbers has likely been limited to solutions of equations, particularly quadratic equations. Recall that if $b^2 - 4ac$ is negative in

$$x = \frac{-b \pm \sqrt{b^2 - 4ac}}{2a}$$

then the solutions to the quadratic equation $ax^2 + bx + c = 0$ are complex. This is easy to verify, since a square root of a negative number can be written in the form

$$\sqrt{-k} = i\sqrt{k} \quad \text{for } k > 0$$

To check this last equation, we square $i\sqrt{k}$ to see if we get $-k$.

$$(i\sqrt{k})^2 = i^2(\sqrt{k})^2 = -k$$

[*Note:* We write $i\sqrt{k}$ instead of $\sqrt{k}i$ so that i will not mistakenly be included under the radical.]

◆   EXAMPLE 2   Converting Square Roots of Negative Numbers to Complex Form

Write in the form $a + bi$.

(A) $\sqrt{-4}$ (B) $4 + \sqrt{-4}$

(C) $\dfrac{-3 - \sqrt{-7}}{2}$ (D) $\dfrac{1}{1 - \sqrt{-9}}$

SOLUTIONS   (A) $\sqrt{-4} = i\sqrt{4} = 2i$

(B) $4 + \sqrt{-4} = 4 + i\sqrt{4} = 4 + 2i$

(C) $\dfrac{-3 - \sqrt{-7}}{2} = \dfrac{-3 - i\sqrt{7}}{2} = -\dfrac{3}{2} - \dfrac{\sqrt{7}}{2}i$

(D) $\dfrac{1}{1 - \sqrt{-9}} = \dfrac{1}{1 - 3i} = \dfrac{1}{1 - 3i} \cdot \dfrac{1 + 3i}{1 + 3i} = \dfrac{1 + 3i}{1 - 9i^2}$

$$= \dfrac{1 + 3i}{10} = \dfrac{1}{10} + \dfrac{3}{10}i$$ ◆

MATCHED PROBLEM 2   Write in the form $a + bi$.

(A) $\sqrt{-16}$ (B) $5 + \sqrt{-16}$

(C) $\dfrac{-5 - \sqrt{-2}}{2}$ (D) $\dfrac{1}{3 - \sqrt{-4}}$

Answers to   **1.** (A) $5 - i$ (B) $1 - i$ (C) $4 - 7i$ (D) $\frac{8}{5} - \frac{1}{5}i$
Matched Problems   **2.** (A) $4i$ (B) $5 + 4i$ (C) $-\frac{5}{2} - (\sqrt{2}/2)i$ (D) $\frac{3}{13} + \frac{2}{13}i$

EXERCISE A.2

Perform the indicated operations and write each answer in the standard form $a + bi$.

1. $(3 - 2i) + (4 + 7i)$ **2.** $(4 + 6i) + (2 - 3i)$

3. $(3 - 2i) - (4 + 7i)$ **4.** $(4 + 6i) - (2 - 3i)$

5. $(6i)(3i)$ **6.** $(5i)(4i)$

7. $2i(3 - 4i)$ **8.** $4i(2 - 3i)$

9. $(3 - 4i)(1 - 2i)$ **10.** $(5 - i)(2 - 3i)$

11. $(3 + 5i)(3 - 5i)$ **12.** $(7 - 3i)(7 + 3i)$

13. $\dfrac{1}{2 + i}$ **14.** $\dfrac{1}{3 - i}$

15. $\dfrac{2 - i}{3 + 2i}$ **16.** $\dfrac{3 + i}{2 - 3i}$

17. $\dfrac{-1 + 2i}{4 + 3i}$ **18.** $\dfrac{-2 - i}{3 - 4i}$

Convert square roots of negative numbers to complex forms,

perform the indicated operations, and express answers in the standard form $a + bi$.

19. $(3 + \sqrt{-4}) + (2 - \sqrt{-16})$

20. $(2 + \sqrt{-9}) + (3 - \sqrt{-25})$

21. $(5 - \sqrt{-1}) - (2 - \sqrt{-36})$

22. $(2 + \sqrt{-9}) - (3 - \sqrt{-25})$

23. $(-3 - \sqrt{-1})(-2 + \sqrt{-49})$

24. $(3 - \sqrt{-9})(-2 - \sqrt{-1})$

25. $\dfrac{5 - \sqrt{-1}}{2 + \sqrt{-4}}$ **26.** $\dfrac{-2 + \sqrt{-16}}{3 - \sqrt{-25}}$

Evaluate:

27. $(1 - i)^2 - 2(1 - i) + 2$

28. $(1 + i)^2 - 2(1 + i) + 2$

29. $\left(-\dfrac{1}{2} + \dfrac{\sqrt{3}}{2}i\right)^3$ **30.** $\left(-\dfrac{1}{2} - \dfrac{\sqrt{3}}{2}i\right)^3$

A.3 SIGNIFICANT DIGITS

◆ **Scientific Notation**

◆ **Significant Digits**

◆ **Calculation Accuracy**

Many calculations in the real world deal with figures that are only approximate. After a series of calculations with approximate measurements, what can be said about the accuracy of the final answer? It seems reasonable to assume that a final answer cannot be any more accurate than the least accurate figure used in the calculation. This is an important point, since calculators tend to give the impression that greater accuracy is achieved than is warranted. In this section we introduce *scientific notation* and we use this concept in a discussion of *significant digits*. We will then be able to set up conventions for indicating the accuracy of the results of certain calculations involving approximate quantities. We will be guided by these conventions throughout the text when writing a final answer to a problem.

◆ Scientific Notation

Work in science and engineering often involves the use of very large numbers. For example, the distance that light travels in 1 yr is called a **light-year**. This distance is approximately

9,440,000,000,000 km

Very small numbers are also used. For example, the mass of a water molecule is approximately

0.000 000 000 000 000 000 000 03 g

It is generally troublesome to write and work with numbers of this type in standard decimal form. In fact, these two numbers cannot even be entered into most calculators as they are written. Fortunately, it is possible to represent any decimal form as the product of a number between 1 and 10 and an integer power of 10; that is, in the form

$$a \times 10^n, \qquad 1 \le a < 10, n \text{ an integer, } a \text{ in decimal form}$$

A number expressed in this form is said to be in **scientific notation**.

◆ EXAMPLE 1   Using Scientific Notation

Each number is written in scientific notation.

$$4 = 4 \times 10^0 \qquad\qquad 0.36 = 3.6 \times 10^{-1}$$
$$63 = 6.3 \times 10 \qquad\qquad 0.0702 = 7.02 \times 10^{-2}$$
$$805 = 8.05 \times 10^2 \qquad\qquad 0.005\ 32 = 5.32 \times 10^{-3}$$
$$3{,}143 = 3.143 \times 10^3 \qquad\qquad 0.000\ 67 = 6.7 \times 10^{-4}$$
$$7{,}320{,}000 = 7.32 \times 10^6 \qquad\qquad 0.000\ 000\ 54 = 5.4 \times 10^{-7}$$

Can you discover a rule that relates the number of decimal places a decimal point is moved to the power of 10 used?

$$7,320,000 \quad = 7.320\ 000. \times 10^6 \quad = 7.32 \times 10^6$$

6 places left

Positive exponent

$$0.000\ 000\ 54 \quad = 0.000\ 000\ 5.4 \times 10^{-7} \quad = 5.4 \times 10^{-7}$$

7 places right

Negative exponent

MATCHED PROBLEM 1 Write in scientific notation:

(A) 450   (B) 360,000   (C) 0.0372   (D) 0.000 001 43

Most scientific calculators express very large and very small numbers in scientific notation. Read the instruction manual for your calculator to see how numbers in scientific notation are entered into your calculator. Numbers in scientific notation are displayed in most calculators as follows.

Calculator display	Number represented
3.207418 −13	$3.207\ 418 \times 10^{-13}$
5.002193 12	$5.002\ 193 \times 10^{12}$

◆ Significant Digits

Suppose we wish to compute the area of a rectangle with the dimensions shown in Figure 1. As it often happens with approximations, we have one dimension to one decimal place accuracy and the other to two decimal place accuracy.

FIGURE 1

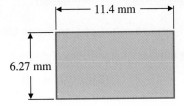

Using a calculator and the formula for the area of a rectangle, $A = ab$, we have

$$A = (11.4)(6.27) = 71.478$$

How many decimal places are justified in this calculation? We will answer this question later in this section. First we must introduce the idea of *significant digits*.

Whenever we write a measurement such as 11.4 mm, we assume that the measurement is accurate to the last digit written. Thus, 11.4 mm indicates that the measurement was made to the nearest tenth of a millimeter—that is, the actual length is between 11.35 mm and 11.45 mm. In general, the digits in a number that indicate the accuracy of the number are called **significant digits**. If (going from left to right) the first digit and the last digit of a number are not 0, then all the digits are significant. Thus, the measurements 11.4 and 6.27 in Figure 1 have three significant digits, the number 100.8 has four significant digits, and the number 10,102 has five significant digits.

If the last digit of a number is 0, then the number of significant digits may not be clear. Suppose we are given a length of 23.0 cm. Then we assume the measurement has been taken to the nearest tenth and say that the number has three significant digits. However, suppose we are told that the distance between two cities is 3,700 mi. Is the stated distance accurate to the nearest hundred, ten, or unit? That is, does the stated distance have two, three, or four significant digits? We cannot really tell. In order to resolve this ambiguity, we give a precise definition of significant digits using scientific notation.

SIGNIFICANT DIGITS

If a number x is written in scientific notation as

$$x = a \times 10^n, \qquad 1 \le a < 10, n \text{ an integer}$$

then the number of significant digits in x is the number of digits in a.

Thus,

3.7×10^3	has two significant digits
3.70×10^3	has three significant digits
3.700×10^3	has four significant digits

All three of these measurements have the same decimal representation, 3,700, but each represents a different accuracy.

◆ **EXAMPLE 2**   Determining the Number of Significant Digits

Indicate the number of significant digits in each of the following numbers.

(A) 9.1003×10^{-3} (B) 1.080×10
(C) 5.92×10^{22} (D) 7.9000×10^{-13}

SOLUTIONS   In all cases the number of significant digits is the number of digits in the number to the left of the multiplication sign (as stated in the definition).

(A) Five   (B) Four   (C) Three   (D) Five   ◆

MATCHED PROBLEM 2   Indicate the number of significant digits in each of the following numbers.

(A) 4.39×10^{12} (B) 1.020×10^{-7}
(C) 2.3905×10^{-1} (D) 3.00×10

The definition of significant digits tells us how to write a number so that the number of significant digits is clear, but it does not tell us how to interpret the accuracy of a number that is not written in scientific notation. We will use the following convention for numbers that are written as decimal fractions.

SIGNIFICANT DIGITS IN DECIMAL FRACTIONS

The number of significant digits in **a number with no decimal point** is found by counting the digits from left to right, starting with the first digit and ending with the last *nonzero* digit.

The number of significant digits in **a number containing a decimal point** is found by counting the digits from left to right, starting with the first *nonzero* digit and ending with the last digit (which may be 0).

Applying this convention to the number 3,700, we conclude that this number (as written) has two significant digits. If we want to indicate that it has three or four significant digits, we must use scientific notation. The significant digits in the following numbers are underlined.

34,007   920,000   25.300   0.0063   0.000 430

◆ Calculation Accuracy

When performing calculations, we want an answer that is as accurate as the numbers used in the calculation warrant, but no more. In calculations involving multiplication, division, powers, and roots, we adopt the following accuracy convention (which is justified in courses in numerical analysis).

ACCURACY OF CALCULATED VALUES

The number of significant digits in a calculation involving multiplication, division, powers, and/or roots is the same as the number of significant digits in the number in the calculation with the smallest number of significant digits.

Applying this convention to the calculation of the area of the rectangle in Figure 1, we have

$$A = (11.4)(6.27)$$ Both numbers have three significant digits.

$$= 71.478$$ Calculator computation

$$= 71.5$$ The computed area is only accurate to three significant digits.

◆   EXAMPLE 3   Determining the Accuracy of Computed Values

Perform the indicated operations on the given approximate numbers. Then use the accuracy conventions stated above to round each answer to the appropriate accuracy.

(A) $\dfrac{(204)(34.0)}{120}$ (B) $\dfrac{(2.50 \times 10^5)(3.007 \times 10^7)}{2.4 \times 10^6}$

SOLUTIONS   (A) $\dfrac{(204)(34.0)}{120}$ 120 has the least number of significant digits (two).

$= 57.8$ Calculator computation: answer must have the same number of significant digits as the number with the least number of significant digits (two).

$= 58$

(B) $\dfrac{(2.50 \times 10^5)(3.007 \times 10^7)}{2.4 \times 10^6}$ 2.4×10^6 has the least number of significant digits (two).

$= 3.132291\ldots \times 10^6$ Calculator computation: answer must have the same number of significant digits as the number with the least number of significant digits (two).

$= 3.1 \times 10^6$

◆

MATCHED PROBLEM 3   Perform the indicated operations on the given approximate numbers. Then use the rounding conventions stated above to write each answer with the appropriate accuracy.

(A) $\dfrac{2.30}{(0.0341)(2.674)}$

(B) $\dfrac{1.235 \times 10^8}{(3.07 \times 10^{-3})(1.20 \times 10^4)}$

We complete this section with two important observations:

1. Many formulas have constants that represent exact quantities. Such quantities are assumed to have infinitely many significant digits. For example, the formula for the circumference of a circle, $C = 2\pi R$, has two constants that are exact, 2 and π. Consequently, the final answer will have as many significant digits as in the measurement R.

2. How do we round a number one place if the last nonzero digit is 5? For example, the following product should have only two significant digits. Do we change the 2 to a 3 or leave it alone?

$(1.3)(2.5) = 3.25$ Calculator result: should be rounded to two significant digits

We will round to 3.2 following the conventions given in the box.

ROUNDING NUMBERS ONE PLACE WHEN LAST NONZERO DIGIT IS 5

(A) If **the digit preceding 5 is odd**, round up 1 to make it even.
(B) If **the digit preceding 5 is even**, do not change it.

We have adopted these conventions in order to prevent the accumulation of round-off errors with numbers having the last nonzero digit 5. The idea is to round up 50% of the time. (There are a number of other ways to accomplish this—flipping a coin, for example.)

EXAMPLE 4   Rounding Numbers When Last Nonzero Digit Is 5

Round each number to three significant digits.

(A) 3.1495   (B) 0.004 135
(C) 32,450   (D) $4.314\ 764\ 09 \times 10^{12}$

SOLUTIONS   (A) 3.15   (B) 0.004 14   (C) 32,400   (D) 4.31×10^{12}

MATCHED PROBLEM 4   Round each number to three significant digits.

(A) 43.0690   (B) 48.05
(C) 48.15   (D) $8.017\ 632 \times 10^{-3}$

Answers to   **1.** (A) 4.5×10^2 (B) 3.6×10^5 (C) 3.72×10^{-2}
Matched Problems   (D) 1.43×10^{-6}
2. (A) Three   (B) Four   (C) Five   (D) Three
3. (A) 25.2   (B) 3.35×10^6
4. (A) 43.1   (B) 48.0   (C) 48.2   (D) 8.02×10^{-3}

EXERCISE A.3

A   *Write in scientific notation.*

1. 640   **2.** 384

3. 5,460,000,000   **4.** 38,400,000

5. 0.73   **6.** 0.004 93

7. 0.000 000 32   **8.** 0.0836

9. 0.000 049 1   **10.** 435,640

11. 67,000,000,000   **12.** 0.000 000 043 2

Write as a decimal fraction.

13. 5.6×10^4 **14.** 3.65×10^6

15. 9.7×10^{-3} **16.** 6.39×10^{-6}

17. 4.61×10^{12} **18.** 3.280×10^9

19. 1.08×10^{-1} **20.** 3.004×10^{-4}

Indicate the number of significant digits in each number.

21. 12.3   **22.** 123

23. 12,300   **24.** 0.001 23

25. 0.012 30   **26.** 12.30

27. 6.7×10^{-1} **28.** 3.56×10^{-4}

29. 6.700×10^{-1} **30.** 3.560×10^{-4}

31. 7.090×10^5 **32.** 6.0050×10^7

B *Round each to three significant digits.*

33. 635,431

34. 4,089,100

35. 86.85

36. 7.075

37. 0.004 652 3

38. 0.000 380 0

Write in scientific notation, rounding to two significant digits.

39. 734

40. 908

41. 0.040

42. 700

43. 0.000 435

44. 635.468 13

Indicate how many significant digits should be in the final answer.

45. (32.8)(0.2035)

46. (0.002 30)(25.67)

47. $\dfrac{(7.21)(360)}{1,200}$

48. $\dfrac{(0.0350)(621)}{8,543}$

49. $\dfrac{(5.03 \times 10^{-3})(6 \times 10^{4})}{8.0}$

50. $\dfrac{3.27(1.8 \times 10^{7})}{2.90 \times 10}$

Using a calculator, compute each of the following, and express each answer with appropriate accuracy.

51. $\dfrac{6.07}{0.5057}$

52. (53,100)(0.2467)

53. $(6.14 \times 10^{9})(3.154 \times 10^{-1})$

54. $\dfrac{7.151 \times 10^{6}}{9.1 \times 10^{-1}}$

55. $\dfrac{6,730}{(2.30)(0.0551)}$

56. $\dfrac{63,100}{(0.0620)(2,920)}$

In the following formulas, the indicated constants are exact. Compute the answer to each problem to an accuracy appropriate for the given approximate values of the variables. (Use $\pi \approx 3.141\ 59$ or π to the accuracy given by your calculator.)

57. *Circumference of a circle* $C = 2\pi R$; $R = 25.31$ cm

58. *Area of a circle* $A = \pi R^{2}$; $R = 2.5$ in.

59. *Area of a triangle* $A = \frac{1}{2}bh$; $b = 22.4$ ft and $h = 8.6$ ft.

60. *Area of an ellipse* $A = \pi ab$; $a = 0.45$ cm and $b = 1.35$ cm

61. *Surface area of a sphere* $S = 4\pi r^{2}$; $r = 1.5$ mm

62. *Volume of a sphere* $S = \frac{4}{3}\pi r^{3}$; $r = 1.85$ in.

C *In the following formulas, the indicated constants are exact. Compute the value of the indicated variable to an accuracy appropriate for the given approximate values of the other variables in the formula. (Use $\pi \approx 3.141\ 59$ or π to the accuracy given by your calculator.)*

63. *Volume of a rectangular parallelepiped (a box)* $V = lwh$; $V = 24.2$ cm³, $l = 3.25$ cm, $w = 4.50$ cm, $h = ?$

64. *Volume of a right circular cylinder* $V = \pi r^{2}h$; $V = 1,250$ ft³, $h = 6.4$ ft, $r = ?$

65. *Volume of a right circular cone* $V = \frac{1}{3}\pi r^{2}h$; $V = 1,200$ in.³, $h = 6.55$ in., $r = ?$

66. *Volume of a pyramid* $V = \frac{1}{3}Ah$; $V = 6,000$ m³, $A = 1,100$ m, $h = ?$

FUNCTIONS AND INVERSE FUNCTIONS

B.1 Functions

B.2 Inverse Functions

B.1 FUNCTIONS

- ◆ **Definition of a Function—Rule Form**
- ◆ **Definition of a Function—Set Form**
- ◆ **Function Notation**
- ◆ **Function Classification**

Seeking correspondences between various types of phenomena is undoubtedly one of the most important aspects of science. A physicist attempts to find a correspondence between the current in an electrical circuit and time; a chemist looks for a correspondence between the speed of a chemical reaction and the concentration of a given substance; an economist tries to determine a correspondence between the price of an object and the demand for the object; and so on. The list could go on and on.

Establishing and working with correspondences among various types of phenomena—whether through tables, graphs, or equations—is so fundamental to pure and applied science that it has become necessary to describe this activity in the precise language of mathematics.

◆ Definition of a Function—Rule Form

What do all the examples cited above have in common? Each describes the matching of elements from one set with the elements in a second set. Consider Charts 1–3, which list values for the cube, square, and square root, respectively.

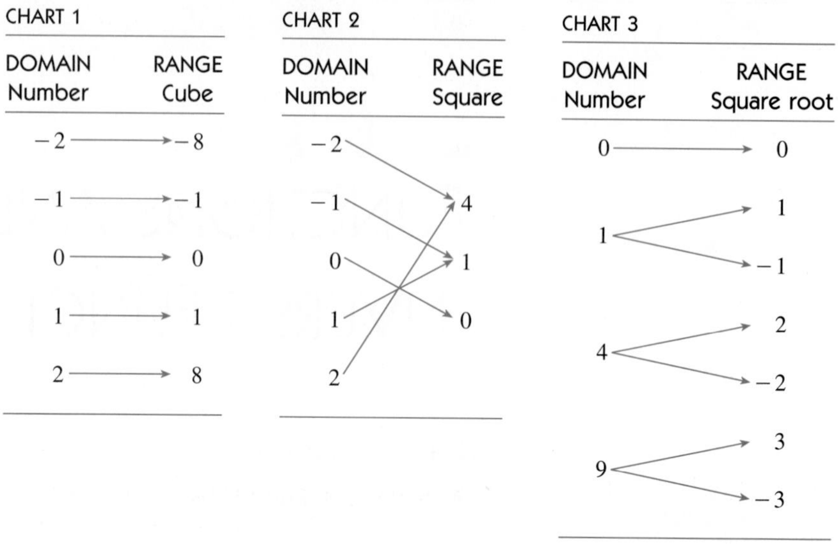

CHART 1

DOMAIN Number	RANGE Cube
−2	−8
−1	−1
0	0
1	1
2	8

CHART 2

DOMAIN Number: −2, −1, 0, 1, 2
RANGE Square: 4, 1, 0

CHART 3

DOMAIN Number: 0, 1, 4, 9
RANGE Square root: 0, 1, −1, 2, −2, 3, −3

Charts 1 and 2 define functions, but Chart 3 does not. Why? The following definition of *function* will explain.

FUNCTION—RULE FORM

A **function** is a rule that produces a correspondence between two sets of elements such that to each element in the first set there corresponds *one and only one* element in the second set.

The first set is called the **domain**. The set of all corresponding elements in the second set is called the **range**.

A variable representing an arbitrary element from the domain is called an **independent variable**. A variable representing an arbitrary element from the range is called a **dependent variable**.

Charts 1 and 2 define functions, since to each domain value there corresponds exactly one range value. For example, the cube of −2 is −8 and no other number. On the other hand, Chart 3 does not specify a function, since to at least one domain value there corresponds more than one range value. For example, to the domain value 4 there corresponds −2 and 2, both square roots of 4.

Some equations in two variables define functions. If in an equation in two variables, say x and y, there corresponds exactly one range value y for each domain value x (x is independent and y is dependent), then the correspondence established by the equation is a function.

♦ **EXAMPLE 1**   Determining If an Equation Defines a Function

Determine which of the following equations define functions with independent variable x and domain all real numbers.

(A) $y - x = 1$ (B) $y^2 - x^2 = 1$

SOLUTIONS   (A) Solving for the dependent variable y, we have

$$y - x = 1 \tag{1}$$
$$y = 1 + x$$

Since $1 + x$ is a real number for each real number x, equation (1) assigns exactly one value of the dependent variable, $y = 1 + x$, to each value of the independent variable x. Thus, equation (1) defines a function.

(B) Solving for the dependent variable y, we have

$$y^2 - x^2 = 1$$
$$y^2 = 1 + x^2 \tag{2}$$
$$y = \pm\sqrt{1 + x^2}$$

Since $1 + x^2$ is always a positive real number and since each positive real number has two real square roots, each value of the independent variable x corresponds to two values of the dependent variable, $y = -\sqrt{1 + x^2}$ and $y = \sqrt{1 + x^2}$. Thus, equation (2) does not define a function.   ◆

MATCHED PROBLEM 1   Determine which of the following equations define functions with independent variable x and domain all real numbers.

(A) $y = x + 3$ (B) $y^2 = x^2 + 3$

It is very easy to determine whether an equation defines a function by examining the graph of the equation. The two equations considered in Example 1 are graphed in Figure 1.

FIGURE 1
Graphs of equations and the vertical line test

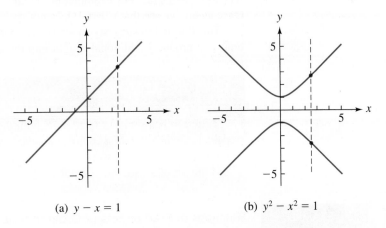

(a) $y - x = 1$ (b) $y^2 - x^2 = 1$

In Figure 1(a), each vertical line intersects the graph in exactly one point, illustrating graphically the fact that each value of the independent variable x corresponds to exactly one value of the dependent variable y. On the other hand, Figure 1(b) shows that there are vertical lines that intersect this graph in two points. This indicates that there are values of the independent variable x that correspond to two different values of the dependent variable y. These observations form the basis for the vertical line test stated below.

VERTICAL LINE TEST

An equation defines a function if and only if each vertical line in the rectangular coordinate system passes through at most one point on the graph of the function.

◆   **Definition of a Function—Set Form**

Since elements in the range of a function are paired with elements in the domain by some rule or process, this correspondence (pairing) can be illustrated by using **ordered pairs** of elements where the first component represents a domain element and the second component represents a corresponding range element. Thus, we can write functions 1 and 2 specified in Charts 1 and 2 as sets of pairs as follows:

Function 1: $\{(-2, -8), (-1, -1), (0, 0), (1, 1), (2, 8)\}$
Function 2: $\{(-2, 4), (-1, 1), (0, 0), (1, 1), (2, 4)\}$

In both cases, notice that no two ordered pairs have the same first component and different second components. On the other hand, if we list the set S of ordered pairs determined by Chart 3, we have

$$S = \{(0, 0), (1, 1), (1, -1), (4, 2), (4, -2), (9, 3), (9, -3)\}$$

In this case, there are ordered pairs with the same first component and different second components. For example, $(1, 1)$ and $(1, -1)$ both belong to the set S. Once again, we see that Chart 3 does not specify a function.

This discussion suggests an alternative but equivalent way of defining functions that produces additional insight into this concept.

FUNCTION—SET FORM

A **function** is a set of ordered pairs with the property that no two ordered pairs have the same first component and different second components. The set of all first components in a function is called the **domain** of the function and the set of all second components is called the **range**.

◆ **EXAMPLE 2**   Functions Defined as Sets of Ordered Pairs

Given the sets,

$H = \{(1, 1), (2, 1), (3, 2), (3, 4)\}$
$G = \{(2, 4), (3, -1), (4, 4)\}$

(A) Which set specifies a function?

(B) Give the domain and range of the function.

SOLUTIONS (A) Set *H* does not specify a function, since (3, 2) and (3, 4) both have the same first components. The domain value 3 corresponds to more than one range value. Set *G* does specify a function, since each domain value corresponds to exactly one range value.

(B) Domain of $G = X = \{2, 3, 4\}$; Range of $G = Y = \{-1, 4\}$ ◆

MATCHED PROBLEM 2   Repeat Example 2 for the following sets.

$$M = \{(3, 4), (5, 4), (6, -1)\}$$
$$N = \{(-1, 2), (0, 4), (1, 2), (0, 1)\}$$

◆ Function Notation

If *x* represents an element in the domain of a function *f*, then we will often use the symbol $f(x)$ in place of *y* to designate the number in the range of *f* to which *x* is paired. It is important not to be confused by this new symbol and think of it as a product of *f* and *x*. The symbol is read "*f* of *x*" or "the value of *f* at *x*." The correct use of this function symbol should be mastered early. Thus, if

$$f(x) = 2x + 3$$

then

$$f(5) = 2(5) + 3 = 13$$

That is, the function *f* assigns the range value 13 to the domain value 5. Thus, (5, 13) belongs to *f*. Can you find another ordered pair that belongs to *f*?

The correspondence between domain values and range values is usually illustrated in one of the two ways shown in Figure 2.

FIGURE 2
Function notation

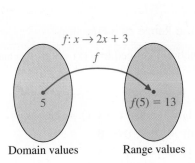

(a) Map

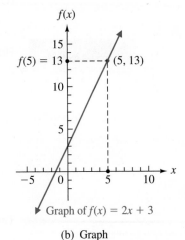

(b) Graph

◆ EXAMPLE 3   Evaluating Functions

If $f(x) = (x/2) - 1$ and $g(x) = 1 - x^2$, find

(A) $f(4)$ (B) $g(-3)$ (C) $f(2) - g(0)$
(D) $f(2 + h)$ (E) $[f(3 + h) - f(3)]/h$ (F) $f[g(3)]$

SOLUTIONS   (A) $f(4) = \dfrac{4}{2} - 1 = 2 - 1 = 1$

(B) $g(-3) = 1 - (-3)^2 = 1 - 9 = -8$

(C) $f(2) - g(0) = \left(\dfrac{2}{2} - 1\right) - (1 - 0^2) = 0 - 1 = -1$

(D) $f(2 + h) = \dfrac{2 + h}{2} - 1 = \dfrac{2 + h - 2}{2} = \dfrac{h}{2}$

(E) $\dfrac{f(3 + h) - f(3)}{h} = \dfrac{\left(\dfrac{3 + h}{2} - 1\right) - \left(\dfrac{3}{2} - 1\right)}{h}$

$= \dfrac{\dfrac{1 + h}{2} - \dfrac{1}{2}}{h} = \dfrac{1}{2}$

(F) $f[g(3)] = f(1 - 3^2) = f(-8) = \dfrac{-8}{2} - 1 = -5$ ◆

MATCHED PROBLEM 3   If $f(x) = 2x - 3$ and $g(x) = x^2 - 2$, find

(A) $f(3)$ (B) $g(-2)$ (C) $f(0) + g(1)$
(D) $f(3 + h)$ (E) $[f(2 + h) - f(2)]/h$ (F) $f[g(2)]$

◆ EXAMPLE 4   Finding the Range of a Function

If the function f defined by $f(x) = x^2 - x$ has domain $X = \{-2, -1, 0, 1, 2\}$, what is the range of f?

SOLUTION $f(-2) = (-2)^2 - (-2) = 6$
$f(-1) = (-1)^2 - (-1) = 2$
$f(0) = 0^2 - 0 = 0$
$f(1) = 1^2 - 1 = 0$
$f(2) = 2^2 - 2 = 2$

Thus, the range of $f = Y = \{0, 2, 6\}$.   ◆

MATCHED PROBLEM 4   Repeat Example 4 for $g(x) = x^2 - 4$, $X = \{-2, -1, 0, 1, 2\}$.

◆ Function Classification

Functions are classified in special categories for more efficient study. You have already had some experience with many *algebraic functions* (defined by means of the algebraic operations addition, subtraction, multiplication, division, powers, and roots), *exponential functions,* and *logarithmic functions.* In this book, we add two more classes of functions to this list, namely, the *trigonometric functions* and the *inverse trigonometric functions.* These five classes of functions are called the **elementary functions**. They are related to each other as follows:

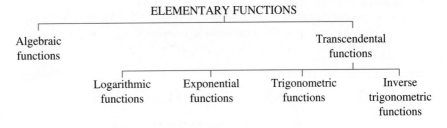

Answers to Matched Problems

1. (A) Defines a function   (B) Does not define a function

2. (A) *M* is a function   (B) $X = \{3, 5, 6\}$, $Y = \{-1, 4\}$

3. (A) 3   (B) 2   (C) -4 (D) $3 + 2h$ (E) 2   (F) 1

4. $Y = \{-4, -3, 0\}$

EXERCISE B.1

For f(x) = 4x − 1, find each of the following.

1. $f(1)$ **2.** $f(2)$ **3.** $f(-1)$

4. $f(-2)$ **5.** $f(0)$ **6.** $f(5)$

If g(x) = x − x², find each of the following.

7. $g(1)$ **8.** $g(3)$ **9.** $g(5)$

10. $g(4)$ **11.** $g(-2)$ **12.** $g(-3)$

For f(x) = 1 − 2x and g(x) = 4 − x², find each of the following.

13. $f(0) + g(0)$ **14.** $g(0) - f(0)$

15. $\dfrac{f(3)}{g(1)}$ **16.** $[g(-2)][f(-1)]$

17. $2f(-1)$ **18.** $\frac{1}{5}g(-3)$

19. $f(2 + h)$ **20.** $g(2 + h)$

21. $\dfrac{f(2 + h) - f(2)}{h}$ **22.** $\dfrac{g(2 + h) - g(2)}{h}$

23. $g[f(2)]$ **24.** $f[g(2)]$

Which of the following equations specify functions, given that x is the independent variable?

25. $x^2 + y^2 = 25$ **26.** $y = 3x + 1$

27. $2x - 3y = 6$ **28.** $y = x^2 - 2$

29. $y^2 = x$ **30.** $y = x^2$

31. $y = |x|$ **32.** $|y| = x$

33. If the function f, defined by $f(x) = x^2 - x + 1$, has domain $X = \{-2, -1, 0, 1, 2\}$, find the range Y of f.

34. If the function g, defined by $g(x) = 1 + x - x^2$, has domain $X = \{-2, -1, 0, 1, 2\}$, find the range Y of g.

35. Indicate which set specifies a function and write down its domain and range.

$$F = \{(-2, 1), (-1, 1), (0, 0)\}$$
$$G = \{(-4, 3), (0, 3), (-4, 0)\}$$

36. Repeat Problem 35 for the following sets.

$$H = \{(-1, 3), (2, 3), (-1, -1)\}$$
$$L = \{(-1, 1), (0, 1), (1, 1)\}$$

Applications

Calculus-Related   *Problems 37–39 pertain to the following relationship: The distance d (in meters) that an object falls in a vacuum in t seconds is given by*

$$d = s(t) = 4.88t^2$$

37. Find $s(0)$, $s(1)$, $s(2)$, and $s(3)$ to two decimal places.
38. Calculator Exercise $[s(2 + h) - s(2)]/h$ represents the average speed of the falling object over the time interval from $t = 2$ to $t = 2 + h$. Use a calculator to compute

each of the following to four significant digits. Then guess the speed of a free-falling object at the end of 2 sec.

(A) $\dfrac{s(3) - s(2)}{1}$ (B) $\dfrac{s(2.1) - s(2)}{0.1}$

(C) $\dfrac{s(2.01) - s(2)}{0.01}$ (D) $\dfrac{s(2.001) - s(2)}{0.001}$

(E) $\dfrac{s(2.0001) - s(2)}{0.0001}$

39. Find $[s(2 + h) - s(2)]/h$ and simplify. What happens as h gets closer and closer to 0? Interpret physically.

B.2   INVERSE FUNCTIONS

◆   **One-to-One Functions**
◆   **Inverse Functions**
◆   **Geometric Relationships**

In this section we develop techniques for determining whether the *inverse function* exists, some general properties of inverse functions, and methods for finding the rule of correspondence that defines the inverse function.

An inverse function is formed by reversing the correspondence in a given function. However, the reverse correspondence for a given function may or may not specify a function. As we will see, only functions that are *one-to-one* have inverse functions.

Many important functions are formed as inverses of existing functions. Logarithmic functions, for example, are inverses of exponential functions, and inverse trigonometric functions are inverses of trigonometric functions with restricted domains.

◆   One-to-One Functions

A **one-to-one correspondence** exists between two sets if each element of the first set corresponds to exactly one element of the second set and each element of the second set corresponds to exactly one element of the first set.

> **ONE-TO-ONE FUNCTION**
>
> A function f is **one-to-one** if each element in the range corresponds to exactly one element from the domain. (Since f is a function we already know that each element in the domain corresponds to exactly one element in the range.)

Thus, if a function f is one-to-one, then there exists a one-to-one correspondence between the domain elements and the range elements of f. To illustrate these concepts, consider the two functions f and g given in the following charts.

FUNCTION f		FUNCTION g	
Domain	Range	Domain	Range
0 ⟶ 0		0 ⟶ 0	
1 ⟶ 1		−2 ⟶	
2 ⟶ 8		2 ⟶ 4	

Function f is one-to-one, since each range element corresponds to exactly one domain element. Function g is not one-to-one, since the range element 4 corresponds to two domain elements, -2 and 2.

Geometrically, if a horizontal line intersects the graph of the function in two or more points, then the function is not one-to-one (see Figure 1a). However, if each horizontal line intersects the graph of a function in at most one point, then the function is one-to-one (see Figure 1b). These observations form the basis for the *horizontal-line test*.

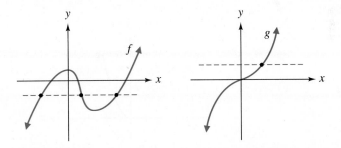

(a) Function f is not one-to-one   (b) Function g is one-to-one

FIGURE 1
Graphs of functions and the horizontal line test

HORIZONTAL-LINE TEST

A function is one-to-one if and only if each horizontal line intersects the graph of the function in at most one point.

◆ **EXAMPLE 1**   Determining Whether a Function Is One-to-One

Graph each function and determine which is one-to-one by the horizontal-line test.

(A) $f(x) = x^2$ (B) $g(x) = x^2$, $x \geq 0$

SOLUTIONS   (A) We graph the function f (Figure 2) and note that it fails the horizontal-line test. Therefore, function f is not one-to-one.

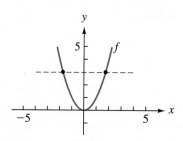

FIGURE 2

(B) We graph the function g (which is function f with a restricted domain) and find that g passes the horizontal-line test (Figure 3). Therefore, function g is one-to-one.

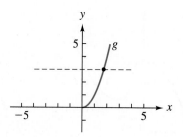

FIGURE 3   ◆

Example 1 illustrates an important point: A function may not be one-to-one, but by restricting the domain it can be made one-to-one. This is exactly what is done to form the inverse trigonometric functions from the trigonometric functions.

MATCHED PROBLEM 1   Graph each function and determine which is one-to-one by the horizontal-line test.

(A) $f(x) = (x - 1)^2$ (B) $g(x) = (x - 1)^2$, $x \geq 1$

◆ Inverse Functions

As we mentioned at the beginning of this section, we are interested in forming new functions by reversing the correspondence in a given function—that is, by reversing all the ordered pairs in the given function. The concept of a one-to-one function plays a critical role in this process. If we reverse all the ordered pairs in a function that is not one-to-one, the resulting set does not define a function. For example, reversing the ordered pairs in the function

$$g = \{(-2, 4), (0, 0), (2, 4)\} \qquad \text{Function } g \text{ is not one-to-one.}$$

produces the set

$$h = \{(4, -2), (0, 0), 4, 2)\} \qquad \text{Set } h \text{ is not a function.}$$

which does not define a function. However, reversing all the ordered pairs in a one-to-one function f always produces a new function, called the *inverse function,* which is denoted by the symbol f^{-1}. For example, reversing the ordered pairs in the function

$$f = \{(0, 2), (1, 3), (2, 4)\} \qquad \text{Function } f \text{ is one-to-one.}$$

produces the inverse function

$$f^{-1*} = \{(2, 0), (3, 1), (4, 2)\} \qquad \text{Set } f^{-1} \text{ is a function.}$$

Furthermore, we see that

$$\text{Domain of } f = \{0, 1, 2\} = \text{Range of } f^{-1}$$
$$\text{Range of } f = \{2, 3, 4\} = \text{Domain of } f^{-1}$$

In other words, reversing all the ordered pairs also reverses the domain and range. This discussion is summarized in the following definition.

INVERSE FUNCTION

The **inverse** of a one-to-one function f, denoted by f^{-1}, is the function formed by reversing all the ordered pairs in the function f. Symbolically,

$$f^{-1} = \{(b, a) \,|\, (a, b) \text{ is an element of } f\}$$
$$\text{Domain of } f^{-1} = \text{Range of } f$$
$$\text{Range of } f^{-1} = \text{Domain of } f$$

---

* *Note:* f^{-1} is a special function symbol used to represent the inverse of the function f. It does not mean $1/f$.

Immediate consequences of the definition of an inverse function are the function–inverse function identities:

FUNCTION–INVERSE FUNCTION IDENTITIES

If f is a one-to-one function, then f^{-1} exists and

$$f^{-1}[f(x)] = x \qquad \text{for all x in the domain of f}$$

and

$$f[f^{-1}(x)] = x \qquad \text{for all x in the domain of } f^{-1}$$

These identities are illustrated schematically in Figure 4.

FIGURE 4
Function–inverse function identities

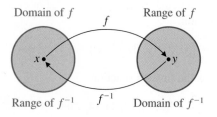

Domain of f f Range of f

x y

Range of f^{-1} f^{-1} Domain of f^{-1}

If the function f is one-to-one and if f maps x into y, then f^{-1} maps y back into x

◆   **EXAMPLE 2**   Finding the Inverse of a Function

For the one-to-one function $f(x) = 2x - 3$, find $f^{-1}(x)$, and check by showing that $f[f^{-1}(x)] = x$ and $f^{-1}[f(x)] = x$.

SOLUTION   For both f and f^{-1} we keep x as the independent variable and y as the dependent variable.

Step 1   Replace $f(x)$ with y.

    f: $y = 2x - 3$ x is independent.

Step 2   Interchange the variables x and y to form f^{-1}.

    f^{-1}: $x = 2y - 3$ x is independent.

Step 3   Solve the equation for f^{-1} for y in terms of x.

$$x = 2y - 3$$
$$y = \frac{x + 3}{2}$$

Step 4   Replace y with $f^{-1}(x)$.

$$f^{-1}(x) = \frac{x + 3}{2}$$

Step 5   Check by showing that $f[f^{-1}(x)] = x$ and $f^{-1}[f(x)] = x$.

$$f[f^{-1}(x)] = 2[f^{-1}(x)] - 3 \qquad f^{-1}[f(x)] = \frac{f(x) + 3}{2}$$

$$= 2\left(\frac{x+3}{2}\right) - 3 \qquad\qquad = \frac{(2x-3) + 3}{2}$$

$$= x + 3 - 3 = x \qquad\qquad = \frac{2x}{2} = x \qquad\qquad ◆$$

MATCHED PROBLEM 2   For the one-to-one function $g(x) = 3x + 2$, find $g^{-1}(x)$, and check by showing that $g[g^{-1}(x)] = x$ and $g^{-1}[g(x)] = x$.

◆   Geometric Relationship

We conclude this discussion of inverse functions by observing an important relationship between the graph of a one-to-one function and its inverse. As an example, let f be the one-to-one function

$$f(x) = 2x - 2 \tag{1}$$

Its inverse is

$$f^{-1}(x) = \frac{x+2}{2} \tag{2}$$

Any ordered pair of numbers that satisfies (1), when reversed in order, will satisfy (2). For example, (4, 6) satisfies (1) and (6, 4) satisfies (2). (Check this.) The graphs of f and f^{-1} are given in Figure 5.

FIGURE 5
Symmetry property of the graphs of a function and its inverse

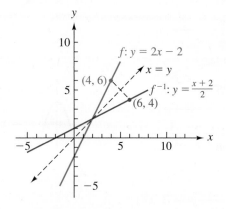

Notice that we sketched the line $y = x$ in Figure 5 to show that if we fold the paper along this line, then the graphs of f and f^{-1} will match. Actually, we can graph f^{-1} by drawing f with wet ink and folding the paper along $y = x$ before the ink dries; f will print f^{-1}. [To prove this, we need to show that the line $y = x$ is the perpendicular bisector of the line segment joining (a, b) to (b, a).]

Knowing that the graphs of f and f^{-1} are symmetric relative to the line $y = x$ makes it easy to graph f^{-1} if f is known, and vice versa.

Answers to
Matched Problems

1. (A) f is not one-to-one   (B) g is one-to-one

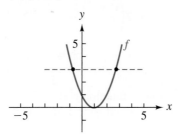   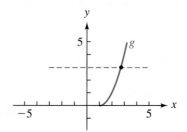

2. $g^{-1}(x) = \dfrac{x-2}{3}$

EXERCISE B.2

In Problems 1–12 indicate which functions are one-to-one.

1.

Domain	Range
-2	$\rightarrow -4$
-1	$\rightarrow -2$
0	$\rightarrow 0$
1	$\rightarrow 2$
2	$\rightarrow 4$

2.

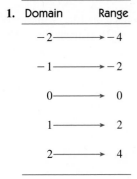

Domain   Range

-2

-1 $\rightarrow 3$

$0 \longrightarrow 0$

1

2 $\rightarrow 5$

3.

Domain   Range

0
1
$2 \longrightarrow 9$
3
4

4.

Domain	Range
0	$\rightarrow 5$
1	$\rightarrow 3$
2	$\rightarrow 1$
3	$\rightarrow 2$
4	$\rightarrow 4$

5.

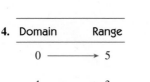

6.

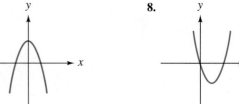

7.

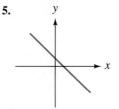

8.

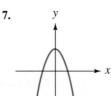

9.

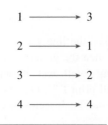

10.

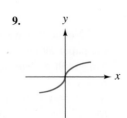

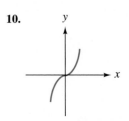

11.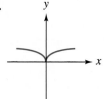

12.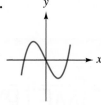

13. Which of the given functions is one-to-one? Write the inverse of the function that is one-to-one as a set of ordered pairs of numbers, and indicate its domain and range.

$$f = \{(-1, 0), (1, 1), (2, 0)\}$$
$$g = \{(-2, -8), (1, 1), (2, 8)\}$$

14. Which of the given functions is one-to-one? Write the inverse of the function that is one-to-one as a set of ordered pairs of numbers, and indicate its domain and range.

$$f = \{(-2, 4), (0, 0), (2, 4)\}$$
$$g = \{(9, 3), (4, 2), (1, 1)\}$$

15. Given $H = \{(-1, 0.5), (0, 1), (1, 2), (2, 4)\}$, graph H, H^{-1}, and $y = x$ on the same coordinate system.

16. Given $F = \{(-5, 0), (-2, 1), (0, 2), (1, 4), (2, 7)\}$, graph F, F^{-1}, and $y = x$ on the same coordinate system.

In Problems 17–20 find the inverse for each function in the form of an equation.

17. $f(x) = 2x - 7$

18. $g(x) = \dfrac{x}{2} + 1$

19. $h(x) = \dfrac{x + 3}{3}$

20. $f(x) = \dfrac{x - 2}{3}$

In Problems 21 and 22 graph the indicated function, its inverse, and $y = x$ on the same coordinate system.

21. f and f^{-1} in Problem 17

22. g and g^{-1} in Problem 18

23. For $f(x) = 2x - 7$, find $f^{-1}(x)$ and $f^{-1}(3)$.

24. For $g(x) = (x/2) + 1$, find $g^{-1}(x)$ and $g^{-1}(-3)$.

25. For $h(x) = (x/3) + 1$, find $h^{-1}(x)$ and $h^{-1}(2)$.

26. For $m(x) = 3x + 2$, find $m^{-1}(x)$ and $m^{-1}(5)$.

27. Find $f[f^{-1}(4)]$ for Problem 23.

28. Find $g^{-1}[g(2)]$ for Problem 24.

29. Find $h^{-1}[h(x)]$ for Problem 25.

30. Find $m[m^{-1}(x)]$ for Problem 26.

31. Find $h[h^{-1}(x)]$ for Problem 25.

32. Find $m^{-1}[m(x)]$ for Problem 26.

PLANE GEOMETRY: SOME USEFUL FACTS

The following four sections include a brief list of plane geometry facts that are of particular use in studying trigonometry. They are grouped together for convenient reference.

C.1 LINES AND ANGLES

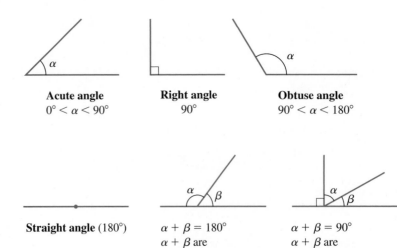

Acute angle	Right angle	Obtuse angle
$0° < \alpha < 90°$	$90°$	$90° < \alpha < 180°$

Straight angle (180°)

$\alpha + \beta = 180°$
$\alpha + \beta$ are
supplementary angles

$\alpha + \beta = 90°$
$\alpha + \beta$ are
complementary angles

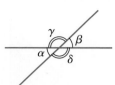

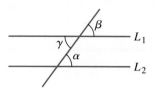

A **straight angle** divided into equal parts forms two **right angles**.

$\alpha = \beta$
$\gamma = \delta$
$\alpha + \delta = \beta + \gamma = 180°$
$\alpha + \delta + \beta + \gamma = 360°$

If $L_1 \parallel L_2$, then $\alpha = \beta = \gamma$.
If $\alpha = \beta = \gamma$, then $L_1 \parallel L_2$.

C.2   TRIANGLES

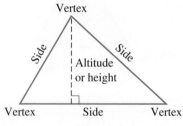

Triangle

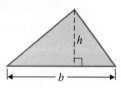

Area: $A = \frac{1}{2}bh$

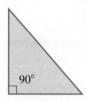

Right triangle
One right angle

(a) **Acute triangle**
All acute angles

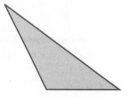

(b) **Obtuse triangle**
One obtuse angle

Oblique triangles
No right angles

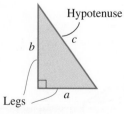

$c^2 = a^2 + b^2$
Pythagorean theorem

$\alpha + \beta = 90°$

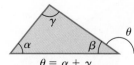

$\theta = \alpha + \gamma$

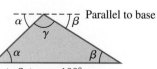

$\alpha + \beta + \gamma = 180°$
The sum of angle measures of all angles in a triangle is 180°.

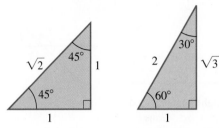

Special triangles

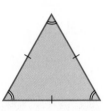

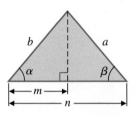

Isosceles triangle
At least two equal sides
At least two equal angles

Equilateral triangle
All sides equal
All angles equal

If $a = b$, then $\alpha = \beta$ and $m = n/2$.
If $\alpha = \beta$, then $a = b$ and $m = n/2$.

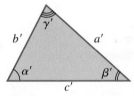

$$\alpha = \alpha', \beta = \beta', \gamma = \gamma'$$

$$\frac{a}{a'} = \frac{b}{b'} = \frac{c}{c'}$$

Similar triangles
Two triangles are similar if two angles of one
triangle are equal to two angles of the other.

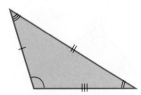

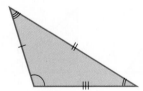

Congruent triangles
Corresponding parts of congruent triangles are equal.

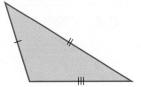

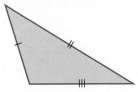

SSS
If three sides of one triangle are equal to the corresponding
sides of another triangle, the two triangles are congruent.

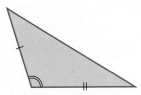

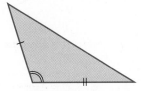

SAS
If two sides and the included angle of one triangle are
equal to the corresponding parts of another triangle, the
two triangles are congruent.

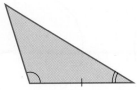

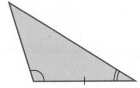

ASA
If two angles and the included side of one triangle are
equal to the corresponding parts of another triangle, the
two triangles are congruent.

C.3   QUADRILATERALS

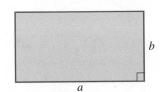

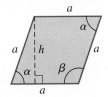

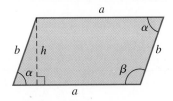

Square
$A = a^2$
$P = 4a$

Rectangle
$A = ab$
$P = 2a + 2b$

Rhombus
Opposite sides are parallel.
$A = ah$
$P = 4a$
$\alpha + \beta = 180°$

Parallelogram
Opposite sides are parallel.
$A = ah$
$P = 2a + 2b$
$\alpha + \beta = 180°$

C.4   CIRCLES

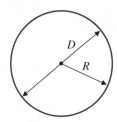

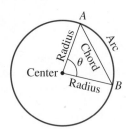

R = Radius
D = Diameter
$D = 2R$
$A = \pi R^2$ (Area)
$C = 2\pi R = \pi D$ (Circumference)
$C/D = \pi$ (For all circles)

$\widehat{AB}$ = Arc AB
AB = Chord AB
$\angle\theta$ subtends AB
$\angle\theta$ subtends $\widehat{AB}$
$\widehat{AB}$ subtends $\angle\theta$
AB subtends $\angle\theta$

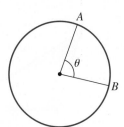

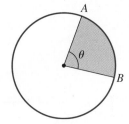

$$\frac{\widehat{AB}}{C} = \frac{\theta}{360°}$$

(C = Circumference)

Circular sector

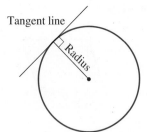

A radius of a circle
is $\perp$ to a tangent line
at the point of tangency.

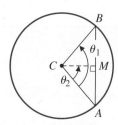

If $CM \perp AB$, then
$$AM = \tfrac{1}{2}AB$$
$$\theta_2 = \tfrac{1}{2}\theta_1$$

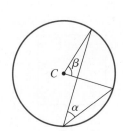

$\beta = 2\alpha$

SELECTED ANSWERS

CHAPTER 1

Exercise 1.1

1. 180° **3.** 45° **5.** 240° **7.** Obtuse
9. Straight **11.** Acute **13.** None **15.** 633′
17. 4,212″ **19.** 1°12′ **21.** 43.351° **23.** 2.213°
25. 103.295° **27.** 13°37′59″ **29.** 83°1′1″
31. 187°12′14″ **33.** $\alpha < \beta$ **35.** $\alpha > \beta$
37. $\alpha < \beta$ **39.** 100 cm **41.** 450 km
43. 240,000 mi **45.** 1,440 cm **47.** 262 cm^2
49. 71.1° **51.** 5.58 mm **53.** 11.5 mm
55. 679 mi **57.** 553 mi **59.** 590 nautical mi
61. 480 nautical mi **63.** 34.9 m **65.** 865,000 mi

Exercise 1.2

1. $b' = 6$ **3.** $c = 110$ **5.** $c' = 2.8$
7. $b = 21$ m, $c = 24$ m **9.** $a = 24$ in., $c = 56$ in.
11. $a = 8.5 \times 10^4$ km, $b = 1.8 \times 10^5$ km
13. $b = 50$ m, $c = 55$ m **15.** $a = 1.2 \times 10^9$ yd,
$c = 2.7 \times 10^9$ yd **17.** $a = 3.6 \times 10^{-5}$ mm,
$b = 7.6 \times 10^{-5}$ mm **19.** $c = 108$ ft **21.** 19.5 ft
23. 63 ft **25.** 1.1 km

Exercise 1.3

1. a/c **3.** b/a **5.** c/a **7.** $\sin \theta$ **9.** $\tan \theta$
11. $\csc \theta$ **13.** 0.432 **15.** 0.709 **17.** 1.41
19. 0.294 **21.** 0.703 **23.** 1.08 **25.** 53.44°
27. 44°20′ **29.** 62°50′ **31.** 22°9′37″
33. $90° - \theta = 31°20′$, $a = 7.80$ mm, $b = 12.8$ mm
35. $90° - \theta = 6.3°$, $a = 0.354$ km, $c = 3.23$ km
37. $90° - \theta = 18.5°$, $a = 4.28$ in., $c = 13.5$ in.
39. $\theta = 56°20′$, $b = 33.6$ cm, $c = 40.4$ cm
41. $\theta = 28°30′$, $90° - \theta = 61°30′$, $a = 118$ ft
43. $\theta = 50.7°$, $90° - \theta = 39.3°$, $c = 171$ mi
49. $90° - \theta = 52.54°$, $a = 6.939$ cm, $c = 8.742$ cm
51. $90° - \theta = 6°48′$, $b = 199.8$ mi, $c = 201.2$ mi
53. $\theta = 37°6′$, $90° - \theta = 52°54′$, $c = 70.27$ cm
55. $\theta = 44.11°$, $90° - \theta = 45.89°$, $a = 36.17$ cm

57. $(\sin \theta)^2 + (\cos \theta)^2 = \left(\dfrac{b}{c}\right)^2 + \left(\dfrac{a}{c}\right)^2 = \dfrac{a^2 + b^2}{c^2} = \dfrac{c^2}{c^2} = 1$

61. (A) $\sin \theta = \dfrac{AD}{OD} = \dfrac{AD}{1} = AD$

  (B) $\tan \theta = \dfrac{AD}{OA} = \dfrac{DC}{OD} = \dfrac{DC}{1} = DC \quad (\angle OED = \theta)$

  (C) $\csc \theta = \dfrac{OE}{OD} = \dfrac{OE}{1} = OE$

63. (A) $\sin \theta$ approaches 1
  (B) $\tan \theta$ increases without bound
  (C) $\csc \theta$ approaches 1
65. (A) $\cos \theta$ approaches 1
  (B) $\cot \theta$ increases without bound
  (C) $\sec \theta$ approaches 1

Exercise 1.4

1. 7.0 m **3.** 211 m **5.** 1.1 km **7.** 29,400 m, or
29.4 km **9.** 22° **11.** (A) 5.1 ft (B) 2.6 ft
13. 3.5 m **15.** 5.69 cm **17.** 134 ft
19. 8.4 ft, 21 ft

21. (A) $\cos \alpha = \dfrac{r}{r + h}$ (B) $r = \dfrac{h \cos \alpha}{1 - \cos \alpha}$

(C) 3,960 mi
23. 8.4 mi **25.** 19,600 mi

27. (A) $T = \dfrac{d - c \cot \theta}{p} + \dfrac{c \csc \theta}{q}$

  (B) 110 sec (C) 360 m
29. \$169,000 **31.** 0.97 km **33.** 386 ft apart; 484 ft
high **35.** $g = 32.0$ ft/sec^2 **37.** $\sin \theta = \frac{3}{5}$

Chapter 1 Review Exercise

1. 7,280″ [1.1] **2.** 60° [1.1] **3.** 8,000 [1.2]
4. 36.38° [1.1] **5.** 560 ft [1.2]
6. (A) b/c (B) c/a (C) b/a (D) c/b (E) a/c
  (F) a/b [1.3]
7. $90° - \theta = 54.8°$, $a = 16.5$ cm, $b = 11.6$ cm [1.3]
8. 144° [1.1] **9.** 4.19 in. [1.1]
10. 74°16′23″ [1.1] **11.** $\alpha > \beta$ [1.1]

12. 7.1×10^{-6} mm [*1.2*]

13. (A) $\cos \theta$ (B) $\tan \theta$ (C) $\sin \theta$ (D) $\sec \theta$
(E) $\csc \theta$ (F) $\cot \theta$ [*1.3*]

14. $90° - \theta = 27°40'$, $b = 7.63 \times 10^{-8}$ m,
$c = 8.61 \times 10^{-8}$ m [*1.3*]

15. (A) $58.97°$ (B) $55°50'$ (C) $0°47'27''$ [*1.3*]

16. $\theta = 40.3°$, $90° - \theta = 49.7°$, $c = 20.6$ mm [*1.3*]

17. $\theta = 40°20'$, $90° - \theta = 49°40'$ [*1.3*]

18. 8.7 ft [*1.2*]   **19.** 940 ft [*1.1*]

20. 107 ft² [*1.1*]

21. $\theta = 66°17'$, $a = 93.56$ km, $b = 213.0$ km [*1.3*]

22. $\theta = 60.28°$, $90° - \theta = 29.72°$, $b = 4,241$ m [*1.3*]

23. 1.0496 [*1.3*]   **24.** 8.8 ft [*1.2*]

25. 24.5 ft, 24.1 ft [*1.4*]   **26.** 2.3°, 0.07 or 7% [*1.4*]

27. 954 mi [*1.1*]   **28.** 46° [*1.4*]

29. 830 m [*1.4*]   **30.** 1,240 m [*1.4*]

31. 760 mph [*1.4*]

32. (A) $\beta = 90° - \alpha$ (B) $r = h \tan \alpha$
(C) $H - h = (R - r) \cot \alpha$ [*1.4*]

CHAPTER 2

Exercise 2.1

1. $\pi/2$ rad   **3.** $\pi/3$ rad   **5.** $2\pi/3$ rad   **7.** 45°

9. 30°   **11.** 150°   **13.** $\pi/6$ rad, $2\pi/6$ rad or $\pi/3$ rad,
$3\pi/6$ rad or $\pi/2$ rad, $4\pi/6$ rad or $2\pi/3$ rad, $5\pi/6$ rad, $6\pi/6$
rad or π rad

15.

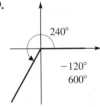

17.

19.

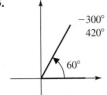

21. (A) 2 rad (B) 1.5 rad   **23.** $\pi/10$ rad ≈ 0.3142 rad

25. $3\pi/20$ rad ≈ 0.4712 rad   **27.** $13\pi/18$ rad $\approx$
2.269 rad   **29.** $(288/\pi)° \approx 91.67°$ **31.** 15° (exact)

33. 3° (exact)

35.

37.

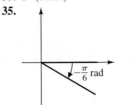

39.

41. (A) $476°$, $-665°$ (B) 9.83 rad, -21.5 rad

43. (A) 58.2 m (B) 8.29 m (C) 20.5 m
(D) 47.1 m

45. (A) 46.4 cm² (B) 42.8 cm² (C) 1.00×10^2 cm²
(D) 192 cm²

47. I   **49.** III   **51.** II   **53.** 1.0008 rad

55. 18.2430°   **57.** 0.4605 rad   **59.** $7\pi/12$ rad $\approx$
1.83 rad   **61.** 12 cm   **63.** 19°   **65.** 1.4×10^6 km

67. 175 ft   **69.** 161 mi   **71.** 843 mi   **73.** 326 mi

75. $\pi/26$ rad ≈ 0.12 rad   **77.** 6,800 mi   **79.** 31 ft

81. 10π rad, 7.2π rad, $2\pi n$ rad   **83.** 6.5 revolutions,
40.8 rad   **85.** 859°

Exercise 2.2

1. 3 mm/sec   **3.** 17 rad/sec   **5.** 3.7 rad/hr
7. 0.593 rad/sec   **9.** 75 m/sec   **11.** 4.65 rad/hr
13. 223.5 rad/sec or 35.6 rps   **15.** $\pi/4,380$ rad/hr;
66,700 mph   **17.** (A) 0.634 rad/hr (B) 28,100 mph
19. 6,900 mph   **21.** 1.61 hr   **23.** $a = 15 \tan 2\pi t$

Exercise 2.3

1. $\sin \theta = \frac{4}{5}$, $\csc \theta = \frac{5}{4}$, $\cos \theta = \frac{3}{5}$, $\sec \theta = \frac{5}{3}$, $\tan \theta = \frac{4}{3}$,
$\cot \theta = \frac{3}{4}$; same values for $Q(6, 8)$ (Why?)

3. $\sin \theta = -\frac{3}{5}$, $\csc \theta = -\frac{5}{3}$, $\cos \theta = \frac{4}{5}$, $\sec \theta = \frac{5}{4}$, $\tan \theta$
$= -\frac{3}{4}$, $\cot \theta = -\frac{4}{3}$; same values for $Q(12, -9)$ (Why?)

5. $\sin \theta = \frac{4}{5}$, $\tan \theta = \frac{4}{3}$, $\csc \theta = \frac{5}{4}$, $\sec \theta = \frac{5}{3}$, $\cot \theta = \frac{3}{4}$

7. $\sin \theta = -\frac{4}{5}$, $\tan \theta = -\frac{4}{3}$, $\csc \theta = -\frac{5}{4}$, $\sec \theta = \frac{5}{3}$,
$\cot \theta = -\frac{3}{4}$

9. $\sin \theta = -\frac{4}{5}$, $\cos \theta = -\frac{3}{5}$, $\tan \theta = \frac{4}{3}$, $\sec \theta = -\frac{5}{3}$,
$\cot \theta = \frac{3}{4}$

11. $\sin \theta = -\frac{4}{5}$, $\cos \theta = \frac{3}{5}$, $\tan \theta = -\frac{4}{3}$, $\sec \theta = \frac{5}{3}$, $\cot \theta = -\frac{3}{4}$

13. 57.29   **15.** -0.9900 **17.** 3.236   **19.** -2.904

21. 0.4202   **23.** -0.4577 **25.** 5.824   **27.** 0.9272

29. 0.8439   **31.** 4.331

33. $\sin \theta = \frac{1}{2}$, $\cos \theta = \sqrt{3}/2$, $\tan \theta = 1/\sqrt{3}$, $\csc \theta = 2$, $\sec \theta = 2/\sqrt{3}$, $\cot \theta = \sqrt{3}$

35. $\sin \theta = -\sqrt{3}/2$, $\cos \theta = \frac{1}{2}$, $\tan \theta = -\sqrt{3}$, $\csc \theta = -2/\sqrt{3}$, $\sec \theta = 2$, $\cot \theta = -1/\sqrt{3}$

37. $\sin \theta = -\sqrt{2}/2$, $\cos \theta = \sqrt{2}/2$, $\tan \theta = -1$, $\csc \theta = -\sqrt{2}$, $\sec \theta = \sqrt{2}$, $\cot \theta = -1$

39. I, IV   **41.** I, III   **43.** I, IV

45. III, IV   **47.** II, IV   **49.** III, IV

51. $\cos \theta = -\sqrt{5}/3$, $\tan \theta = 2/\sqrt{5}$, $\csc \theta = -\frac{3}{2}$, $\sec \theta = -3\sqrt{5}$, $\cot \theta = \sqrt{5}/2$

53. $\cos \theta = \sqrt{5}/3$, $\tan \theta = -2/\sqrt{5}$, $\csc \theta = -\frac{3}{2}$, $\sec \theta = 3/\sqrt{5}$, $\cot \theta = -\sqrt{5}/2$

55. $\sin \theta = -\sqrt{2}/\sqrt{3}$, $\cos \theta = 1/\sqrt{3}$, $\tan \theta = -\sqrt{2}$, $\csc \theta = -\sqrt{3}/\sqrt{2}$, $\cot \theta = -1/\sqrt{2}$

57. 0.6191   **59.** 4.938   **61.** $-0.081\ 69$

63. -1.553 **65.** -0.8812 **67.** 1.079   **69.** 1.778

71. 0.7112   **73.** -0.9800 **75.** Tangent and secant

77. (A) $\theta = 1.2$ rad
   (B) $(a, b) = (5 \cos 1.2, 5 \sin 1.2) = (1.81, 4.66)$

79. (A) $\theta = 2$ rad
   (B) $(a, b) = (1 \cos 2, 1 \sin 2) = (-0.416, 0.909)$

81. 3.22 units   **83.** k, $0.94k$, $0.77k$, $0.50k$, $0.17k$

85. Summer solstice: $E = 0.97k$; winter solstice: $E = 0.45k$

87. (A) 2.598 08, 2.938 93, 3.139 53, 3.141 57, 3.141 59
   (B) $\pi = 3.141\ 592\ 65 \ldots$ (C) No

91. $I = 24$ amp   **93.** $L = 12 \csc \theta + 8 \sec \theta$

95. (A) 2.01; -0.14 (B) $y = -0.75x + 3.75$

Exercise 2.4

1. 29.3°   **3.** 25.1°   **5.** 24.4°   **7.** Closer

9. 40 km/hr   **11.** $\theta = 60°$

13. 3×10^{10} cm/sec   **15.** $-6°$

Exercise 2.5

1. $\alpha = \theta = 60°$

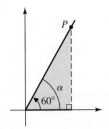

3. $\alpha = |-60°| = 60°$ **5.** $\alpha = |-\pi/3| = \pi/3$

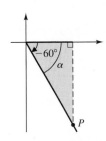   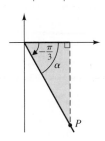

7. $\alpha = \pi - 3\pi/4 = \pi/4$

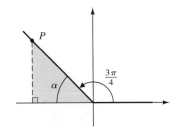

9. $\alpha = 210° - 180° = 30°$

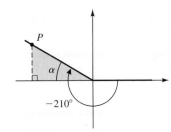

11. $\alpha = 5\pi/4 - \pi = \pi/4$

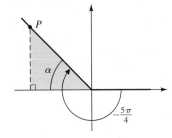

13. 1   **15.** $\frac{1}{2}$ **17.** 1   **19.** 1   **21.** $1/\sqrt{3}$, or $\sqrt{3}/3$ **23.** 0   **25.** $-\frac{1}{2}$ **27.** 0   **29.** $-\sqrt{3}$

31. $\sqrt{3}/2$ **33.** Not defined   **35.** $-\sqrt{3}$ **37.** $-\frac{1}{2}$

39. -1 **41.** $-1/\sqrt{2}$, or $-\sqrt{2}/2$ **43.** -1

45. $-\sqrt{3}/2$ **47.** $-\sqrt{3}$ **49.** $1/\sqrt{2}$, or $\sqrt{2}/2$

51. $2/\sqrt{3}$ **53.** 0   **55.** (A) 90°, 270° (B) $\pi/2$, $3\pi/2$

57. (A) 0°, 180° (B) 0, π **59.** (A) 30° (B) $\pi/6$

61. (A) $120°$ (B) $2\pi/3$ **63.** (A) $120°$ (B) $2\pi/3$
65. $240°, 300°$ **67.** $5\pi/6, 11\pi/6$
69. (A) $x = 14, y = 7\sqrt{3}$
     (B) $x = 4/\sqrt{2}, y = 4/\sqrt{2}$
     (C) $x = 10/\sqrt{3}, y = 5/\sqrt{3}$
71. $3\sqrt{3}$ cm^2 **73.** $150\sqrt{3}$ in.^{2}

Exercise 2.6

1. (A) π (B) $3\pi/2$
3. (A) $(1, 0)$ (B) $(0, 1)$ (C) $(0, 1)$
    (D) $(-1, 0)$ (E) $(1, 0)$ (F) $(0, 1)$
5. (A) 0 to 1   (B) 1 to 0   (C) 0 to -1
    (D) -1 to 0   (E) 0 to 1
7. (A) 1 to 0   (B) 0 to -1 (C) -1 to 0
    (D) 0 to 1   (E) 1 to 0
9. $\pi/2, 5\pi/2$ **11.** $0, \pi, 2\pi, 3\pi, 4\pi$
13. $0, \pi, 2\pi, 3\pi, 4\pi$ **15.** $3\pi/2, 7\pi/2$
17. $-2\pi, 0, 2\pi$ **19.** $-3\pi/2, -\pi/2, \pi/2, 3\pi/2$
21. $\pi/2, 3\pi/2, 5\pi/2, 7\pi/2$ **23.** $0, \pi, 2\pi, 3\pi, 4\pi$
25. 0.7   **27.** -0.7 **29.** -0.8 **31.** -2 **33.** 1
35. -2 **37.** -0.2088 **39.** 7.228   **41.** 119.2
43. -0.04334 **45.** -29.99 **47.** -12.04
49. $-1/\sqrt{2}$ or $-\sqrt{2}/2$ **51.** $-\sqrt{2}$ **53.** -1
55. $-1/\sqrt{3}$ or $-\sqrt{3}/3$ **57.** -2 **59.** Not defined
61. All 0.9525
63. (A) Both 1.6   (B) Both -1.5 (C) Both 0.96
65. (A) Both -0.14 (B) Both 0.23   (C) Both 0.99
67. (A) 1.0   (B) 1.0   (C) 1.0   **69.** 1
71. csc x **73.** csc x **75.** cos x
77. (A) Identity (4)   (B) Identity (9)   (C) Identity (2)
79. 2π **81.** $s_1 = 1, s_2 = 1.540\ 302, s_3 = 1.570\ 792,$
$s_4 = 1.570\ 796, s_5 = 1.570\ 796; \pi/2 = 1.570\ 796$

Chapter 2 Review Exercise

1. (A) $\pi/3$ (B) $\pi/4$ (C) $\pi/2$ [2.1]
2. (A) $30°$ (B) $90°$ (C) $45°$ [2.1]
3. (A) $874.3°$ (B) -6.793 rad   [2.1]
4. 185 ft/min   [2.2]   **5.** 80 rad/hr   [2.2]
6. $\sin \theta = \frac{3}{5}; \tan \theta = -\frac{3}{4}$ [2.3]
7. (A) 0.7355   (B) 1.085   [2.3, 2.6]
8. (A) 0.9171   (B) 0.9099   [2.3, 2.6]
9. (A) 0.9394   (B) 5.177   [2.3, 2.6]
10. (A) $\alpha = 60°$

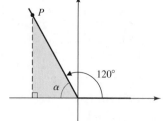

(B) $\alpha = \dfrac{\pi}{4}$

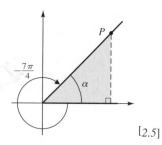

[2.5]

11. (A) $\sqrt{3}/2$ (B) $1/\sqrt{2}$, or $\sqrt{2}/2$ (C) 0   [2.5]
12. (A) $(1, 0)$ (B) $(-1, 0)$ (C) $(0, 1)$
    (D) $(0, 1)$ (E) $(-1, 0)$ (F) $(0, -1)$ [2.6]
13. (A) 0 to 1   (B) 1 to 0   (C) 0 to -1
    (D) -1 to 0   (E) 0 to 1   (F) 1 to 0   [2.6]
14. (A) 1 to 0   (B) 0 to -1 (C) -1 to 0
    (D) 0 to 1   (E) 1 to 0   (F) 0 to -1 [2.6]
15. $42°$ [2.1]   **16.** 3 rad   [2.1]   **17.** 6 cm   [2.1]
18. $53\pi/45$ [2.1]   **19.** $15°$ [2.1]
20. 0; not defined; 0; not defined   [2.5]
21. (A) I   (B) IV   [2.3]
22. (A) $7.4°$ (B) $76°40'$ (C) $37°40'$ [2.5]
23. (A) 0.75 rad   (B) 1.28 rad   (C) 0.86 rad   [2.5]
24. -0.992 [2.3, 2.6]   **25.** 0.130   [2.3, 2.6]
26. 0.973   [2.3, 2.6]   **27.** -4.34 [2.3, 2.6]
28. -1.30 [2.3, 2.6]   **29.** 1.26   [2.3, 2.6]
30. 0.683   [2.3, 2.6]   **31.** -1.07 [2.3, 2.6]
32. 0.284   [2.3, 2.6]   **33.** -3.38 [2.3, 2.6]
34. 0.757   [2.3, 2.6]   **35.** -0.864 [2.3, 2.6]
36. (A) $\pi/6$ (B) $\pi/4$ (C) $\pi/3$ [2.5]
37. $-\sqrt{3}/2$ [2.5]   **38.** $-1/\sqrt{3}$ [2.5]
39. $-1/\sqrt{2}$ [2.5]   **40.** -1 [2.5]
41. -1 [2.5]   **42.** 0   [2.5]   **43.** $\sqrt{3}/2$ [2.5]
44. -2 [2.5]   **45.** -1 [2.5]
46. Not defined   [2.5]   **47.** $\sqrt{3}/2$ [2.5]
48. $\frac{1}{2}$ [2.5]   **49.** 0.407 24   [2.3, 2.6]
50. $-0.338\ 84$ [2.3, 2.6]   **51.** 0.646 92   [2.3, 2.6]
52. 0.496 39   [2.3, 2.6]
53. $\cos \theta = \frac{3}{5}, \tan \theta = -\frac{4}{3}$ [2.3]   **54.** $7\pi/6$ [2.5]
55. $\cos \theta = \sqrt{21}/5, \sec \theta = 5/\sqrt{21}, \csc \theta = -\frac{5}{2},$
    $\tan \theta = -2/\sqrt{21}, \cot \theta = -\sqrt{21}/2$ [2.3]
56. $135°, 315°$ [2.5]   **57.** $5\pi/6, 7\pi/6$ [2.5]
58. (A) 20.3 cm   (B) 4.71 cm   [2.1]
59. (A) 618 ft^2 (B) 1,880 ft^2 [2.1]
60. 215.6 mi   [2.1]   **61.** 0.422 rad/sec   [2.2]
62. All 0.754   [2.6]
63. (A) Both -0.871 (B) Both -1.40
    (C) Both -3.38 [2.6]
64. D [2.6]   **65.** cos x [2.6]   **66.** cos x [2.6]
67. $(\cos 1.4, \sin 1.4) \approx (0.170, 0.985)$ [2.3]
68. 57 m   [2.1]   **69.** 5.74 units   [2.3]

70. 200 rad, $100/\pi \approx 31.8$ revolutions   *[2.1]*
71. 7.5 rev, 15 rev   *[2.1]*   **72.** 62 rad/sec   *[2.2]*
73. 16,400 mph   *[2.2]*   **74.** -17.6 amp   *[2.3]*
75. (A) $L = 10 \csc \theta + 2 \sec \theta$ *[2.3]*

(B) θ	0.9	1.0	1.1	1.2	1.3
L	16.0	15.6	15.6	16.2	17.9

76. $\beta = 23.3°$ *[2.4]*   **77.** $\alpha = 41.1°$ *[2.4]*
78. 11 mph   *[2.4]*

CHAPTER 3

Exercise 3.1

1. 2π, 2π, π **3.** (A) 1 unit   (B) Indefinitely far
(C) Indefinitely far   **5.** $-3\pi/2$, $-\pi/2$, $\pi/2$, $3\pi/2$
7. -2π, $-\pi$, 0, π, 2π **9.** No x intercepts
11. (A) None   (B) -2π, $-\pi$, 0, π, 2π
(C) $-3\pi/2$, $-\pi/2$, $\pi/2$, $3\pi/2$

13.
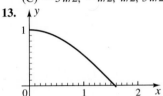

15.
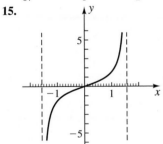

17.
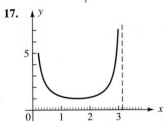

19.
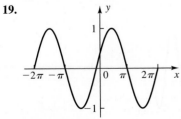

21.
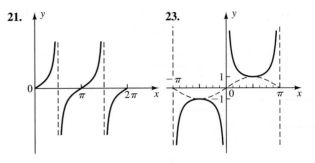

23.

25. 0, $\pi/2$, and π are not in the domains of the respective functions. Either an error message will occur or some very large number will appear because of round-off error.

27. (A)
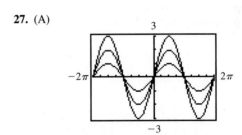

(B) Highest point: 1, 2, 3; lowest point: -1, -2, -3
(C) Highest point: p; lowest point: $-p$

29. (A)
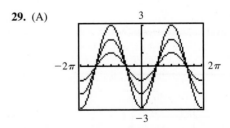

(B) Highest point: 1, 2, 3; lowest point: -1, -2, -3
(C) Highest point: $-p$; lowest point: p

31. (A)
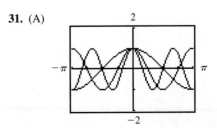

(B) 1, 2, 3   (C) n

33. (A)

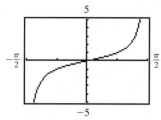

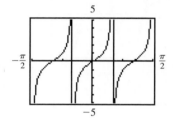

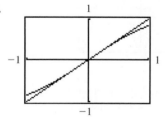

(B) 1, 2, 3
(C) *n*

35.

Exercise 3.2

1.

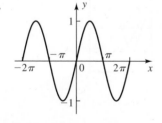

3. Amplitude = 3, Period = 2π

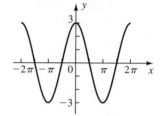

5. Amplitude = 2, Period = 2π

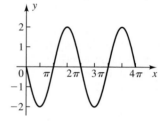

7. Amplitude = $\frac{1}{2}$, Period = 2π

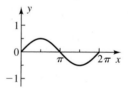

9. Amplitude = 1, Period = π

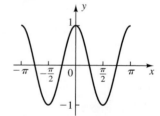

11. Amplitude = 1, Period = 1

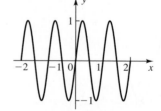

13. Amplitude = 1, Period = 8π

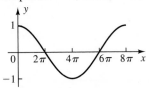

15. Amplitude = 2, Period = $\pi/2$

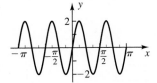

17. Amplitude = $\frac{1}{3}$, Period = 1

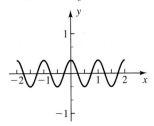

19. Amplitude = $\frac{1}{4}$, Period = 4π

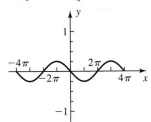

21.

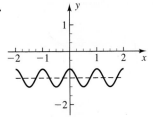

23.

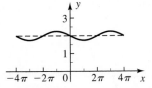

25. $y = 5 \sin 2x$ **27.** $y = -4 \sin(\pi x/2)$

29. $y = 8 \cos(x/4)$ **31.** $y = -\cos(\pi x/3)$

33. $y = 0.5 \sin 2x$

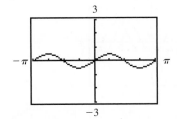

35. $y = 1 + \cos 2x$

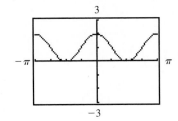

37. $y = 2 \cos 4x$

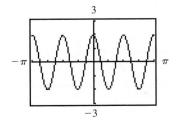

39. (A)

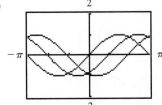

(B) The graph of $y = \sin(x - a)$ is the same as the graph of sin x shifted a units to the right.

41. (A)

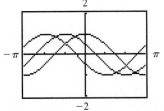

(B) The graph of $y = \cos(x + a)$ is the same as the graph of cos x shifted a units to the left.

43. (A)

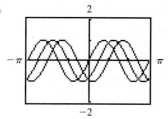

(B) The graph of $y = \sin(2x - a)$ is the same as the graph of $\sin 2x$ shifted $a/2$ units to the right.

45. Amplitude $= 110$, Period $= \frac{1}{60}$ sec, Frequency $= 60$ Hz

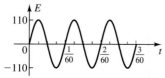

47. $E = 12 \cos 80\pi t$

49. (A) 2,220 kg
(B) $y = 0.2 \sin 2\pi t$
(C)

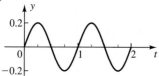

51.

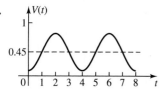

53.

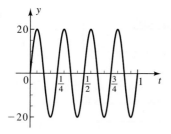

Exercise 3.3

1. Phase shift $= -\pi/2$

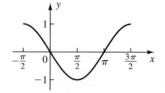

3. Phase shift $= \pi/4$

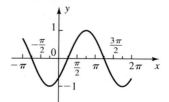

5. Phase shift $= \pi/2$

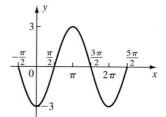

7. Amplitude $= 1$, Period $= 1$, Phase shift $= \frac{1}{2}$

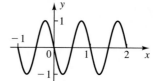

9. Amplitude $= 4$, Period $= 2$, Phase shift $= -\frac{1}{4}$

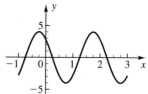

11. Amplitude $= 2$, Period $= \pi$, Phase shift $= -\pi/2$

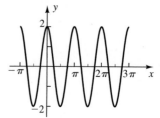

13.

15.

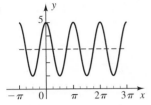

17. Amplitude = 2, Period = $2\pi/3$, Phase shift = $\pi/6$

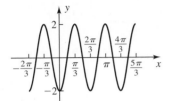

19.

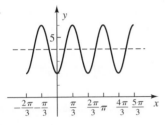

21. Both have the same graph; thus, $\cos(x - \pi/2) = \sin x$ for all x.

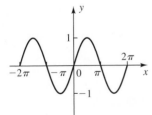

23. $y = 5 \sin(\pi x/2 - \pi/2)$ **25.** $y = 2 \cos(x/2 - 3\pi/4)$
27. x intercept: -1.047; $y = 2 \sin(x + 1.047)$

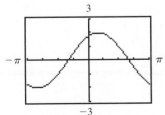

29. x intercept: 0.785; $y = 2 \sin(x - 0.785)$

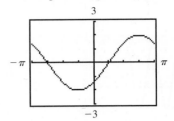

31. x intercept: -0.644; $y = 5 \sin(2x + 1.288)$

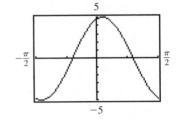

33. x intercept: 1.682; $y = 3 \sin(x/2 - 0.841)$

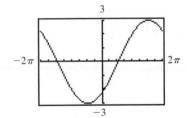

35. Amplitude = 5m, Period = 12 sec,
Phase shift = -3 sec

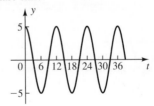

37. Amplitude = 30, Period = $\frac{1}{60}$, Frequency = 60 Hz,
Phase shift = $\frac{1}{120}$

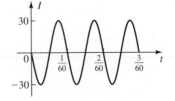

39. $x = 20 \sin(8\pi t + \pi/2)$; Amplitude = 20, Period = $\frac{1}{4}$, Phase shift = $-\frac{1}{16}$

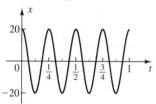

41. (A) 85

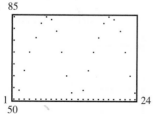

(B) $y = 67.5 + 17.5 \sin\left(\dfrac{\pi}{6}x - 2\right)$

(C) 85

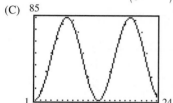

Exercise 3.4

1. Amplitude = 10, Frequency = 60 Hz, Phase shift = $\frac{1}{2}$
3. $I = 20 \cos 60\pi t$ **5.** 30 ft, 1,311 ft, 82 ft/sec
7.

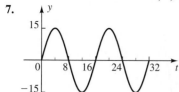

9. (A) $y = 2 \sin \dfrac{\pi}{75} r$

(B) $T = 393$ sec

11.

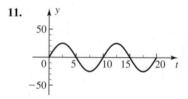

13. Period = 10^{-8} sec, $\lambda = 3$ m   **15.** $B = 2\pi \times 10^{18}$
17. Period = 10^{-5} sec, Frequency = 10^5 Hz, no
19.

21.

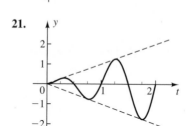

Exercise 3.5

1.

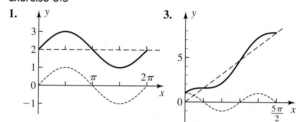

3.

5.

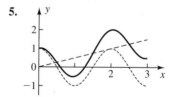

7.

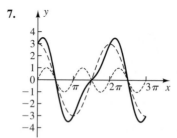

9.

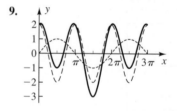

11.

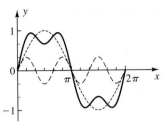

13.

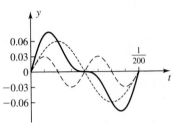

15.

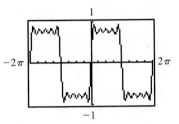

17. (A)

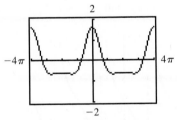

(B)

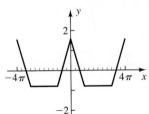

19.

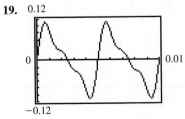

21. $V = 0.45 - 0.35 \cos(\pi t/2),\ 0 \le t \le 8$

23. (A)

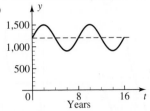

(B) 1,500 deer
(C) 900 deer

25. (A)

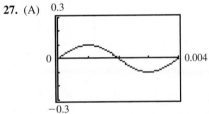

(B) $11.5 million
(C) $4 million

27. (A)

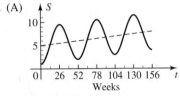

(B)

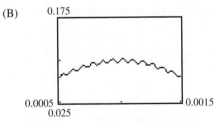

Exercise 3.6

1.   **3.**

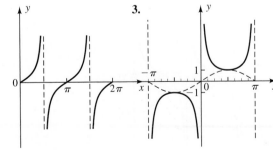

5. Period $= \pi/2$

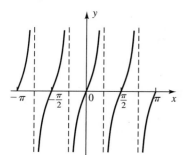

7. Period $= \frac{1}{2}$

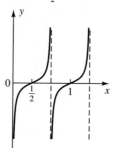

17. Period $= \pi/2$, Phase shift $= \pi/2$

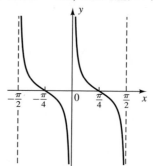

9. Period $= 2$

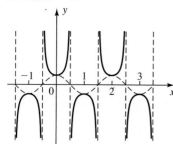

11. Period $= 2\pi$

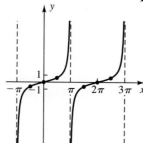

19. Period $= 2$, Phase shift $= \frac{1}{2}$

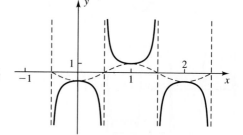

13. Period $= 4\pi$

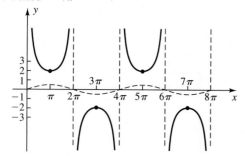

21. Period $= \pi/2$, Phase shift $= -\pi/2$

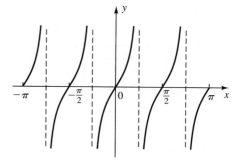

15. Period $= \pi$, Phase shift $= \pi/2$

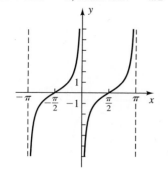

23. Period $= 1$, Phase shift $= 1$

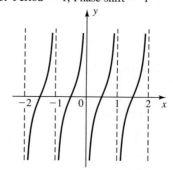

25. Period $= 2$, Phase shift $= \frac{1}{2}$

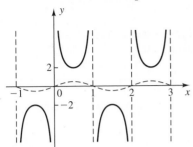

27. $y = \tan \dfrac{x}{2}$

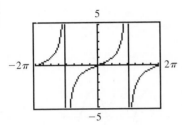

29. $y = 2 \csc 2x$

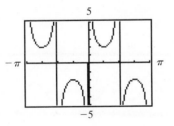

31. $y = \sec 2x$

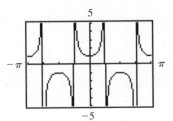

33. $y = \cot 3x$

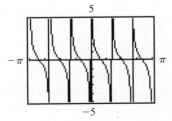

35. (A) $a = 15 \tan 2\pi t$
(B)

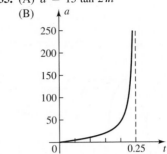

Chapter 3 Review Exercise

1.

[3.1]

2.

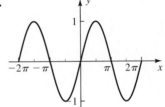

[3.1]

3.

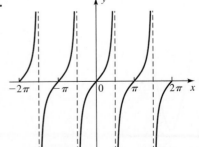

[3.1]

4.

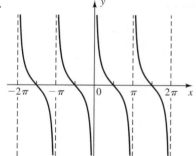

[3.1]

5.

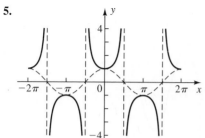

[3.1]

6.

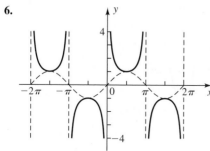

[3.1]

7.

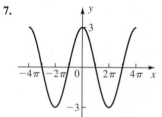

[3.2]

8.

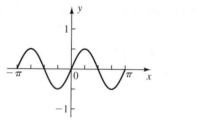

[3.2]

9.

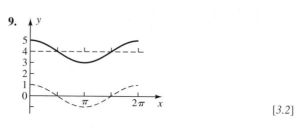

[3.2]

10.

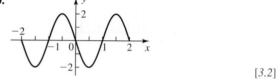

[3.2]

11.

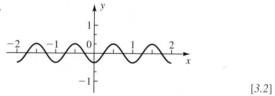

[3.2]

12.

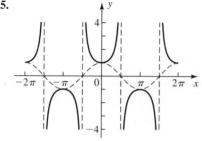

[3.2]

13.

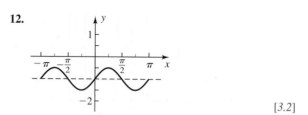

[3.2]

14.
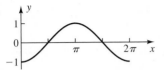

[3.3]

15.
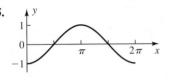

[3.3]

16.
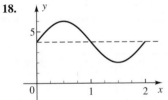

[3.3]

17.
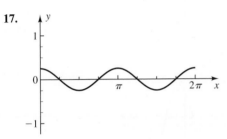

[3.3]

18.

[3.3]

19.
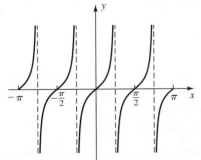

[3.6]

20.
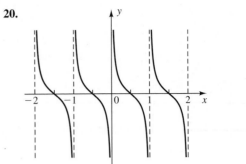

[3.6]

21.
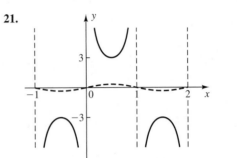

[3.6]

22.
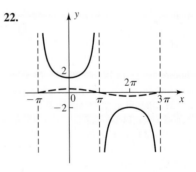

[3.6]

23.
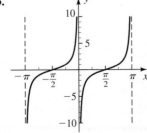

[3.6]

24.

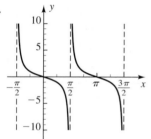

[3.6]

25.

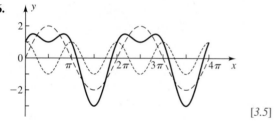

[3.5]

26.

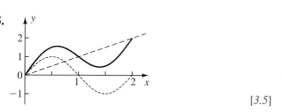

[3.5]

27. Period = 1, Amplitude = $\frac{1}{3}$ [3.2]   **28.** 1 [3.6]
29. $\pi/2$ [3.3]   **30.** Amplitude = 3, Period = 2, Phase
shift = -1 [3.3]   **31.** Period = 2, Phase shift =
-1 [3.6]   **32.** Period = π, Phase shift = $\pi/2$ [3.6]
33. $y = 4 \sin \pi x$ [3.2]   **34.** $y = -0.5 \cos 2x$ [3.2]

35.

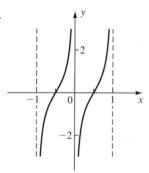

[3.6]

36.

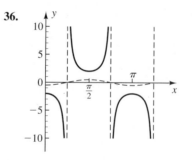

[3.6]

37. $y = -\sin(\pi x - \pi/4)$ [3.3]
38. $\frac{1}{2} + \frac{1}{2} \cos 2x$

[3.2]

39. x intercept: -0.464, $y = 2 \sin(2x + 0.928)$

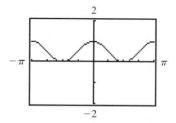

[3.3]

40.

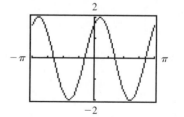

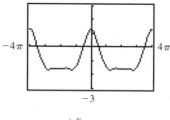

[3.5]

41. (A) $y = \sec x$

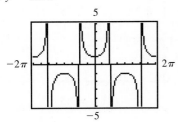

(B) $y = \csc x$

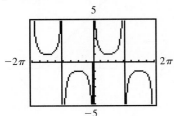

(C) $y = \cot x$

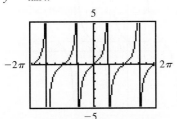

(D) $y = \tan x$

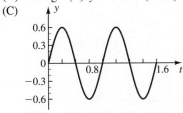

[3.6]

42. $y = -4 \cos 16\pi t$ [3.2]
43. $P = 1 + \cos(\pi n/26), 0 \le n \le 104$ [3.3]
44. (A) 179 kg   (B) $y = 0.6 \sin(2.5\pi t)$
   (C)

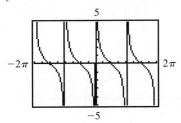

[3.2]

45. (A) Period $= \frac{1}{280}$ sec, $B = 560\pi$

    (B) Frequency $= 400$ Hz, $B = 800\pi$

    (C) Period $= \frac{1}{350}$ sec, frequency $= 350$ Hz   [3.2]

46. $E = 18 \cos 60 \pi t$ [3.2]
47. Amplitude $= 6$ m, Period $= 20$ sec, Phase shift $= 5$ sec

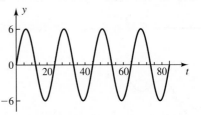

[3.3]

48. E: Amplitude $= 20$, Period $= \frac{1}{50}$; I: Amplitude $= 15$, Period $= \frac{1}{50}$, Phase shift $= \frac{1}{200}$

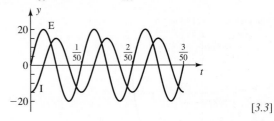

[3.3]

49. Wave height $= 24$ ft, Wavelength $= 184$ ft, Speed $= 31$ ft/sec   [3.4]   **50.** Period $= 10^{-15}$ sec, Wavelength $= 3 \times 10^{-7}$ m   [3.4]

51. (A) Simple harmonic motion

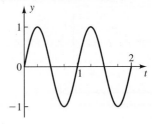

(B) Resonance

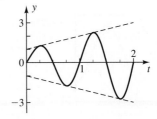

(C) Damped harmonic motion

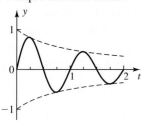

[3.4]

52.

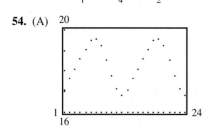

[3.5]

53. (A) $h = 1,000 \tan \theta$

(B)

[3.6]

54. (A)

20

16 1   24

(B) $y = 18.22 + 1.37 \sin\left(\dfrac{\pi x}{6} - 1.7\right)$

(C)

20

16 1   24

[3.3]

Chapters 1–3 Cumulative Review Exercise

1. 90°; $\pi/2$ rad [1.1, 2.1]   **2.** 21.78° [1.1]

3. 95.68° [2.1]   **4.** -12.48 rad [2.1]

5. 65°, 14 in., 31 in. [1.3]

6. $\sin \theta = -\dfrac{15}{17}$, $\tan \theta = -\dfrac{15}{8}$ [2.3]

7. (A) 0.3939 (B) -1.212 (C) 0.8267 [2.3, 2.6]

8. (A) $\alpha = \dfrac{\pi}{6}$ (B) $\alpha = 45°$

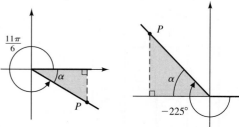

[2.5]

9. (A)

(B)

(C)

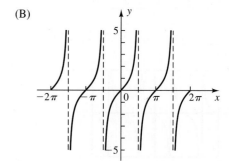
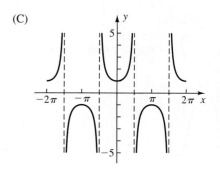

[3.1]

10. 43°30′ [*1.3*]   **11.** 20.94 cm [*1.1, 2.1*]

12. 31.4 cm/min [*2.2*]   **13.** 27°31′22″ [*1.3*]

14. 8 [*1.2*]   **15.** 4π/15 [*2.1*]   **16.** 5.1 ft [*2.1*]

17. (A) 57.6° (B) 25°30′ (C) 0.42 (D) π/3 [*2.5*]

18. (A) $-1/\sqrt{2}$ (B) $-\sqrt{3}/2$ (C) $\sqrt{3}$
(D) Not defined [*2.5*]

19. $\sin\theta = \sqrt{5}/3$, $\tan\theta = -\sqrt{5}/2$, $\sec\theta = -\frac{3}{2}$,
$\csc\theta = 3/\sqrt{5}$, $\cot\theta = -2/\sqrt{5}$ [*2.3*]

20. 13.4 cm² [*2.1*]

21. Period $= \pi$, Amplitude $= \frac{1}{2}$

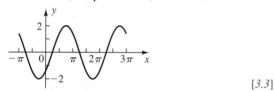

[*3.2*]

22. Period $= 2\pi$, Amplitude $= 2$, Phase shift $= \pi/4$

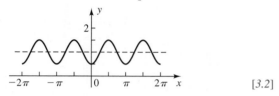

[*3.3*]

23. Period $= \pi/4$

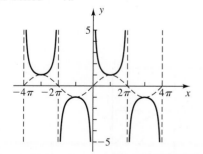

[*3.6*]

24. Period $= 4\pi$

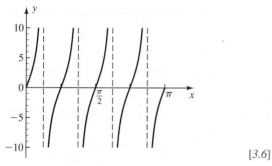

[*3.6*]

25. Period $= 2$

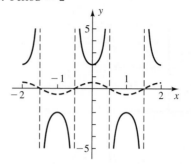

[*3.6*]

26. Period $= 1$, Phase shift $= -\frac{1}{2}$

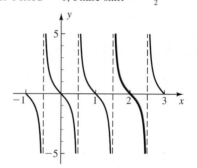

[*3.6*]

27.

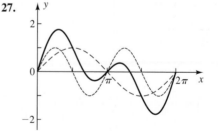

[*3.5*]

28. $-2\sin(2\pi x)$ [*3.2*]   **29.** $\sec x$ [*2.6*]

30. 210°, 330° [*2.5*]   **31.** 44.1 in., 57.8°, 32.2° [*1.3*]

32. 57°50′, 32°10′ [*1.1*]

33. $(-0.666, 0.746)$ [*2.3, 2.6*]

34. 18.37 units [*2.3*]

35. $\sin\theta = \dfrac{a}{\sqrt{1+a^2}}$, $\cos\theta = \dfrac{1}{\sqrt{1+a^2}}$, $\cot\theta = \dfrac{1}{a}$,

$\csc\theta = \dfrac{\sqrt{1+a^2}}{a}$, $\sec\theta = \sqrt{1+a^2}$ [*2.3, 2.6*]

36. $y = 4\sin(\pi x - \pi/3)$ [*3.3*]

37. $y = \frac{1}{2} - \frac{1}{2}\cos 2x$

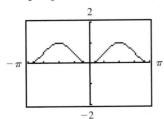

[3.2]

38. x intercept: 1.287, $y = 3\sin\left(\frac{x}{2} - 0.6435\right)$

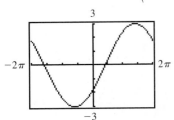

[3.3]

39.

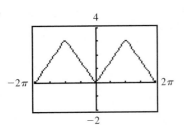

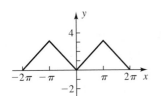

[3.5]

40. (A) $y = \tan x$

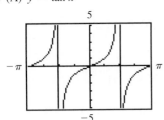

(B) $y = \sec x$

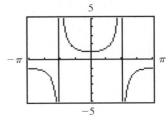

(C) $y = \csc x$

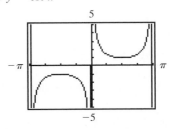

(D) $y = \cot x$

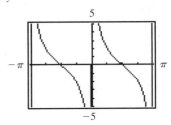

[3.6]

41. 773 mi   [*1.1, 2.1*]   **42.** 45 cm^3 [*1.2*]
43. 41°, 36 ft   [*1.4*]   **44.** $\frac{4}{3}$ ft   [*1.2, 1.4*]
45. 1.7°, 5%   [*1.4*]
46. Building: 51 m high; street: 17 m wide   [*1.4*]
47. (A) 1.7 mi   (B) 1.3 mi   [*1.4*]   **48.** 16 ft   [*2.1*]
49. (A) 942 rad/min   (B) 64,088 in./min   [*2.2*]
50. 27.8°   [*2.4*]   **51.** 84°   [*2.4*]
52. Wavelength = 123 ft, Period = 4.9 sec   [*3.4*]
53. A1 = $\frac{1}{2}\sin x$ [*2.6*]   **54.** A2 = $\frac{1}{2}x$ [*2.1, 2.6*]
55. A3 = $\frac{1}{2}\tan x$ [*2.6*]
57.

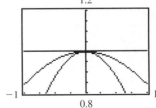

[2.6]

58. Amplitude = 5, Period = $\pi/5$ sec,
Frequency = $5/\pi$ Hz

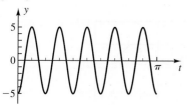

[3.2]

59. Amplitude = 12, Period = $\frac{1}{30}$, Frequency = 30 Hz,
Phase shift = $\frac{1}{60}$

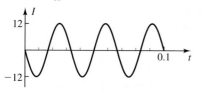

[3.3]

60. Period = 2×10^{-13} sec   [3.4]
61. (A) $y = 100 \csc \theta + 70 \sec \theta$
    (B)

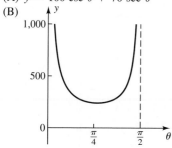

[1.4, 2.3, 3.5]

62. (A) $d = 50 \tan \theta$ (B) $\theta = 40\pi t$
    (C) $d = 50 \tan 40\pi t$
    (D)

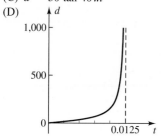

[3.6]

63. (A)

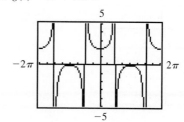

(B) $y = 45 + 26 \sin\left(\dfrac{\pi x}{6} - 2.2\right)$

(C)

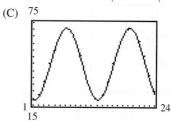

[3.3]

CHAPTER 4

Exercise 4.1

1. $\csc x = -\frac{3}{2}$, $\sec x = 3/\sqrt{5}$, $\tan x = -2/\sqrt{5}$,
$\cot x = -\sqrt{5}/2$ **3.** $\cot x = \frac{1}{2}$, $\cos x = -1/\sqrt{5}$,
$\csc x = -\sqrt{5}/2$, $\sec x = -\sqrt{5}$ **5.** 1   **7.** $\sec x$
9. $\tan x$ **11.** $\sec \theta$ **13.** $\cot^2 \beta$ **15.** 2
17. $\cos x = -\sqrt{15}/4$, $\csc x = 4$, $\sec x = -4/\sqrt{15}$,
$\tan x = -1/\sqrt{15}$, $\cot x = -\sqrt{15}$
19. $\cot x = -\frac{1}{2}$, $\sec x = \sqrt{5}$, $\cos x = 1/\sqrt{5}$,
$\sin x = -2/\sqrt{5}$, $\csc x = -\sqrt{5}/2$
21. $\sin x = \frac{2}{3}$, $\cos x = -\sqrt{5}/3$, $\sec x = -3/\sqrt{5}$,
$\tan x = -2/\sqrt{5}$, $\cot x = -\sqrt{5}/2$
23. $-\cot y$ **25.** $\csc x$ **27.** 0   **29.** $\csc^2 x$
31. $\cot w$ **33.** (A) 1   (B) 1   **35.** I, II   **37.** II, III
39. All   **41.** I, IV   **43.** $a \cos x$ **45.** $a \sec x$
47. $\dfrac{x^2}{25} + \dfrac{y^2}{4} = 1$

Exercise 4.2

1. A   **3.** E   **5.** H   **7.** J   **9.** L   **11.** K
99. $g(x) = \tan x$

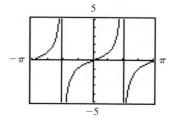

101. $g(x) = 1 + \sec x$

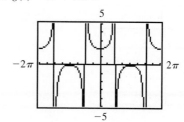

103. $g(x) = 2 \sin x$

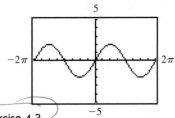

Exercise 4.3

13. $\dfrac{\sqrt{2}}{2}(\sin x - \cos x)$ **15.** $-\cos x$

17. $\dfrac{1 - \tan x}{1 + \tan x}$ **19.** $\dfrac{\sqrt{3} + 1}{2\sqrt{2}}$ **21.** $\dfrac{\sqrt{3} + 1}{2\sqrt{2}}$

23. $\sqrt{3}/2$ **25.** $\sqrt{3}$

27. $\sin(x - y) = \dfrac{-2 - \sqrt{75}}{12}$; $\tan(x + y) =$

$\dfrac{\sqrt{75} - 2}{\sqrt{5} + 2\sqrt{15}}$

29. $\sin(x - y) = \dfrac{-2\sqrt{8} - 1}{3\sqrt{5}}$; $\tan(x + y) = \dfrac{1 - 4\sqrt{2}}{2 + 2\sqrt{2}}$

45.

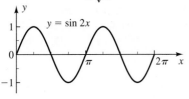

47.

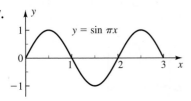

51. $y2 = \dfrac{1}{2}\sin x + \dfrac{\sqrt{3}}{2}\cos x$

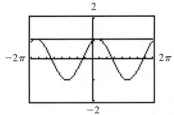

53. $y2 = -\dfrac{\sqrt{3}}{2}\cos x + \dfrac{1}{2}\sin x$

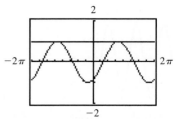

55. $y2 = \dfrac{\tan x + 1}{1 - \tan x}$

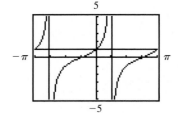

59. (C) 3,940 ft

Exercise 4.4

1. $\sqrt{3}/2 = \sqrt{3}/2$ **3.** $-\sqrt{3} = -\sqrt{3}$

5. $\sqrt{2} + \sqrt{3}/2$ **7.** $2 - \sqrt{3}$

33.

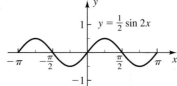

35.

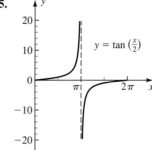

37.

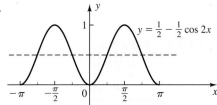

$y = \frac{1}{2} - \frac{1}{2}\cos 2x$

39. $\sin 2x = \frac{24}{25}$, $\cos 2x = \frac{7}{25}$, $\tan 2x = \frac{24}{7}$

41. $\sin 2x = -\frac{24}{25}$, $\cos 2x = \frac{7}{25}$, $\tan 2x = -\frac{24}{7}$

43. $\sin 2x = -\frac{120}{169}$, $\cos 2x = -\frac{119}{169}$, $\tan 2x = \frac{120}{119}$

45. $\sin(x/2) = \sqrt{3}/3$, $\cos(x/2) = \sqrt{6}/3$

47. $\sin(x/2) = \sqrt{\dfrac{3 + 2\sqrt{2}}{6}}$, $\cos(x/2) = -\sqrt{\dfrac{3 - 2\sqrt{2}}{6}}$

49. $\sin(x/2) = -2\sqrt{5}/5$, $\cos(x/2) = \sqrt{5}/5$

51. $\sin x = 1/\sqrt{10}$, $\cos x = 3/\sqrt{10}$, $\tan x = \frac{1}{3}$

53. $\sin x = -2/\sqrt{5}$, $\cos x = 1/\sqrt{5}$, $\tan x = -2$

55. $\sin x = -2/\sqrt{5}$, $\cos x = -1/\sqrt{5}$, $\tan x = 2$

65.

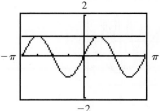

67.

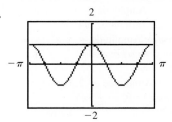

69. $0 \le x \le 2\pi$

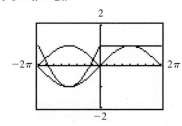

71. $-2\pi \le x \le -\pi$, $\pi \le x \le 2\pi$

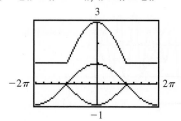

73. $g(x) = \cot \dfrac{x}{2}$

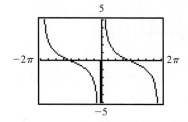

75. $g(x) = \csc 2x$

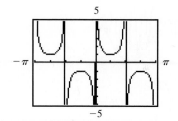

77. $g(x) = 2\cos x - 1$

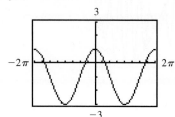

79. $d = \dfrac{v_0^2 \sin 2\theta}{32 \text{ ft/sec}^2}$ **81.** $x = 2\sqrt{3} \approx 3.464$ cm,

$\theta = 30.000°$ **83.** $\cos \theta = \frac{4}{5}$

Exercise 4.5

1. $\frac{1}{2}\cos 12A + \frac{1}{2}\cos 2A$ **3.** $\frac{1}{2}\sin 5\theta + \frac{1}{2}\sin \theta$

5. $2\cos 6\theta \cos \theta$ **7.** $-2\cos 3u \sin 2u$

9. $(2 - \sqrt{3})/4$ **11.** $\frac{1}{4}$ **13.** $\sqrt{2}/2$ **15.** $\sqrt{2}/2$

31. (A)

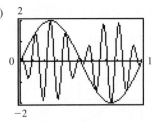

(B) $y1 = \sin(18\pi x) - \sin(14\pi x)$

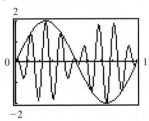

33. (A)

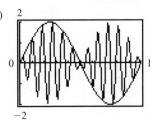

(B) $y1 = \cos(22\pi x) - \cos(26\pi x)$

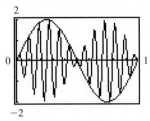

35. $y = 2k \sin 517\pi t \cos 5\pi t; f_b = 5$ Hz

37. (A)

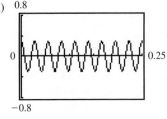

(B)

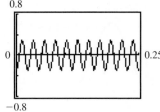

(C)

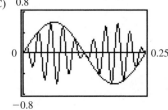

(D) $y2 = 0.6 \sin 80\pi t \sin 8\pi t$

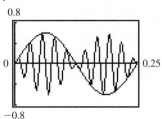

Exercise 4.6

1. $y = \sqrt{2} \sin(t + \pi/4)$; Amplitude $= \sqrt{2}$, Period $= 2\pi$, Frequency $= 1/2\pi$, Phase shift $= -\pi/4$
3. $y = \sqrt{2} \sin(t + 5\pi/4)$; Amplitude $= \sqrt{2}$, Period $= 2\pi$, Frequency $= 1/2\pi$, Phase shift $= -5\pi/4$
5. $y = 2 \sin(t + 5\pi/6)$; Amplitude $= 2$, Period $= 2\pi$, Frequency $= 1/2\pi$, Phase shift $= -5\pi/6$
7. $y = 2 \sin(t + 4\pi/3)$; Amplitude $= 2$, Period $= 2\pi$, Frequency $= 1/2\pi$, Phase shift $= -4\pi/3$
9. $y = \sqrt{2} \sin(t - \pi/4)$; Amplitude $= \sqrt{2}$, Period $= 2\pi$, Frequency $= 1/2\pi$, Phase shift $= \pi/4$

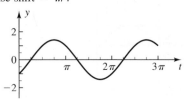

11. $y = \sqrt{2}\,\sin(\pi t + \pi/4)$; Amplitude $= \sqrt{2}$,
Period $= 2$, Frequency $= \frac{1}{2}$, Phase shift $= -\frac{1}{4}$

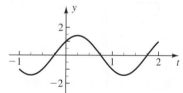

13. $y = 2\sin(\pi t - \pi/6)$; Amplitude $= 2$,
Period $= 2$, Frequency $= \frac{1}{2}$, Phase shift $= \frac{1}{6}$

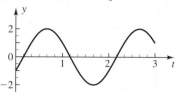

15. $y = \sqrt{2}\,\sin(2\pi t + \pi/4)$; Amplitude $= \sqrt{2}$,
Period $= 1$, Frequency $= 1$, Phase shift $= -\frac{1}{8}$

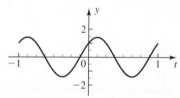

17. $y = 2\sin(2\pi t - 2\pi/3)$; Amplitude $= 2$,
Period $= 1$, Frequency $= 1$, Phase shift $= \frac{1}{3}$

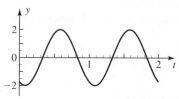

19. $y = 5\sin(\pi t - 0.64)$; Amplitude $= 5$, Period $= 2$,
Frequency $= \frac{1}{2}$, Phase shift $\approx 0.64/\pi \approx 0.20$

21. $y = \sqrt{34}\,\sin(3t + 2.60)$; Amplitude $= \sqrt{34}$,
Period $= 2\pi/3$, Frequency $= 3/2\pi$, Phase shift $= -0.87$

23. Intercepts: 0.20, -0.80, Phase shift: 0.20

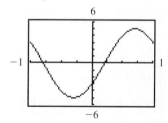

25. Intercepts: 0.18, -0.87, Phase shift: -0.87

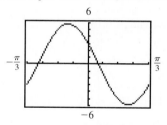

27. $y = 5\sin(8t + 4.07)$; Amplitude $= 5$ cm,
Period $= \pi/4$ sec, Frequency $= 4/\pi$ Hz,
Phase shift $= -0.51$ sec

Chapter 4 Review Exercise

1. $1/(\csc x)$ [4.1]   **2.** $1/(\cos x)$ [4.1]
3. $(\cos x)/(\sin x)$ [4.1]   **4.** $(\sin x)/(\cos x)$ [4.1]
5. $1 - \cos^2 x$ [4.1]   **6.** $\cos x$ [4.1]
18. $\frac{1}{2} = \frac{1}{2}$ [4.4]   **19.** $1/\sqrt{2} = 1/\sqrt{2}$ [4.4]
20. $\frac{1}{2}\cos 3t - \frac{1}{2}\cos 13t$ [4.5]
21. $2\sin 3w \cos 2w$ [4.5]   **22.** $y = 2\sin(t + 7\pi/6)$,
Amplitude $= 2$, Period $= 2\pi$, Frequency $= 1/2\pi$,
Phase shift $= -7\pi/6$ [4.6]

42. $\dfrac{1}{2} - \dfrac{\sqrt{3}}{4}$ [4.5]   **43.** $-\sqrt{6}/2$ [4.5]

44. $\sec x = -\frac{3}{2}$, $\sin x = \sqrt{5}/3$, $\csc x = 3/\sqrt{5}$,
$\tan x = -\sqrt{5}/2$, $\cot x = -2/\sqrt{5}$ [4.1]

45. $\sin 2x = \frac{24}{25}$, $\cos 2x = -\frac{7}{25}$, $\tan 2x = -\frac{24}{7}$ [4.2]
46. $\sin(x/2) = -3/\sqrt{13}$, $\cos(x/2) = 2/\sqrt{13}$,
$\tan(x/2) = -\frac{3}{2}$ [4.2]   **47.** $y = \sqrt{2}\,\sin(\pi t + 3\pi/4)$,
Amplitude $= \sqrt{2}$, Period $= 2$, Frequency $= \frac{1}{2}$,
Phase shift $= -\frac{3}{4}$

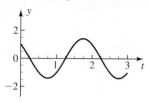

[4.6]

48. $y = 2\sin(2\pi t + \pi/3)$, Amplitude $= 2$, Period $= 1$,
Frequency $= 1$, Phase shift $= -\frac{1}{6}$

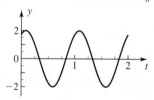

[4.6]

49. $\sin x = -5/\sqrt{26}$, $\cos x = 1/\sqrt{26}$, $\tan x = -5$ [4.4]
57. $y = 2 \sin(4t - 0.64)$, Amplitude $= 2$, Period $= \pi/2$,
Frequency $= 2/\pi$, Phase shift $= 0.16$ [4.6]
58. $g(x) = 4 + 2 \cos x$

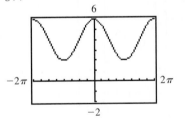

[4.2]

59. $g(x) = \tan 2x$

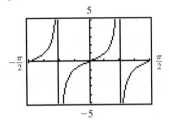

[4.4]

60. $g(x) = 2 - \cos 2x$

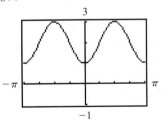

[4.4]

61. $g(x) = -2 + \sec 2x$

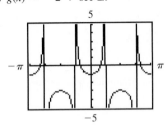

[4.4]

62. $g(x) = 2 + \cot(x/2)$

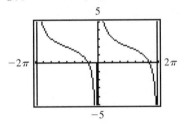

[4.4]

63. $-2\pi \le x < -\pi, 0 \le x < \pi$

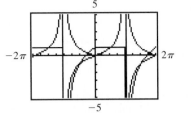

[4.4]

64. $-\pi < x \le 0, \pi < x \le 2\pi$

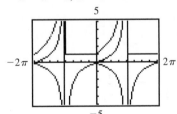

[4.4]

65.

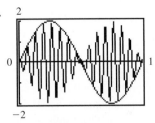

[4.5]

66. $y1 = \sin 32\pi x - \sin 28\pi x$

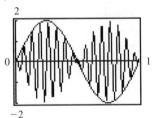

[4.5]

67. Intercepts: -0.62, 0.16; Phase shift $= 0.16$

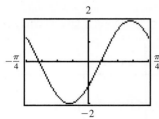

[4.6]

68. $a \tan x$ [4.1]   **69.** $57.5°$ [4.3]

70. $x = \frac{445}{39} \approx 11.410$, $\theta \approx 32.005°$ [4.4]

71. $y = 0.6 \sin 130\pi t \sin 10\pi t$, Beat frequency $=$ 10 Hz   [4.5]

72. (A)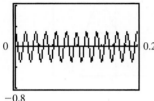

(B)

(C)

(D)

[4.5]

73. $y = 10 \sin(3t + 3.79)$, Amplitude $= 10$ cm, Period $= 2\pi/3$ sec, Frequency $= 3/2\pi$ Hz, Phase shift $= -1.26$ sec   [4.6]

74. x intercepts: -1.26, -0.21, 0.83, 1.88; Phase shift $= -1.26$

[4.6]

CHAPTER 5

Exercise 5.1

1. 0   **3.** $\pi/6$ **5.** $\pi/4$ **7.** $\pi/3$ **9.** $\pi/6$
11. 0   **13.** 1.155 **15.** 1.548 **17.** Not defined
19. $2\pi/3$ **21.** $-\pi/4$ **23.** $-\pi/3$ **25.** $5\pi/6$
27. -0.6 **29.** -1.5 **31.** $\sqrt{3}$ **33.** $\sqrt{2}/2$
35. $-1/\sqrt{3}$ **37.** -1.328 **39.** 1.001
41. -1.029 **43.** 2.456
45.

47. $120°$ **49.** $-45°$ **51.** $-60°$ **53.** $71.80°$
55. $-54.16°$ **57.** $-86.69°$ **59.** $\frac{24}{25}$ **61.** $-\frac{1}{2}$
63. $-\frac{24}{25}$ **65.** $2/\sqrt{10}$ **67.** $\sqrt{1 - x^2}$

69. $\dfrac{x}{\sqrt{1 - x^2}}$ **71.** $\dfrac{2x}{x^2 + 1}$ **73.** $1 - 2x^2$

75. $x = \dfrac{1}{b} \sin^{-1} \dfrac{y}{a} - \dfrac{c}{b}$, $-a \le y \le a$

77.

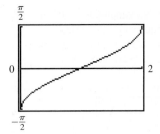

Since $\sin^{-1} x$ is defined only for $-1 \le x \le 1$, no graph results from using values outside this interval. Some calculators may give an error message in this situation.

79.

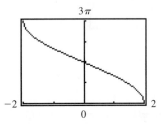

81.

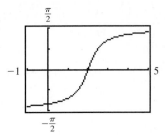

85. (A)

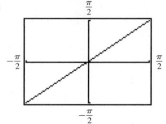

(B)

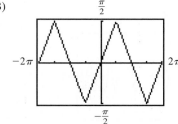

For any real number x, $\sin x$ is a real number between -1 and 1, thus $\sin^{-1}(\sin x)$ is defined for all real numbers x. Enlarging the graphing interval extends the graph of this periodic function.

87. (A) $\theta = 2 \sin^{-1}(1/M)$, $M > 1$ (B) 72°, 52°
89. (B) 248 ft^3
91. 2.6 ft

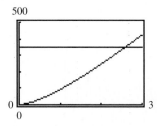

93. 35.8 in.
95. 12.1 ft

83. (A)

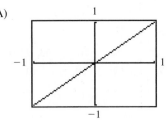

(B)

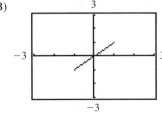

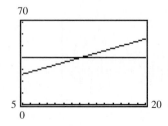

97. (B) 46.4 ft

Exercise 5.2

1. $\pi/6$ **3.** $\pi/2$ **5.** $\pi/4$ **7.** 1   **9.** $\frac{4}{3}$
11. $3\pi/4$ **13.** $2\pi/3$ **15.** $-\pi/6$ **17.** Not defined
19. $\frac{4}{5}$ **21.** $-\frac{4}{3}$ **23.** $-\frac{1}{2}$ **25.** 33.4   **27.** -4
29. 0.315   **31.** 2.105   **33.** 1.022   **35.** 2.948
37. 120°   **39.** 135°   **41.** $-60°$ **43.** 71.80°
45. $-54.16°$ **47.** 93.31°   **49.** $-\frac{8}{19}$ **51.** $\frac{24}{7}$
53. $\dfrac{1}{\sqrt{x^2+1}}$ **55.** $\dfrac{|x|}{\sqrt{x^2-1}}$ **57.** $\dfrac{2x}{x^2+1}$

61.

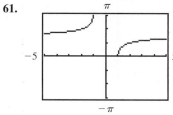

63.

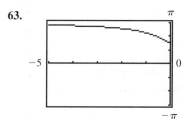

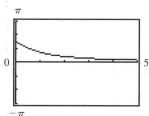

Exercise 5.3

1. $\pi/4, 3\pi/4$ **3.** $2\pi/3, 4\pi/3$ **5.** $-\pi/3$
7. 60°, 300°   **9.** 240°, 300°   **11.** $-30°$
13. $\pi/4 + 2k\pi, 3\pi/4 + 2k\pi$, k any integer
15. $2\pi/3 + 2k\pi, 4\pi/3 + 2k\pi$, k any integer
17. $-\pi/3 + k\pi$, k any integer
19. 60° + $k360°$, 300° + $k360°$, k any integer
21. 240° + $k360°$, 300° + $k360°$, k any integer
23. $-30° + k180°$, k any integer   **25.** 0.64, 2.50
27. 2.05, 4.23   **29.** -1.53 **31.** 74.93°, 285.07°
33. 237.14°, 302.86°   **35.** $-83.96°$
37. 0.64 + $2k\pi$, 2.50 + $2k\pi$, k any integer

39. 2.05 + $2k\pi$, 4.23 + $2k\pi$, k any integer
41. $-1.53 + k\pi$, k any integer
43. 74.93° + $k360°$, 285.07° + $k360°$, k any integer
45. 237.14° + $k360°$, 302.86° + $k360°$, k any integer
47. $-83.96° + k180°$, k any integer
49. $k\pi$, k any integer   **51.** $\pi + 2k\pi$, k any integer
53. $\pi/6, 5\pi/6$ **55.** 101.537°, 258.463°   **57.** 0.446,
2.695, 6.729, 8.979   **59.** $-2.915, -0.226, 3.368, 6.057$
61. $-283.762°, -103.762°, 76.238°, 256.238°$
63. (A) $a > 1$ or $a < -1$ (B) $a = 1$ or $a = -1$
        (C) $0 < a < 1$ or $-1 < a < 0$ (D) $a = 0$
65. $-2.930, -0.212$ **67.** $-1.318, 1.318$
69. $-2.034, 1.107$

Exercise 5.4

1. $\pi/4$ **3.** 450°, 630°   **5.** $\pi/6$
7. $\pi/6, 5\pi/6, 7\pi/6, 11\pi/6$ **9.** 0°, 90°, 180°
11. $\pi/6, \pi/3$ **13.** $\pi/4$ **15.** $\pi/2$
17. $2k\pi, \pi + 2k\pi, \pi/3 + 2k\pi, 5\pi/3 + 2k\pi$, k any integer
19. 90° + $k360°$, 270° + $k360°$, 45° + $k360°$,
225° + $k360°$, k any integer   **21.** $\pi/6, \pi/2, 5\pi/6$ **23.** 180°
25. 90°, 270°   **27.** $\pi/6, 5\pi/6$ **29.** $\pi/2, 3\pi/2$
31. 104.5°   **33.** 0.9987, 5.284
35. 0.2524 + $k\pi$, 1.318 + $k\pi$, k any integer
37. 0.5752 + $k\pi/2$, 0.9956 + $k\pi/2$, k any integer
39. 0.9987 + $2k\pi$, $-0.9987 + 2k\pi$, k any integer
41. 1.823 + $2k\pi$, 4.460 + $2k\pi$, k any integer
43. $\pi/4 + 2k\pi, 5\pi/4 + 2k\pi$, k any integer
45. $2k\pi, \pi/3 + 2k\pi, -\pi/3 + 2k\pi$, k any integer
47. 0, $\pi/2$ **49.** 0   **51.** 0, $\pi/2, \pi$
53. 0, $\pi/12, 5\pi/12$ **55.** $t = 2k\pi, \pi/2 + 2k\pi$,
k an integer   **57.** $t = \frac{1}{2} + 2k, 1 + 2k$, k an integer
59. 0.739   **61.** 0.257, 2.029   **63.** 1.844, 2.564,
3.720, 4.439   **65.** 0.002 613 sec   **67.** 33.21°
69. 1.78 rad   **71.** 35.64 ft^2
73. (A) $L = 12.4575$ mm   (B) $y = 2.6495$ mm
75. 64.1°   **77.** $(r, \theta) = (1, 30°), (1, 150°)$
79. $\theta = 45°$

Chapter 5 Review Exercise

1. $\pi/6$ *[5.1]*   **2.** $\pi/6$ *[5.1]*   **3.** $-\pi/2$ *[5.1]*
4. π *[5.1]*   **5.** $-\pi/3$ *[5.1]*   **6.** $2\pi/3$ *[5.1]*
7. $\pi/4$ *[5.1]*   **8.** $-\pi/3$ *[5.1]*   **9.** $\pi/2$ *[5.1]*
10. Not defined *[5.1]*   **11.** $2\pi/3$ *[5.2]*
12. $2\pi/3$ *[5.2]*   **13.** $-\pi/2$ *[5.2]*   **14.** $3\pi/4$ *[5.2]*
15. -0.9750 *[5.1]*   **16.** 0.9188 *[5.1]*   **17.** Not
defined *[5.1]*   **18.** 1.557 *[5.1]*   **19.** 0.1757 *[5.2]*
20. -0.2857 *[5.2]*   **21.** $\pi/6, 11\pi/6$ *[5.4]*
22. 0°, 30°, 150°, 180° *[5.4]*   **23.** $\pi/6, 5\pi/6$,
$7\pi/6, 11\pi/6$ *[5.4]*   **24.** 120°, 180°, 240° *[5.4]*   -

25. $\pi/16, 3\pi/16$ *[5.4]*   **26.** $-120°$ *[5.4]*

27. 0.315 *[5.1]*   **28.** -1.5 *[5.1]*   **29.** $-\frac{3}{5}$ *[5.1]*

30. $-2/\sqrt{5}$ *[5.1]*   **31.** $\sqrt{5}/2$ *[5.1]*   **32.** 4 *[5.2]*

33. $\sqrt{10}/3$ *[5.2]*   **34.** $2\sqrt{6}/5$ *[5.2]*

35. -1.192 *[5.1]*   **36.** Not defined *[5.1]*

37. -0.9975 *[5.1]*   **38.** 1.095 *[5.1]*

39. 1.142 *[5.1]*   **40.** 0.9423 *[5.2]*

41. 0.9894 *[5.2]*   **42.** 1.001 *[5.2]*

43. $135°$ *[5.1]*   **44.** $30°$ *[5.1]*   **45.** $-90°$ *[5.1]*

46. $69.78°$ *[5.1]*   **47.** $-85.41°$ *[5.1]*

48. $0.14°$ *[5.1]*   **49.** $90°, 270°$ *[5.4]*

50. $\pi/12, 5\pi/12$ *[5.4]*   **51.** $-\pi$ *[5.4]*

52. $k360°, 180° + k360°, 30 + k360°, 150° + k360°,$ k any integer *[5.4]*   **53.** $2k\pi, \pi + 2k\pi,$ $\pi/6 + 2k\pi, -\pi/6 + 2k\pi, k$ any integer *[5.4]*

54. $120°, 240°$ *[5.4]*   **55.** $7\pi/12, 11\pi/12$ *[5.4]*

56. $0.7878 + 2k\pi, 2.354 + 2k\pi, k$ any integer *[5.3]*

57. $1.690 + 2k\pi, 4.593 + 2k\pi, k$ any integer *[5.3]*

58. $-1.343 + k\pi, k$ any integer *[5.3]*

59. $0.2441 + 2k\pi, 2.898 + 2k\pi, k$ any integer *[5.3]*

60. $0.6259 + 2k\pi, 2.516 + 2k\pi, k$ any integer *[5.4]*

61. $1.178 + k\pi, -0.3927 + k\pi, k$ any integer *[5.4]*

62. $0, \pi/2$ *[5.4]*   **63.** $0, \pi/3, 2\pi/3$ *[5.4]*

64. $\pi/12, 5\pi/12, 0, \pi/3$ *[5.4]*

65. $0.375, 2.77$ *[5.4]*   **66.** $-\frac{24}{25}$ *[5.1]*

67. $\frac{24}{25}$ *[5.1]*   **68.** $\frac{3}{2}$ *[5.2]*   **69.** $\dfrac{x}{\sqrt{1 - x^2}}$ *[5.1]*

70. $\dfrac{1}{\sqrt{x^2 + 1}}$ *[5.1]*   **71.** $\dfrac{\sqrt{1 - x^2} - x^2}{\sqrt{x^2 + 1}}$ *[5.1]*

72. $\dfrac{2 - x^2}{x^2}$ *[5.2]*

73.

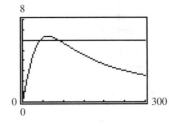

[5.1]

74.

[5.1]

75.

[5.1]

76. $0.253, 2.889$ *[5.3]*   **77.** $-0.245, 2.897$ *[5.3]*

78. ± 1.047 *[5.3]*   **79.** ± 0.824 *[5.4]*

80. 0 *[5.4]*   **81.** $4.227, 5.197$ *[5.4]*

82. $0.604, 2.797, 7.246, 8.203$ *[5.4]*

83. $0.228, 1.008$ *[5.4]*   **84.** $0.000\ 24$ sec *[5.4]*

85. $0.001\ 936$ sec *[5.4]*

86. (A) $\theta = 2 \arctan (500/x)$ (B) $45.2°$ *[5.1]*

87. $\theta = \arctan \dfrac{14x}{x^2 + 3,920}; 4.54°$ *[5.1]*

88. 43.9 ft or 89.3 ft

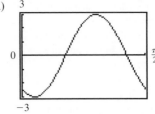

[5.1]

89. (B) 0.7165 in.^{2} *[5.1]*   **90.** 0.5743 in. *[5.1]*

91. 2.82 ft *[5.4]*

92. (A)

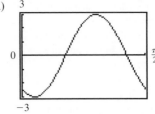

(B) 0.74 sec, 1.16 sec (C) 0.37 sec, 1.52 sec [5.4]
93. (A) $y = 3 \sin(4t + 4.07)$ (B) 0.74 sec, 1.16 sec
(C) 0.37 sec, 1.52 sec [4.6, 5.4]

Chapters 1–5 Cumulative Review Exercise
1. 241.22° [2.1] **2.** 8.83 rad [1.1, 2.1]
3. 54°, 36°, 6.1 m [1.3]
4. $\cos \theta = -\frac{5}{13}$, $\cot \theta = \frac{5}{12}$ [2.3]
5. (A) 0.3786 (B) 15.09 (C) −0.5196 [2.3, 2.6]
6. (A)

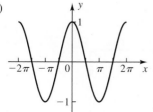

(B)

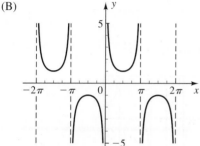

(C)

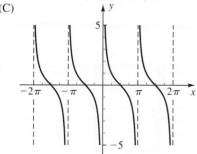

[3.1, 3.2, 3.6]

11. $\frac{1}{2}$ [2.5] **12.** $1/\sqrt{2}$ [2.5] **13.** $\sqrt{3}$ [25]
14. Undefined [2.5] **15.** $\pi/4$ [5.1]
16. 0 [5.1] **17.** $5\pi/6$ [5.1] **18.** Undefined [5.1]
19. $5\pi/6$ [5.2] **20.** $\pi/3$ [5.2]
21. 0.0505 [1.3, 5.1] **22.** 2.379 [1.3, 5.1]
23. −1.460 [1.3, 5.1] **24.** 0.3218 [5.2]
25. 1.176 [5.2] **26.** Undefined [5.2]
27. 30°, 150° [5.3] **28.** $-\pi/6$ [5.3]
29. $-\pi$ [5.3] **30.** $\frac{1}{2}(\sin 10u + \sin 4u)$ [4.5]
31. $-2 \sin 3w \sin 2w$ [4.5]

32. $2 \sin(t + \pi/6)$; Amplitude = 2, Period = 2π,
Frequency = $1/(2\pi)$, Phase shift = $-\pi/6$ [4.6]
33. 92°27′43″ [1.1] **34.** 144° [1.1]
35. $x = 6$, $\theta = 33.7°$ [1.3] **36.** $2\pi/5$ [2.1]
37. $\alpha = 45°$

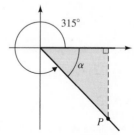

[2.5]

38. $\sin \theta = -1/\sqrt{5}$, $\cos \theta = -2/\sqrt{5}$, $\cot \theta = 2$,
$\sec \theta = -\sqrt{5}/2$, $\csc \theta = -\sqrt{5}$ [2.3]
39. Period = 4π, Amplitude = 2

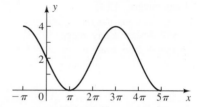

[3.2]

40. Period = π, Amplitude = 3, Phase shift = $\pi/2$

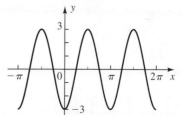

[3.3]

41. Period = 1, Phase shift = $\frac{1}{4}$

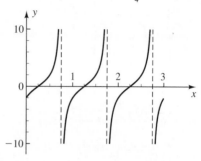

[3.6]

42. $y = 1 + 3 \cos(\pi x)$ [3.2]
43. Hypotenuse: 46.0 cm, Angles: 25°0′, 65°0′ [1.3]
50. $\sin(x/2) = 4/\sqrt{17}$, $\cos 2x = \frac{161}{289}$ [2.3, 4.4]

51. 0   *[4.5]*

52. $y = 2\sin(\pi t - \pi/3)$, Amplitude $= 2$, Period $= 2$,
Frequency $= \frac{1}{2}$, Phase shift $= \frac{1}{3}$

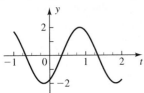

                                       [4.6, 3.3]

53. (A) $\sqrt{21}/5$ *[5.1]*   (B) $\sqrt{6}$ *[5.1]*
      (C) 3   *[5.1]*   (D) $\sqrt{15}$ *[5.2]*

54. (A) -1.016 *[5.1]*   (B) Not defined   *[5.1]*
      (C) 0.7828   *[5.1]*   (D) 0.9762   *[5.2]*

55. $60°$ *[1.3, 5.1]*   **56.** $64.01°$ *[1.3, 5.1]*

57. $7\pi/6, 3\pi/2, 11\pi/6$ *[5.4]*

58. $90° + k180°$, k any integer   *[5.4]*

59. $\pi/3 + k\pi, 2\pi/3 + k\pi$, k any integer   *[5.4]*

60. $3.745 + 2k\pi, 5.679 + 2k\pi$, k any integer   *[5.3]*

61. $1.130 + 2k\pi, 5.153 + 2k\pi$, k any integer   *[5.3]*

62. $1.823 + 2k\pi, 4.460 + 2k\pi$, k any integer   *[5.4]*

63. 9.27   *[2.3]*

64. $\sin\theta = -\sqrt{1 - a^2}$, $\tan\theta = \dfrac{-\sqrt{1 - a^2}}{a}$,

$\cot\theta = \dfrac{-a}{\sqrt{1 - a^2}}$, $\sec\theta = \dfrac{1}{a}$,

$\csc\theta = \dfrac{-1}{\sqrt{1 - a^2}}$ *[2.3, 2.6]*

65.

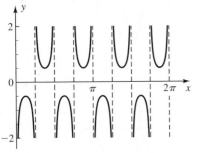

                                       [3.6]

67. $\frac{7}{9}$ *[4.4, 5.1]*   **68.** $\dfrac{\sqrt{1 - x^2} - x^2}{\sqrt{1 + x^2}}$ *[5.1]*

69. $\pi/2 + 2k\pi, \pi + 2k\pi$, k any integer   *[5.4]*

70. 0.6662, 2.475   *[5.4]*

71. $y = 4\sin(t/2 + 2.21)$, Amplitude $= 4$, Period $= 4\pi$,
Frequency $= 1/(4\pi)$, Phase shift $= -4.42$ *[4.6]*

72. $g(x) = 3 - \cos x$

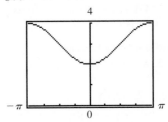

                                       [4.2]

73. $g(x) = 4 + 2\cos 2x$

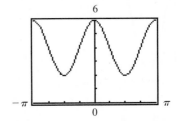

                                       [4.4]

74. $g(x) = 1 + \sec 2x$

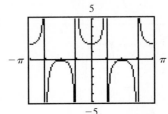

                                       [4.4]

75. $g(x) = 3 - \cot(x/2)$

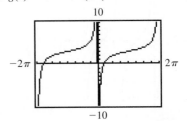

                                       [4.4]

76.

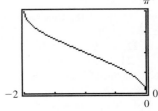

                                       [5.1]

77.

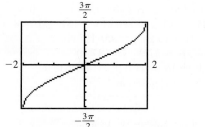

[5.1]

78.

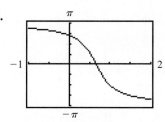

[5.1]

79. −1.893, 1.249   [5.3]   **80.** 0.642   [5.4]
81. 0, 3.895, 5.121, 9.834, 12.566   [5.4]
82. Intercepts: −4.43, 1.85, Phase shift = −4.43
(≈ −4.42 from Problem 71)

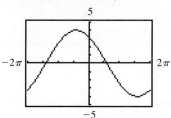

[4.6]

83. 200 mi   [1.1]   **84.** 1,429 m   [1.4]
85. a cot x   [4.1]   **86.** x = 2√3, θ = 30°   [4.4]
87. 0.443 sec   [5.4]
88. 0.529, 0.985

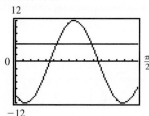

[5.4]

89. (A) y = 12 sin(5t + 4.069)
       (B) 0.409 sec, 1.105 sec   [4.6, 5.4]

90. Amplitude = 60, Period = $\frac{1}{45}$, Frequency = 45 Hz,
Phase shift = $\frac{1}{180}$

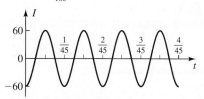

[3.4]

91. t = 0.007758   [5.4]
92. (A) d = 200 cot θ
       (B)

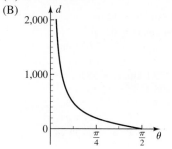

[3.6]

93. (B) 12.5°   [5.1]
94. 64.9 ft or 308.3 ft

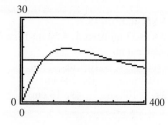

[5.1]

95. 49.5π rad   [2.1]
96. 11,550π ≈ 36,300 cm/min   [2.2]
97. 148 cm   [5.1]   **98.** 40.8 cm   [5.1]
99. (A)

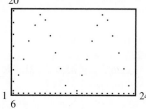

(B) y = 12.435 + 6.835 sin(πx/6 − 1.6)

(C)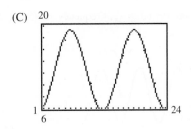

[3.3]

CHAPTER 6

Exercise 6.1

1. $\gamma = 106°$, $a = 14$ mi, $b = 12$ mi
3. $\alpha = 101°$, $b = 64$ cm, $c = 55$ cm
5. $\alpha = 98.0°$, $b = 4.32$ mm, $c = 7.62$ mm
7. $\gamma = 22.9°$, $a = 27.3$ cm, $c = 12.6$ cm
9. $\alpha = 67°20'$, $a = 55.1$ km, $c = 58.8$ km
11. $\alpha = 31.7°$, $\gamma = 96.3°$, $c = 15.1$ cm
13. No solution
15. $\beta = 44.76°$, $\gamma = 116.3°$, $c = 133.7$ yd
17. $\beta = 135.24°$, $\gamma = 25.84°$, $c = 64.99$ yd
19. No solution
21. $\alpha = 17°40'$, $\gamma = 128°30'$, $c = 1{,}740$ ft
23. Triangle 1: $\alpha = 124.0°$, $\gamma = 28.7°$, $c = 141$ ft;
Triangle 2: $\alpha' = 56.0°$, $\gamma' = 96.7°$, $c' = 292$ ft
25. $1.20 \approx 1.46$
29. $k = a \sin \beta = 66.8 \sin 46.8° \approx 48.7$
31. $AC = 12.8$ mi   **33.** 8.37 mi from A; 4.50 mi from B
35. 230 m   **37.** 2.8 nautical mi
39. 109 ft   **41.** $\beta = 113.4°$, $c = 10.8$ m
43. $r = 9.73$ mm, $s = 10.7$ mm   **45.** 284 mi
47. 4.42×10^7 km, 2.39×10^8 km   **49.** 16 cm

Exercise 6.2

1. $a = 6.00$ mm, $\beta = 65°0'$, $\gamma = 64°20'$
3. $c = 26.2$ m, $\alpha = 33.3°$, $\beta = 12.7°$
5. $\alpha = 62.7°$, $\beta = 36.3°$, $\gamma = 81.0°$
7. $\alpha = 29°12'$, $\beta = 63°26'$, $\gamma = 87°22'$
9. $\alpha = 30.3°$, $a = 46.2$ ft, $c = 27.2$ ft
11. $b = 19.3$ m, $\alpha = 40.6°$, $\gamma = 72.9°$ **13.** No solution
15. $a = 20.4$ yd, $\alpha = 47.2°$, $\beta = 66.4°$
17. $\alpha = 84.1°$, $\beta = 29.7°$, $\gamma = 66.2°$
19. $a = 24.5$ cm, $\beta = 22.6°$, $\gamma = 45.0°$ **21.** No solution
23. $\beta = 28.2°$, $a = 984$ m, $c = 1{,}310$ m
25. No solution
27. Triangle 1: $\beta = 65.3°$, $\gamma = 68.0°$, $c = 23.1$ yd;
Triangle 2: $\beta = 114.7°$, $\gamma = 18.6°$, $c = 7.93$ yd
29. No solution

31. $b^2 = a^2 + c^2 - 2ac \cos 90° = a^2 + c^2 - 0 = a^2 + c^2$ **33.** $-0.872 \approx -0.873$ **37.** $CB = 613$ m
39. 113.3°   **41.** 710 km   **43.** 121 cm
45. $\alpha = 65°20'$, $\beta = 27°0'$, $\gamma = 87°40'$
47. $AC = 23.0$ ft, $AB = 17.0$ ft   **49.** 638 mi
51. 74.4°   **53.** $DC = 166$ m

Exercise 6.3

1. 102 m^2 **3.** 12 cm^2 **5.** 11.6 in.^{2}
7. $40{,}900$ ft^2 **9.** 45.4 cm^2 **11.** 129 m^2

15. $A_1 = A_3 = \dfrac{ab}{2} \sin(180° - \theta) = \dfrac{ab}{2} \sin \theta = A_2 = A_4$

Exercise 6.4

1. $|\mathbf{u} + \mathbf{v}| = 71$ km/hr, $\theta = 29°$
3. $|\mathbf{u} + \mathbf{v}| = 57$ lb, $\theta = 33°$
5. $|\mathbf{u} + \mathbf{v}| = 154$ knots, $\theta = 21.9°$
7. $\mathbf{H} = 35$ lb, $\mathbf{V} = 23$ lb   **9.** $\mathbf{H} = 11$ knots, $\mathbf{V} = 20$ knots   **11.** $\mathbf{H} = 178$ km/hr, $\mathbf{V} = 167$ km/hr
13. $|\mathbf{u} + \mathbf{v}| = 190$ lb, $\alpha = 18°$
15. $|\mathbf{u} + \mathbf{v}| = 9.1$ knots, $\alpha = 11°$
17. $|\mathbf{u} + \mathbf{v}| = 699$ mi/hr, $\alpha = 7.4°$
19. Magnitude $= 5.0$ km/hr; Direction $= 140°$
21. 250 mi/hr, 350°   **23.** Magnitude $= 2{,}500$ lb; Direction $= 13°$ (relative to $\mathbf{F_1}$)   **25.** 32°
27. $2{,}500 \sin 15° = 650$ lb; $2{,}500 \cos 15° = 2{,}400$ lb
29. To the left

Exercise 6.5

1.

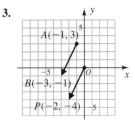

3.

5. $\langle 4, 2 \rangle$ **7.** $\langle 4, -4 \rangle$ **9.** 5   **11.** $\sqrt{29}$
13. (A) $\langle -2, 6 \rangle$ (B) $\langle 4, 2 \rangle$
       (C) $\langle 11, 2 \rangle$ (D) $\langle 6, 18 \rangle$

15. (A) $\langle 1, -6 \rangle$ (B) $\langle 3, 0 \rangle$
(C) $\langle 7, 3 \rangle$ (D) $\langle 3, -6 \rangle$
17. $\mathbf{u} = \langle \frac{4}{5}, -\frac{3}{5} \rangle$ **19.** $\mathbf{u} = \langle 2/\sqrt{13}, -3/\sqrt{13} \rangle$
21. $\mathbf{v} = 3\mathbf{i} - 2\mathbf{j}$ **23.** $\mathbf{v} = 4\mathbf{j}$ **25.** $\mathbf{v} = 2\mathbf{i} + 3\mathbf{j}$
27. $-\mathbf{i} - 7\mathbf{j}$ **29.** $-17\mathbf{j}$ **31.** $-4\mathbf{i} - \mathbf{j}$
41. Left side: 676 lb; Right side: 677 lb **43.** For AB, a compression of 9,000 lb; for CB, a tension of 7,500 lb

Exercise 6.6

1. -1 **3.** -19 **5.** 0 **7.** 0 **9.** 56.3°
11. 113.2° **13.** 90.0° **15.** Orthogonal **17.** Not orthogonal **19.** $2/\sqrt{10} = 0.632$ **21.** 0
23. $5/\sqrt{2} = 3.54$ **31.** $\langle 3, 0 \rangle$ **33.** $-\frac{36}{13}\mathbf{i} - \frac{24}{13}\mathbf{j}$
35. $\frac{6}{17}\mathbf{i} + \frac{24}{17}\mathbf{i}$ **37.** 4,900 ft-lb **39.** 85 ft-lb
41. 9 ft-lb **43.** 20 ft-lb

Chapter 6 Review Exercise

1. Infinitely many [6.1] **2.** 1 [6.1] **3.** 0 [6.1]
4. 1 [6.1] **5.** 0 [6.1] **6.** 1 [6.1] **7.** 2 [6.1]
8. 1 [6.1] **9.** 0 [6.1] **10.** 1 [6.2]
11. 1 [6.2] **12.** 0 [6.2]
13. $\beta = 22°$, $a = 90$ cm, $c = 110$ cm [6.1]
14. $\gamma = 82°$, $a = 21$ m, $c = 22$ m [6.1]
15. $a = 21$ in., $\beta = 53°$, $\gamma = 78°$ [6.2]
16. $\beta = 49°$, $\gamma = 69°$, $c = 15$ cm [6.1]
17. 224 in.² [6.3] **18.** 79 cm² [6.3]
19. 9.4 at 32° [6.4] **20.** $\mathbf{H} = 9.8$, $\mathbf{V} = 6.9$ [6.4]
21. $\langle 2, -5 \rangle$ [6.5] **22.** 13 [6.5]
23. -8 [6.6] **24.** 4 [6.6]
25. 36.9° [6.6] **26.** 123.7° [6.6]
27. $a = 97.7$ m, $\beta = 72.8°$, $\gamma = 42.2°$ [6.2]
28. $\beta = 43°30'$, $\gamma = 101°10'$, $c = 22.4$ in. [6.1]
29. $\beta = 136°30'$, $\gamma = 8°10'$, $c = 3.24$ in. [6.1]
30. $\alpha = 50°$, $\beta = 59°$, $\gamma = 71°$ [6.2]
31. 3,380 m² [6.3] **32.** 980 mm² [6.3]
33. $|\mathbf{u} + \mathbf{v}| = 23.3$, $\theta = 15.9°$ [6.4]
34. (A) $\langle 2, -3 \rangle$ (B) $\langle 6, 3 \rangle$ (C) $\langle 16, 6 \rangle$
(D) $\langle 15, 8 \rangle$ [6.5]
35. (A) $5\mathbf{i} - 4\mathbf{j}$ (B) $\mathbf{i} + 2\mathbf{j}$ (C) $5\mathbf{i} + 3\mathbf{j}$
(D) $5\mathbf{j}$ [6.5]
36. $\mathbf{u} = \langle -\frac{8}{17}, \frac{15}{17} \rangle$ [6.5]
37. (A) $\mathbf{v} = -5\mathbf{i} + 7\mathbf{j}$ (B) $\mathbf{v} = -3\mathbf{j}$
(C) $\mathbf{v} = -4\mathbf{i} - \mathbf{j}$ [6.5]
38. (A) Orthogonal (B) Not orthogonal [6.6]
39. (A) $17/\sqrt{10} = 5.38$ (B) $-7/\sqrt{10} = -2.21$ [6.6]
40. $-2.68 \approx -2.70$ (Checks) [6.1]
41. $k = 12.7 \sin 52.3°$ [6.1]

48. 5.9 cm, 17 cm [6.2] **49.** 252 m [6.1]
50. 25 cm [6.1] **51.** 540 ft [6.2]
52. 220,000 ft² [6.3] **53.** 4.00 km [6.2]
54. 660 mi [6.2] **55.** 325 mi [6.1]
56. 260 km/hr at 57° [6.4]
57. Magnitude = 489 lb; Direction = 13.4° [6.4]
58. 10,800 ft-lb [6.6]
59. Compression: 150 lb; Tension: 300 lb [6.5]
60. (A) $|\mathbf{u}| = |\mathbf{w}| \cos \theta$; $|\mathbf{v}| = |\mathbf{w}| \sin \theta$
(B) $|\mathbf{u}| = 40$ lb; $|\mathbf{v}| = 124$ lb (C) $\theta = 80°$ [6.5]
61. 56 ft-lb [6.6]

CHAPTER 7
Exercise 7.1
1.

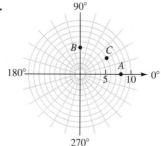

3.

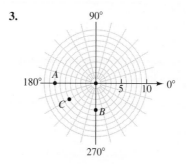

5.

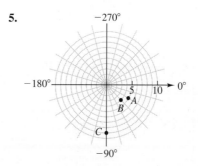

7.

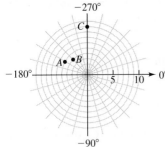

9.

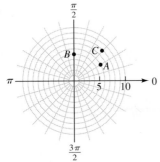

11.

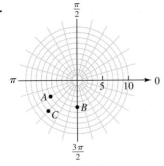

13.

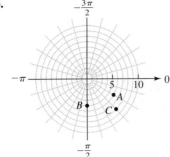

15.

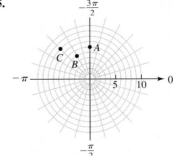

17. $(4, 4\sqrt{3})$ **19.** $(0, -9)$ **21.** $(-2\sqrt{2}, -2\sqrt{2})$
23. $(-5\sqrt{3}, 5)$ **25.** $(-3\sqrt{3}, 3)$ **27.** $(-2\sqrt{3}, 2)$
29. $(4, \pi/6)$ **31.** $(8, 3\pi/4)$ **33.** $(8, 4\pi/3)$
35. $(7, 3\pi/2)$

37.

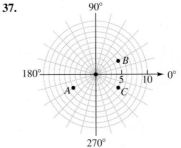

39.

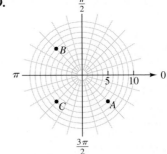

41. $r = 6 \cos \theta$ **43.** $r(2 \cos \theta + 3 \sin \theta) = 5$

45. $r^2 = 9$, or $r = \pm 3$ **47.** $r^2 = \dfrac{1}{\sin 2\theta}$

49. $r^2 = \dfrac{4}{4 - 5 \sin^2\theta}$ **51.** $2x + y = 4$

53. $x^2 + y^2 = 8x$ **55.** $x^2 - y^2 = 4$

57. $x^2 + y^2 = 16$ **59.** $y = \dfrac{1}{\sqrt{3}}x$

61. $y - 3 = -2\sqrt{x^2 + y^2}$, or $(y - 3)^2 = 4(x^2 + y^2)$
63. 1.615 units

Exercise 7.2

1.

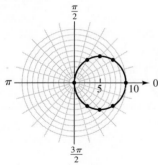

3.

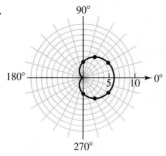

5.

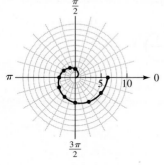

7.

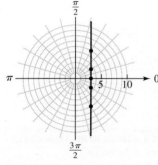

9.

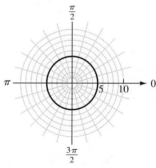

11.

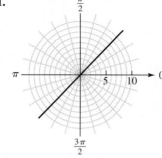

13.

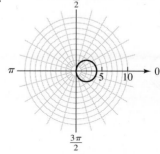

15.

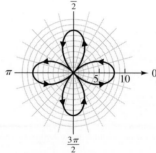

17.

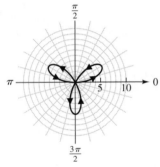

19.

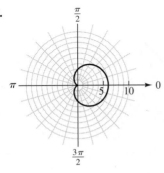

21.

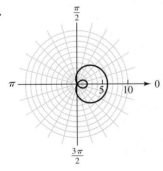

23.

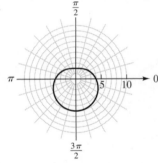

25.

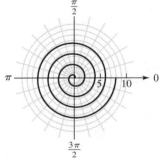

27.

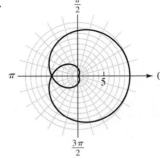

29.

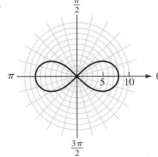

31.

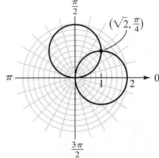

$(r, \theta) = (\sqrt{2}, \pi/4)$
[Note that $(0, 0)$ is not a solution to the system even
though the graphs cross at the origin.]

33.

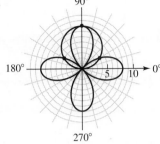

$(r, \theta) = (4, 30°), (4, 150°), (-8, 270°)$
[Note that $(0, 0)$ is not a solution to the system even
though the graphs cross at the origin.]

35. $n = 2$

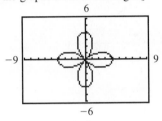

$n = 4$

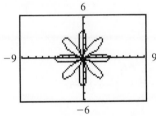

$n = 6$

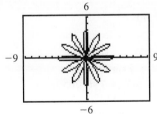

37. $n = 3$

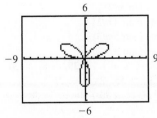

$n = 5$

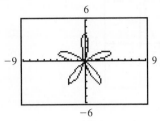

$n = 7$

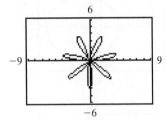

39.

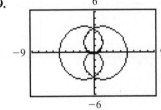

41.

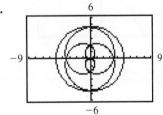

43.

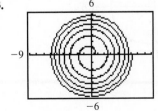

45.

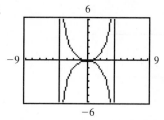

47. (A) 9.5 k (B) 12.0 k (C) 13.5 k (D) 12.0 k
49. (A) Ellipse

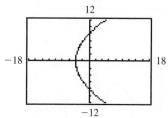

    (B) Parabola

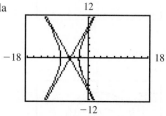

    (C) Hyperbola

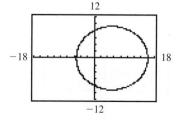

51. (A) Perihelion = 2.85×10^7 mi
Aphelion = 4.33×10^7 mi

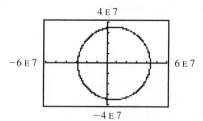

    (B) Faster at perihelion

Exercise 7.3
1.   **3.**

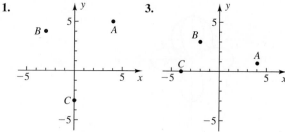

5.

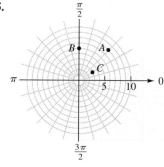

7.

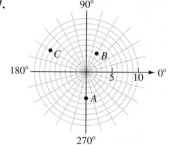

9. $z_1 z_2 = 50(\cos 77° + i \sin 77°)$,
$z_1/z_2 = 2(\cos 13° + i \sin 13°)$
11. $z_1 z_2 = 21(\cos 265° + i \sin 265°)$,
$z_1/z_2 = \frac{7}{3}(\cos 61° + i \sin 61°)$
13. $\sqrt{2}(\cos \pi/4 + i \sin \pi/4)$
15. $2(\cos 5\pi/6 + i \sin 5\pi/6)$
17. $4(\cos \pi/2 + i \sin \pi/2)$
19. $2(\cos 4\pi/3 + i \sin 4\pi/3)$
21. $4(\cos 5\pi/3 + i \sin 5\pi/3)$
23. $8(\cos 3\pi/2 + i \sin 3\pi/2)$
25. $1 + i$ **27.** $-1 + i$ **29.** -8 **31.** $12i$
33. $-3 - i3\sqrt{3}$ **35.** $2\sqrt{3} - 2i$
37. $2i, 2(\cos 90° + i \sin 90°)$

39. $4i$, $4(\cos \pi/2 + i \sin \pi/2)$
41. $\sqrt{34}(\cos 59.0° + i \sin 59.0°)$
43. $\sqrt{58}(\cos 156.8° + i \sin 156.8°)$
45. $\sqrt{61}(\cos 320.2° + i \sin 320.2°)$ **47.** $7.16 + 5.46i$
49. $-4.78 - 1.48i$ **51.** $8.59 - 6.87i$
57. (A) $(8 + 0i) + (3\sqrt{3} + 3i) = (8 + 3\sqrt{3}) + 3i$
(B) $13.5(\cos 12.8° + i \sin 12.8°)$
(C) 13.5 lb at an angle of 12.8°

Exercise 7.4

1. $27(\cos 45° + i \sin 45°)$
3. $32(\cos 450° + i \sin 450°) = 32(\cos 90° + i \sin 90°)$
5. $64(\cos 180° + i \sin 180°)$ **7.** -4
9. $16\sqrt{3} + 16i$ **11.** 1
13. $2(\cos 15° + i \sin 15°)$, $2(\cos 195° + i \sin 195°)$
15. $2(\cos 30° + i \sin 30°)$, $2(\cos 150° + i \sin 150°)$,
$2(\cos 270° + i \sin 270°)$
17. $2^{1/10}(\cos 27° + i \sin 27°)$,
$2^{1/10}(\cos 99° + i \sin 99°)$,
$2^{1/10}(\cos 171° + i \sin 171°)$,
$2^{1/10}(\cos 243° + i \sin 243°)$,
$2^{1/10}(\cos 315° + i \sin 315°)$
19. $1(\cos 0° + i \sin 0°)$, $1(\cos 60° + i \sin 60°)$,
$1(\cos 120° + i \sin 120°)$, $1(\cos 180° + i \sin 180°)$,
$1(\cos 240° + i \sin 240°)$, $1(\cos 300° + i \sin 300°)$
21. 2, $-1 + i\sqrt{3}$, $-1 - i\sqrt{3}$

23. $-3, \dfrac{3}{2} - \dfrac{3\sqrt{3}}{2}i, \dfrac{3}{2} + \dfrac{3\sqrt{3}}{2}i$

27. $1, 0.309 + 0.951i, -0.809 + 0.588i$,
$-0.809 - 0.588i, 0.309 - 0.951i$
29. $0.855 + 1.481i, -1.710, 0.855 - 1.481i$

Chapter 7 Review Exercise

1.

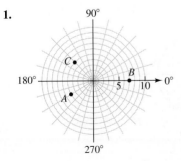

[7.1]

2.

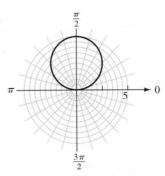

[7.2]

3.

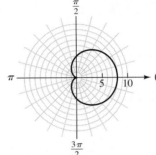

[7.2]

4.

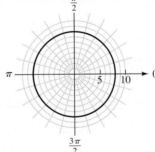

[7.2]

5. $(2, 2)$ [7.1]   **6.** $(2, 5\pi/6)$ [7.1]
7.

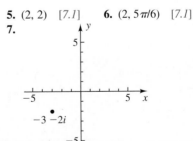

[7.1]

8.

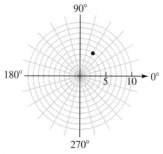

[7.3]

9. $z_1z_2 = 27(\cos 79° + i \sin 79°)$,
$z_1/z_2 = 3(\cos 5° + i \sin 5°)$ [7.3]
10. $16(\cos 40° + i \sin 40°)$ [7.4]
11.

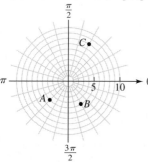

[7.1]

12.

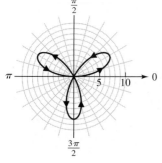

[7.2]

13.

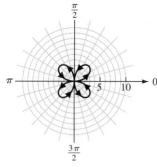

[7.2]

14.

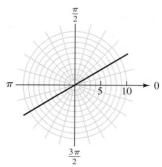

[7.2]

15. $r^2 = 8r \cos \theta$, or $r = 8 \cos \theta$ [7.1]
16. $3x - 2y = -2$ [7.1]   **17.** $x^2 + y^2 = -3x$ [7.1]
18. $2(\cos 210° + i \sin 210°)$ [7.3]
19. $-3 + 3i$ [7.3]
20. $8(\cos 195° + i \sin 195°)$ [7.3]
21. $\frac{1}{2}(\cos 75° + i \sin 75°)$ [7.3]
22. -4 [7.4]
23. $4, -2 + i2\sqrt{3}, -2 - i2\sqrt{3}$ [7.4]
24. $2(\cos 70° + i \sin 70°), 2(\cos 190° + i \sin 190°)$,
$2(\cos 310° + i \sin 310°)$ [7.4]
25. $[2(\cos 30° + i \sin 30°)]^2 = 4(\cos 60° + i \sin 60°) =$
$2 + i2\sqrt{3}$ [7.3]
26.

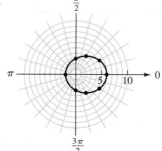

[7.2]

27.

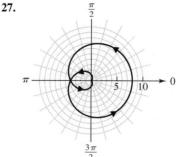

[7.2]

28. $y - 3 = -2\sqrt{x^2 + y^2}$, or
$(y - 3)^2 = 4(x^2 + y^2)$ [7.1]
29. $2.289, -1.145 + 1.983i, -1.145 - 1.983i$ [7.4]
31.

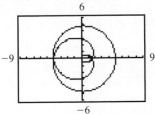

[7.2]

32.

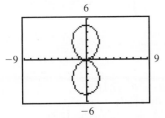

[7.2]

33. $n = 1$

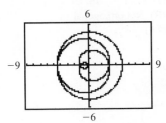

$n = 2$

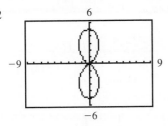

$n = 3$

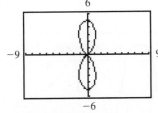

The graph always has two leaves. [7.2]

34. (A) Hyperbola

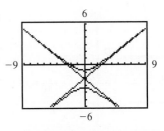

(B) Parabola

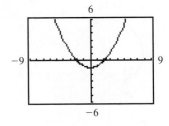

(C) Ellipse

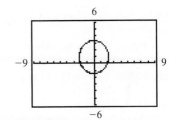

[7.3]

Chapters 1–7 Cumulative Review Exercise

1. α [2.1]   **2.** 4.82 cm, 74.7°, 15.3° [1.3]
3. $\sec \theta = \frac{25}{7}$, $\tan \theta = -\frac{24}{7}$ [2.3]
5. -0.5 [2.5]   **6.** $\sqrt{3}$ [2.5]   **7.** $2\pi/3$ [5.1]
8. $\pi/4$ [5.2]   **9.** 0.6867 [2.3]
10. 0.3249 [2.6]   **11.** 0.9273 [5.1]
12. 1.347 [5.2]   **13.** $2 \sin 2t \cos t$ [4.5]
14. $a = 22$ cm, $b = 21$ cm, $\gamma = 81°$ [6.1]
15. $\alpha = 28°$, $a = 34$ m, $c = 48$ m [6.1]
16. $\alpha = 40°$, $\gamma = 106°$, $b = 14$ in. [6.2]
17. $\alpha = 33°$, $\beta = 45°$, $\gamma = 102°$ [6.2]
18. 110 in.^{2} [6.3]   **19.** 200 ft^2 [6.3]
20. $\mathbf{H} = 12$, $\mathbf{V} = 5.5$ [6.4]   **21.** 7.5 at 31° [6.4]
22. $\langle -7, 9 \rangle$; $\sqrt{130}$ [6.5]   **23.** 143.5° [6.6]

24.

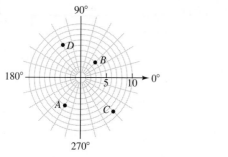

[7.1]

25.

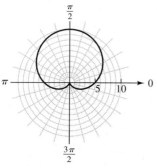

[7.2]

26. $(-3, 3)$ [7.1]   **27.** $(4, 5\pi/6)$ [7.1]
28. $2\sqrt{2}[\cos(7\pi/4) + i \sin(7\pi/4)]$ [7.3]
29. $-3i$ [7.3]
30. $z_1 z_2 = 15(\cos 65° + i \sin 65°)$,
$z_1/z_2 = 0.6(\cos 35° + i \sin 35°)$ [7.3]
31. $81(\cos 100° + i \sin 100°)$ [7.4]
32. $x = 6$, $\theta = 33°40'$ [1.3]
33. $\csc \theta = -\sqrt{17}$, $\cos \theta = -4/\sqrt{17}$ [2.3]
34. Amplitude $= 2$, Period $= \pi$, Frequency $= 1/\pi$,
Phase shift $= -\pi/2$

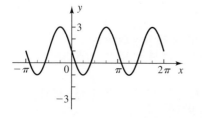

[3.3]

38. $\tan(x/2) = \frac{1}{7}$, $\sin 2x = \frac{336}{625}$ [2.3, 4.4]
39. $y = \sqrt{2} \sin(2\pi t + 3\pi/4)$; Amplitude $= \sqrt{2}$,
Period $= 1$, Frequency $= 1$, Phase shift $= -\frac{3}{8}$

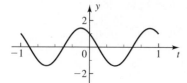

[3.3, 4.6]

40. $4/\sqrt{7}$ [5.1]
41. (A) 1.033 [5.1]   (B) -1.052 [5.1]
      (C) 0.2767 [5.2]
42. $k\pi$, $2\pi/3 + 2k\pi$, $4\pi/3 + 2k\pi$, k any integer [5.4]
43. $0.8481 + 2k\pi$, $2.294 + 2k\pi$, k any integer [5.4]
44. (A) No triangle exists
      (B) $\beta = 64°10'$, $\gamma = 66°20'$, $c = 17.7$ cm;
           $\beta' = 115°50'$, $\gamma' = 14°40'$, $c = 4.89$ cm
      (C) $\beta = 38°50'$, $\gamma = 91°40'$, $c = 27.7$ cm [6.1]
45. $|\mathbf{u} + \mathbf{v}| = 41.9$, $\theta = 11.0°$ [6.4]
46. (A) $\langle 3, -18 \rangle$ (B) $10\mathbf{i} - 11\mathbf{j}$ [6.5]
47. $\mathbf{u} = \langle 0.28, -0.96 \rangle$ [6.5]   **48.** $2\mathbf{i} + 3\mathbf{j}$ [6.5]
49. (A) Orthogonal (B) Not orthogonal
      (C) Orthogonal [6.6]
50.

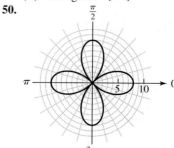

[7.2]

51. $r = 6 \tan \theta \sec \theta$ [7.1]
52. $x^2 + y^2 = 4y$ [7.1]
53. $6\sqrt{2}(\cos 165° + i \sin 165°)$ [7.3]
54. $(\sqrt{2}/3)(\cos 75° + i \sin 75°)$ [7.3]
55. $8i$ [7.4]

56. i, $-\frac{1}{2}\sqrt{3} - \frac{1}{2}i$, $\frac{1}{2}\sqrt{3} - \frac{1}{2}i$ [7.4]

57. $\theta = 4$ rad, $(a, b) = (-1.307, -1.514)$ [2.3]
58. $\theta = 3.785$, $s = 7.570$ [2.3]
59.

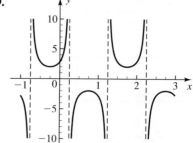

[3.6]

60. $\dfrac{1 + x^2}{1 - x^2}$ [4.4, 5.1]

62. $x - 1 = -\sqrt{x^2 + y^2}$ or
$(x - 1)^2 = x^2 + y^2$ [7.1]
63.

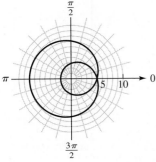

[7.2]

64. 1.587, $-0.794 + 1.375i$, $-0.794 - 1.375i$ [7.4]
65. (A) $\cos 3\theta = \cos^3 \theta - 3 \cos \theta \sin^2 \theta$,
$\sin 3\theta = 3 \cos^2 \theta \sin \theta - \sin^3 \theta$ [7.4]
66. $g(x) = -1 + 3 \cos 2x$

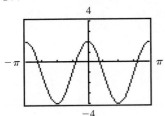

[4.4]

67. $g(x) = \sec 2x - 3$

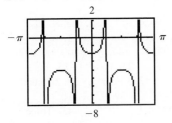

[4.4]

68. $g(x) = -1 + \cot(x/2)$

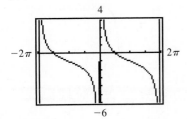

[4.4]

69.

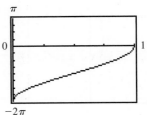

[5.1]

70.

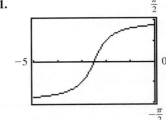

[5.1]

71.

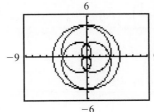

[5.1]

72. $-1.768, 1.373$ [5.3]   **73.** 0.582   [5.4]
74. 3.909, 4.313, 5.111, 5.516   [5.4]
75.

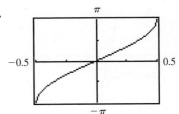

[7.2]

76. (A) Ellipse

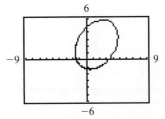

[7.2]

(B) Parabola

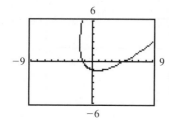

(C) Hyperbola

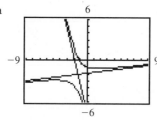

[7.2]

77. 650 ft   [6.1]   **78.** 550 ft   [6.2]
79. (A) 48 ft   (B) 41 ft   [1.4, 6.1]
80. Station A: 8.7 mi; Station B: 6.3 mi   [1.4, 6.1]
81. (A) Period = $\frac{1}{70}$ sec, B = 140π

    (B) Frequency = 80 Hz, B = 160π

    (C) Period = $\frac{1}{50}$ sec, Frequency = 50 Hz   [3.2]

82. Wave height = 4 ft, Wavelength = 82 ft,
Speed = 20 ft/sec   [3.4]
83. Area = $\frac{1}{2}\theta + \frac{1}{2}\sin\theta\cos\theta$ [2.1, 2.6]
84. Area = $\frac{1}{2}\sin^{-1}x + \frac{1}{2}x\sqrt{1-x^2}$ [2.6, 5.1]
85. $\theta = 0.553$ [5.4]   **86.** x = 0.412   [5.4]
87. (A) R = r + (H − h) tan α

    (B) $\beta = \tan^{-1}\left(\dfrac{H-h}{R-r}\right)$ [1.4]

88. 18°; 252 mph   [6.4]
90. (A) 574 ft   (B) 2.9°   [6.1, 6.2]
91. (A) 1,250 ft   (B) 4.6°   [1.4, 2.1, 6.2]
92. $A = r^2(\pi - \theta + \sin\theta)$ [5.4]
93. 2.6053   [5.4]
94. (A)   5

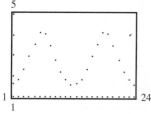

    (B) y = 2.845 + 1.275 sin(πx/6 − 1.8)

(C)   5

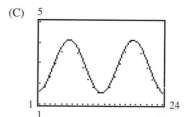

[3.3]

APPENDIX A
Exercise A.1
1. −3, 0, 5 (answers vary)   **3.** $\frac{2}{3}$ (answers vary)
5. (A) True   (B) False   (C) True
7. (A) 0.363 6$\overline{36}$, rational   (B) 0.777$\overline{7}$, rational
    (C) 2.645 751 31. . . , irrational
    (D) 1.625 0$\overline{0}$, rational
9. (A) 2 and 3   (B) −4 and −3
    (C) −5 and −4
11. y + 3   **13.** (3 · 2)x   **15.** 7x
17. (5 + 7) + x   **19.** 3m   **21.** u + v   **23.**(2 + 3)x
25. (A) True   (B) False; 4 − 2 ≠ 2 − 4
    (C) True   (D) False; 8/2 ≠ 2/8

Exercise A.2
1. 7 + 5i   **3.** −1 − 9i   **5.** −18   **7.** 8 + 6i
9. −5 − 10i   **11.** 34   **13.** $\frac{2}{5} - \frac{1}{5}i$
15. $\frac{4}{13} - \frac{7}{13}i$ **17.** $\frac{2}{25} + \frac{11}{25}i$ **19.** 5 − 2i
21. 3 + 5i   **23.** 13 − 19i   **25.** $1 - \frac{3}{2}i$
27. 0   **29.** 1

Exercise A.3
1. 6.4×10^2 **3.** 5.46×10^9 **5.** 7.3×10^{-1}
7. 3.2×10^{-7} **9.** 4.91×10^{-5} **11.** 6.7×10^{10}
13. 56,000   **15.** 0.0097   **17.** 4,610,000,000,000
19. 0.108   **21.** Three   **23.** Five   **25.** Four
27. Two   **29.** Four   **31.** Four   **33.** 635,000
35. 86.8   **37.** 0.004 65   **39.** 7.3×10^2
41. 4.0×10^{-2} **43.** 4.4×10^{-4} **45.** Three
47. Two   **49.** One   **51.** 12.0   **53.** 1.94×10^9
55. 53,100   **57.** 159.0 cm   **59.** 96 ft^2
61. 28 mm^2 **63.** 1.65 cm   **65.** 13 in.

APPENDIX B
Exercise B.1
1. 3   **3.** −5   **5.** −1   **7.** 0   **9.** −20
11. −6   **13.** 5   **15.** $-\frac{5}{3}$ **17.** 6
19. −3 − 2h   **21.** −2
23. g[f(2)] = g(−3) = −5   **25.** Not a function

27. Function   **29.** Not a function   **31.** Function
33. $Y = \{1, 3, 7\}$ **35.** F; $X = \{-2, -1, 0\}$, $Y = \{0, 1\}$
37. 0 m; 4.88 m; 19.52 m; 43.92 m
39. $19.52 + 4.88h$; the ratio tends to 19.52 m/sec, the speed at $t = 2$

Exercise B.2

1. One-to-one   **3.** Not one-to-one   **5.** One-to-one
7. Not one-to-one   **9.** One-to-one
11. Not one-to-one
13. g is one-to-one; $g^{-1} = \{(-8, -2), (1, 1), (8, 2)\}$;
Domain: $\{-8, 1, 8\}$; Range: $\{-2, 1, 2\}$
15.

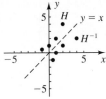

17. $f^{-1}(x) = \dfrac{x + 7}{2}$ **19.** $h^{-1}(x) = 3x - 3$

21.

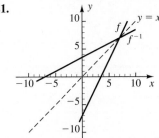

23. $f^{-1}(x) = \dfrac{x + 7}{2}$; $f^{-1}(3) = 5$
25. $h^{-1}(x) = 3x - 3$; $h^{-1}(2) = 3$
27. 4   **29.** x **31.** x

GRAPHING TRIGONOMETRIC FUNCTIONS (3.1–3.3)

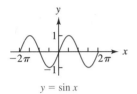

$y = \sin x$

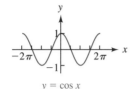

$y = \cos x$

$y = A \sin(Bx + C)$ $y = A \cos(Bx + C)$

Amplitude $= |A|$ Period $= \dfrac{2\pi}{B}$

Frequency $= \dfrac{B}{2\pi}$

Phase shift $= -\dfrac{C}{B} \begin{cases} \text{left if } -C/B < 0 \\ \text{right if } -C/B > 0 \end{cases}$

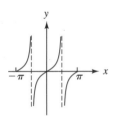

$y = \tan x$

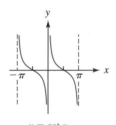

$y = \cot x$

$y = A \tan(Bx + C)$ $y = A \cot(Bx + C)$

Period $= \dfrac{\pi}{B}$

Phase shift $= -\dfrac{C}{B} \begin{cases} \text{left if } -C/B < 0 \\ \text{right if } -C/B > 0 \end{cases}$

LAW OF SINES (6.1)

$$\frac{\sin \alpha}{a} = \frac{\sin \beta}{b} = \frac{\sin \gamma}{c}$$

LAW OF COSINES (6.2)

$a^2 = b^2 + c^2 - 2bc \cos \alpha$
$b^2 = a^2 + c^2 - 2ac \cos \beta$
$c^2 = a^2 + b^2 - 2ab \cos \gamma$

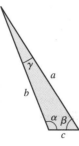

HERON'S FORMULA FOR AREA (6.2)

If the semiperimeter s is

$$s = \frac{a + b + c}{2}$$

then $A = \sqrt{s(s - a)(s - b)(s - c)}$.

INVERSE TRIGONOMETRIC FUNCTIONS (5.1, 5.2)

$y = \sin^{-1} x$ means $x = \sin y$

where $-\dfrac{\pi}{2} \le y \le \dfrac{\pi}{2}$ and $-1 \le x \le 1$

$y = \cos^{-1} x$ means $x = \cos y$
where $0 \le y \le \pi$ and $-1 \le x \le 1$

$y = \tan^{-1} x$ means $x = \tan y$

where $-\dfrac{\pi}{2} < y < \dfrac{\pi}{2}$

and x is any real number

$y = \cot^{-1} x$ means $x = \cot y$
where $0 < y < \pi$ and x is any real number

$y = \sec^{-1} x$ means $x = \sec y$

where $0 \le y \le \pi$, $y \ne \dfrac{\pi}{2}$,

and $x \le -1$ or $x \ge 1$

$y = \csc^{-1} x$ means $x = \csc y$

where $-\dfrac{\pi}{2} \le y \le \dfrac{\pi}{2}$, $y \ne 0$,

and $x \le -1$ or $x \ge 1$

The ranges for $\sec^{-1}$ and $\csc^{-1}$ are sometimes selected differently.

VECTORS (6.4–6.6)

Vector $\mathbf{v} = \overrightarrow{AB}$